INVESTIGATIONS INTO AMBER

BADANIA BURSZTYNU

Materiały z międzynarodowego symposium interdisciplinarnego:
Bursztyn bałtycki i inne żywice kopalne
997 Urbs Gyddanyzc – 1997 Gdańsk
2–6 wrzesień 1997, Gdańsk

pod redakcją
Barbary Kosmowskiej-Ceranowicz
i Henryka Panera

Muzeum Archeologiczne w Gdańsku
Muzeum Ziemi, Polska Akademia Nauk
Gdańsk, 1999

INVESTIGATIONS INTO AMBER

Proceedings of the International Interdisciplinary Symposium:
Baltic Amber and Other Fossil Resins
997 Urbs Gyddanyzc – 1997 Gdańsk
2–6 September 1997, Gdańsk

Edited by
Barbara Kosmowska-Ceranowicz
& Henryk Paner

The Archaeological Museum in Gdańsk
Museum of the Earth, Polish Academy of Sciences
Gdańsk, 1999

The Archeological Museum in Gdańsk wishes to thank the following companies and individuals — members of the Amber Association of Poland — for their generosity in supporting the publication of this volume:

- Wojciech Kalandyk — ART-7, Gdańsk
- Jacek Leśniak — VENUS, Gdańsk
- Adam & Leszek Dulińscy — BALT S.C., Gdańsk
- Wojciech Klucznik — PRACOWNIA WOYTEK KLUCZNIK, Gdynia
- Danuta & Mariusz Gliwińscy — ARTISTIC JEWELRY, Sopot
- Anna Klucznik — KLUCZNIK GALLERY, Gdańsk
- Mariusz Drapikowski — ART-MODERN, Gdańsk
- Stanisław Jacobson — Gdańsk
- Gabriela Gierłowska — PRACOWNIA BIŻUTERII, Gdańsk
- Adam Pstrągowski — SILVER & AMBER, Gdynia
- Wiesław Gierłowski — PRACOWNIA KONSERWACJI ZABYTKÓW, Gdańsk
- Jarosław Lis — KOLIA, Gdańsk

INVESTIGATIONS INTO AMBER

ISBN 83-85824-01-4

Editorial assistance Alicja Pielińska, Joanna Popiołek

Design and typesetting Barbara M. Gostyńska

Cover design Jerzy Kamrowski

Cover photographs Andy Chopping

Translation & proof-reading

- **German texts** Siegfried Ritzkowski, Brigitte Krumbiegel, Günter Krumbiegel and contributors
- **English texts** Barbara M. Gostyńska and contributors
- **Polish summaries** Barbara Kosmowska-Ceranowicz and contributors

FOREWORD

In 1997 the latest in a series of international amber conferences took place at the Archaeological Museum in Gdańsk, Poland, as one of many special events marking the City of Gdańsk's millennium. The wealth of diverse research results presented at the symposium far outweighed expectations and proved to be of great interest to the attending delegates, who represented a broad range of disciplines.

The sudden surge of new developments in amber studies, which began in the 1970s, was repeated in the early 1990s, when research workers were granted access to amber deposits in Sambia. During the Russian communist era the Sambian peninsula was strictly out of bounds to foreign scientists, whilst literature relating to it was only available in Russian and, hence, read almost exclusively in Eastern Bloc countries. The instigation of an annual international amber conference by the Museum of the Earth, and the founding of a Russian-German-Polish working group inspired a large number of research projects, the results of which were presented at the International Interdisciplinary Symposium: Baltic Amber and Other Fossil Resins, held in Gdańsk on 2–4 September 1997. The growing interest in amber among academics has also led to an increase in its popularity in the economic and public sectors; an International Amber Fair known as AMBERIF has been staged in Gdańsk every year since 1994. The amber-working industry has also flourished, whilst organic inclusions in fossil resins have become popular collector's items.

The papers published in this volume cover a wide variety of topics. Deposits in Sambia and the Ukraine are given comprehensive coverage, whilst earlier debate about the origins and age of Bitterfeld deposits is resumed. The latest results of research into Palaeogene amber-bearing sediments in southern Poland also feature in this publication. Arthropod inclusions are given slightly less attention than in previous years. Articles about scale insects in Cretaceous amber and a new species of Heteroptera are accompanied by communiqués about two unique finds: a lizard in Baltic amber and mammalian ichnites. Of interest to comparative analyses is a study on the capture of modern plant inclusions in resin. Archaeology is also well represented in these post-conference proceedings. Papers include evidence not only from Poland, but also from Latvia and Slovakia. The use of amber by Roman period societies in Poland's Kuiavia region is dealt with, the results of a long-standing Italian-Polish collaborative project are presented, and a progress report is given on the production of an up-to-date catalogue of Neolithic amber artefacts from the former collection of the Albert University in Königsberg. The latest chemico-physical methods used in the examination of amber are outlined by authors presenting the results of research into Baltic amber, fossil resins from Austria, beckerite from the Goitsche mine, cedarite and other fossil resins from Canada, as well as resins from the historic Otto Helm collection and the ethnographic collection of the Israel Museum. There is even a paper which deals with the analysis of gas inclusions in amber. The volume is brought to an end by contributions on amber-working at the Kaliningrad Amber Factory from 1945 to 1996, and in twentieth-century Poland, followed by by two ethnographic studies — one being a consideration of the use of amber in Polish folk art, the other pertaining to Sicilian amber.

Unfortunately, papers and posters presented at the symposium by members of the Russian Academy of Sciences, Kaliningrad (E.M. Emelyanov, V.V. Sivkov & S.M. Isachenko and V.V. Sivkov & A.D. Krylov) were not submitted for publication. Neither, regretfully, does this volume include J. Okulicz-Kozaryn's work offering a new insight into the history of the amber-bearing Baltic coast in late antiquity. J. Grabowska's account of the development of amber-working in Gdańsk's heyday is also missing as are G. Jabłoński's observations on contemporary amber art and A. A. Żuravlev's report on the reconstruction of the amber room. Some of the aforementioned papers can be found in summary form in the volume of abstracts printed to accompany the Symposium*. We sincerely hope that the contributors to INVESTIGATIONS INTO AMBER as well as other researchers working on this subject will persevere for many years to come in their efforts to fathom the mysteries of amber.

Barbara KOSMOWSKA-CERANOWICZ
Henryk PANER

**Bursztyn bałtycki...* Muzeum Ziemi (Konferencje naukowe) Streszczenia 8, 1997, Warszawa or **Baltic Amber...* Museum of the Earth (Scientific conferences) Abstracts 9, 1997, Warsaw.

CONTENTS

AMBER DEPOSITS

ANIMAL AND PLANT INCLUSIONS IN AMBER

AMBER IN THE ARCHAEOLOGICAL RECORD

PHYSICO-CHEMICAL ANALYSIS OF FOSSIL RESINS

AMBER PROCESSING

NATURWISSENSCHAFTLICHE FORSCHUNGEN ÜBER BERNSTEIN IN POLEN

Barbara KOSMOWSKA-CERANOWICZ

Kurzfassung

Seit den 50iger Jahren des XX. Jahrhunderts entwikkeln sich die Bernstein-Forschungen in Polen zunehmend, während gleichzeitig die Anzahl von Bernstein-Sammlungen amwächst. Bedeutende Forschungsergebnisse erscheinen im letzten Viertel des XX. Jahrhunderts. Die Veröffentliechung wurde aufgrund der Bibliographie und wissenschaftlichen Arbeit am Sammlungsmaterial der Bernstein Abteilung des Museums der Erde in Warschau erarbeitet.

Die ersten Mitteilungen

Die in den Jahren 1993–1994 zusammengestellte und kommentierte *Bibliografie des polnischen Schrifttums und der Arbeiten polnischer Autoren in der Weltliteratur* geht bis auf das Jahr 1534 zurück[1].

In dieser Zeit wird Bernstein in den Kräuterbüchern erwähnt.

Beim Studium der frühen Manuskripte zeigt sich, daß sich die Autoren aus dem XVI. Jh. an viel frühere Werke anlehnten, zum Beispiel an das vom Thomas, einem Arzt des Breslauer Fürsten HEINRICH DEM VIERTEN[2] oder an das von Johann STANKO[3], dem Königsarzt.

Vereinzelte Mitteilungen aus dem XVI. bis XVIII. Jh. betrafen vor allem Funde des Rohstoffes Bernstein, seine physikalisch-chemischen Eigenschaften, Sorten und Bemerkungen über organische Inklusen. K. KLUK hat im Jahre 1781 auf verwandte Harze und Kunstbernstein aufmerksam gemacht. Obwohl so frühzeitige Mitteilungen über Bernstein existieren, begann in Polen erst in den 70-er Jahren des XIX. Jhs. die bedeutende Entwicklung der Untersuchungen. Allerdings waren wir nicht die einzigen; da die Deutschen den besten Zugang zu den größten Lagerstätten hatten, wurde von ihnen bereits Mitte des XIX. Jhs. Bernstein wissenschaftlich bearbeitet.

Das weitere Interesse an Bernstein und die ersten Monographien

Im XIX. Jh. sind in Polen 84 Werke und Notizen erschienen, die über Bernstein unter verschiedenen Gesichtspunkten berichten. Schon S. STASZIC, Vater der polnischen Geologie, schrieb in seinem Hauptwerk (1815) über Bernstein. Er wies auf seine Verbreitung hin und verglich den Bernstein aus Pokucie (später karpatischer Bernstein genannt), mit anderen ähnlichen Funden aus den Karpaten.

Von J. FREYER (1833) und J. HACZEWSKI (1838) erschienen die ersten Monographien, mit für die damalige Zeit aktuellem Wissen über Bernstein, der Beschreibung polnischer Bernsteingruben und -bearbeitungswerkstätten, sowie der Terminologie der polnischen Bernsteinarbeiter. Teilweise war es noch die polonisierte deutsche Terminologie, z. B. **klar, bastard** (die Deutschen übernahmen vom Französischen), **knoch**, die A. CHĘTNIK (1981) erst in XX. Jh. durch den polnischen Wortschatz zu ersetzen versuchte.

Zum wachsenden Interesse an Bernstein trug zweifellos die stürmische Zunahme individueller Sammler und Handwerker nach der Einführung des Regierungsgesetzes im Jahre 1840 bei. Es wurden Steuererleichterungen beim Rohstoffhandel eingeführt und alle Gebühren für die Erzeugnisse aufgehoben (nach SZACKI Maschinenschrift, nicht datiert).

Das Interesse an Bernstein in Hinblick auf seine Verbreitung und den Bergbau zeigten J. POŁCZYŃSKI

---

[1] Bursztyn bałtycki i inne żywice kopalne... 1993 (Baltic amber and other fossil resins in Polish literature and works by Polish authors in world literature. An annotated bibliography 1534–1993 (collective work under the guidance of KOSMOWSKA-CERANOWICZ). Part I: Amber in nature, culture and art (PIETRZAK & RÓŻYCKA). Muzeum Ziemi, Opracowania dokumentacyjne, **12**, Warszawa.

[2] *Aggregatum*, Kopie ca. 1450 (siehe KOWALCZYK 1991).

[3] J. STANKO: *Antobolomeum*, Manuskript Nr 225 (1472) in Archiv und Bibliothek des Krakauer Kathedral Kapitels des Wawelschloßes. Siehe auch ROSTAFIŃSKI 1900.

(1850, 1851) und L. ZEJSZNER (1844). M. RACIBORSKI schrieb in den Jahren 1891–1903 über pflanzliche Inklusen im Bernstein, in dem er die Umwelt des bernsteinführenden Waldes nachzubilden versuchte. Bei Bernsteinbeschreibungen und der Absicht, ihn in der Mineraliensystemathik unterzubringen, richtete sich ZEJSZNER nach den Werken von J. J. BERZELIUS 1833) und G. ROSE (1861). Es wurden Arbeiten fremder Autoren, übersetzt und besprochen (SCHOEDLER wurde überzetzt bei F. BERDAU 1871; HÄPKE wurde rezensiert bei JANOTA 1876; GOEPPERT — bei MORAWSKI 1883; HELM — bei KRAMSZTYK 1887 & SIEMIRADZKI 1888; TSCHIRCH, SCHMID und ERDÖS von SIMONBERG 1935).

J. SIEMIRADZKI (1888) schrieb über die Kollektion von 600 Käfern in der Bernsteinsammlung von O. HELM (aufgegliedert in 43 Familien, darunter 27% Springkäfer und Chrysomelidae, 4% Staphylinidae und 3% Curculionidae; von den Wasserkäfern (so von SIEMIRADZKI benannt) wurden in der Sammlung *Pyrinus* [?] und *Laccophilus* [Fam. Dytyscidae] erwähnt). Nach dem *Bericht der Danziger Gesellschaft für Naturkunde* aus dem Jahre 1902 zählte die Käfersammlung schon 800 Objekte. Derselbe Verfasser beschrieb etwas später (1909) die samländische Bernstein-Formation in Preußen. Auch danach behandelte er in seinen geologischen Schriften Bernstein zusammen mit anderen „Fossilen Früchten Polens" (nach 1922).

Zu den Ausnahmen gehören Originalwerke aus dem Bereich der Paläoenthomologie von A. WAGA (1826). WAGA hat als erster (1883) *Paleognathus succini* (LEUTHNER) WAGA beschrieben. Erst 1922 erschien die Beschreibung von J. STACH der neuen Gattung des Springkäfers (*Sminthurus*). Trotz weniger Originalwerke wurde in Polen die Entdeckung der gestreiften Muskulatur bei Diptera in Bernstein von A. PETRUNKIEWICZ (1958) und kontroverse Mitteilungen von G. KIRCHNER (KOZŁOWSKI 1951; BERNATT 1955) veröffentlicht.

Erst nach über 50 Jahren zeigten die polnischen Enthomologen deutliches Interesse an Bernstein, damals nach dem II. Weltkrieg änderte sich die Standortverteilung der Untersuchungszentren.

Anfänge der Forschungen und die Entstehung von Kollektionen

In der ersten Hälfte des XX. Jhs. ist eine Veröffentlichung von J. NIEDŹWIEDZKI (1907) erschienen. Er beschrieb zum ersten mal den karpatischen Bernstein und verlieh ihm den Namen Delatynit nach dem Fundort Delatyń. Heute wissen wir, daß es ein fossiles Harz aus der Gruppe des Rumänits ist, das seine Eigenschaften während orogenetischer Prozesse erlangte.

Die einzigen aber zahlreichen Werke über Bernstein aus den Jahren 1913–1964 (und nachgelassene bis 1985 herausgegeben) von A. CHĘTNIK betreffen die Ethnographie der Kurpie-Region (nördlicher Teil des Narew-Gebiets).

Trotz vieler Ungenauigkeiten, denen Mangel an Wissen zugrunde liegt, schuf der Verfasser neben zahlreichen ethnographischen Veröffentlichungen auch eine Dokumentation zur Verbreitung und Gewinnung des Bernsteins in dieser Region. Bezeichnend ist der Satz von P. SZACKI, der nicht nur den Untersuchungsstand in den frühen Nachkriegsjahren, sondern auch die Eigenart des Kurpielandes im Zusammenhang mit dem Wissen über Bernstein aufzeigt: „Das heutzutage nur noch geringe Wissen über Bernstein, Orte seiner Lagerstätte, Art der Ablagerungen und Gattungen des Bernsteins war in der Blütezeit der Bernsteinkunst in Kurpien beträchtlich. Die Kenntnis von der Regelmäßigkeit der Bernsteinlagerung und umfangreiche Terminologie brachten das zum Ausdruck" (Maschinenschrift, nicht datiert im Archiv der Bernstein Abteilung, Museum der Erde, Warschau).

Die Bernsteinkollektionen von CHĘTNIK wurden zur Grundlage der Sammlungen im Bezirkmuseum in Łomża und im Museum der Erde der Polnischen Akademie der Wissenschaften in Warschau, wo er in den Jahren 1951–1958 gearbeitet hat. Eine kleine Populärmonographie von Z. MULICKI (1951) wurde für viele Jahre zum notwendiger Lehrbuch bei den Gesellen- und Bernstein-Meisterprüfungen in der Handwerkerzunft. Das Buch von MULICKI informiert genau über den damaligen Wissensstand. Es enthält jedoch eine zweifellos falsche Nachricht über einen Vogel der im Bernstein beim Ausbaggern des Presidentenbeckens im neugebauten Hafen in Gdynia gefunden worden sein soll. Illustrationen und Literatur sind unvollkommen zitiert.

Nachdem Polen nach dem zweiten Weltkrieg ein wichtiger Zugang zum Meer zuerkannt wurde, nahmen die Chancen beträchtlich zu sowohl für die Entwicklung der Forschung in Zusammenhang mit der Erkundung des Geländes und der Bernsteingewinnung, als auch für die Gewerbeentwicklung mit eigenem Material und für die Entwicklung des Handels. Leider gab es in den ersten 25 Jahren keine besonderen Erfolge.

Das Bernsteinvorkommen in Tertiär- als auch in Quartärsedimenten an der Ostseeküste wurde zum erstenmal in Anlehnung an die geologischen Untersuchungen von E. PASSENDORFER & J. ZABŁOCKI (1946) dokumentiert. Die kleine aber bedeutende Bernsteinsammlung von ZABŁOCKI befindet sich heute noch in der Kollektion der Universität von Toruń. Bernstein wurde auch im Niederungsteil des Odergebiets gefunden (PASSENDORFER 1948). Wichtige Arbeiten und die Landkarte von E. SUKERTOWA-BIEDRAWINA (1950, 1961, siehe auch KOSMOWSKA-CERANOWICZ 1997), angelehnt an deutsche Literatur über Ostpreußen, zeigen ähnliche Probleme über die in Polen seinerzeit unbekannte Bernsteinverbreitung auf dem Warmia- und Mazury-Gebiet.

Einen internationalen Ruf gewann die Arbeit von H. CZECZOTT (1961) über die Flora im Bernstein. Die Analyse aller paläobotanischen Werke und die kritischen Synthesen der Pflanzengemeinschaften aus dem Tertiär finden noch heute Anwendung in Untersuchungen der pflanzlichen Inklusen (vergleiche POINAR 1992 u. a.).

Eine großartige Errungenschaft war die Erfassung der Bernsteinkollektionen und deren Popularisierung durch Ausstellungen erstenmals in Polen. Eine bedeutende Rolle bei den Untersuchungen, bei der Popularisierung und bei der Einrichtung von Bernsteinkollektionen spielte Z. ZALEWSKA (Geologin und Botanikerin), in den Jahren 1958–1974 Leiterin der Bernstein Abteilung im Museum der Erde. In den Ausstellungsführern (1964, 1968, 1974) nach eigenem Konzept und Szenar, stellte sie Aktuelles für das Wissen über Bernstein der damaligen Zeit zusammen. Ihre Arbeiten betreffend Genese und Stratigraphie der Lagerstätte des Baltischen Bernsteins werden oft in der Literatur zitiert, sie haben bestimmt zum Interesse der Geologen an Bernstein beigetragen, obwohl sie verschiedene Ungenauigkeiten enthalten. Ihr Forschungsaufenthalt in der Ukraine und ihre Dokumentation der Aufschlüsse im Dniepr-Tal sind erst nach ihr Tode erschienen (1995).

Die Bernsteinerzeugnisse betreffende Fragen

Es ist sowohl bezeichnend als auch logisch, daß man zu Beginn der Nachkriegsuntersuchungen versuchte, eine genaue Dokumentation der Erzeugnisse und des Rohbernsteins, der in Museen vorhanden war, mit Hilfe von Umfragen zusammenzustellen.

War es lediglich das Ziel, sich über den Besitz der Untersuchungsmaterialien klar zu werden, oder war es nur die alltägliche Tätigkeit auf dem Gebiet des Museumsweens, die zur Bearbeitung eines Katalogs führen sollte? Es steht dagegen fest, daß es zur Veröffentlichung des Materials beigetragen hat und die Entstehung zahlreicher interessanter Bearbeitungen veranlaßte.

Im Jahre 1954 fragte S. BERNATT (Mariner und Journalist) bei dem Zentralen Vorstand der Museen und des Kunstwerkeschutzes (eigentlich ohne Erfolg) nach dem Verzeichnis der Bernsteinerzeugnisse in polnischen Museen. Er bekam die Erlaubnis eine Umfrage nach altertümlichen Erzeugnissen durchzuführen. Er fragte nach der Charakteristik der Kunstwerke, nach Fotographien und nach den Erzeugnissen, die sich entweder in privaten oder kirchlichen Besitz befinden. Einen Teil der Ergebnisse veröffentlichte er in *Tygodnik Morski* (Meereswochenzeitschrift 1961, 1962), ein Teil blieb in der Handschrift *Über die Madonna in Oliwa* (1966, nach seinem Nachlass, der in Museum der Erde sich befindet).

BERNATT sammelte auch Daten in ausländischen Museen: in Gotha, Weimar, Braunschweig, Brighton, Neapel, St. Petersburg und in Dänemark (Museum der Kunstindustrie). Nach M. PELCZAR (1946) wurden die Sammlungen, unter anderem mit Kunsterzeugnissen aus Bernstein, vom Stadtmuseum in Danzig (Rzeźnicza Str. 25/28), in die Umgebung von Gotha verschleppt.

Nach dem Katalog von A. ROHDE (1937, 73–78) gab es in der Kollektion von Danzig folgende sieben Kunstwerke: 1 — „Anna Selbdritt in vergoldeter Bronzefassung", Anfang XVII Jh., 11,5 cm ohne Einfassung; 2 — „Fußschale", erste Hälfte des XVII Jhs., 10,8 cm; 3 — „Reliefbildnis Kaiser Konstantins", um 1725, 5,7 cm; 4 — „Scheren-Etui" von W. KRÜGER (?), um 1720, 11 cm; 5 — „Dose mit mytologischer Szene auf dem Deckel", um 1620, 8,5 x 5 cm; 6 — „Dose mit Liebespaar auf dem Deckel", um 1680, 7 cm; 7 — „Ovale Dose mit Aktäonczene auf dem Deckel", XVII-XVIII Jh., 8,3 x 6 cm.

Unabhängig von BERNATT arbeitete seit 1957 A. CHĘTNIK an der Dokumentation der artiger Erzeugnisse. Er besuchte persönlich Museen in den einzelnen Städten. Die Ergebnisse der Rundfrage, ergänzt mit aktuellen Daten, sind erst nach 25 Jahren erschienen (KOSMOWSKA-CERANOWICZ & POPIOŁEK 1981).

Im Museum der Erde bearbeitete Z. ZALEWSKA eine Umfrage im Jahre 1966. Ihre Ergebnisse, dargestellt auf der Landkarte *Bernstein in Museen von Polen*, wurden auf der Ausstellung *Bernstein in Polen* gezeigt. Auf dieser Karte kamen im Vergleich zur früheren (von A. CHĘTNIK) etwa 20 neue Museen und Institutionen hinzu.

Auch in den 60-er Jahren arbeitete J. GRABOWSKA — damalige Verwalterin der Sammlungen von Malbork und zweifenllos die beste Kennerin dieses Themas — an der wissenschaftlichen Bearbeitung auf Grund der versandten Umfrage. Eine Monographie der polnischen Erzeugnisse, die Grabowska noch in 1969 (vom H. RACZYNIEWSKI im Brief an BERNATT verfaßt) vorbereitete, ist leider nie erschienen. Das zusammengestellte Material verarbeitete die Autorin in dem Buch *Polnischer Bernstein* (1982).

Bedeutender Fortgang der Untersuchungen in den 70-er Jahren und der gegenwärtige Stand

Die erste Konferenz zum Thema Bernsteinforschungen fand am 27.01.1951 im Museum der Erde in Warschau statt. R. KOZŁOWSKI, einer der führenden polnischen Paläontologen hielt das Referat *Ueber Wissenschaftliche Bernsteinforschungen* (1951). A. CHĘTNIK sprach über den Bergbau und die Bernsteinindustrie in Kurpien.

Der Forschungsplan, der damals im Museum der Erde entstand, wurde völlig, um nicht zu sagen einzig und allein, im wichtigen Bereich der Sammlung und der Dokumentation der Kollektionen realisiert.

Ohne diese Unterlagen wären die Untersuchungsarbeiten kaum möglich. R. KOZŁOWSKI schrieb 1970 in seinem Brief an BERNATT: „In unserem Land ist das Interesse an mit Bernstein verbundenen Problemen sehr ungenügend, obwohl wir ohne Zweifel eine Menge wertvolles Material für die Wissenschaft besitzen. Ihr Artikel kann also unsere Wissenschaftler aus verschiedenen Gebieten zu Untersuchungen anregen (...)"[4]. Es war ein Antwortbrief auf den zugesandten Artikel, in dem BERNATT die Methoden zur Identifizierung von Succinit popularisierte, die früher von HELM und seit den 60-er Jahren von BECK angewandt wurden. Er erwähnte auch die Kollektion von 360 Bernsteinstücken mit tierischen Inklusen, die 1939 dem Institut für Paläontologie der Warschauer Universität geschenkt worden war. Aus dieser Sammlung bearbeitete und bestimmte A. SKALSKI *Epiborkhausenites obscurotrimaculatus* (1973). Diese Bezeichnung veröffentlichte BERNATT in einen Artikel, früher als der Verfasser selbst.

Die Untersuchungen von Lepidopteren im baltischen Bernstein und anderen fossilen Harzen, die A. SKALSKI zusammen mit BERNATT aufnahm, dauerten bis zu seinem Tode im Jahre 1996. Zum Thema fossile Lepidoptera veröffentlichte SKALSKI etwa 15 Arbeiten.

Nicht nur die ausführliche Arbeit von BERNATT trug zur Belebung des Interesses bei. Hilfreich waren auch zwei Monografien über Bernstein von S. S. SAWKIEWICZ (1970) und dem Litauer V. KATINAS (1971), die in der damaligen UdSSR herausgegeben und in Polen als Rezensionen erschienen. Besonders die Monografie von SAWKIEWICZ bot eine unfangreiche Bibliografie, die Nachforschungen in der Literatur erleichterte.

Ob überhaupt und in welchem Umfang war die in Polen seit Mitte der 70-er Jahre begonnene rasche Entwicklung der Bernsteinforschungen von der wachsenden Nachfrage auf dem Markt abhängig? Der Bernstein ist doch vor allem ein Schmuckstein. In den ersten 25 Nachkriegsjahren war die Einfuhr des Rohstoffes aus dem Samland für den Betrieb der Bernsteinerzeugnisse in Gdańsk-Wrzeszcz nicht größer als 3 Tonnen pro Jahr.

Private bearbeitungsbereite Hersteller, konnten zunächst den Widerstand der Behörden und die Steuerrestriktionen nicht überwinden. Günstige Änderungen für die „nicht volkseigene Wirtschaft" erfolgten nach den Macht antritts GIEREKS. Im Jahre 1970 hörte der formale Import des Bernsteins aus dem Samland auf. Als Antwort auf den Rohstoffmangel begann seit Anfang der 70-er Jahre, gemäß der zehn Konzessionen, aber auch illegal, die Gewinnung des Bernsteins mit der hydraulischen Methode. Beträchtliche Bernsteinlager in fossilen Stränden wurden während der Bauarbeiten am Nordhafen entdeckt. Innerhalb von zwanzig Jahren (1970–1990) stieg der Rohstoffverbrauch bis auf 60 t jährlich an. Nach W. GIERŁOWSKI (1995): bis 6 t jährlich durch Auffangen, 4 t mit Hilfe der hydraulischen Methode durch Konzessionsfirmen, 35 t durch illegale Förderung, 15 t durch private Einfuhr aus der GUS. Die freie Marktwirtschaft ab 1990 war Anlaß für erhöhten Rohstoffbedarf: im Jahre 1990 — 54 t, 1991 — 86 t, 1992 — 123 t, 1993 — 126 t, 1994 — 122 t.

Gleichzeitig mit dem Ansteigen des Bernsteinbedarfs verlief die Intensivierung von Forschungen und geologischen Untersuchungen der Bernstein-

[4] In Nachlass von BERNATT.

lagerstätten. Wurde sie durch die Erzeugung angeregt? Oder wirkte die Popularisierung des Bernsteins seit Mitte der 70-er Jahre als polnischer Schmuckstein, tief in der polnischen Tradition verwurzelt, auf das steigende der Interesse? Und war es vielleicht eine Rückkoppellung?

Den Rang eines Forschungszentrums erlangte das Museum der Erde aufgrund der Untersuchungen, seiner Inspiration, Popularisierung (durch Sonderausstellungen und seit 1977 auch ausländische, Kataloge- und Museumsführerveröffentlichungen) und vor allem seit 1978 infolge der interdisziplinären Organisation von in der Regel jährlichen Landes- und Auslandstreffen. Solche Zentren gab es früher in Danzig und Königsberg.

Auf Anregung des Museums der Erde wurde 1995 in Kaliningrad eine russisch-deutsch-polnische Arbeitsgruppe zur Bernsteinforschung gegründet. Eine weitere Tagung fand auf deutsche Initiative 1996 in Bochum am Deutschen Bergbaumuseums statt (die Vorträge wurden in der Zeitschrift des Deutschen Bergbaumuseums in Bochum *METALLA* 1997 veröffentlicht). Die Autorin des Artikels ist überdies die Vertreterin Polens in der Internationalen Mineralogischen Gesellschaft (IMA) in WGOM (Working Group on Organic Minerals).

Bernstein-Kollektionen des Museums der Erde in Warschau

Die im Museum der Erde seit 1950 zuerst von A. CHĘTNIK und dann von Z. ZALEWSKA gesammelten Kollektionen schufen eine gute Grundlage für den Beginn von Mehrzweckforschungen. Mitte der 70-er Jahre konnten aus allen vorhandenen Sammlungen 6 Themenkollektionen gebildet werden, die Gegenstand der entomologischen, paläobotanischen, chemischen, archäologischen und allgemeingesehen naturwissenschaftlichen Forschungen sind.

Die größte Sammlung TIERISCHER INKLUSEN, beträgt heute 17 674 Objekte. Zu den wichtigsten Stücken gehören vor allem 66 Holotypen, aber auch die Stücke mit den seltensten Gruppen der Gliederfüßer, wie z. B. die größte Kollektion der Welt: 6 Stücke der Fächerflügler (Strepsiptera), ein Fersenspringer (Embioptera — sehr selten in Bernstein) und eine Megaloptera. Die Grundbestimmungen werden von unserer Mitarbeiterin der Entomologin Frau Dr. R. KULICKA laufend durchgeführt, während Spezialuntersuchungen der einzelnen Tierfamilien von Spezialisten vorgenommen werden. In den vergangenen 25 Jahren sind auf der Grundlage dieser Sammlung 22 Arbeiten erschienen, die taxonomische, ökologische, biochemische und paläohistologische Forschungen betreffen. Insekten oder andere Gliederfüßer erwecken den Anschein als ob sie noch lebendig sind. P. MIERZEJEWSKI (zwei Arbeiten in 1976), vom Museum der Erde hat ihre inneren Organe, wie Lunge, Spinndrüse, Teile der Augen herauspräpariert und ist selbst an Sehzellen und Bestandteile der einzelnen Zellen, wie z. B. Hornhautzellen des Auges, gelangt.

Dank einem Teil der Kollektion der TIERISCHEN INKLUSEN, die von dem Fundort Gdańsk-Stogi stammt, konnte man eine statistische Zusamenstellung der tierischen Systematik erstellen, die typisch für Baltischen Bernstein ist (Anteile von zehn ausgewählten Insekten-Ordnungen der Bernsteinfauna und anderer fossiler Harze).

Die Kollektion PFLANZLICHER INKLUSEN beträgt 1172 Objekte. Da die Untersuchung von pflanzlichen Resten sehr schwierig ist, ist das Interesse der Botaniker dafür sehr gering. Vorangekommen sind lediglich die Untersuchungen an Lebermoosen (Hepaticae), die von Dr. R. GROLLE in Jena durchgeführt wurden. Er hat drei neue Holotypen aus dieser Kollektion bestimmt und in zwei Arbeiten beschrieben (1985).

Die Blüten und Blütenstände der Fagaceae (43 Stück) unserer Kollektion bearbeitet Dr. CREPET (1989) aus den USA.

Es ist eigenartig und paradox, dass der Baltische Bernstein, der heute mit physikalischen und chemischen Methoden sehr leicht zu identifizieren ist, pflanzenkundlich nicht bis zu Ende bearbeitet wurde. Wir können etwas über die Mutterbäume des Dominikanischen Bernsteins, des Glessits, des Siegburgits und des Bernsteins aus Borneo aussagen, aber nicht über die des Succinits! Succinit hat dieselbe IR Kurve die V. KATINAS (aus Wilna) vom rezenten Harz der Zeder — *Cedrus atlantica* MANETTI erhielt[5]. Leider kann erst die botanische Bestätigung die Herkunft des Succinit endgültig klären.

Die REGIONALE KOLLEKTION enthält (692 St. und Proben) neben dem Succinit (= Baltischer Bernstein) aus der baltischen Region, solchen aus der Ukraine, aus Weissrussland und von Bitterfeld auch die ihn begleitenden Harze und außerdem andere fossile, subfossile und rezente Harze aus der ganzen Welt. Die Harze werden mit physikalischen und

[5] Gezeigt auf „Sixth meeting on amber and amber-bearing sediments" in Museum der Erde 1988 in Warschau.

chemischen Methoden; im Museum der Erde, vor allem mit Infrarot-Spektroskopie untersucht, oft unter Mitarbeit ausländischer Forscher (S. SAWKIEWICZ, G. KRUMBIEGEL, N. VÁVRA, C. BECK, und anderen). In anderen polnischen Instituten wurden auch Stücke des Bernsteins mit Hilfe gaschromatographischen Analysen erforscht. Letztens haben die Chemiker und Mineralogen aus Wrocław angefangen offene Fragen des undurchsichtigen Bernsteins mit Positron Anihilation Spectroscopy zu erforschen (CHOJCAN & SACHANBIŃSKI 1993).

Über die Eigenschaften des Bernsteins und anderer fossiler Harze wurden in letzten 25 Jahren ca. 45 Arbeiten veröffentlicht.

Alle fossile Harze wurden im Museum der Erde seit 1985 auch intensiv gesammelt und mit IR-Licht erforscht. Heute hat die Sammlung 31 Arten, und das Material bildet die Grundlage für einen Atlas der IR-Kurven der fossilen Harze, der in Bearbeitung ist.

Besonders intensiv hat das Museum der Erde, zusammen mit Dr. G. KRUMBIEGEL aus dem Geiseltalmuseum Halle seit einigen Jahren gearbeitet, um die neuentdeckten Harzarten im Tagebau Goitsche (teilweise bei BARTHEL und HETZER), Königsaue und anderen Fundorten in Sachsen-Anhalt zu bestimmen. Es waren Gedanit, Gedano-Succinit, Glessit, Siegburgit, Goitschit, sog. Schwarzer Bernstein, Beckerit sowie Retinit, Krantzit und Oxykrantzit.

Am Strand des Baltikums wie auch im Meer kann man nicht nur Bernstein, sondern auch verschiedene andere subfossile oder rezente Harze, die leicht und gelb sind, ziemlich oft finden. Diesen Harzen wurden mehr oder wenigerer treffende Namen verliehen. Für Leute, die sich mit Bernstein beschäftigen, ist es sog. junger Bernstein. Die Händler verkaufen den Kollektionären solche Stücke fälschich als Kopal, ganz falsch als Gedanit und nur manchmal exakt richtig als Kolophonium. Die Folgen finden wir in den musealen Sammlungen, in denen unbestimmte oder falschbestimmte Stücke viele Jahre auf eine richtige Identifizierung warten. Die Kolophonium-Kollektion im Museum der Erde enthält zwei grosse Klumpen (7 und 3,3 kg!) sowie ein Fragment des 238 kg schweren Stückes, das ein Schwede, L. BROST, 1988 aus der Ostsee gefischt hat. Das Harz hat die Form einer Tonne. Die Bestimmung des Kolophoniums habe ich mit den amerikanischen Autoren (BECK *et al.* 1993) gemacht. Derzeit prüfe ich das Alter des Kalophoniums mit den ukrainischen Kollegen aus Kiev mit Hilfe der Radiokarbon — ^{14}C Methode (KOSMOWSKA-CERANOWICZ *et al.* 1996).

Die Regionalsammlung gibt auch Einblick in die Entdeckungen und Untersuchungen der Lagerstätten und Funde fossiler Harze aus Polen und 24 anderen Ländern. In Polen gibt es ca. 600 Fundorte Baltischen Bernsteins in Quartär-Sedimenten und in 3 Tertiär-Lagerstätten: Chłapowo in der Danziger Region, Górka Lubartowska in Mittel-Ostpolen und Możdżanowo bei Stolp.

In den 70 ger Jahren würden vom Staatlichen Geologischen Institut in Warschau, teilweise gemeinsam mit dem Museum der Erde, die Chłapowo Lagerstätte durch drei Aufschlussbohrungen entdeckt und beschrieben sowie deren Vorräte berechnet.

In drei Profilen von Bohrungen in Chłapowo (Chlapau), finden sich in grosser Tiefe (ca. 120 m), dieselben geologischen Schichten, die in Samland-Profil anzutreffen sind. Hier stellte man auch fest, dass die Sedimentation des Bernsteins, ähnlich wie im Samland, unter den gleichen Verhältnissen wie in einem Flachmeer zur Ablagerung gelangten. Eine Analyse der Sedimente beweist die vorherrschende lithologische Deltafazies eines Flußmündungsgebietes. Die grösseren Mächtigkeiten der Schichten (Latdorf) in dem westlichen Teil des Deltas (Chłapowo Gebiet) im Gegensatz zur Halbinsel Samland, kann man mit einer grössenen Subsidenz (Absenkung) des Untergrundes dieses Sedimentationsbeckens interpretieren.

Das Delta des Bernsteinsflusses ERIDANOS ist wenigstens 115 km breit und liegt zwischen Karwia (Karwen) in Polen und Jantarnyj (Palmnicken) in Russland. Im mittleren Teil des Eridanos-Deltas, dem Żuławy (Weichsel-Werder) und der Zatoka Gdańska (Danziger Bucht), wurde das bernsteinführende Tertiär infolge Erosion völlig abgetragen. Diese Erosionsprozesse (glaziale Abtragung) fanden während des gesamten Tertiärs und des Quartärs statt. Die holozäne Abtragung war während der Litorina-Zeit die größte und intensivste. Im Quartär wurden Tausende Tonnen Bernstein ins Binnenland umgelagert und dort in wenigen Metern Tiefe wieder sedimentiert.

Über die Lagerstätten in Polen haben die Forscher ca. 40 mal geschrieben; über die Lagerstätten im Ausland 24 mal.

Die Sammlung der NATURFORMEN von Baltischen Bernstein im Museum der Erde enthält etwas mehr als 4047 Stücke und ist die reichhaltigste Kollektion dieser Art in den Museen der Welt. Ähnliche aber kleinere Sammlungen besitzt das Bernsteimuseum in Kaliningrad. Besondere Stücke sind auch in Göttingen in dem Teil der ehemaligen Königsberger Sammlung zu finden. Das größte Stück wiegt 4800 g und hat eine gut erhaltene glatte Oberfläche, die auf ehemals flüssiges Harz hindeutet. In Göttingen befindet sich auch ein besonderes Bernsteinfragment, ein Tropfsteinvorgang (Draperie) der aus Zapfen und Stalaktiten gebildet wurde. Die Naturformen-Kollektion zeigt die Menge sowie die Art und Weise der Harzabsonderung und die Größe der Bäume in den tertiären Wäldern und hat deshalb besonderen Ausstellungswert.

Kleine natürliche Fließformen von Krantzit sind Seltenheiten.

Die VARIANTEN-Kollektion des Baltischen Bernsteins (2846 Stücke) läßt die Unterschiedlichkeit der Innenstruktur des Bernsteins, die Wirkung der Verwitterung und die eingeschlossenen organischen Substanzen erkennen. In Polen gibt es sehr viel Literatur über die Bernstein-Varianten, die von A. CHĘTNIK durch etnographische Untersuchungen gegründet wurde. Wir haben fast 200 Volksnamen von Bernsteinvarianten, die von CHĘTNIK im Form eines kleinen Lexikons zusammengestellt wurden (1981). So eine etnographische Seltenheit habe ich noch nie in der Weltliteratur gesehen!

Die Sammlung von BERNSTEINERZEUGNISSEN vom Neolithikum bis zur Gegenwart (1086 Stücke) wird im Museum der Erde für Archäologen, Kunsthistoriker, aber auch für die Bernsteinausstellungen gesammelt. Das Schlossmuseum Malbork ist für Bernsteinkunst spezialisiert. In den letzten Jahren wurde diese Sammlung dank dem Sammler C. WÓJCIAK durch einen Satz Objekte neolithischer Erzeugnisse aus dem Weichsel Werder (Niedźwiedziówka Ort) bereichert, der sehr wertvoll für archäologische Untersuchungen ist.

Besonders interessante Einzelteile sind Kopien von zwei Bernsteinlinsen, hergestellt um 1700 von PORSCHIN in Königsberg, (die Originale sind in Schloß Rosenborg in Dänemark) sowie Kopien von zwei Feldern des Schachspiels des Bernstein-Zimmers.

Obwohl uns nicht alles befriedigen konnte, könnte man viele Lücken im Vergleich zu dem weltweiten Fortschritt des Wissens über dieses organische Mineral zeigen (wir haben es aufgegeben paläohistologische Untersuchungen durchzuführen, wir erforschen nicht die Mikroorganismen), Polen wird bekannt und weltweit genannt; wir nehmen Teil in internationalen Forschungsgruppen (WGOM der IMA, IUPPS und anderen). Ausländische Forscher veröffentlichen gern Ergebnisse ihrer Untersuchungen in polnischen wissenschaftlichen Zeitschriften. Im Bereich der Popularisierung der Bernstein-Kentnisse und -Erkenntnisse, ist Polen im Ausland durch bedeutende Ausstellungen (Prag, Venedig, San Francisko, Passau, Konstanz, Bielefeld u.a.) und Kataloge (*Ambra oro del Nord*, *Spuren des Bernsteins* u.a.) bekannt.

Barbara KOSMOWSKA-CERANOWICZ
Muzeum Ziemi PAN
Aleja Na Skarpie 27
00-488 Warszawa, Poland

BADANIA PRZYRODNICZE NAD BURSZTYNEM W POLSCE

Barbara KOSMOWSKA-CERANOWICZ

Streszczenie

Zestawiona w latach 1993–1994 bibliografia komentowana polskiego piśmiennictwa i prac autorów polskich w literaturze światowej na temat bursztynu sięga roku 1534. Mimo tak długiej tradycji, znaczący rozwój badań rozpoczął się w Polsce dopiero w latach 70. dwudziestego wieku.

W XIX w. w Polsce ukazały się 82 prace, recenzje, tłumaczenia i omówienia prac autorów obcych. Rozwój zainteresowań badawczych był niewątpliwie związany ze wzrostem liczby indywidualnych poszukiwaczy i rzemieślników. Do publikacji najważniejszych zaliczamy pierwsze monografie Jana FREYERA (1833) i Józefa HACZEWSKIEGO (1838). Do wyjątków należą trzy oryginalne prace z zakresu paleoentomologii (WAGA 1826; 1883; STACH po 1922).

W pierwszej połowie XX w. ukazała się odkrywcza praca NIEDŹWIEDZKIEGO (1907), który po raz pierwszy opisał **delatynit**.

W latach 1913–1964 najliczniejsze były prace A. CHĘTNIKA, które dotyczą etnografii regionu Kurpiowszczyzny. Jego kolekcje bursztynu stały się zalążkiem zbiorów i podstawą pierwszych wystaw Muzeum Okręgowego w Łomży i Muzeum Ziemi PAN w Warszawie. Niewielka popularna monografia Z. MULICKIEGO (1951), na wiele lat stała się podręcznikiem, którego znajomości wymagano przy egzaminach w cechu rzemieślniczym.

Po przyznaniu Polsce znacznego dostępu do morza po II wojnie światowej, szanse rozwoju badań były znaczne. Szczególnych efektów niestety w pierwszych 25 latach nie było wiele.

Jedna z pierwszych prac geologicznych dotyczy bursztynu w osadach trzeciorzędowych i czwartorzędowych w brzegu Bałtyku (PASSENDORFER, ZABŁOCKI 1946). Ważne prace i mapa E. SUKERTOWEJ-BIEDRAWINY, oparte na niemieckiej literaturze Prus Wschodnich, udokumentowały rozprzestrzenienie bursztynu na Ziemi Warmińsko-Mazurskiej (1950). Międzynarodowy zasięg uzyskała praca H. CZECZOTTOWEJ (1961) dotycząca składu i wieku flory w bursztynie. Osiągnięciem Z. ZALEWSKIEJ były zbiory bursztynu w Muzeum Ziemi. Jej prace (1971, 1974) są wielokrotnie cytowane.

Dzięki powojennym badaniom metodą ankietowania S. BERNATT, J. GRABOWSKA, A. CHĘTNIK i Z. ZALEWSKA przeprowadzili dokładną dokumentację wyrobów i kolekcji surowego bursztynu pozostających w muzeach.

Pierwsza ważna konferencja na temat badań bursztynu odbyła się w Muzeum Ziemi w 1951 roku. Plan badań, jaki wówczas powstał, realizowany był najpełniej w zakresie gromadzenia i dokumentacji zbiorów, bez których żadne prace badawcze nie byłyby możliwe.

Do ożywienia zainteresowań przyczyniło się niewątpliwie rosnące zapotrzebowanie rynku po zmianach wprowadzonych do gospodarki przez GIERKA w 1970 roku. W tych latach ustał formalny import bursztynu z Sambii, a w odpowiedzi na brak surowca rozpoczęła się na polskim wybrzeżu eksploatacja bursztynu z osadów kopalnych plaż. W dwudziestoleciu 1970–1990 zużycie surowca wzrastało aż do osiągnięcia 60 t rocznie, a w 1996 — wyniosło około 200 t.

S. BERNATT popularyzował wyniki dawnych badań O. HELMA i najnowszych C. W. BECKA, a także A. SKALSKIEGO.

Muzeum Ziemi PAN choćby z racji posiadania jednych z największych zbiorów bursztynu na świecie, inspirowaniu badań i ich popularyzacji, organizacji od 1978 roku interdyscyplinarnych, z reguły corocznych krajowych i międzynarodowych spotkań badaczy bursztynu, przez wielu zaczynało być uznawane za centrum badawcze.

W połowie lat 70. z wszystkich zbiorów bursztynu Muzeum Ziemi można było utworzyć sześć kolekcji tematycznych: inkluzji roślinnych, zwierzęcych, form naturalnych, odmian bursztynu bałtyckiego, zbiorów regionalnych i wyrobów. Wszystkie te kolekcje są przedmiotem badań przyrodniczych.

Do najważniejszych osiągnięć ostatniego 25-lecia należą poszukiwania i badania trzeciorzędowych złóż bursztynu w rejonie Chłapowa, Parczewa, Możdżanowa), a także dokumentacja wszelkich znalezisk. Badania dotyczyły również złóż bitterfeldzkich i ostatnio sambijskich. Rozpoczęły się również badania nagromadzeń wtórnych w osadach czwartorzędowych.

Już od końca lat 60. w Polsce prowadzi się badania żywic kopalnych metodami fizycznymi i chemicznymi (IRS, ESR — T. URBAŃSKI, Politechnika Warszawska). W Muzeum Ziemi zgromadzono kolekcję żywic kopalnych częściowo identyfikowanych metodą IRS. Katalog wyników obejmuje około 500 krzywych IRS, które publikowane są w licznych pracach. Również w Muzeum Ziemi prowadzono badania na zawartość siarki i próby określenia wieku bezwzględnego żywic subfosylnych metodą ^{14}C (we współpracy z Ukrainą). Są również prace dotyczące zawartości mikroelementów (L. KOZIOROWSKA) oraz wyniki badań metodami elektronowego rezonansu paramagnetycznego (EPR), anihilacji pozytonów i spektrografii masowej (zespół wrocławski skupiony wokół M. SACHANBIŃSKIEGO).

Pracami A. SKALSKIEGO rozpoczął się ogromny rozwój badań paleoentomologicznych. Badania paleobotaniczne prowadzone są konsekwentnie, chociaż w mniejszym zakresie, co wynika nie tylko z większego stopnia trudności, jaki materiał ten przedstawia, ale i z mniejszej kolekcji.

Na 680 prac objętych wyżej wspomnianą bibliografią około 440 reprezentuje dorobek badaczy — przyrodników i chemików w ostatnich 25 latach.

Literatur

BECK C. W., STOUT E. C., KOSMOWSKA-CERANOWICZ B.
1993 A large find of supposed amber from Baltic Sea, *Geologiska Föreningens i Stockholm Förhandlinger*, **115**, 2, 145–150.

CHOJCAN J. & SACHANBIŃSKI M.
1993 Freevolumes in amber probed by positron annihilation spectroscopy, [*in*:] *Proceedings of XXVIII Zakopane School of Physics*, 296–298, Kraków.

CREPET W. L.
1989 History and implications of the early North American fossil record of Fagaceae. Evolution, [*in*:] *Sistematics and fossil history of the Hamamelidae 2, The Systematics Association Special Vol.* 40B, 45–66, Oxford.

GIERŁOWSKI W.
1995 Stan obecny i perspektywy rozwoju polskiego bursztynnictwa. *Muzeum Ziemi — Konferencje naukowe — Streszczenia referatów*, **4**, 26–32.

KOSMOWSKA-CERANOWICZ B.
1997 Bursztyn na mapach dawnych Prus Wschodnich, [*in*:] Ziemie Dawnych Prus Wschodnich w kartografii, *Z dziejów kartografii*, **8**, 67–80, Olsztyn.

KOSMOWSKA-CERANOWICZ B., KOVALIUK N. & SKRIPKIN V.
1996 Sulphur content and radiocarbon dating of fossil and sub-fossil resins, *Prace Muzeum Ziemi*, **44**, 47–50.

KOWALCZYK M.
1991 Aggregatum Tomasza biskupa Sarepty w rękopisie Biblioteki Jagiellońskiej 777, [*in*:] *Kultura średniowieczna i staropolska*, 117–126, PWN, Warszawa.

PETRUNKEVITCH A.
1958 Amber spiders in European collections, *Trans. Conn. Acad. Arts Sci.*, **41**.

POINAR G.
1992 *Life in amber*, Stanford.

ROHDE A.
1937 *Bernstein ein deutscher Werkstoff*, Berlin.

ROSTAFIŃSKI J.
1900 *Średniowieczna historia naturalna. Systematyczne zestawienie roślin, zwierząt, minerałów oraz wszystkich innego rodzaju leków prostych używanych w Polsce od XII do XVI wieku*, Kraków .

WAGA A.
1883 Note sur un Lucanide incrusté dans le succin (Paleognatus *Leuthenr succini* Waga), *Annales Societatis Ent. Fr.*, **6** (3), 191–194.

EOCENE PALAEOGEOGRAPHY AND SEDIMENTATION IN THE BALTIC DEEP AND ADJONING AREAS

Alexander I. BLAZCHISHIN

Abstract

The main subject of this paper is presented from a palaeooceanographic point of view, with particular attention being paid to the origin of amber-bearing deposits.

Source materials

This article is based on general data relating to the formation of the Baltic Deep (BLAZCHISHIN 1991). Palaeogene basins occupied only the most southern part of the Baltic Deep (Fig. 1) and the adjoning palaeoshelf: the Polish-Lithuanian depression and the eastern part of the Central European basin. Palaeogene deposits were determined in the West Baltic (Arkona Deep and Mecklenburg Bay) by seismoacoustic data (SVIRIDOV & LITVIN 1978) and in the central part of the Bay of Gdańsk, where they lie imbedded in ancient valleys at the fracture zones of latitude direction (SVIRIDOV 1990).

Palaeocene-Eocene deposits were exposed in six marine borehole samples taken by the "Petrobaltic" company along Poland's coast from Bornholm to the Hel spit (KRAMARSKA 1992). Palaeocene deposits have only been found in cores from Hanoe Bay near the south pirolitycznej coast of Sweden (KUMPAS 1978). The unbroken cover of Palaeogene deposits is more characteristic of western areas — the islands of Denmark and the Jutland Peninsula. Cores taken from the underwater slope of the Samland Peninsula revealed a rather wide distribution of amber-bearing Eocene deposits (BLAZCHISHIN 1974; 1997).

Early Eocene

The reconstruction of the Early Eocene (Fig. 1) is very similar to that of the Late Palaeocene (BLAZCHISHIN 1991). The palaeogeography of both periods was determined by the same geodynamic events. The progressive and differentiated spreading of the North Atlantic caused a multitude of transgressions and regressions in the Early Palaeogene of the European palaeoshelf. The separation of Greenland from Europe probably began in the *Vetzeliella varielongituda* Zone (COX *et al.* 1986) at the beginning of Early Eocene. This event was preceded by regional rises in the North Sea basin and powerful volcanic activity in the region of Scotland-Ireland (Thule) (JACQUE & THOUVENIN 1975). Volcanic activity was also prevalent in the Skagerrak street (AM 1973) and the ash which this produced spread far east, up to Sambia.

The aforementioned pre-spreading movements can be compared with the inverse rises in the southern part of the North Sea (ZIEGLER 1982). In the east, the inversion of the Polish-Danish Trough caused the formation of an orogenic barrier — the Kujawy-Pomeranian ridge, which isolated the Polish-Lithuanian basin from the North Sea basin (KOSMOWSKA-CERANOWICZ 1979). This event ocurred just after the uplift of the Lower San anticlinorium. The termination of inversive movement coincided with the beginning of spreading in the Norway-Greenland basin (KNOX & HARLAND 1979). Thus, it is clear that Alpine movements are not directly connected with ocean crust formation in the North-Atlantic region.

Early Eocene transgression was limited by western Poland areas (CIUK 1983). At this time the rest of the Polish-Lithuanian basin was existing which was connected with the North Sea trough Baltic Deep (Fig.1). In East Germany the sea penetrated to the western Lausitz (TRÖGER 1984).

Estuarine-continental coal-bearing Early Eocene units can be retraced along the boundary of the sea basin from Szczecin to Potsdam Berlin and further in the Lower Saale river and to south-eastern Harz Foreland. The widespread development of swamp landscapes is characteristic for the lower half of the Ypresian. The Early Eocene basin differed by a disperse composition (London Clay and its analogies in Western Europe). Some authors have attributed this not only to the deep of the basin and very slack

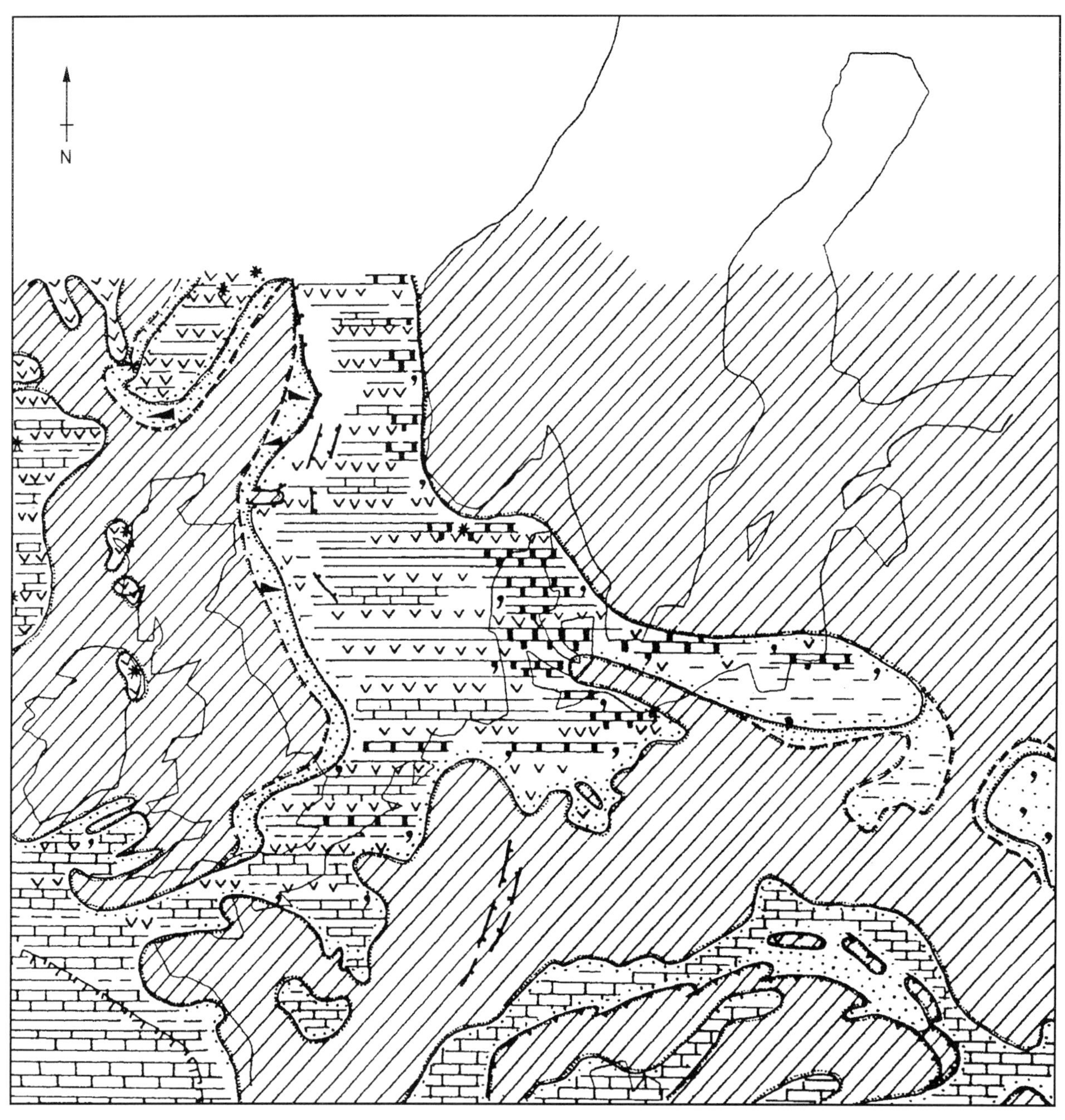

Fig. 1. Early Eocene palaeogeography and sedimentation in the Baltic Deep and adjoining areas. Conventional signs — see Fig. 3.

circulation, up to stagnation, but also to the wide development of lateritic weathering indicating hot humid climatic conditions. This explains the red colour of the upper part of London Clay (KNOX & HARLAND 1979). The basin was not very deep. This is attested to by the presence of clays and storm-originated layers of yellow sand. In general, wave action played an important part in the formation of London Clay and Tarras Clays (Germany). In the Tarras Clays the good preservation of fish skeletons, crayfish and other organisms was noted, in particular crinoids, which characterize very still water conditions (GRIPP 1964). Among clay minerals in continental deposits kaolinite (in marine deposits — smectite) predominates. The smectite component is undoubtedly associated with pyroclastic decomposition which is represented everywhere in abundance in the lower part of Ypresian sequences.

Thin clay layers (Mo-Clay formation) appear with London Clay in northern Denmark and the southern

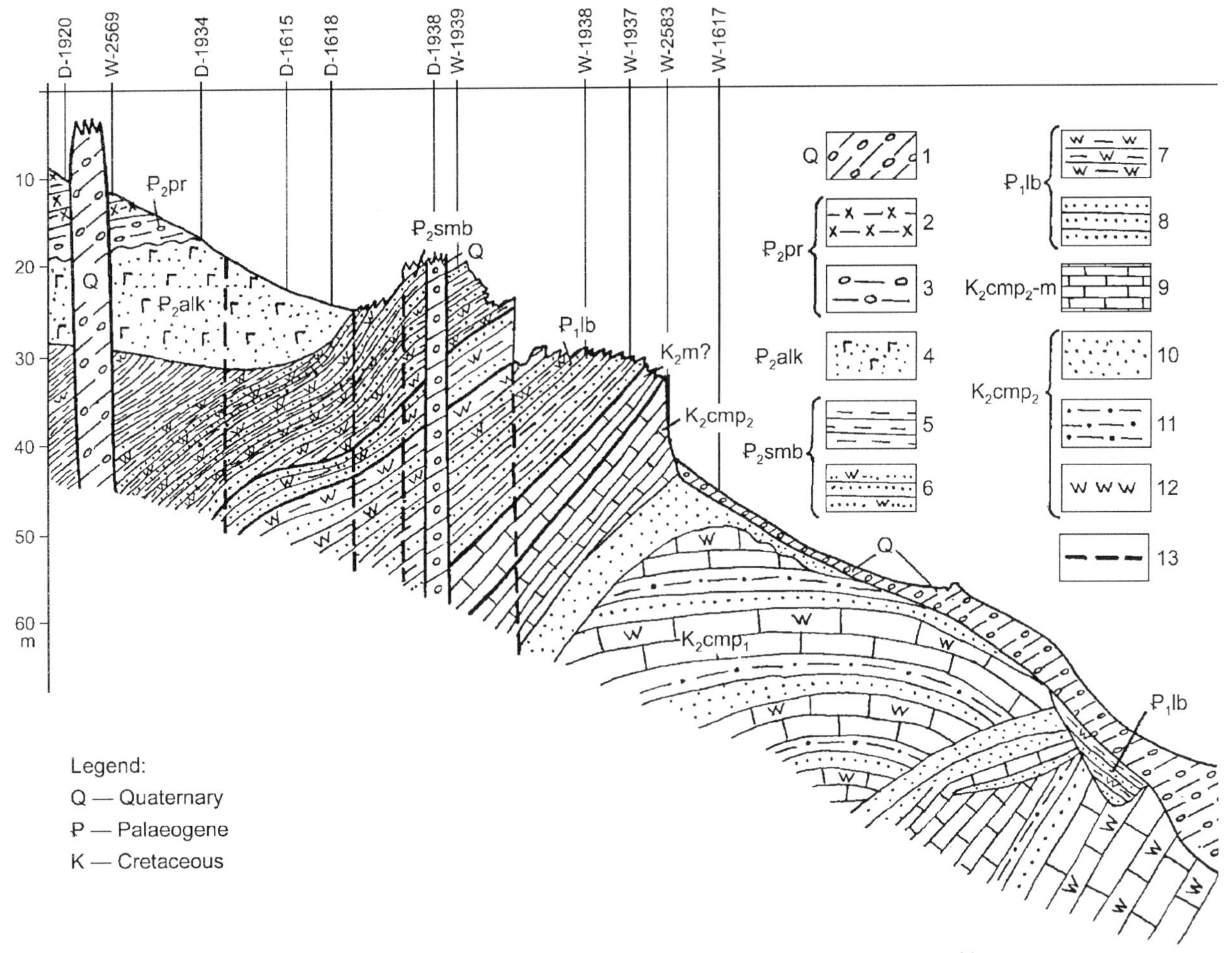

Fig. 2. Geological sequence of submarine slope of the Sambian Peninsula (seismoacoustic profile B-2). 1 — moraine; 2 — "blue earth"; 3 — "wild earth"; 4 — "lower shifting sand" with glauconite; 5 — silty clays; 6 — sands with silizites; 7 — siliceous clays; 8 — silts and sands; 9 — limestones, marls; 10 — sandstones; 11 — siltstones; 12 — flints; 13 — fractures.

North Sea. In the upwelling zone they interlaminated with diatomites. In the Limfjord area the Eocene part of this sequence contains 140 ash layers. The diatomites alternate with ash layers of three characteristic texture types (from laminated to bioturbated), which show at the change of the oxic-anoxic conditions. Transitory stagnation periods occurred in all areas and a period of diatomite accumulation (1–2 mln years) was noted in 12 cases (PEDERSEN 1981).The sedimentation of London Clay and its analogues came to an end in the middle of the Early Eocene in connection with a new short regression which can, according to PLEZIAT (1981), be identified by the epeirogenetic deformations in the Alpine faulting region.

The reliable Early Eocene deposits of the south-east Baltic Deep associated with the Sambian suite include contemporary non-calcareous micaceous clays, silts with *Angulerina muralix* and *Globorotalia acuta* as well as glauconitic-quartz sands of a total thickness of up to 60 m (on average 25–30 m) (KATINAS 1971). The peculiarity of the Sambian deposits lies in their laminated character, volcanic ash admixture and essential silicification. At submarine slope the Early Eocene silicites of the Sambian suite the stony banks along the western boundary of the Sambian Peninsula were formed (Fig. 2). The Sambian suite is comparable to the Olsztyn layers in north-eastern Poland.

Middle Eocene

During the Middle Eocene a progressive rise in the sea level took place at the same time as the degradation of orogenic barriers in inversion troughs. Ultimately, this paved the way for the Late Eocene wide marine expansion. The global Eocene transgression had a eustatic character brought on by geodynamic causes. In their number the decrease of capacity of sea floor spreading in phase of mid-ocean ridge formation be called and also the increase of the spreading rate (KASTING & RICHARDSON 1985). High rates of spreading brought about the metamorphosis of carbonates with plenty of CO_2 being released into the atmosphere. Palaeopedological and isotope data indicate that this period of warming (with an average temperature increase of 2–3°C) reached its peak by the beginning of the Early Eocene (ALBERS 1981; BURCHARDT 1978). During the Middle Eocene the average temperature gradually decreased as the sea level rose. The Late Eocene transgression maximum coincided with a stable tendency towards a colder climate. Moreover, the climate of the Late Palaeocene–Early Eocene was characterized by a long dry period. Eastern passat winds, similar to those in modern-day subtropical zones, were dominant and seasonal upwellings in the area near the coasts of Norway and Denmark were developing. It is possible that these upwelling conditions characterize the transgressive Eocene deposits (glauconitic-quartz sands and silts) in the marginal eastern part of the Polish-Lithuanian basin.

Some authors (MCKENNA 1983) believe that the Faeroe-Iceland land bridge, which separated the Norway-Greenland basin from the North Atlantic, still existed during the Eocene. In the Middle Eocene only its eastern part was exposed to destruction. Thus, the epicontinental Central-European basin was probably still isolated both from the Arctic and Atlantic during the Early Eocene. The North Sea basin was connected with the Atlantic through the English Channel, but already by the middle Lutetian period the uplift of anticline D'Artois isolated it from the ocean (POMEROL 1978). The problematic periodical connections leaved with Tethys across the Moravian street (POŻARYSKA 1977). It is possible that this isolation occurred as a result of the hot continental climate during the Middle Eocene. Whatever the case may be, the Middle-Late Eocene is the last period in the Mesozoic-Cenozoic history of the European basin when the carbonate plateau was still intact. With the exclusion of the Paris basin, reef and chalk limestones are known from the Upper Lutetian-Barthonian of Germany and the English-Belgian Basin, whilst marls occur in the Middle Eocene of Denmark.

The Lillebelt suite in Denmark contains layers with siliceous clays (DIENESEN *et al.* 1978). This shows that the upwelling conditions continued here in the Middle Eocene. Siliceous (spiculose) sandstones are known from border (Early–Middle Eocene) sequences of Schleswig-Holstein (GRIPP 1964). Unlike the preceding stage, the Middle–Late Eocene basins in north Germany had an open connection both with the North Ocean (general widespreading of radiolarians sediments) as well as with southern seas (the presence of a Tethyan Nummulites fauna). These basins were well-ventilated. This is reflected in the abundance of authigenic glauconite and small remains of carbonate organisms.

Late Eocene

The reconstruction of the Late Eocene (Fig. 3) reflects the maximal extent of transgression. The connection with the Dnieper-Donec basin was restored along the border of the East-European platform. The wide Moravian street connecting the Central-European basin with the Carpathian Flysh basin was open. The foraminifera complex of the Lublin area (*Globigerina semiinvoluta* Zone or NP-17 of standard scale) dates from the upper part of the Barthonian. This complex is similar to the fauna of the Kiev suite (Ukraine) which indicates that the sea was approaching from the east. Despite extensive research efforts no large foraminifera have been found in the Kiev suite, which indicates that there was no direct connection with the late Eocene basin of the Carpathians nor with those of Silesia (POŻARYSKA 1977). The widespread presence of Nummulites Calauer facies is typical for the Palaeogene of the Alpine-Carpathian region in the Moravian street area (TRÖGER 1984). The Middle-Polish anticlinorium which divided the Polish-German basin into eastern and western parts existed for a long time; in the south-east the Little-Polish massive was uplifted. At the same time, judging by the distribution of heavy mineral complexes, the Kujawy-Pomeranian ridge was blocked during the Late Eocene by a barrier of transgressive waters and sediment transport from the west (KOSMOWSKA-CERANOWICZ 1979). The connection of the European basin with the Atlantic was renewed during La-Manche and probably during Faeroe-Iceland threshold (the eastern part of Thulean bridge having been destroyed). Regressive tendencies (an indication of arid climate) occurred in the second half of the Pryabonian. These tendencies were related to the Pyrenees orogenic phase. In accordance with new data the Early Pyrenees stadium of geodynamic activation took place about 45 million years ago, yet in the Barthonian, the Late

Pyrenees stadium — 35 million years ago, during the Lower Oligocene (ZIEGLER 1982).

***

The most widespread of Middle-Late Eocene deposits are represented by the transgressive series (sands, silts) of terrigenous-glauconitic formation. Deposits of the Alkian suite are imbedded (up to 30 m thick) in the Early Eocene deposits of the Sambian Peninsula and its submarine slope with a hiatus. The amber-bearing deposits of the Prussian suite lie 20 to 30 metres above the transgressive boundary. The structure of both suites is strikingly similar — in the bottom of the basal horizon are glauconitic-quartz sands and clayey silts with phosphorite nodules ("wild earth"). The productive "blue earth" horizon (glauconitic micaceous silts with amber and phosphorites) lies directly above. This horizon is overlain by gravel and sands of varying grain size ("shifting sand").

In accordance with recent K-Ar absolute dates (RITZKOWSKI 1997) the "wild earth" of the Prussian suite was formed in the Ypresian period and "blue earth" — in the Lutetian. Thus, the Alkian suite and the lower part of the Prussian suite date from the Early Eocene, whilst the upper part of the Prussian suite dates from the Middle Eocene. Similarly, all amber-bearing horizons of Sambia were formed between the Early and Middle Eocene. This took place during the first stage of transgression when the Eocene basin covered only a small area in the form of a deep bay. The Lutetian basin was somewhat larger than the Ypresian basin. In general, it is clear why amber deposits were formed at this time. The wide Late Eocene street (Fig. 3) most probably

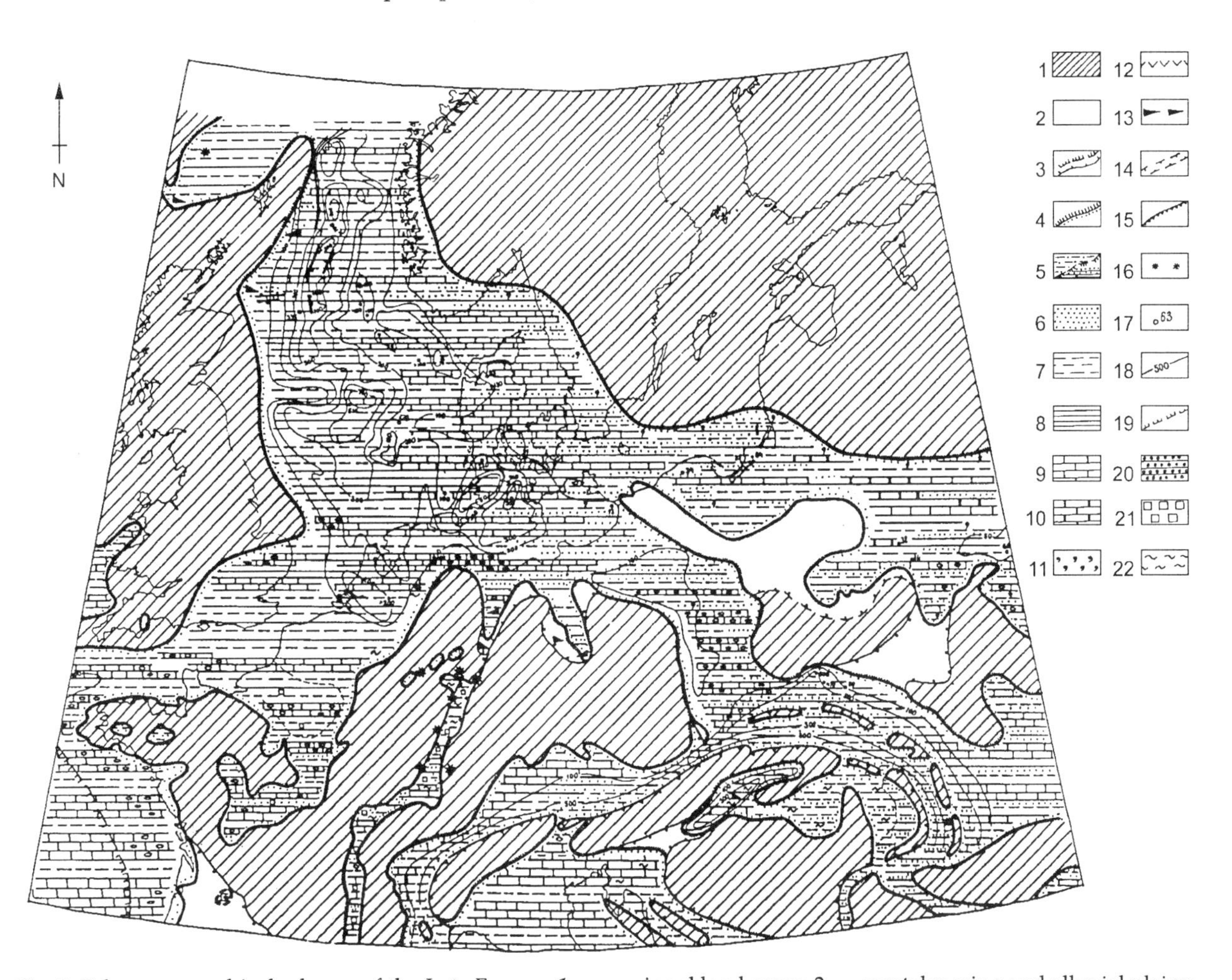

Fig. 3. Palaeogeographical scheme of the Late Eocene. 1 — erosional land areas; 2 — coastal marine and alluvial plains; 3 — sedimentation basin boundaries; 4 — Late Eocene coastline; 5 — coastal plains periodically submerged by sea; 6 — sands, sandstones; 7 — silts, siltstones; 8 — clays, mudstones, shales; 9 — carbonate deposits (limestones, marls); 10 — siliceous deposits; 11 — glauconite; 12 — volcanoclastics; 13 — peat, lignite, coal; 14 — fractures; 15 — alpine orogenic front; 16 — volcanoes; 17 — basic sections and deposit thickness (in metres); 18 — isopach, m; 19 — continental slope; 20 — conglomerates; 21 — evaporites; 22 — redbeds.

promoted the distant drift of amber from the deposits where it formed.

The Pomeranian horizon, below the Mosina beds and Kiev suite of Belarus and the Ukraine, also appear by stratigraphical analogies of Alkian and Prussian suites in Poland. Amber-bearing deposits (Połczyno member) were discovered in north Poland (Chłapowo area) by PIWOCKI & OLKOWICZ-PAPROCKA (1987). However, their stratigraphic position suggests that they are younger than Sambian "blue earth" (RITZKOWSKI 1997).

Late Eocene deposits have been revealed on the sea floor between the Polish coast and the Słupsk Trough. Their thickness ranged from 3.6 up to 11.5 m in 4 core samples taken by the "Petrobaltic" company (KRAMARSKA 1992). The composition of deposits (grey-green glauconitic-quartz sands and others) corresponds with the afore mentioned terrigenous-glauconitic formation of Sambia. Their foraminifera fauna (*Globomalina micra* Zone) are comparable with those of the lower Mosina beds of North Poland, Schönewalde layers in eastern Germany, the Prussian suite of Sambia and the Kiev suite of the Ukraine and Belarus.

What is certain is that there is an absence of information on the genesis of amber-bearing "blue earth". The avandeltaic hypothesis (KATINAS 1971) was considered more grounded in the long time. However, this hypothesis incorrectly combined many reliable lithologic-geochemical factors (KRASNOV & KAPLAN 1976). The most important of these factors was the evidence of glauconite authigenesis, which has recently been confirmed by K-Ar absolute dates. Other factors included: low sedimentation rates which indicate intensive bioturbation, and also the low content of organic carbon (on average 0.32%), as well as phosphorite and shark tooth accumulations. The marine genesis of "blue earth" has been confirmed by the correlation of iron and sulphur forms, higher boron content in adsorbed complex, sapropel composition of organic matter and the presence of zeolites.

Further investigations are necessary to gain a more precise picture of the Eocene sedimentation environment of the Baltic Deep and adjoining areas.

Alexander I. BLAZCHISHIN
Atlantic Branch
of P. P. Shirshov Institute of Oceanology RAS
Prospect Mira 1,
236000 Kaliningrad, Russia

PALEOGEOGRAFIA I AKUMULACJA OSADÓW W EOCENIE NIECKI BAŁTYCKIEJ I OKOLIC

Alexander I. BLAZCHISHIN

Streszczenie

Rekonstrukcja wczesnego eocenu odzwierciedla ważniejsze geodynamiczne wydarzenia — stopniowe i dyferencjalne rozprzestrzenianie się północnego Atlantyku, oddzielenie Grenlandii od Europy, rozwój inwersyjnych ruchów w basenie Środkowej Europy i w basenie Morza Północnego. Nasiliła się działalność wulkaniczna w prowincji szkocko-irlandzkiej i w cieśninie Skagerrak, skąd popioły wulkaniczne dotarły na wschód aż do Sambii. Transgresja wczesnoeoceńska kończyła się na zachodnich obszarach Polski, pozostały basen polsko-litewski łączył się z Morzem Północnym przez nieckę bałtycką. W dolnym eocenie (iprez) szeroko rozciągały się obszary błotniste, gdzie tworzył się węgiel brunatny. W centralnych częściach basenów formowały się czerwone iły londyńskie, analogiczne iły Tarras w Niemczech i iły z Mo w Danii ze znaczną zawartością smektytu. Warstwowane łupki ilaste i diatomity limfordzkie w Danii formowały się w warunkach upwellingu.

W środkowym eocenie na tle degradacji barier orogenicznych i ruchów inwersyjnych miało miejsce stopniowe podnoszenie się poziomu morza, co poprzedziło rozległą morską ekspansję w górnym eocenie. Globalna eoceńska transgresja miała charakter eustatyczny (długookresowych zmian poziomu oceanu światowego), uwarunkowany zwiększeniem prędkości spredingu w fazie rozwoju grzbietu śródoceanicznego. Szczyt ocieplenia klimatu przypadł we wczesnym eocenie, w późnym eocenie klimat oziębił się. Sezonowy upwelling wyrażony w facjach glaukonitowo-kwarcowych rozwijał się we wschodniej marginalnej części basenu polsko-litewskiego, w Szlezwiku-Holsztynie i w Jutlandii.

Transgresywne facje górnego eocenu w Danii, Sambii, Polsce, na dnie niecki bałtyckiej, a także na Ukrainie i Białorusi zaznaczyły się podobnymi prowincjami faunistycznymi, co wskazuje na szeroką łączność basenu z rejonami Tetydy na południu i arktycznymi na północy. W górnym priabonie — w związku z orogenezą pirenejską — zauważa się tendencje regresywne i ślady osuszania klimatu. Zgodnie z nowym datowaniem metodą potasowo-argonową (RITZKOWSKI 1997), dolna część bursztynonośnego pokładu Sambii jest datowana na wczesny eocen (iprez), produktywna niebieska ziemia (fm. pruska) na środkowy eocen (lutet).

Jeśli uwzględni się paleogeograficzny schemat wczesnego eocenu (fig. 1), można zrozumieć warunki formowania się ważniejszych w świecie pokładów bursztynu. We wczesnym stadium transgresji eoceńskiej basen morski stanowił wąski, głęboko wcięty w granice lądu obszar wodny, w którym osadzały się bursztynowe złoża okruchowe. Szeroka późnoeoceńska cieśnina (fig. 3) przyczyniła się raczej do dalszego rozprzestrzeniania bursztynu niż do jego koncentracji.

Bibliography

ALBERS H.
1981 Neue Daten zum Klima des nordwesteuropaeischen Alttertiars, *Fortschritte in der Geologie von Rheinland und Westfalen*, **29**, 483–503.

AM K.
1973 Geophysical indications of Permiam and Tertiary igneous activity in the Skagerrak, *Norges Geologiske Undersökele*, **287**, 1–25.

BLAZCHISHIN A. I.
1974 Geological structure of submarine coast slope of Sambian Peninsula, *Regionale geology of Baltic land* [in Russian], 165–172, "Zinatne", Riga.

1991 *Baltic Sea and North Sea. Palaeogeographical Atlas of Shelf Region of Eurasia for Mesozoic and Cenozoic, Part 1. Expalanatory note*, 1–11 & 15, Manchester.

1997 *New data on wide-spread of Paleogene amber-bearing deposits at submarine slope of Sambian Peninsula. The ecological problems of Kaliningrad region* [in Russian], 75–80, Russian Geogr. Society, Kaliningrad State University.

BURCHARDT B.
1978 Oxygen isotope paleotemperatures for the Tertiary period in the North Sea area, *Nature*, **275**, 291–292.

CIUK E.
1983 Paleogen i podłoże mezozoizne w otworze Goleniów 18-2 w Zielonczynie, woj. szczecińskie, *Przegląd Geologiczny*, **31** (7), 415–420.

DIENESEN A., MICHELSEN O. & LIEBERKIND K.
1977 A survey of the Paleocene and Eocene deposits of Jylland Fyn, *Danmarks geologiske Undersřgelse, B*, **1**, 1–15.

GRIPP K.
1964 *Erdgeschichte von Schleswig-Holstein*, 1–680, K. Wachholtzverlag, Neumuenster.

JACQUE M. & THOUVENIN J.
1975 Lower Tertiary tuffs and volcanic activity in the North Sea, [*in*:] Woodland A. W. (ed.), *Petroleum and the Continental Shelf of North-West Europe, Part 1, Geology*, 477–485, Applied Science Publishers, Barking, Essex.

KATINAS V.
1971 *Amber and amber-bearing deposits of Southern Baltic Land* [in Russian], 1–155, "Mintis", Vilnius.

KASTING J. & RICHARDSON S.
1985 Seafloor hydrotermal activity and spreading rates: the Eocene carbon dioxide greenhouse revisited, *Geochimica et Cosmochimica Acta*, A, **75**, 465–487.

KNOX R. W. & HARLAND K.
1979 Stratigraphical relationships of the Early Palaeogene ash-serries of NW Europe, *Journal of the Geological Society*, **136**, 463–470, London.

KOSMOWSKA-CERANOWICZ B.
1979 Zmienność litologiczna i pochodzenie okruchowych osadów trzeciorzędowych wybranych rejonów północnej i środkowej Polski w świetle wyników analizy przezroczystych minerałów ciężkich, *Prace Muzeum Ziemi*, **30**, 3–73.

KRAMARSKA R.
1992 Paleogene deposits from the southern Baltic. Meerwissenschaftliche Berichte, *Proceedings of the Second Marine Geol. Conference "The Baltic", Warnemünde*, 86–83.

KRASNOV S. G. & KAPLAN A. A.
1976 On genesis of amber-bearing Paleogene deposits in Kaliningrad region by lithological investigations [in Russian], *Lithology and useful minerals*, **4**, 95–106.

KUMPAS M.
1978 Distribution of sedimentary rocks the Hanoe bay and S. of Oeland, South Baltic, *Stockholm Contributions in Geology*, **31** (3), 95–103.

MCKENNA M.
1983 *Cenozoic paleogeography of North Atlantic land bridges. Structure and Development of Greenland-Scotland Ridge*, 351–399, New York-London.

PEDERSEN G. K.
1981 Anoxic events during sedimentation of a Paleogene diatomite in Denmark, *Sedimentology*, **28** (4), 487–504.

PIWOCKI M. & OLKOWICZ-PAPROCKA I.
1987 Litostratygrafia paleogenu, perspektywy i metodyka poszukiwań bursztynu w Północnej Polsce, *Biuletyn Instytutu Geologicznego*, **356**, 1–22.

PLEZIAT J-C.
1981 Late Cretaceous to Late Eocene palaeogeographic evolution of southwest Europe, *Palaeogeography, Palaeoclimatology, Palaeoecology*, **36**, 263–320.

POMEROL CH.
1978 Evolution paleogeographie et structurale du basin de Paris du Precambrian a l'actuel, en relation avec les regions avoisinantes, *Geologie en Mijnbouw*, **57**, 533–543.

POŻARYSKA K.
1977 Upper Eocene Foraminifera of East Poland and their paleogeographical meaning, *Acta Paleontologica Polonica*, **22**, 1.

RITZKOWSKI S.
1997 K-Ar-Alterbestimmungen der bernsteinführenden Sedimente des Samlands (Paläogen, Bezirk Kaliningrad), Sondernheft *Metalla*, **66**, 19–23.

SVIRIDOV N. I.
1990 *The seismoacoustic character of work area. Geoacoustic and gase-lithogeochemical investigations in Baltic Sea* [in Russian], 34–47, Institute of Oceanologie RAS, Moscow.

SVIRIDOV N. I. & LITVIN V. M.
1978 The bottom structure of south-western part of Baltic Sea [In Russian], *Soviet Geology*, **4**, 27–41.

TRÖGER K. A.
1984 *Abriss der historischen Geologie*, 1–718, Akademie-Verlag, Berlin.

ZIEGLER P.
1982 Geological Atlas of Western and Central Europe, *Elsevier*, 1–130, Maatsch: Shell Intern. Petrol, Amsterdam.

CRITERIA FOR THE RECOGNITION OF THE PRUSSIAN SUITE (FORMATION) AND PRUSSIAN HORIZON

Nadezhda P. LUKASHINA & Gennadij S. KHARIN

Abstract

On the basis of indicators which are characteristic of the Prussian suite in the Sambian stratotype section (North-West Russia), correlation of the local stratigraphic subdivisions of the Upper Palaeogene amber-bearing deposits of Denmark, Northern Germany, Poland, Belarus and the Ukraine was carried out.

Palaeogene sediments in the western part of the Kaliningrad region and in the adjacent aquatoria of the Baltic Sea, the Curonian Bay and the Bay of Gdańsk are represented by terrigenous sandy-argillaceous facies. These occur at different horizons of the Upper Cretaceous and are covered by Neogene or Quaternary deposits. Upper Eocene sediments, containing amber and phosphorites, are of the greatest commercial value. The Prussian suite, which was first recognized at the end of the last century by ZADDACH (1868), belongs to the most productive mass of the Upper Eocene.

This work aims to correlate local stratigraphic subdivisions of amber-bearing deposits, as well as to identify a regional stratigraphic subdivision — the Prussian horizon. Published materials form the basis of this paper, supplemented by the authors' own data about foraminifera from the amber opencast mine located on the Sambian Peninsula (western part of the Kaliningrad region).

In a stratotype section on the Sambian Peninsula, the Prussian suite occupies a thickness of more than 40 m. Proceeding upwards from the bottom, it consists of four layers:

1. „wild earth" — clayey, glauconitic-quartz sands of different grain size with lumps of clay, concretions of phosphorites and shark's teeth washed out from the underlying Alkas suite;
2. „blue earth" — greenish-grey, clayey, micaceous, glauconitic-quartz sands and silts, including amber pieces, concretions of phosphorite and pyrite, wood remnants and pyritic tubes, which serve as traces of worm activity.
3. "floating earth" — greenish-grey, glauconitic-quartz sands of different grain size with concretions of siderite, pyrite and phosphorite, with amber and sometimes with ferrugineous sands, clay balls and concretions of phosphorites in the basis — so-called "krant";
4. "white wall" — strongly micaceous silts with carbonised remnants of plants and small-sized amber pieces (KATINAS 1971).

The largest deposits of amber in the world are to be found in the north-western part of the Kaliningrad region, confined to a layer of "blue earth". The amber-bearing sediments of these deposits appear to have been formed in the delta of a river which drained Scandinavia (KATINAS 1971; JAWOROWSKI 1984; 1987). Frequently found attributes of anaerobic conditions, i.e. pyritisation, a smell of hydrogen sulphide, and the black colour of sediments made it possible to consider them as lagoon or lagoon-deltaic deposits (KHARIN 1995). A river of the Sambian Peninsula entered into a strait, which connected the North Atlantic and the Tethys (or Paratethys). The waters of this river carried debris material and non-fossil resins from coniferous trees. Here, in the river delta, resin accumulated and matured, transforming into amber. This is how the primary deposits of amber were formed. These deposits were eroded and the amber distributed by coastal currents in a south-easterly direction in winter and north-west in summer (CINCURA 1989). Amber is considered as a placer mineral because it is carried large distances from its primary place of deposition and forms secondary accumulations (TROPHIMOV 1974). The distribution of amber across the waters of the Eocene Sea is associated with the formation of secondary deposits in Poland, Germany[1], Denmark[2], Belarus

[1] Upper Oligocene–Lower Miocene. (Ed.)

[2] In Mo-clay from Middle Eocene. (Ed.)

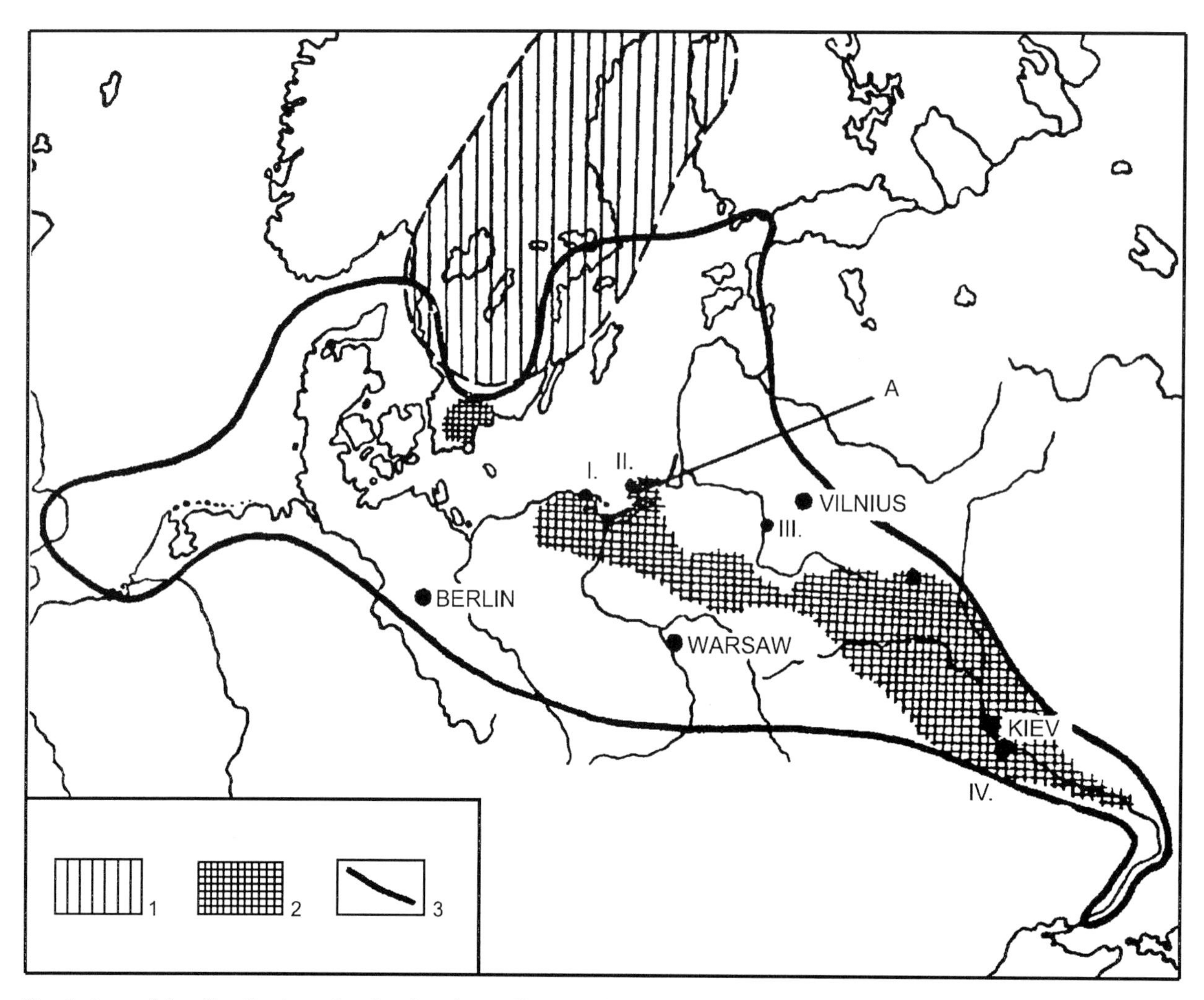

Fig. 1. Area of the distribution of amber-bearing sediments (KATINAS 1971): 1 — primary deposits of amber; 2 — amber of Palaeogene deposits; 3 — boundary of distribution of redeposited amber; A — Kaliningrad region, I–IV — position of sections (I: Chłapowo, II: Sambia, III: Grodno, IV: Obuchov)

and the Ukraine[3] (Fig. 1; KATINAS 1971). The properties and structure of this amber are similar to amber from the Sambian Peninsula. When the strait was closed by the Mazuro-Belarus Rise and existed as a bay, amber accumulated on the northern slope of this elevation. As a result, we can expect to find large quantities of loose amber in this area. Currents transported not only amber but also other minerals, in particular glauconite (KATINAS 1971; KHARIN1995).

Amber-bearing deposits outside the Kaliningrad area are represented by other facies. The Chłapowo area, in north-eastern Poland, is characterised by a lithological-facial composition of amber-bearing deposits similar to those of the Sambian Peninsula (KOSMOWSKA-CERANOWICZ & MÜLLER 1985). Taking this assumption further led to the conclusion that the amber deposits of Sambia and Chłapowo (Fig. 1: I–II) came from one delta. However, detailed prospecting has revealed that the productive layer of "blue earth" could not be traced to the west and south-west of the Sambian Peninsula. These were most probably different marine lagoons of the same age, connected by tributaries of an ancient river (the Pra-Neva[4]). In the Ukraine and Belarus amber

[3] This theory of currents has opponents: the present of amber in Ukraine is interpreted with transport of material from South, not with migration in Eocene basin (see KOSMOWSKA-CERANOWICZ *et al.* 1990; TUTSKIJ 1997). (Ed.)

[4] After KOSMOWSKA-CERANOWICZ & KONART 1989: Eridanus. (Ed.)

debris occurs among terrigenic-glauconitic deposits (KATINAS 1987), but can also occur among carbonaceous and clayey-carbonaceous (marly) deposits.

The Prussian suite, as a unit of the local stratigraphic system, was distinguished by BALTAKIS (1967) on the basis of certain attributes and the lithological structure of its constituent layers, flora and fauna, availability of amber, glauconite, and phosphorite nodule content. However, the occurrence of "blue earth" could not be accepted as a commonly applied criterion, because amber was transported by currents of the Upper Eocene sea and accumulated outside "blue earth" facies. Amber is normally found, not only in Upper Eocene deposits, but also in terrigenous deposits of the Oligocene and even Quaternary periods, where redeposition of the amber took place. In this case, the Prussian suite should be identified primarily on the basis palaeontological data.

Abundant occurrence of fauna in Prussian suite sediments is associated with "krant" facies. NOETLING (1885) found 152 species of organism here — from bryozoan to vertebrates. This faunal complex served as the basis for comparing the Prussian suite with the Lattorfian stage of Northern Germany, which belonged at that time to the Lower Oligocene. The microphitoplankton complex which was classified by ZATULA (1973) on the basis of the Prussian suite, revealed its similarity with the Late Eocene microflora of the Ukraine. KAPLAN *et al.* (1977) demonstrated that balls of "blue earth" roof incorporate large bivalvia complexes, including considerable amounts of *Cubitostrea plicata Sol.*, which is described from Bartonian and Latdorfian stages.

Palynocomplexes of "blue earth" and "floating earth" have been studied. It transpires that their composition is similar to that of complexes from the lower half of the Kharkov series (Belarus) and the Obukhov suite of the Kiev-Dnieper region, as well as to the Almin horizon of the Crimea and the Kuban section in the North Caucasus (GRIGELIS *et al.* 1988; DIDKOVSKY *et al.* 1979). The occurrence of lithological composition and fauna of the same stratotype made it possible to identify the Prussian suite in south-western Lithuania (Tab.1). It is represented here by glauconite-quartz sands of different grain size, without amber pieces and phosphorite nodules. The absence of these could be explained by the fact that only the upper layers of the Prussian suite — "floating earth" and "white wall" — were deposited here. This area lies in close proximity to the Mazurian-Belarus Swell and up until recent times was subjected to washing out processes. On the basis of its foraminifera, the Prussian suite in the Sambian Peninsula can be assigned to planktonic foraminifera zones P 15, P 16 and in part to P 17 (HARLAND *et al.* 1985). The Prussian suite in Lithuania yielded planctonic foraminifera *Globoquadrina corpulenta* (Subb.), which confirms this connection. Foraminifera are represented here predominantly by benthonic species (GRIGELIS *et al.* 1971). Benthonic foraminifera characterise Upper Eocene deposits in Belarus, the Grodno region and Gomel (PHURSENKO & PHURSENKO 1961), and in Poland in the vicinities of Chłapowo and in the amber-bearing Połczyno strata (ODRZYWOLSKA-BIEŃKOWA 1987). Foraminifera complexes in Upper Eocene marly deposits in the north-eastern part of the Polish Lowland are more abundant and contain more planktonic species (POŻARYSKA 1977). Our study has shown that in Prussian suite deposits from the amber quarry on the Sambian Peninsula planktonic foraminifera are predominant, indicating the existence of a boreal sea of normal salinity. Varieties of these foraminifera are present in layers of "wild earth" and "blue earth", whilst they are virtually absent from "floating earth" and "white wall", which shows that the basin gradually became shallower over time. The aforementioned species are predominantly Upper Eocene, but younger species of Recent face, which were subjected to redeposition, were also found. The differences between Upper Eocene foraminifera complexes from various areas is explained by facial conditions: the domination of benthonic species indicates shallow-water sedimentation conditions, while planktonic species are indicative of deep-water conditions, which reached up to about 200 m in the vicinitiy of the Sambian Peninsula.

The systematic similarity of foraminifera in the amber-bearing deposits of Belarus, the Ukraine, Poland and Germany, as well as the wide incidence of glauconite and phosphorites in these sediments and sediments of the Sambian Peninsula is indicative of their synchrony. There are some differences with regard to the age of amber-bearing deposits. Thus, in Denmark amber-bearing deposits of Mo-clay formation are Middle Eocene (KATINAS 1987); on the Sambian Peninsula, according to absolute dating established by KAPLAN *et al.* (1977), the age of "blue earth" is about 37.7 mln. years — Upper Eocene, this age is corroborated by the results of research into spore-pollens and phitoplankton in "wild earth" and "blue earth" (KOSMOWSKA-CERANOWICZ *et al.*

SERIES		Stages	Planktonic-foraminifera zones	Nannoplankton zones	Formation / Suite	Denmark	N Poland	Chłapowo	Sambian Peninsula	S-W Lithuania	N-W Belarus	N Ukraine
OLIGOCENE	LOWER	Rupelian	P 20	NP					Palvesk fm		Ckarkkov fm	Mezhigorsk fm
			P 19	23								
			P 18	NP 22			Czempin fm					
			P 17	NP			Lower Mosina fm	Chłapowo mb				
EOCENE	UPPER	Priabonian		21	Prussia fm			Połczyno mb	Prussia fm	Prussia fm	Obuchov fm	Obuchov fm
			P 16	NP 20			Pomerania fm	Upper Mieroszyno mb				
			P 15	NP 18								
	MIDDLE	Bar.	P 14	NP 17		Mo-clay fm		Lower Mieroszyno mb	Alk fm		Kiev fm	Kiev fm
		Lutetian	P 13	NP 16								
			P 12	NP 15								
			P 11	-								Buchak fm
			P 10	NP 14								
	LOWER	Ypresian	P 9						Sambia fm			Kanev fm
			P 8	-								
			P 7	NP 12								
			P 6	NP 10								

Table 1. Correlation of the local stratigraphic schemes of amber-bearing deposits of Middle und Upper Palaeogene.

1997), but according to RITZKOWSKI (1997) it is nearer 44.1 mln. years — Middle Eocene. Proceeding from the composition of foraminifera and nannoplankton, the amber-bearing Połczyno member belongs to nanno-zone NP 21 — the transition between the Upper Eocene and Lower Oligocene (ODRZYWOLSKA-BIEŃKOWA 1987; KOSMOWSKA-CERANOWICZ & MÜLLER 1985). In the Ukraine amber-bearing deposits of the Mezhigorsk suite are Lower Oligocene in age. This topic remains open to debate and requires further research to reach a satisfactory conclusion.

Discoveries of amber in Palaeogene sediments in Belarus, the Ukraine, Poland, Denmark, Germany and on the Sambian Peninsula (Russia) could be used for correlating local stratigraphic subdivisions, as well as for identifying regional Prussian horizons. We believe that this horizon incorporates the Prussian suite proper, a stratotype of which is to be found in the amber open cast mines of the Sambian Peninsula, Obuchov and the Charkov suites of Belarus, the Mezhigorsk suite of the Ukraine, and the lower part of the Mosina formation in northern Poland.

Acknowledgements

The work has been carried out by the Geology Laboratory of the Atlantic, with the support of RFFI, grant 96-15-98336.

The authors wish to express their thanks to Dr V. SIVKOV for his assistance in the selection of material from the amber qaurry, to Dr A. KRYLOV and Dr V. SIVKOV (President and Director of the "Sea Venture Bureau") for their financial support and to Prof. E. EMELYANOV for his valuable remarks made after looking through this paper.

Nadezhda P. LUKASHINA & Gennadij S. KHARIN
Atlantic Branch
of P. P. Shirshov Institute of Oceanology RAS
pr. Mira 1, 236000 Kaliningrad, Russia
e-mail: kharin@geology.ioran.kern.ru

KRYTERIA ROZPOZNANIA SUITY (FORMACJI) PRUSKIEJ I POZIOMU PRUSKIEGO

Nadezhda P. LUKASHINA,
Gennadij S. KHARIN

Streszczenie

Bursztynonośne osady różnych facji są rozprzestrzenione na rozległej powierzchni, która w starszym paleogenie była częścią cieśniny łączącej Północny Atlantyk z Tetydą. Na teren Półwyspu Sambijskiego (Rejon Kaliningradzki) rzeki niosły nie do końca utwardzone żywice sosnowych lasów ze Skandynawii i osadzały je w lagunowych deltach, gdzie ulegały procesom przemiany i formowały pierwotne złoża bursztynu. Ruchy tektoniczne i oscylacje poziomu morza doprowadzały do rozmywania pierwotnych bursztynonośnych osadów. Prądy w cieśninach o różnych kierunkach (w powiązaniu z monsunami) roznosiły bursztyn na południowy-wschód i północny-zachód. Dzięki tym procesom, tworzyły się na obecnych obszarach Danii, Północnych Niemiec, Polski, Białorusi i Ukrainy wtórne bursztynonośne złoża[1].

Główny bursztynonośny horyzont na Półwyspie Sambijskim to górnoeoceńska suita (formacja) pruska, wydzielona na podstawie kryteriów litologicznych i paleontologicznych, obecności bursztynu, ziarn glaukonitu i konkrecji fosforytowych. Kryteria te poza Sambią nie zawsze są miarodajne. Bursztyn w warunkach redepozycji może być spotykany w coraz młodszych osadach (aż do czwartorzędowych). Stąd potrzeba oddzielenia w profilu stratygraficznym, jednoczesnego w czasie z sedymentacją pruskiej formacji, **horyzontu** pruskiego. Horyzont obejmuje następujące lokalne podziały stratygraficzne: Obuchowską i Charkowską suitę Białorusi, Mezogorską Ukrainy, dolną część formacji mosińskiej (ogniwo Połczyna) w północnej Polsce, piętro bartońskie w Północnych Niemczech i formację Mo w Danii.

Wiek pruskiego horyzontu opiera się na faunistycznym i florystycznym kompleksie. Systematyczne podobieństwo otwornic i innych mikroorganizmów w bursztynonośnych osadach Danii, Północnych Niemiec, Polski, Białorusi, Ukrainy i Półwyspu Sambijskiego potwierdza ich równowiekowość. Ponadto i inne wskazówki mogą być użyte dla korelacji górnego paleogenu i wydzielenia pruskiego horyzontu.

[1] Teoria prądów jest dyskusyjna. Jak wynika z badań, obecność bursztynu na Ukrainie tłumaczy się transportem materiału z południa, a nie migracją w basenie eoceńskim (por. KOSMOWSKA-CERANOWICZ *et al.* 1990; TUTSKIJ 1997). (Red.)

Bibliography

BALTAKIS V.
1970 On the stratigraphy and lithostratigraphical correlation of the paleogene deposits of Sambia [in Russian], *Paleontologija i stratigrafija Pribaltiki i Belorussii*, sb. **2**, 325–340, Vilnius.

CINCURA L.
1989 Paleoclimatic problems of Fennoscandia from the view-point of the Tethys realm, *Terra*, **101**, (1), 42–45.

DIDKOVSKY W. J., ZELINSKAJA W. A. & ZERNETCKY B. F.
1979 *The biostratigraphical substantation of borders in Paleogene and Neogene of Ukraine*, Nauk. Dumka, Kiev.

GRIGELIS A. A., BALTAKIS V. & KATINAS V.
1971 Stratigraphy of Paleogene deposits of Pribaltica [in Russian], *Izwiestija Akademii Nauk SSSR*, ser. geol., **3**, 107–116, Moskwa.

GRIGELIS A. A., BURLAK A. F., ZOSIMOWICH W. U., IVANIK M. M., KRAEWA E. J., LULJEWA S. J. & STOTLAND A. B.
1988 New data on stratigraphy and paleogeography of Paleogene deposits of the West European part of USSR [in Russian], *Sowietskaja geologija*, **12**, 41–55, Moskwa.

HARLAND W. B., COX A. V., LLEWELLYN P. G., PICTON C. A. G., SMITH A. G. & WALTERS R.
1985 A geologic time scale, 1–140, "Mir".

JAWOROWSKI K.
1984 Warunki sedymentacji bursztynonośnych osadów trzeciorzędowych z okolic Chłapowa, *Przegląd Geologiczny*, **32** (4),194–196.

1987 Geneza bursztynonośnych osadów paleogenu w okolicach Chłapowa, *Biuletyn Instytutu Geologicznego*, **356**, 86–102.

KAPLAN A. A., GRIGELIS A. A., STRELNIKOVA N. I. & GLIKMAN L. S.
1977 Stratigraphy and correllation of Paleogene deposits of South-Western Pribaltic region [in Russian], *Sowietskaja geologija*, **4**, 30–43.

KATINAS V.
1971 Amber and amber-bearing deposits of the Southern Pribaltic [in Russian], 1–156 "Mintis", Vilnius.

1987 Amber-bearing terrigenic-glauconitic formation of Paleogene of Pribaltica and Byelorussia, [*in:*] *Tectonica, facies and formations of west of West-European platform*, 184–189, Minsk.

KHARIN G. S.
1995 Geological conditions of the amber-bearing deposits originating in the Baltic Region, *Amber & Fossils*, **1**, 47–54.

KOSMOWSKA-CERANOWICZ B., KOCISZEWSKA-MUSIAŁ G., MUSIAŁ T. & MÜLLER C.
1990 Bursztynonośne osady trzeciorzędowe okolic Parczewa (The amber-bearing Tertiary sediments near Parczew), *Prace Muzeum Ziemi*, **41**, 21–35, Warszawa.

KOSMOWSKA-CERANOWICZ B., KOHLMAN-ADAMSKA A. & GRABOWSKA I.
1997 Erste Ergebnisse zur Lithologie und Palynologie der bernsteinfürenden Sedimente im Tagebau Primorskoje, *Metalla*, **66**, 5–17, Bochum.

KOSMOWSKA-CERANOWICZ B. & KONART T.
1989 Tajemnice bursztynu, *Sport i Turystyka*, Warszawa.

KOSMOWSKA-CERANOWICZ B. & MÜLLER C.
1985 Lithology and calcareous nannoplankton in amber-bearing Tertiary sediments from boreholes Chłapowo Northern Poland), Bulletin of the Polish Academy of Science, *Terre*, **33** (3–4), 119–129.

NOETLING F.
1885 Die Fauna des samländischen Tertiärs. Teil 1 (Gastropoda, Pecycypoda, Bryozoa, Geologischer Teil), *Abhandlungen zur geologischen Specialkarte von Preußen und den Thüringischen Staaten*, **6** (4), I–VIII & 1–109, Tafelband, Berlin.

ODRZYWOLSKA-BIEŃKOWA E.
1987 Biostratygrafia paleogenu w okolicach Chłapowa na podstawie mikrofauny (Summary: Biostratigraphy of the Paleogene in the area of Chłapowo, based on microfauna), *Biuletyn Instytutu Geologicznego*, **356**, 52–63.

PHURSENKO F. W. & PHURSENKO K. B.
1961 Foraminifera of the top of Eocene of Byelorussia and their stratigraphical significance [in Russian], *Paleontology and stratigraphy of BSSR*, sb. III, ŔS BSSR, 246–354, Minsk.

POŻARYSKA K.
1977 Upper Eocene Foraminifera of East Poland and their palaeogeographical meaning, *Acta Palaeontologica Polonica*, **22** (1), 3–50.

RITZKOWSKI S.
1997 K-Ar-Alterbestimmungen der bernsteinfürenden Sedimente des Samlandes (Paläogen, Bezirk Kaliningrad), *Metalla*, **66**, 19–23, Bochum.

TROPHIMOV W. S.
1974 Amber [in Russian], 1–184, "Nauka".

TUTSKIJ W.
1997 Geologie und Entstehung der Bernsteinvorkommen im Nordwesten der Ukraine, *Metalla*, **66**, 57–62, Bochum.

ZADDACH G.
1867 Das Tertiärgebirge Samlands, *Schriften der Physikalisch-Ökonomischen Gesellschaft zu Königsberg in Pr.*, **8**, 85–197, Königsberg.

ZATULA K. F.
1973 Kompleksy gistrichosfer iz jantarenosnych otlozenij Pribaltiki, *Doklady Akademii Nauk SSSR*, **212** (4), 981–983.

DAS GEOLOGISCHE ALTER DER BERNSTEINFÜHRENDEN SEDIMENTE IN SAMBIA (BEZIRK KALININGRAD), BEI BITTERFELD (SACHSEN-ANHALT) UND BEI HELMSTEDT (SE-NIEDERSACHSEN)

Siegfried RITZKOWSKI

Kurzfassung

Die bernsteinführenden Sedimente in östlichen Mitteleuropa besitzen sehr unterschiedliche geologische Alter. Die ältesten sind möglicherweise im Paläozän gebildet, die mit Bitterfelder Bernstein im jüngsten Oberoligozän. Die Sedimentation von Bernstein war kein Kurzzeit-Ereignis, sondern erfolgte während eines Zeitraums von mindestens 23 Ma (47–24 Ma B.P.).

Die K-Ar-Alter der Glimmer geben Hinweis auf die Herkunft des Sediments. Die K-Ar-Alter der Glimmer von Sambia und Bitterfeld betragen etwa 1 Ga. Sie weisen auf ein Liefergebiet in SW-Schweden und S-Norwegen und schließen somit eine Herkunft des Sediments sowohl vom Variszikums des südlichen Festlandes als auch aus dem alten Kristallin um den Bottnischen Meerbusen aus. Die Glimmer untereozäner Glaukonitsande von Helmstedt/ Niedersachsen, die um 346 Ma B.P. während der variszischen Orogenese gebildet wurden, stammen von dem südlichen Festland des oberoligozänen Meeres.

Die radiometrischen Alter der bernsteinführenden Folgen und ihrer Glimmer sind nur zwei, wenngleich grundlegende Befunde zur Entschlüsselung der Bernsteinlagerstätten. Es fehlt bislang eine grenzüberschreitende Synopse der Vorkommen in Mittel- und Osteuropa und ihrer paläogeographischen Situation.

Problemstellung

Bernstein ist ein fossiles Harz. Das Harz, das in Wäldern eines Festlandes von den Bäumen tropfte, legte einen weiten Weg zurück und erfuhr Veränderungen, ehe es etwa in der Blauen Erde, so beim Baltischen Bernstein, oder im Glimmerschluff des Tagebaus Goitsche bei Bitterfeld, so beim Bitterfelder Bernstein, in marinen Sedimenten eingelagert wurde. Die bernsteinführenden Sedimente sind gleichsam die ersten geologischen Zeugen über das fossile Harz. Sie liefern die ersten und grundlegenden Informationen über den Bernstein, soweit er sie nicht selbst enthüllt.

Das geologische Alter der bernsteinführenden Schichten

Die erste und wichtigste Information ist das geologische Alter der bernsteinführenden Horizonte. Die Feststellung der Gleichzeitigkeit oder Ungleichzeitigkeit bedarf verläßlicher Methoden und identer Bezugsskalen. Stratigraphische Begriffe besitzen bei verschiedenen Autoren unterschiedliche Bedeutung. Sie verlieren im Laufe der Zeit auch ihre Eindeutigkeit. Es ist deshalb notwendig, chronostratigraphischen Angaben früherer Autoren gleichsam „umzurechnen" auf den heutigen Stand der Kenntnis. Im Folgenden wird auf den geochronologische und chronostratigraphische Skala von BERGGREN *et al.* (1995) Bezug genommen.

Die Ungenauigkeit chronostratigraphischer Zeitangaben lässt sich am „Unter-Oligozän" demonstrieren. Dieser Zeitabschnitt, den BEYRICH (1856) definiert und im wesentlichen A. v. KOENEN (1865; 1889–1894) mit konkretem, paläontologischen Inhalt erfüllt hatte, umfaßt einen Zeitraum vom Mitteleozän bis zum Mittel-Oligozän (MARTINI & RITZKOWSKI 1968; 1970), also die Stufen Lutetium, Bartonium, Priabonium und Rupelium der chronostratigraphischen Skala nach BERGGREN *et al.* (1995). Diese Stufen decken zwischen ca. 47 bis ca. 30 Ma B.P. einen Zeitraum von etwa 17 Mio Jahren. In diesem Zeitabschnitt der Erdgeschichte liegen die bernsteinführenden Folgen im östlichen Mitteleuropa. Deshalb bedarf die stratigraphische Korrelation im Eozän und Oligozän zwischen Deutschland, Nordpolen, Samland und Ukraine weitergehender Präzisierung.

Epoch* (Ma)	Stage* (Ma)	Sambia (Ma)	Chłapowo	Bitterfeld/ Helmstedt
MIOCENE	Aquitanian	? Flußsand (N1)		
23,8				
	Chattian (4,7)			Bitterfelder Glimmersand
OLIGOCENE	28,5			
	Rupelian (5,2)		Chłapowo	
33,7				
	Priabonian (3,3)		Chłapowo?	
	37,0			
	Bartonian (4,3)	Grüne Mauer 38,1+/-1,4 38,8+/-1,2		
EOCENE	41,3			
	Lutetian (7,7)	Blaue Erde 44,1+/-1,1 Wilde Erde 47,0+/-1,2		
	49,0			
	Ypresian (5,5/5,8)			Spurensand
54,5 (54,8)				
	Thanetian (3,4/3,1)			
	57,9			
PALEOCENE 10,5 (10,2)	Selandian (3,0)			
	60,9			
	Danian (4,1)			
65,0				
CRETACEOUS	Maastrichtian			

* Nach BERGGREN *et al.* 1995.

Tabelle 1. Chronologische und chronostratigraphische Tabelle des Paläogens.

Das Alter der bernsteinführenden Folgen im Samland

Die stratigraphische Einstufung der bernsteinführenden Horizonte des Samlands in das Unter-Oligozän gründen sich vornehmlich auf die Molluskenfaunen (BEYRICH 1853; MAYER 1861; NOETLING 1883; 1885; 1888). Die stratigraphische Auswertung der Dinoflagellaten-Zysten im Profil des Tagebaus Primorskoje/Sambia bei KOSMOWSKA-CERANOWICZ *et al.* (1997) führt zur Einstufung in die Zonen D 12 für die „Wilde Erde" und die „Blaue Erde" und in die Zonen D 12/13 für die „Grüne Mauer" und den schokoladenfarbenen Ton darüber in die Zonnen D 12/13 ein. Die Zone D 12 wird von KÖTHE (1990) dem Priabonum, die Zone D 13 bereits dem Rupelium zugeordnet. Einstufungen mit Hilfe des kalkigen Nannoplanktons oder planktonischer Foraminiferen, die zu den verläßlichsten Angaben hätten führen können, sind infolge der Karbonatfreiheit der Sedimente nicht möglich. Die radiometrischen Datierungen von KAPLAN *et al.* (1977), die jedoch keine Erläuterungen der Meßstandards und der Labormethoden enthalten und deshalb einer kritischen Beurteilung sich entziehen, deuten auf ein eozänes Alter der bernsteinführenden Horizonte „Untere Blaue Erde" (41+/-3,5 Ma), „Blaue Erde" (37,7+/-3,0 Ma) und „Grüne Mauer" (34,6+/-3,0 Ma).

Eigene Untersuchungen an Glaukoniten (AHRENDT *et al.* 1995) ergaben folgende K-Ar-Alter und chronostratigraphische Einstufungen:

Formation/Member	K-Ar-Alter (Ma)	Chronostratigraphie/Chronologie
Grüne Mauer	38,1+/-1,4 38,8+/-1,2	Bartonium/Middle Eocene (41,3–37,0 Ma BP)
Blaue Erde	44,1+/-1,1	Lutetium/Middle Eocene (49,0–41,3 Ma BP)
Wilde Erde	47,0+/-1,5	Lutetium/Middle Eocene (49,0–41,3 Ma BP)
Untere Blaue Erde	Ohne Befund	?Ypresium/Early Eocene (54,5–49,0 Ma BP)

Tabelle 2. Sambia, Tagebau Primorskoje, K-Ar-Alter der Glaukonite (AHRENDT *et al.* 1995). Chronologische Daten aus BERGGREN *et al.* 1995.

Insgesamt tritt Bernstein in Folgen auf, deren Sedimentation in den Zeitraum zwischen 47 und 38 Ma

B.P. fällt und eine Zeitspanne von mindestens 9 Ma (in den Verläßlichkeitsgrenzen 12 Ma) umfaßt. Zählt man den Bernstein der „Unteren Blauen Erde" hinzu, die vermutlich im Unteren Eozän sedimentiert wurde, dann könnte diese Zeitspanne sogar 17 Ma betragen. Die Sedimentation des samländischen Bernsteins ist nicht ein Kurzzeit-Ereignis, sondern erfolgte über einen beträchtlichen geologischen Zeitraum.

Das geologische Alter der bernsteinführenden Folgen von Chłapowo

In dem Profil von Chłapowo westlich Gdańsk liegt der bernsteinführende Horizont gemäß KOSMOWSKA-CERANOWICZ & MÜLLER (1985) über marinen Schichten mit kalkigem Nannoplankton der Zonen NP 19–20 (Priabonium, Late Eocene) und in marinen Schichten mit der NP 21 (Late Eocene/Early Oligocene). Die Obergrenze der NP-Zone 19–20 wird bei BERGGREN *et al.* (1995) mit 34,2 Ma angegeben, die Obergrenze der NP 21 mit 32,8 Ma. Im norddeutschen Raum enthalten die Schichtverbände, die BEYRICH dem Unteroligozän zugerechnet hat, in ihrem unteren Abschnitt kalkiges Nannoplankton der Zone NP 21. Es kann also vereinfacht festgehalten werden, bei Chłapowo liegt der Bernstein auf einer Lagerstätte oligozänen Alters. Die Schichtfolge ist nicht altersgleich der des Samlandes.

Für ein oligozänes Alter sprechen auch die Vergesellschaftungen der fossilen Dinoflagellaten-Zysten. Sie werden der Dino-Zone D 14 zugeordnet (KOSMOWSKA-CERANOWICZ *et al.* 1997), die fast vollständig den Zone NP 23 und NP 24 nach kalkigem Nannoplankton und damit der Rupel-Stufe entsprechen (KÖTHE 1990). Dagegen decken sich die biostratigraphischen Einstufungen für das Samland auf Grund der Dinoflagellaten-Zysten (D 12 Priabonium und D 12/13 Priabonium/Rupelium) nicht mit den radiometrischen Befunden für „Blaue Erde" und „Grüne Mauer" (Lutetium und Bartonium) bei AHRENDT *et al.* (1995).

Das geologische Alter des bernsteinführenden Glimmerschluffs von Bitterfeld

Bei Bitterfeld in Sachsen-Anhalt liegt der Glimmerschluff der Oberen Cottbuser Folge, der den Bitterfelder oder Sächsischen Bernstein geliefert hat, über den Glimmersanden der Chattischen Stufe (Late Oligocene) und über dem Flöz Breitenfeld und unter dem miozänen Bitterfelder Hauptflöz (KRUMBIEGEL [*in*:] KOSMOWSKA-CERANOWICZ & KRUMBIEGEL 1989). Das Flöz Breitenfeld wurde zwar aufgrund von Sporomorphen als unteres Miozän angesehen (KRUTZSCH [*in*:] BARTHEL & HETZER 1982). Die marinen Dinoflagellaten des Bernsteinschluffes jedoch stellen oberoligozäne Vergesellschaftungen dar (BLUMENSTENGEL, frdl. mdl. Mitt. 1997). Marines Plankton und Glaukonit im Bernsteinschluff über Flöz Breitenfeld zeigen an, daß das Flöz Breitenfeld nur eine regressive Phase in dem oberoligozänen Meeresraum markiert.

Gemäß BERGGREN *et al.* (1995) umfaßt die Chattische Stufe den Zeitraum zwischen 28,5 und 23,8 Ma B.P. Der Bitterfelder Bernsteinschluff dürfte also im jüngsten Oligozän, etwa vor 24 Ma gebildet worden sein.

Das geologische Alter der bernsteinführenden Folgen in Ostpolen und in der Ukraine

Die Bernstein-Vorkommen der Region von Parczew (Ostpolen) (KOSMOWSKA-CERANOWICZ 1996, 300), deren Trägergesteine dem Bartonium und Priabonium (POŻARYSKA 1977, 21; KOSMOWSKA-CERANOWICZ & LECIEJEWICZ 1997) zugerechnet oder in der Ukraine, die nach TUTSKIJ & STEPANJUK (1999) ins Oligozän gestellt werden, weiten nicht den Zeitraum der Bildung der Bernsteinvorkommen.

Das Problem der Gleichaltrigkeit von Bernstein und Sediment

Die bernsteinführenden Horizonte des Paläogens in Deutschland, Polen, Samland und Ukraine wurden während einer Zeitspanne von mindestens 23 Ma (47–24 Ma B.P.) gebildet. Diese große Zeitspanne veranlaßt zu der Frage:

- sind diese Bernstein-Lagerstätten und -Vorkommen Bildungen jeweils synchroner Bernsteinwälder, oder
- resultieren die unterschiedlich alten Vorkommen aus der Umlagerung von Bernstein einer Lagerstätte?

Diese einfache Alternative kann in mannigfacher Weise variiert werden.

Bei Betonung der Gemeinsamkeiten in Fauna und Flora von Bitterfelder (Sächsischem) und Baltischem Bernstein kommt WEITSCHAT (1997, 79) zu der Ansicht, daß „derartig große faunistische Übereinstimmungen ... sich nur durch ein gemeinsames und zeitgleiches Ursprungsgebiet erklären" lassen, wobei als zeitgleich durchaus eine Zeitspanne von ca. 10 Ma verstanden wird (*ibid.* S. 81). Die Lagerstätte Bitterfeld soll demzufolge umgelagerten Baltischen Bernstein enthalten. RÖSCHMANN (1997, 86) hingegen betont die Unterschiede in der Zusammensetzung der Dipteren-Fauna (Sciaridae und Ceratopogonidae), die „... gegen die Identität der Ereignisorte der Harzproduktion" und „... gegen eine Gleichbehandlung der Inklusen von Baltischem und Sächsischem Bernstein ..." sprechen.

Bitterfelder Bernstein wird in einem regional und stratigraphisch engbegrenzten Vorkommen gefunden. Unter Baltischem Bernstein dagegen faßt man alle Stücke des Bernsteins der Ostsee-Region zusammen, ungeachtet, ob sie aus eozänen, oligozänen oder gar quartären Schichten stammen. Baltischer Bernstein der Sammlungen stellt somit ein *mixtum compositum* dar, das nicht nach konkreter Fundschichten zugeordnet werden kann.

Die Analyse des Materials Bernstein und seiner Inklusen ist nicht Gegenstand dieser Darstellung. Vielmehr sollen die bernsteinführenden Sedimente durch die Altersbestimmung ihrer Hellglimmer näher charakterisiert werden. Diese Methode wird hier erstmals auf tertiäre Sedimente angewendet. Deshalb gibt es bisher nur wenige Daten. Das Verfahren und die Verläßlichkeit der Daten werden anderenorts veröffentlicht.

Glimmer als Indikator des Liefergebiets

Die Sedimente, die Bernstein führen, sind feinkörnige Sedimente, überwiegend schluffig und feinsandig. Sie enthalten Hellglimmer in beträchtlichen Anteil. Während Sandkörner im Regelfall rollend oder springend transportiert werden, sind Glimmer weithin Bestandteil der Schwebfracht. Bernstein nimmt infolge seiner geringen Dichte eine Zwitterstellung ein.

Glimmer stammen aus kristallinen Gesteinen. Das K-AR-Alter der Glimmer wird massenspektrometrisch bestimmt. Es erlaubt Rückschlüsse auf das Herkunftsgebiet der Glimmer.

Aus dem **Samland** liegen bislang zwei Glimmeralter vor:

Pr. Nr.	Herkunft	Schicht /Alter	Mineral	K-Ar-Alter (Ma)	2s-Fehler (Ma)
14.164	Tagebau Jantarnoje/Sambia leg. KATINAS ca. 1970, ded. CERANOWICZ	Flußsand, Unt. Miozän,	Muskowit	1.010,3	32,4
14.092	Tagebau Primorskoje/Sambia leg. RITZKOWSKI 1995	Blaue Erde Mitteleozän	Muskowit	893,5	23,3

Tabelle 3. Tagebau Primorskoje/Sambia, K-Ar-Alter von Glimmern.

Der **Bitterfelder Glimmerschluff** ist oberoligozänen Alters. Das Oberoligozän war eine Zeit der Regression des Meeres. Glimmersande füllten die große, flache Bucht in Ostdeutschland und Polen aus, die das ausgedehnte Meer der Rupel-Zeit hinterlassen hatte. Gleichsam in der letzten Sekunde des Oberoligozäns, direkt unter den ausgedehnten terrestrischen Flözbildungen liegt die Bernstein-Lagerstätte des Tagebaus Goitsche bei Bitterfeld.

An drei Proben von Hellglimmern aus dem Bitterfelder Bernsteinschluff wurden die K-Ar-Alter bestimmt mit folgendem Ergebnis:

Pr. Nr.	Vorkommen	Mineral	Alter (Ma)	2s-Fehler (Ma)
14.153	Tgb. Goitsche, Bitterfeld, leg. CERANOWICZ 1986	Muskowit	1.038,2	20,5
14.154	Tgb. Goitsche, Bitterfeld, leg. CERANOWICZ 1986	Muskowit	994,3	22,4
14.155	Tgb. Goitsche, Bitterfeld, ded. Museum Ribnitz-Damgarten	Muskowit	969,6	19,7
arithm. Mittel			1.000,7	21,1

Tabelle 4. Tagebau Goitsche b. Bitterfeld, Bernstein-Schluff, Oberoligozän: K-Ar-Alter der Glimmer.

Die braunkohlenführende Schichtfolge in der Region **Helmstedt** enthält einen Glaukonit-Sand, dessen K-Ar-Alter an Glaukoniten mit 52,8 +/-1,4 Ma im (Untereozän, Ypresium) bestimmt wurde (AHRENDT *et al.* 1995). Ein Glimmersand im unteren Teil dieser Folge lässt an Fauna und sedimentologischen Merkmalen auf einen lagunären, brackischen Bildungsraum (Haff) schließen. Die dicken Hell-Glimmer-Pakete zeigen einen kurzen Transportweg an. Vier Analysen haben folgende Schließungsalter der Glimmer ergeben:

Pr. Nr.	Vorkommen	Mineral	Alter (Ma)	2s-Fehler (Ma)
	Tagebau Schöningen	Muskowit	352,9	7,5
	Tagebau Schöningen	Muskowit	345,3	7,8
	Tagebau Schöningen	Muskowit	346,1	7,0
	Tagebau Schöningen	Muskowit	341,0	7,7
arithm. Mittel			346,3	7,5

Tabelle 5. Tagebau Schöningen b. Helmstedt (SE-Niedersachsen), Glimmersand (Untereozän), K-Ar-Alter der Glimmer.

Folgerungen

Die Alter um 1 Ga für die Glimmer der eozänen Glaukonitsande (Blaue Erde) und miozänen Braunkohlensande des Samlands schließen eine Herkunft aus dem variszischen Gebirge aus. Das Ukrainische Massiv kommt seiner archaischen und frühproterozoischen Metamorphosealter wegen (älter als 1,8 Ga; KHAIN 1985) als Liefergebiet nicht in Betracht. Die östlichen und mittleren Teile Skandinaviens (Kola-Karelien-Kraton: 3,2–2,5 Ga) und Svecofinnische Krustenentwicklung (2,1–1,7 Ga; KRAUSS & LINDH 1990) scheiden ebenfalls infolge zu hohen Alters als Liefergebiete aus. Dagegen stellt der Südwestschwedische Gneiskomplex SGC (LINDH & JOHANSON 1995) in Südnorwegen und Westschweden ein ausgedehntes Gebiet dar, das eine sekundäre magmatische Reaktivierung etwa zwischen 1.000 und 900 Ma B.P. erfahren hat (WILSON & SUNDIN 1979; KHAIN 1985; KRAUSS & LINDH 1990; LINDH & JOHANSON 1995). Dies ist die einzige Region, die Glimmer mit einem Alter um 1 Ga liefern konnte. Es ist deshalb hier das Herkunftsgebiet der Glimmer zu vermuten.

Ein K-Ar-Alter der Glimmerum 1 Ga schließt auch das Umfeld des Bottnischen Meerbusen, nämlich die Gesteine der Svecofinnischen Orogenese und des archaischen Karelischen Massiv als Liefergebiet der Glimmer aus. Das bedeutet auch, daß das übrige Sediment und der Bernstein nicht aus dieser Region gekommen sein dürften. Dieser Befund spricht somit gegen die Annahme, daß ein Flußsystem („Eridanos") im Gebiet des Bottnischen Meerbusens der Materiallieferant der bernsteinführenden Sedimente des Samlandes wäre.

Die Glimmer aus dem oberoligozänen Bernsteinschluff bei Bitterfeld ergeben etwa die gleichen Alter wie die Proben aus dem Eozän und Miozän des Samlands. Kristalline Gesteine mit K-Ar-Alter der Glimmer um 1.000 Ma gibt es nicht im variszischen Gebirge im Süden der Lagerstätte Bitterfeld. Die Glimmer von Bitterfeld sind ebenfalls aus Scandinavien (SW-Schweden) herzuleiten.

Weitere Indizien für eine Materialherkunft aus Skandinavien sieht VALETON (1958; 1959) in den Schwermineralien der oberoligozänen und miozänen Glimmertone Norddeutschlands. Gleichfalls macht LOTSCH 1969 auf den skandinavischen Einfluß aufmerksam, der sich in einem Gebiet von Ost-Mecklenburg bis östlich der Oder mit silifizierten Korallen, Kieseloolithen und anderen Gesteinsmerkmalen, die auf Skandinavien hinweisen, äußert. FAY (1986) zeigt an Hand der Schwermineralspektren, daß das gesamte oberoligozäne Becken in Norddeutschland und Polen bis fast zu dessen Südrand von Skandinavien her beliefert wurde. Umfangreiche Schwermineralanalysen sind von KOSMOWSKA-CERANOWICZ *et al.* (zuletzt 1997) veröffentlicht worden.

Die K-Ar-Alter der Glimmer des untereozänen Glimmersandes von Helmstedt, östliches Niedersachsen, die bei 340 Ma B.P. liegen, deuten auf ein thermisches Ereignis der variszischen Orogenese und lassen ein Herkunftsgebiet im variszischen Gebirge, also im Süden von Bitterfeld suchen. Diese Vermutung steht in Übereinstimmung mit der faziellen Zonierung im Helmstedt-Hallieschen Braunkohlenrevier, die eine Küstenebene am südlichen Ufer des eozänen Meeres beschreibt. Während des Eozän trägt das südliche Festland erheblich zur Sedimentbildung bei.

In diesen eozänen Schichtfolgen zwischen Halle und Helmstedt treten jedoch nur Krantzite und

Oxikrantzite als fossile Harze auf, nie Bernstein-Arten wie sie im Samland oder bei Bitterfeld gefunden werden (LIETZOW & RITZKOWSKI 1996). Deshalb kann der Bernstein von Bitterfeld nicht durch Umlagerung aus den eozänen Folgen der Region hergeleitet werden. Gleichfalls läßt die Herkunft der Glimmer aus SW-Skandinavien daran zweifeln, daß Bitterfelder Bernstein ein fossiles Harz ist, das aus einem contemporären, d. i. oberoligozänen Wald auf dem südlichen Festlande stammte.

Es existiert eine einzige Quelle für Glimmer mit einem K-Ar-Alter um 1.000 Ma. Sie kann sowohl die Blaue Erde des Samlandes im Eozän, den Bitterfelder Bernsteinschluff im Oberoligozän (als auch die samländische Braunkohlensande im Untermiozän) beliefert haben. Es ist daher nicht zwingend notwendig, die Bitterfelder Glimmer durch Umlagerung aus der samländischen Bernsteinformation herzuleiten.

Angesichts der Mengen an Glimmer im gesamten oberoligozänen Becken in Norddeutschland und Polen erscheint es wenig realistisch, die Glimmer allein durch Aufarbeitung einer eozänen Bernsteinformation im Umfeld des Samlandes zu beziehen. Angesichts der Homogenität der Analysendaten ist es nicht wahrscheinlich, daß das Analysenalter der Glimmer durch Mischung von Glimmern unterschiedlichen Alters entstanden ist. Ferner läßt sich nach BARTHEL & HETZER (1982, 334) baltischer Bernstein mit einem gutem Abrollungsgrad nicht zu einen Bitterfelder Bernstein umlagern, der einen geringen Abrollungsgrad aufweist.

Die Informationen, die an den Glimmern gewonnen wurden, verringern somit erheblich die Möglichkeiten, Sediment und damit auch den Bitterfelder Bernstein von einem südlichen oder westlichen Festlande oder durch Aufarbeitung eozäner Sedimente herzuleiten, wie dies WEITSCHAT (1997, 74–84) in Erwägung zieht. Vielmehr bedarf es einer differenzierenden, paläogeographischen Rekonstruktion des gesamten östlichen Mitteleuropa im Paläogen, um zu einer Theorie zu gelangen, die die Vielfalt der bekannten Daten über den Bernstein und die bernsteinführenden Sedimente vereint.

Siegfried RITZKOWSKI
Institut und Museum für Geologie und Paläontologie
Universität Göttingen
Goldschmidtstr. 3
D-37077 Göttingen, Germany

WIEK BURSZTYNONOŚNYCH OSADÓW SAMBII (REJON KALININGRADU), BITTERFELDU (SAKSONIA-ANHALT) I HELMSTEDTU (SE DOLNA SAKSONIA)

Siegfried RITZKOWSKI

Streszczenie

Wiek bursztynonośnych osadów wschodniej części centralnej Europy jest bardzo zróżnicowany. Najstarsze tworzyły się prawdopodobnie w paleocenie, podczas gdy bitterfeldzkie osady z bursztynem powstawały w najstarszym oligocenie. Sedymentacja osadów bursztynonośnych odbywała się w okresie nie krótszym niż 23 miliony lat (47–24 Ma).

Geologiczny wiek łyszczyków może wskazać na pochodzenie osadów. Łyszczyki z Sambii i Bitterfeldu datowane metodą potasowo-argonową (K-Ar) mają około 1 Ga. Na podstawie badań łyszczyków, południowo-zachodnią Szwecję i południową Norwegię można uznać jako tereny alimentacji, co jednocześnie wyklucza pochodzenie osadów tak z obszaru południowych waryscydów jak i ze starego krystaliniku wokół zatoki Botnickiej. Łyszczyki dolnoeoceńskich piasków z Helmstedt (Dolna Saksonia), które zostały utworzone podczas orogenezy waryscyjskiej około 346 Ma B.P., pochodzą z lądu położonego na południe od morza późnooligoceńskiego.

Radiometryczny (K-Ar) wiek bursztynonośnych sekwencji i wiek zawartych w nich łyszczyków stanowią jedyne podstawowe dane, które mogą rozszyfrować pochodzenie bursztynu w złożach danego rejonu. Niestety stale jeszcze brakuje uzgodnienia danych dotyczących występowania osadów i ich paleogeograficznej sytuacji wzdłuż linii granic, dla osadów centralnej i wschodniej Europy.

Literatur

AHRENDT H., KÖTHE A., LIETZOW A., MARHEINE D. & RITZKOWSKI S.
1995 Lithostratigraphie, Biostratigraphie und radiometrische Datierung des Unter-Eozäns von Helmstedt (SE-Niedersachsen), *Zeitschrift der deutschen geologischen Gesellschaft*, **146**, 450–457, Hannover.

BARTHEL M. & HETZER H.
1982 Bernstein-Inklusen aus dem Miozän des Bitterfelder Raumes, *Zeitschrift für angewandte Geologie*, **28**, 314–336, Berlin.

BERGGREN W. A., KENT D. V., SWISHER III C. C. & AUBRY M.-P.
1995 *A revised Cenozoic geochronology and chronostratigraphy*, [*in*:] Berggren W. A., Kent D.V., Aubry M.-P. & Hardenbol J. (eds.), Geochronology, time scales and global stratigraphic correlation, *SEPM (Society for Sedimentary Geology), Spec. Publ.*, **54**, 129–212, Tulsa (Okl.).

BEYRICH E.
1853 Die Conchylien des norddeutschen Tertiärgebirges, *Zeitschrift der deutschen geologischen Gesellschaft*, **5**, 273–385 & 5 Taf., Berlin.

1856 Über den Zusammenhang der norddeutschen Tertärbildungen, *Abhandlungen der königlichen Akademie der Wissenschaften Berlin, Physikalische Klasse*, **1855**, 1–20, Berlin

FAY M.
1986 Marine sand deposits of the Northwest German Paleogene: heavy minerals and provenance, [*in*:] Tobien H. (Koord.), Nordwestdeutschland im Tertiär, *Beitrage zur Regionalen Geologie der Erde*, **18**, 92–104, Berlin.

KAPLAN A. A., GRIGIALIS A. A., STRELNIKOVA N. I. & GLIKMAN L. S.
1977 Stratigraphie und Korrelation ..., *Sovietskaja Geologija*, **4**, 30 – 43, Moskau.

KHAIN V. E.
1985 Geology of the USSR, *Beitr. Reg. Geol. d. Erde*, **17**, 1–272, Berlin-Stuttgart.

KOENEN A. V.
1865 Die Fauna der unteroligozänen Tertiärschichten von Helmstedt bei Braunschweig, *Zeitschrift der deutschen geologischen Gesellschaft*, **17**, 459–534, Berlin.

1889–1894 Das norddeutsche Unter-Oligocän und seine Molluskenfauna, *Abhandlungen zur geologischen Specialkarte von Preußen und den Thüringischen Staaten*, **10** (1–7), 1–1458 & 101 Taf., Berlin.

KÖTHE A.
1990 Paleogene Dinoflagellates from Northwest Germany, *Geologisches Jahrbuch*, **A 118**, 3–11, Hannover.

KOSMOWSKA-CERANOWICZ B.
1996 Bernstein — die Lagerstätte und ihre Entstehung, [*in*:] Ganzelewski M. & Slotta R. (Hrsg.), Bernstein — Tränen der Götter, *Veröffentlichungen aus dem Deutschen Bergbau-Museum Bochum*, **64**, 161–168, Bochum.

KOSMOWSKA-CERANOWICZ B. & MÜLLER C.
1985 Lithology and calcareous nannoplankton in amber-bearing Tertiary sediments from boreholes Chłapowo, *Bulletin of the Polish Academy of Sciences. Terre*, **33**, 119–129, Warszawa.

KOSMOWSKA-CERANOWICZ B. & KRUMBIEGEL G.
1989 Geologie und Geschichte des Bitterfelder Bernsteins und anderer fossiler Harze, *Hallesches Jahrbuch für Geowissenschaften*, **14**, 1–25.

KOSMOWSKA-CERANOWICZ B., KOHLMAN-ADAMSKA A. & GRABOWSKA I.
1997 Erste Ergebnmsise zur Lithologie und Palynologie der bernsteinführenden Sedimente im Tagebau Primorskoje, Sonderheft *Metalla*, **66**, 5–17, Bochum.

KOSMOWSKA-CERANOWICZ B. & LECIEJEWICZ K.
1997 Amber beds at the southern shores of the Eocene sea, *Museum of the Earth, Scientific conferences, Summary of lectures*, **7**, 13–17, Amberif '97, Gdańsk.

KRAUSS M. & LINDH A.
1990 Der südliche Baltische Schild — seine tektonische Krustenentwicklung und Beziehungen zum mitteleuropäischen Raum, *Zeitschrift geologischen Wissenschaften*, **18**, 569–586, Berlin.

LIETZOW A. & RITZKOWSKI S.
1996 Fossile Harze in den braunkohlenführenden Schichten von Helmstedt (Paläozän–Eozän, SE-Niedersachsen), [*in*:] Ganzelewski M. & Slotta R. (Hrsg.), Bernstein — Tränen der Götter, *Veröffentlichungen aus dem Deutschen Bergbau-Museum Bochum*, **64**, 83–88, Bochum.

LINDH A. & JOHANSON J.
1995 Grabnitic rocks as a source of granite: The Gösta and Sundsta Granites, south-west Sweden, *Geologische Rundschau*, **84**, 164–174.

MARTINI E. & RITZKOWSKI S.
1968 Was ist das „Unter-Oligozän?", *Nachrrichten der Akademie der Wissenschaften zu Göttingen, II. Math.-Phys. Klasse*, **13**, 231–251, Göttingen.

1970 Stratigraphische Stellung der obereozänen Sande von Mandrikowka (Ukraine) und Paralleliesierungmöglichkeiten mit Hilfe das fossilen Nannoplanktons, *Newsletters on Stratigraphy*, **1** (2), 49–60, Leiden.

MAYER K.
1861 Die Faunula des marinen Sandsteins von Klein-Kuhren, *Vierteljahresschrift der naturforschenden Gesellschaft in Zürich*, **6**, 1–109.

NOETLING F.
1883 Über das Alter der samländischen Tertiärformation, *Zeitschrift der deutschen geologischen Gesellschaft*, **35**, Berlin.

1885 Die Fauna des samländischen Tertiärs. Teil 1 (Vertebrata, Crustacea und Vermes, Echinodrmata), *Abhandlungen zur geologischen Specialkarte von Preußen und den Thüringischen Staaten*, **6** (3), I–VIII & 1–216, Berlin.

1888 Die Fauna des samländischen Tertiärs. Teil 1 (Gastropoda, Pecycypoda, Bryozoa, Geologischer Teil),

Abhandlungen zur geologischen Specialkarte von Preußen und den Thüringischen Staaten, **6** (4), I–VIII & 1–109, Tafelband, Berlin.

POŻARYSKA K.
1977 Upper Eocene Foraminifera of east Poland and their paleogeographical meaning, *Acta Paleontologica Polonica*, **22** (1), 3–54.

RÖSCHMANN F.
1997 Ökofaunistischer Vergleich von Nematoceren-Faunen (Insecta; Diptera: Sciaridae und Ceratopogonidae) des Baltischen und Sächsischen Bernsteins (Tertiär, Oligozän–Miozän), *Paläontologisches Zeitschrift*, **71** (1/2), 79–87.

TUTSKIJ V. & STEPANJUK L.
1999 Geologie und Mineralogie des Bernsteins von Klessow, Ukraine, *Investigations into Amber*, 53–60, Gdańsk.

VALETON I.
1958 Der Glaukonit und seine Begleitminerale aus dem Tertiär von Walsrode, *Mitteilungen aus dem geologischen Staatsinstitut in Hamburg*, **27**, 88–131.
1959 Zur Petrographie der miozänen Glimmerton- und Glimmersandfazies Nordwestdeutschlands, *Mitteilungen aus dem geologischen Staatsinstitut in Hamburg*, **28**, 110–125.

WEITSCHAT W.
1997 Bitterfelder Bernstein — ein eozäner Bernstein auf miozäner Lagerstätte, *Sonderheft Metalla*, **66**, 71–84, Bochum.

WILSON M. R. & SUNDIN N. O.
1979 Isotopic age determinations on rocks and minerals from Sweden 1960–1978, *Severiges Geologiska Undersökning, Rapporter och Meddelanden*, **16**.

AMBER IN THE NORTHERN LUBLIN REGION — ORIGIN AND OCCURRENCE

Jacek Robert KASIŃSKI & Elżbieta TOŁKANOWICZ

Abstract

Prospect drilling in the northern Lublin Region has evidenced numerous amber accumulations within clastic deposits of the Siemień Formation of Middle/Upper Eocene age, particularly in the vicinity of the middle Wieprz river valley near Lubartów and Parczew. Due to the fact that amber nodules are the sedimentary equivalent of fine quartz grains, amber concentrations in this area are associated with fine-grained clastic marine deposits containing glauconite and mollusc, coral and fish remains. Amber of the northern Lublin Region exhibits the same characteristic features as amber from the Ukrainian massif and their common origin is further evidenced by heavy mineral composition. Concentrations of amber originated as a result of the development of small delta systems and local overflood transport at the northern slope of the Metacarpathian Ridge, which have been reworked with littoral barrier facies. The Eocene shoreline was situated much further south of the present-day erosional border of the current extent of Eocene sediments.

Introduction

Work on the preparation of methods for the *in situ* prospecting for amber within Palaeogene sediments began at the Polish Geological Institute in 1992. The question was, how to delimit areas of amber occurrence for the successful identification of prospecting sites? As part of this project, an inventory was drawn up of amber accumulations within Palaeogene deposits in an area NW of the River Vistula and Palaeogene facies analysis was carried out for this area (KASIŃSKI *et al.* 1993a; 1993b). A program of drilling amber prospecting boreholes (KASIŃSKI *et al.* 1994) within seven areas between Dęblin-on-the-Vistula and Kodeń and Sławatycze on the River Bug was prepared based on the results of this work.

During the years 1996–1997, prospect drilling was performed in these areas thanks to the financial support of the National Fund for Environmental Protection and Water Management. The results obtained from 36 core and trenching samples confirmed the potential for amber exploitation in three of the seven designated areas (KASIŃSKI *et al.* 1997); in one of them the presence of amber had not previously been noted. All of the favourable sites are located in the central part of the area studied.

The age of the amber-bearing deposits was established as Middle/Upper Eocene.

Methodology

The assumption that, due to its low density, amber is the sedimentary equivalent of fine sands and silts was the basic foundation for defining the sites of greatest potential for amber extraction. During the processes of transportation and sedimentation, amber fragments are laid down in those areas, where the energy of the sedimentary environment and its motive force decrease to a level making it impossible to maintain fine sand and silt quartz grains in suspension. Such sedimentary environments correspond roughly to areas where the clasticity coefficient has a value of 0.5–4.0. Locations of this type, recognized as potentially amber-bearing, were delimited on a map of the Palaeogene outcrop belt of a sub-Quaternary surface.

Area of study

The area under study is located in southeast Poland, in the northern part of the Lublin Region, and encloses a relatively narrow belt of Palaeogene outcrops of a sub-Quaternary surface between the middle Vistula valley and upper Bug valley. From a palaeogeographical point of view of it lies at the southern margin of the Mid-European Eocene epicontinental basin (Fig. 1). This central part, situated in the middle section of the Wieprz river valley near Lubartów and Parczew, which holds much promise for further amber prospecting, corresponds to an area known as the Parczew delta (KOSMOWSKA-CERANOWICZ *et al.* 1990).

Amber prospecting history

In the northern part of the Lublin area, in the areas surrounding Łuków, Parczew, Lubartów and Puławy, the presence of amber within Palaeogene sediments has been known from core and trenching samples for a long time. Amber has been recorded at a very shallow depth of 0.2–2.3 m below the ground surface near Siemień Pond, 8 km west of Parczew; here, the amber-bearing deposits were situated at a depth of over 2.6 m there (RÜHLE 1955; WOŹNY 1966a; 1966b).

Amber-bearing deposits have also been recognized in some cores; the best report has been compiled for the Luszawa core, taken from the flood terraces of the Wieprz river, 15 km north of Lubartów.

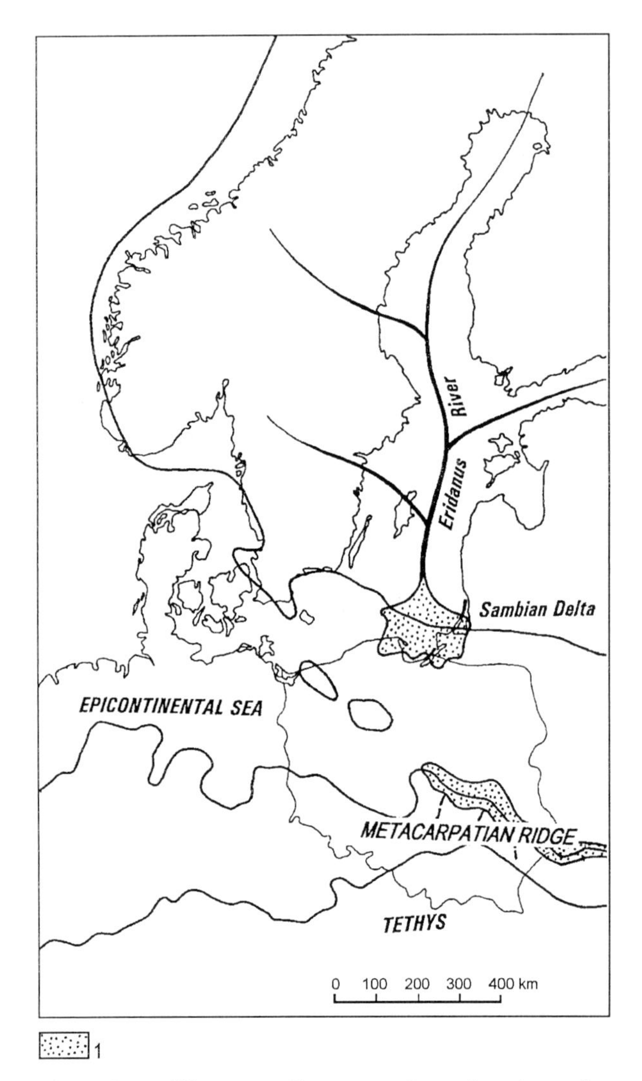

Fig. 1. Central European Eocene marine epicontinental basin (after: VINKEN 1988 & ZIEGLER 1990; JAWOROWSKI 1987 modified): 1 — areas of delta development.

Amber-bearing sediments 2.8 m thick were found there at a depth of 13.2 m (MOJSKI *et al.* 1966).

In the mid-sixties, during seismic investigation carried out by the Oil Prospecting Enterprise from Cracow, large amber nodules ("fist-sized", according to eye-witness accounts) were extracted from a borehole in Leszkowice village, 10 km north of Lubartów (ZALEWSKA 1974). Unfortunately, the exact location of this borehole is unknown today.

In 1989, during elaboration of the construction crush-material deposit Górka Lubartowska by the "Polgeol" Geological Enterprise, 9 km northwest of Lubartów, Palaeogene deposits with numerous amber nodules were found below Quaternary sands and gravels. Estimates were made as to how economically exploitable these amber resources were (STRZELCZYK 1990).

Amber-bearing deposits also occur in similar stratigraphic positions in natural outcrops and at an open amber mine in the Ukraine, where they are observed in deep erosional valleys of the Dnieper and Dnester river systems (SREBRODOLSKIJ 1984; TUTSKIJ 1997). The presence of amber within Palaeogene deposits has also been confirmed in Western Belarus (RÜHLE 1948; 1985; BOGDASAROV 1988).

Stratigraphic position

Amber-bearing associations lie within a few Upper Cretaceous and lower Palaeogene lithostratigraphic units (Fig. 2): (1) the Lower Mastricht unit, which occurs in the form of marls, limestones and decalcified siliceous rocks (gaizes), usually 2–5 m thick, though in some parts it reaches a thickness of up to 10 m (POŻARYSKI 1951); (2) the Sochaczew Formation (Danian), represented by quartz-glauconite sands (partly silicified) with phosphates and marl, gaize and organodetrital limestone intercalations with mollusc and echinoid remains, and (3) the Puławy Formation (Montian), which occurs in the form of clayey/marly sediments, gaizes and organodetrital limestones (HARASIMIUK & HENKIEL 1984; KRASSOWSKA 1990). There are no Upper Palaeocene (Tanetian) and Lower Eocene (Ypresian) deposits in the area under investigation.

The age of the amber-bearing association, defined in the research area as the Siemień Formation, was established following biostratigraphic examinations: (1) micropalaeontological (GIEL 1997), (2) palynological (WAŻYŃSKA 1997), and (3) calcareous nannoplancton analyses (GAŹDZICKA 1997) as Middle/Late Eocene: late Bartonian and early

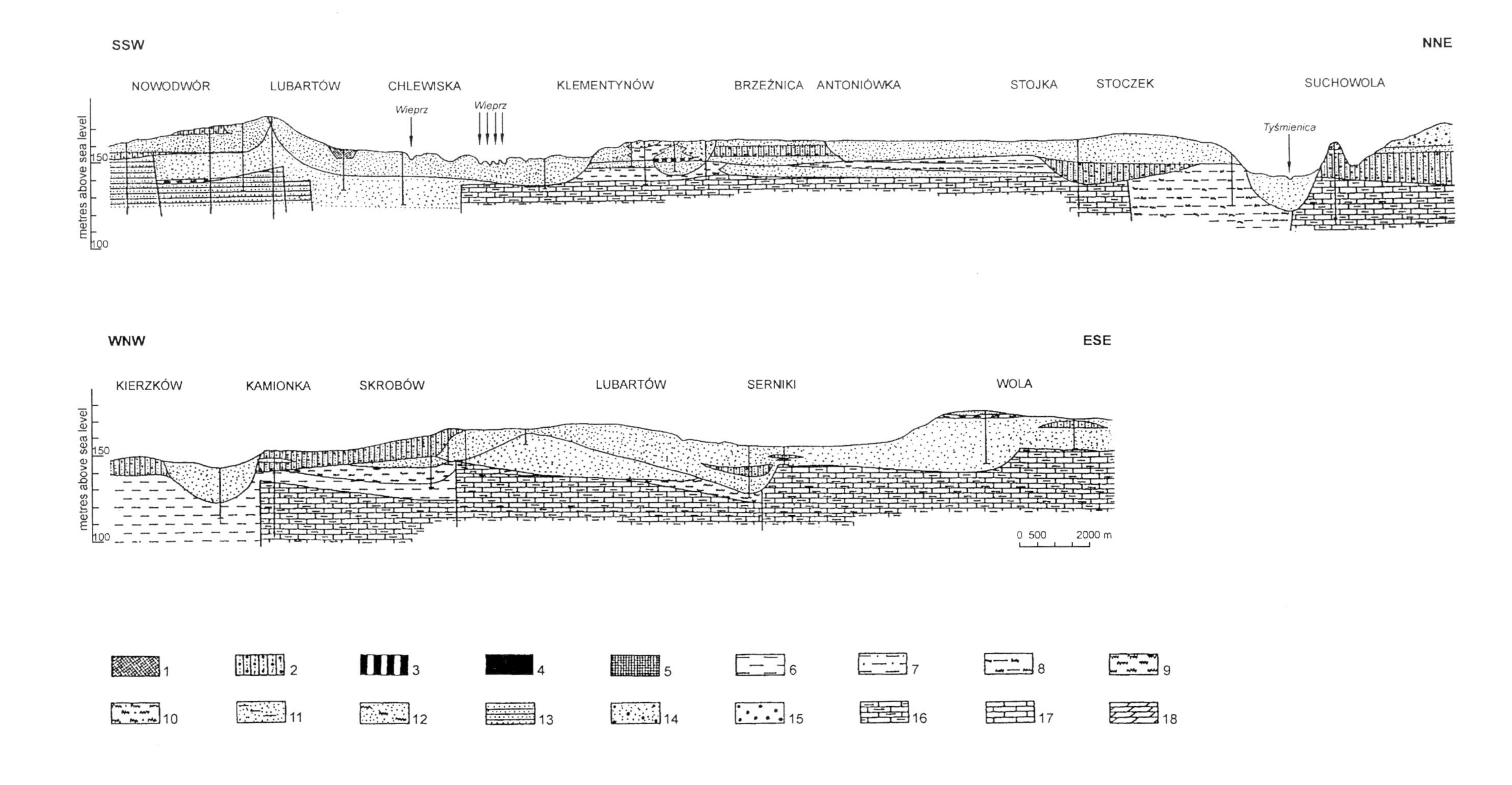

Fig. 2. Geological cross-sections of the Central Area on the middle Wieprz river valley: 1 — anthropogenic deposit; 2 — till; 3 — peat; 4 — lignite; 5 — lime gyttja; 6 — clay; 7 — sandy clay; 8 — clayey silt; 9 — silt; 10 — sandy silt; 11 — clayey sand; 12 — silty sand; 13 — sandstone; 14 — gravelly sand; 15 — gravel; 16 — marl; 17 — limestone; 18 — gaize.

Priabonian. These results correspond to the biostratigraphic data published earlier (WOŹNY 1966; POŻARYSKA & LOCKER 1971; POŻARYSKA 1977; POŻARYSKA & ODRZYWOLSKA-BIEŃKOWA 1977; UBERNA & ODRZYWOLSKA-BIEŃKOWA 1977; UBERNA 1981; KOSMOWSKA-CERANOWICZ & POŻARYSKA 1984; VINKEN 1988; GRABOWSKA 1992; SŁODKOWSKA 1993; 1996; GAŹDZICKA 1994). They are also comparable to the results of absolute age radiometric examination, carried out on the glauconite grains using the K-Ar method, which defined the absolute age of the Siemień Formation as 39.5 ± 3.0 Ma, 41.7 ± 0.4 Ma and 42.2 ± 3.0 Ma BP (BURACZYŃSKI & KRZOWSKI 1994; KRZOWSKI 1997), which corresponds to the late part of the Middle Eocene (Bartonian).

Neogene deposits, overlying the amber-bearing association occur in the form of isolated lobes, mostly in the northern and northwestern part of the area under consideration, where they infill erosional fossil valleys (Fig. 2). The complete lithostratigraphic column of these deposits, preserved rather sparsely, consists of (1) fine quartz sands with carboniferous silt and clay intercalations of the Middle Miocene Adamów Formation, (2) grey/brownish silts and clays with lignite lenses of the Middle Miocene Grey Clay Member of the Poznań Formation, and (3) grey-greenish clays, silts and fine sands of the Green Clay Member and green and bunter clays with silty and sandy intercalations of the Flame Clay Member, representing a stratigraphic column of the upper part of Middle Miocene, the whole Upper Miocene and, probably, also the lowermost Pliocene (CIUK 1970; PIWOCKI & ZIEMBIŃSKA-TWORZYDŁO 1995; 1997). Preglacial deposits of the Kozienice Member: sands with greenish clay and gravel intercalations, lay in the uppermost part of the column (BARANIECKA 1976; 1979; 1981).

Quaternary deposits are represented by Pleistocene deposits, occurring in the form of sands, silty sands and gravely sands with till intercalations (MOJSKI & TREMBACZOWSKI 1975); till content increases towards the North. Varve clays, silts and peats occur less frequently. More thick Holocene deposits, represented by clastic and peat sediments, occur in river valleys, particularly in those of the Bug, Vistula and Wieprz rivers.

Lithology and facies

The amber-bearing association is represented by clastic sediments of the Siemień Formation. Two to five lithological complexes may be established within the lithostratigraphic column of this formation (WOŹNY 1966a; 1966b; 1977; UBERNA & ODRZYWOLSKA-BIEŃKOWA 1977). These are as follows:

- a complex of fine and medium, greenish quartz-glauconite sands with fining-upward sequence with insertions of Cretaceous carbonate weathering crust within the bottom part. Quartz-gravel grains (so-called "beans") up to 2 cm in diameter, lidyte grains and rounded phosphate concretions up to 6 cm in diameter occur in the lowest part of the column, with dark-green sandy silts and clays occurring in the upper part of the complex. Complex thickness: up to 3.8 m;
- a bed of very fine grey-greenish (green at the top) quartz-glauconite sands with silt additions and single phosphate concretions, up to 2.3 m thick;
- a bed of very fine grey-greenish silty and clayey quartz-glauconite sands with very frequent non-rounded phosphate concretions up to 6 cm in diameter, with mollusc (*Chlamys*, *Ostrea*) and coral remains of up to 4 cm in diameter; fish remains and carbonized plant detritus may also be present within this bed. Bed thickness: up to 0.3 m.
- a complex of fine and very fine grey-greenish quartz-calcareous sands containing glauconite and amber additions with mollusc and coral

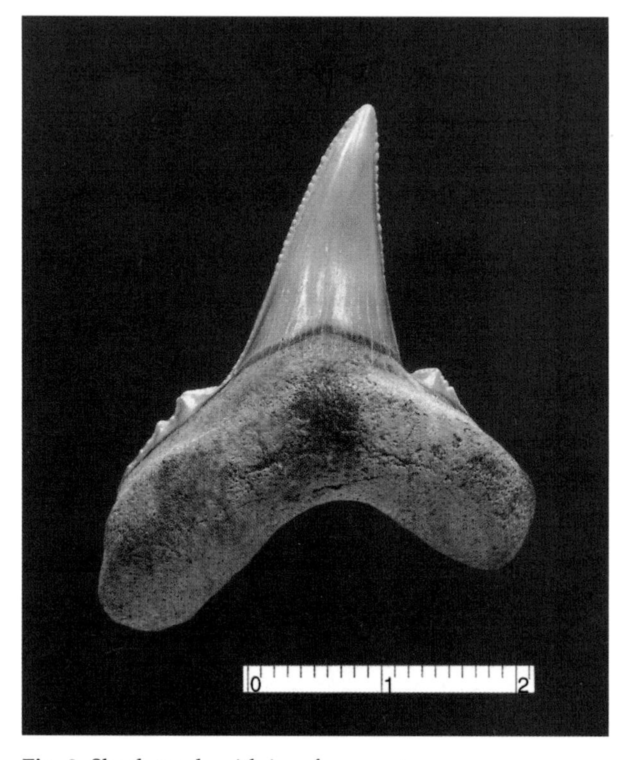

Fig. 3. Shark tooth with jaw fragment.

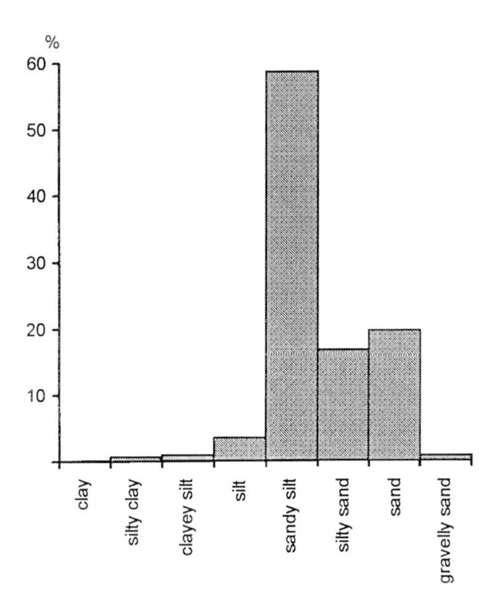

Fig. 4. Statistical diagram of core intervals with amber occurrence related to lithology of sediments.

remains, more fossilized at the top (loose sandstone). A fining-upward sequence may be observed in the uppermost part of the complex, where a level of calcareous siltstone with calcareous gaize intercalation occurs. Complex thickness: up to 3.8 m.

- a complex of dark-green glauconite-bearing differentiated sediments: clayey silts, very fine quartz-glauconite silty sands, clayey gaizes and amber. Thin intercalations of hard gaize with faunal remains (molluscs of *Chlamys* and *Ostrea* groups, fish teeth) (Fig. 3) also occur here. Complex thickness: up to 3.0 m.

In general, the Eocene lithological column is distinctly bipartite. The lower series consists of more coarse-grained sediments: fine and medium quartz-glauconite sands with a substantial content of quartz and lidyte gravel and phosphate concretions and only small silt additions; frequent inclusions of pelecypod and coral fauna occur here. Amber grains are rather rare within this series. The upper series consists of more fine-grained sediments: very fine quartz-glauconite sands and silts (locally also clays) with frequent amber grains and fine detritus of pelecypod/coral fauna. However, in numerous bores the lithological column of Eocene deposits is greatly reduced and limited to the upper series only.

Both series are substantially differentiated with heavy mineral composition (the upper one being much richer) and amber content. The heavy mineral composition of the lower series is similar to the heavy mineral composition of Cretaceous weathering marls (NAWROCKA-MIKLASZEWSKA 1997). The heavy mineral composition of the upper series shows a genetic relationship to the amber-bearing association of the Ukraine (MOJSKI *et al.* 1966; SREBRODOLSKIJ 1980). Amber grains occur mostly in the finest sediments (Fig. 4), due to the low density of amber — about 1.15 Mg/m^3; this makes the amber grains the sedimentary equivalent of very fine quartz grains (KASIŃSKI *et al.* 1993b). The majority of amber grains — nearly 60% of their total weight — are related to sandy silts, and more than 90% occur with fine grain clastic forms of sandy silts to fine and medium grain sands (Fig. 5).

Amber description

Amber from the sample cores consists of frequent, small, sharp-edged grains and occasionally different sized nodules.

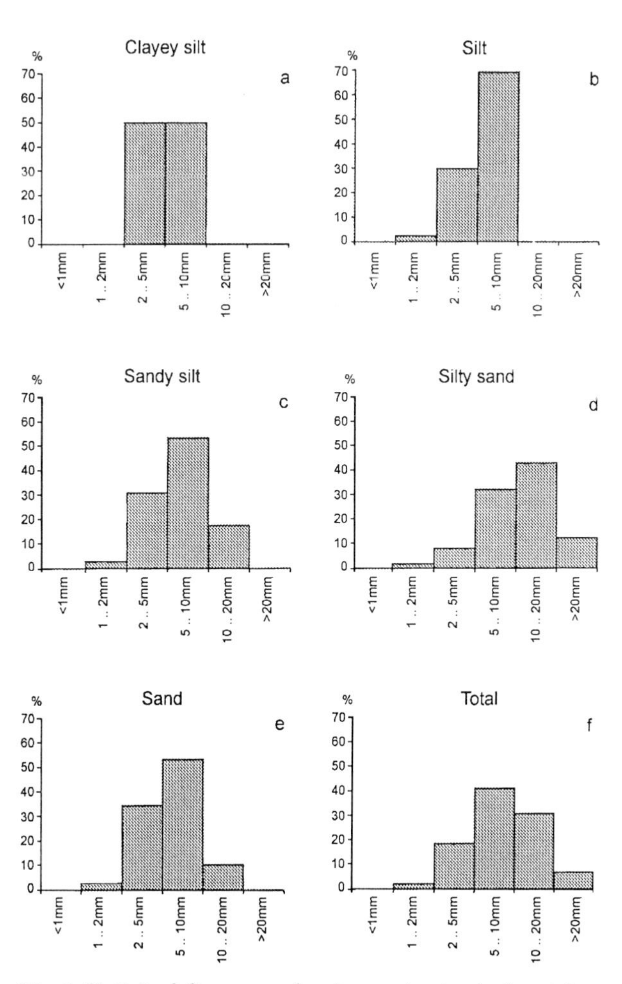

Fig. 5. Statistical diagrams of amber grain-size (a–f, weight-percent) distribution within different lithological members.

The sharp-edged grains originated following the fragmentation of lumps during the drilling process. They are not a valid material for studying the forms and size of amber clasts, but their freshly crushed surfaces with flint-like fractures are useful for the description of amber varieties and colours. Transparent, shiny and opaque varieties are represented in the samples. The colours are deeply differentiated, from very pale yellow through different shades of honey-yellow to dark brown (Fig. 6). Some clasts containing amber grains of atypical colour, such as lemon-yellow (Fig. 6d) and blood-red, have been also evidenced. The dark-brown grains are mostly very fragile as a result of their advanced state of weathering. The sharp-edged grain size is usually less than 10 mm.

The shape of amber nodules found intact were often related to natural forms of resin accumulations, such as icicle, intra-cortex and sub-cortex forms (Figs. 6d, 6e). Desiccation cracks and primary sculpture (Figs. 6a, 6d) were preserved on some nodules. Forms with characteristic "finger-holes", representing twig imprints, were also evidenced. The majority of the lumps had more or less rounded edges. However, typical amber pebbles were not evidenced in the samples. The surface of nodules was dull or covered with a weathered cortex; some lumps had a dark-brown to black cortex. All three basic amber varieties: transparent, shiny and opaque were noted, with honey-yellow and honey-brown being the predominant colours.

The size of more than 90% (by weight) of the amber nodules fell within a range of 2–20 mm (Fig. 5f) with the biggest specimen, identified as a sub-cortex form measuring 40 x 35 x 11 mm (Fig. 6e). Grain-size distribution of the amber nodules was related to the lithology of co-existent rocks (Figs. 5a–5e), with the largest lumps (modal class 10–20 mm) being most frequent in silty sand; more than 80% (by weight) of the amber measured over 10 mm in diameter (Fig. 5d).

Sedimentary environment

Sedimentation of the amber-bearing association began with the transgression of the Eocene sea, which covered large areas of Central Europe during the Middle Eocene (see Fig. 1). A characteristic transgressive level of gravely sediments with phosphate concretions was deposited in the lowermost part of the sequence. Quartz-glauconite sands, frequently with clayey and/or calcareous additions, silts and siltstones with glauconite and gaizes and sandy marls with thin sandy and oolitic limestone intercalations were laid down in the upper part. In the southern and south-eastern part of the area under examination, these deposits were replaced by a complex of silts and clays without glauconite.

Resin grains — the precursors of amber — originated in amber-bearing tropical and sub-tropical forests which once covered large areas, according to some authors (WEITSCHAT 1997) as far as from the Elbe river to the Ural Mountains and from the Scandinavian Penninsula to the Tethys shoreline. These pieces of resin were transported by rivers together with clastic grains and deposited in favourable hydrodynamic conditions within the distal parts of river-mouth deltas in the littoral zone of a shallow shelf sea. A substantial part of the primary amber deposits were eroded and re-transported by sea currents and wave activity and redeposited within barrier facies by coastal currents.

Following analogies with amber concentrations in the Sambian Peninsula and in the vicinity of the Bay of Gdańsk (KATINAS 1971; JAWOROWSKI 1987; GRIGALIS *et al.* 1988), amber accumulations along the southern shore of the Eocene sea have also been interpreted as being related to the development of large river deltas. Two such deltas have been established in this region: the Parczew delta with the deposits of the Siemień Formation (described here) and the Klesów-Gatcza delta which lies in the border zone between the Ukraine and Belarus (KOSMOWSKA-CERANOWICZ *et al.* 1990; KOSMOWSKA-CERANOWICZ & LECIEJEWICZ 1995).

During the course of geological prospecting the only amber (succnite) concentrations related to marine environments were confirmed. This fact was evidenced by the widespread presence of glauconite and frequent inclusions of marine fauna remains within the amber-bearing sediments. An area of recent amber occurrence is probably located relatively far from the coastline of the Eocene sea, which was situated much further south of the recent erosional border of Eocene deposits (Fig. 7). Sediments of the upper part of the Middle Eocene are known in endemical occurrences within the Sołokija Trough in the Roztocze region (GAŹDZICKA 1994), and small amber grains occurring as a secondary deposit within Miocene sediments of the Carpathian Foredeep (PAWŁOWSKI *et al.* 1985) attest to the origins of amber-bearing Eocene sediments over large areas of the Roztocze region, where they were transported by further erosive processes. The shallow marine character of the warm-temperate basin of normal salinity is evidenced by the content of pelecypod fauna occurring there (WOŹNY 1977).

Although a limited number of amber grains is related to the transgressive facies of the lower lithological series, variously sized river deltas were probably also the basic sedimentary environment of primary amber accumulations in the northern Lublin Region. In the described area, it was more a system of numerous small, superimposed deltas, located

Fig. 6. Varieties of Baltic amber from the northern Lublin region: a — amber nodules of the shiny and opaque type, yellow, dark-honey-yellow and brown in colour. Most parts of the nodules present a flinty fracture. The nodule in the lower-left corner is a fragment of an icicle form with a thick, weathered crust; b — sharp-edged amber nodules with fresh flinty fracture: shiny variety, yellow colour, and shiny and partly transparent variety, brown in colour with partly preserved weathered crust. Maximum size 16 mm; c — amber nodules of the opaque and shiny variety, honey-yellow, orange and brown in colour. Part of the nodules present a flinty fracture surface. The amber nodule in the lower-left corner has a natural surface sculpture with small finger holes — twig traces; d — amber nodules of the opaque and partly shiny type, of different colours from light-honey-yellow through orange-brownish to brown. The brown amber nodule on the left has a deeply weathered, dark-brown crust. Maximum size 28 mm; e — the biggest amber nodule — a sub-cortical form measuring 40 x 35 x 11 mm; amber of the shiny type, honey-yellow in colour. The nodule has a thick, dark-brown, weathered crust.

along the Eocene sea shoreline from the present-day Vistula river valley to the Bug river valley and further still to Kiev in the Ukraine. Single isolated amber concentrations within Oligocene deposits in Belarus (RÜHLE 1985) are probably similar in origin. The palaeogeography of alimentary areas, and in particular the relatively narrow continental belt of the Metacarpathian Ridge between the epicontinental Eocene sea and Paratethys induces this conception; the question is, where was a place for

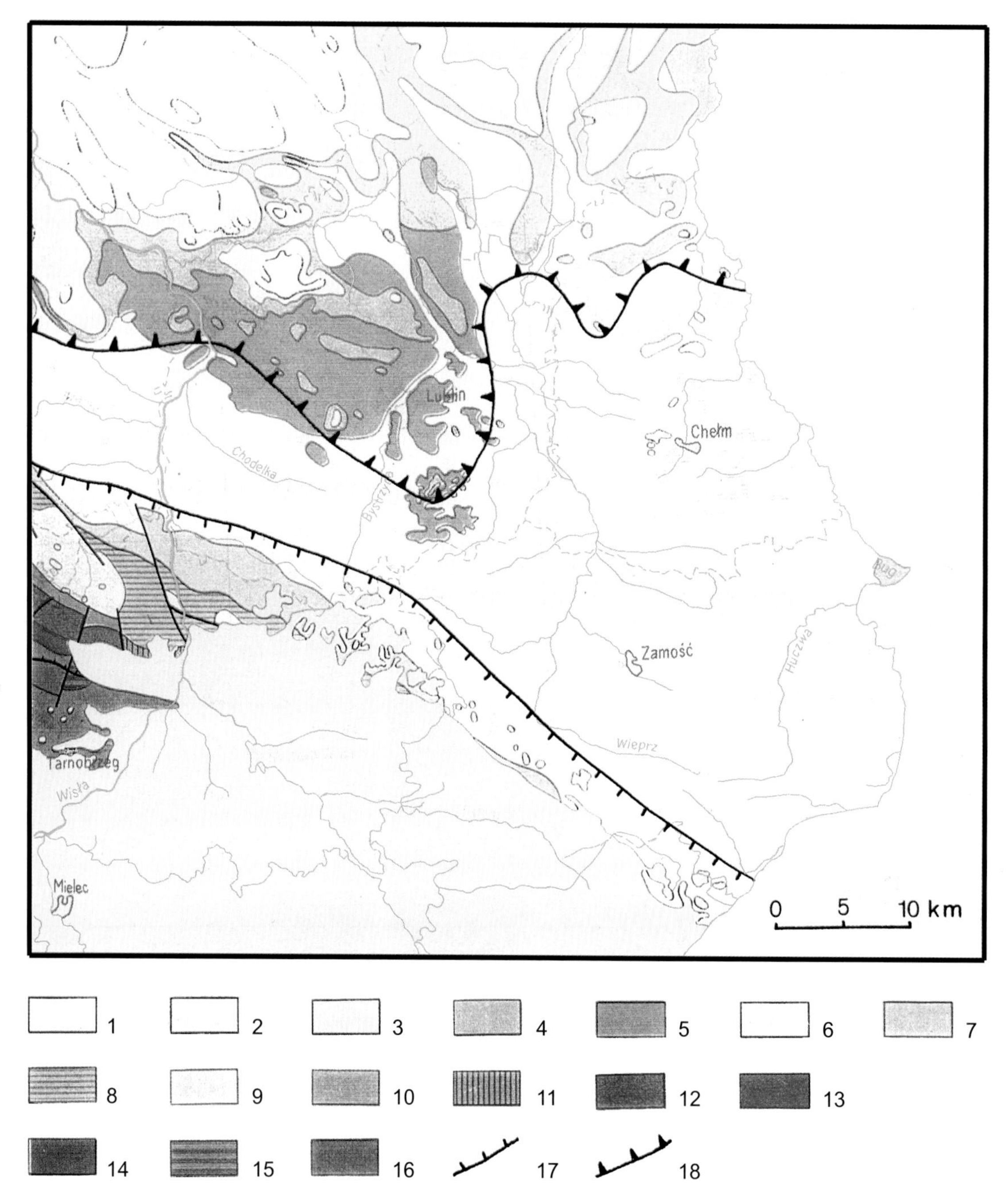

Fig. 7. Palaeogeographical southern shoreline of the Eocene sea and Recent erosional southern border of Eocene deposits (map without Quaternary deposits, geology after PIWOCKI 1990). Coenozoic: 1 — Mio-Pliocene (Poznań Formation), 2 — continental Miocene, 3 — marine Miocene of the Carpathian Foredeep, 4 — Eocene, 5 — Palaeocene; Mesozoic: 6 — Maastrichtian, 7 — Campanian, 8 — Santonian and Cognacian, 9 — Malm, 10 — Dogger, 11 — Liassic, 12 — Jurassic — undivided, 13 — Lower Triassic (Bunter Sandstone); Palaeozoic: 14 — Devonian, 15 — Silurian and Ordovician — undivided, 16 — Cambrian; extent lines: 17 — Eocene shoreline, 18 — recent extent of Eocene deposits.

development of large rivers — like the Eridan one, which shaped the Sambian delta. Amber grains were probably also transported to the sea directly from the forest areas with areal overfloods.

Jacek Robert KASIŃSKI & Elżbieta TOŁKANOWICZ
Państwowy Instytut Geologiczny
ul. Rakowiecka 4
00-975 Warszawa, Poland
e-mail: jkas@pgi.waw.pl; etol@pgi.waw.pl

BURSZTYN NA PÓŁNOCNEJ LUBELSZCZYŹNIE — GENEZA I WYSTĘPOWANIE

Jacek Robert KASIŃSKI,
Elżbieta TOŁKANOWICZ

Streszczenie

W latach 1996–1997 Państwowy Instytut Geologiczny przeprowadził badania geologiczno-rozpoznawcze, mające na celu określenie perspektyw występowania bursztynu w utworach eocenu północnej Lubelszczyzny, w pasie wychodni tych utworów na powierzchnię podczwartorzędową od Dęblina nad Wisłą po Kodeń i Sławatycze nad Bugiem. Prace sfinansowano ze środków Narodowego Funduszu Ochrony Środowiska i Gospodarki Wodnej. Wyniki 36 wierceń i szybików poszukiwawczych pozwoliły na stwierdzenie perspektywiczności trzech spośród rejonów wytypowanych w projekcie badań, w tym jednego zupełnie nowego, gdzie dotychczas nie notowano wystąpień bursztynu. Wszystkie rejony perspektywiczne leżą w centralnej części obszaru badań, w okolicach Lubartowa i Parczewa.

Jako podstawę wyznaczenia obszarów perspektywicznych przyjęto założenie, że bursztyn, ze względu na małą gęstość, stanowi ekwiwalent sedymentacyjny utworów drobnopiaszczystych i mułkowych. W cyklu procesów transportu i sedymentacji okruchy bursztynu osadzają się zatem w momencie, gdy energia środowiska, a co za tym idzie potencjał trakcyjny, obniża się do poziomu uniemożliwiającego utrzymanie w zawiesinie drobniejszej frakcji ziarn piasku i ziarn mułku. Środowiska takie odpowiadają w dostatecznym przybliżeniu obszarom, w których wartość współczynnika klastyczności waha się w granicach 0.5–4.0. Obszary odpowiadające temu przedziałowi, uznane za perspektywiczne dla poszukiwań bursztynu, wykartowano wzdłuż pasa wychodni utworów paleogenu na powierzchnię podczwartorzędową.

Wiek utworów asocjacji bursztynonośnej określono na przełom eocenu środkowego i górnego (barton–priabon). Utwory te zalegają na osadach mastrychtu górnego, danu i montu, wykształconych w postaci margli i wapieni, odwapnionych skał krzemionkowych (gez i opok), oraz piasków kwarcowo-glaukonitowych z fosforytami, miejscami zdiagenezowanych. Reprezentują je dwa kompleksy litologiczne, zbudowane ze skał klastycznych: (1) kompleks dolny, złożony z osadów o grubszej frakcji — piasków kwarcowo-glaukonitowych ze żwirem i konkrecjami fosforytowymi oraz obfitą fauną małżowo-koralowcową i (2) kompleks górny, w skład którego wchodzą osady drobniejszej frakcji — piaski mułkowate i mułki kwarcowo-glaukonitowe.

Okruchy bursztynu występują głównie w kompleksie górnym, choć pewna ich liczba pojawia się także w osadach kompleksu dolnego. Należą one do odmian: przezroczystej, przeświecającej i nieprzezroczystej o rozmaitej barwie — od jasnożółtej, poprzez różne odcienie miodowej do ciemnobrązowej. Kształty bryłek odzwierciedlają często naturalne formy, wśród których zidentyfikowano formy soplopodobne, międzykorowe i podkorowe. Bryłki bursztynu są często pokryte korą wietrzeniową barwy ciemnobrązowej do prawie czarnej.

Występowanie bursztynu związane jest przede wszystkim z mułkami piaszczystymi, piaskami mułkowatymi i piaskami (ponad 90%). Większe bryłki (ponad 10 mm średnicy) występują najczęściej w piaskach mułkowatych.

W toku prac poszukiwawczych stwierdzono, że na omawianym obszarze występowanie bursztynu jest związane wyłącznie ze środowiskiem morskim, na co wskazuje powszechna obecność w osadach asocjacji bursztynonośnej glaukonitu oraz liczne szczątki fauny morskiej, w tym zęby ryb. Obszar pierwotnej akumulacji bursztynu był zapewne dość odległy od brzegu, który rozciągał się znacznie dalej na południe od dzisiejszej granicy utworów eocenu, mającej charakter erozyjny. Bursztyn osadzał się w wielu nakładających się na siebie drobnych deltach; okruchy bursztynu były zapewne także znoszone do morza przez spływy powierzchniowe bezpośrednio z obszarów porośniętych lasem produkującym żywice, z których powstał bursztyn. Osady te w toku dalszej ewolucji zostały w znacznej części rozmyte, a okruchy bursztynu uległy redepozycji w facjach barierowych.

Bibliography

BARANIECKA M. D.

1976 Charakterystyka geologiczna osadów trzeciorzędowych wybranych obszarów Mazowsza, *Prace Muzeum Ziemi*, **25**, 15–28, Warszawa.

1979 Osady plioceńskie Mazowsza jako podłoże czwartorzędu, *Biuletyn Geologiczny Uniwersytetu Warszawskiego*, **23**, 23–36, Warszawa.

1981 Osady trzeciorzędowe południowo-zachodniej części niecki warszawskiej na przykładzie profilu z Kaczorówka, *Kwartalnik Geologiczny*, **25** (2), 365–386, Warszawa.

[BOGDASAROV A. A.] БОГДАСАРОВ А. А.

1988 The amber finds in the Central Part of the Brest-Podlasie Depression and their physico-chemical properties, [*in*:] 6th Meeting on Amber and Amber-Bearing Sediments, 20–21.10.1988, Reports summaries, *Prace Muzeum Ziemi*, **41**, 161, Warszawa.

BURACZYŃSKI J. & Z. KRZOWSKI

1994 Middle Eocene in the Sołokija Graben on Roztocze Upland, *Geological Quarterly*, **38** (4), 739–758, Warszawa.

CIUK E.

1970 Schematy stratygraficzne trzeciorzędu Niżu Polskiego, *Kwartalnik Geologiczny*, **14** (4), 754–771, Warszawa.

GAŹDZICKA E.

1994 Middle Eocene calcareous nannofossils from the Roztocze region (SE Poland) — their biostratigraphic and paleogeographic significance, *Geological Quarterly*, **38** (4), 727–734. Warszawa.

1997 Wyniki analizy nannoplanktonu wapiennego w utworach eocenu lubelszczyzny, [*in*:] Kasiński J. R., Piwocki M., Saternus A., Tołkanowicz E. & Wojciechowski A., *Realizacja projektu prac geologicznych dla określenia perspektyw występowania bursztynu w utworach eocenu Lubelszczyzny*, **11** (3), 1–3, Centralne Archiwum Geologiczne P.I.G., Warszawa.

GIEL M. D.

1997 Wyniki badań mikropaleontologicznych z obszaru północnej Lubelszczyzny, [*in*:] Kasiński J. R., Piwocki M., Saternus A., Tołkanowicz E. & Wojciechowski A., *Realizacja projektu prac geologicznych dla określenia perspektyw występowania bursztynu w utworach eocenu Lubelszczyzny*, **11** (1), 1–13, Centralne Archiwum Geologiczne P.I.G., Warszawa.

GRABOWSKA I.

1992 *Wyniki analiz sporowo-pyłkowych 5 próbek z trzech profili: Narol 1, Piekiełko 4, Łaszczówka 3 (Wyżyna Lubelska)*, Centralne Archiwum Geologiczne P.I.G., Warszawa.

[GRIGALIS A. A. *et al.*] ГРИГАЛИС А. А., БУРЛАК А. Ф., ЗОСИМОВИЧ В. Ю., ИВАНИК М. М., КРАЕВА Е. Ю., ЛЮЕВА С. А. & СТОТЛАНД А. Б.

1988 Новые данные по стратиграфии и палеогеографии отложений запада европейской части СССР, *Стратиграфия и палеогеография*, **12**, 41–54, Москва.

HARASIMIUK M. & A. HENKIEL,

1984 Kenozoik Lubelskiego Zagłębia Węglowego, [*in*:] *Przewodnik 56. Zjazdu Polskiego Towarzystwa Geologicznego, Lublin 6–8.09.1984*, 56–70, Wyd, Geol., Warszawa.

JAWOROWSKI K.

1987 Geneza bursztynodajnych osadów paleogenu w okolicach Chłapowa, *Biuletyn Instytutu Geologicznego*, **356**, 89–102, Warszawa.

KASIŃSKI J. R., PIWOCKI M. & TOŁKANOWICZ E.

1993a *Poszukiwanie bursztynu, piasków kwarcowych i surowców ilastych w osadach trzeciorzędowych wschodniej części Niżu Polskiego*, 1–92, Centralne Archiwum Geologiczne P.I.G., Warszawa.

1993b Upper Paleogene facies setting in North-East Poland and its control of amber distribution, [*in*:] *Abstracts 2. Baltic Stratigraphic Conference, Vilnius 9–14.05.1933*, 39, Lith. Geol. Inst., Vilnius.

KASIŃSKI J. R., SATERNUS A. & TOŁKANOWICZ E.

1994 *Projekt prac geologicznych dla określenia perspektyw występowania złóż bursztynu w utworach eocenu Lubelszczyzny*, 1–40, Centralne Archiwum Geologiczne P.I.G., Warszawa.

KASIŃSKI J. R., PIWOCKI M., SATERNUS A., TOŁKANOWICZ E. & WOJCIECHOWSKI A.

1997 *Realizacja projektu prac geologicznych dla określenia perspektyw występowania bursztynu w utworach eocenu Lubelszczyzny*, 1–59, Centralne Archiwum Geologiczne P.I.G., Warszawa.

[KATINAS V.] КАТИНАС В.

1971 Янтарь и янтареносные отлжения Южной Прибалтики, *Труды Литовского Научного Исследовательного Института*, **20**, Вильнюс.

KOSMOWSKA-CERANOWICZ B., KOCISZEWSKA-MUSIAŁ G. , MUSIAŁ T. & MÜLLER C.

1990 *Bursztynonośne osady trzeciorzędu okolic Parczewa*, Prace Muzeum Ziemi, **41**, 21–35, Warszawa.

KOSMOWSKA-CERANOWICZ B. & LECIEJEWICZ K.

1995 Złoża bursztynu na południowym brzegu morza eoceńskiego, [*in*:] Materiały 2. Seminarium Amberif '95, "Bursztyn bałtycki, złoża i warsztaty", *Muzeum Ziemi /Konferencje naukowe/Streszczenia*, **4**, 18–22, Gdańsk.

KOSMOWSKA-CERANOWICZ B. & POŻARYSKA K.

1984 On new research of Tertiary sediments in Polish Lowlands, *Bulletin de la Academie Polonaise des Sciences, Series Sciences de la Terre*, **29** (1), 81–90, Warszawa.

KRASSOWSKA A.

1990 Utwory młodszej kredy górnej i paleocenu dolnego w głębokich otworach wiertniczych w rejonie Puław, Lublina i Lubartowa, *Przegląd Geologiczny*, **38** (4), 168–173, Warszawa.

KRZOWSKI Z.

1997 Eocene in Mielnik on the Bug River, *Geological Quarterly*, 41 (1), 61–68, Warszawa.

MOJSKI J. E., RZECHOWSKI J. & WOŹNY E.

1966 Górny eocen w Luszawie nad Wieprzem koło Lubartowa, *Przegląd Geologiczny*, **14** (12), 513–517, Warszawa.

MOJSKI J. E. & TREMBACZOWSKI J.
1975 Osady kenozoiczne Polesia Lubelskiego, *Biuletyn Instytutu Geologicznego*, **290**, 97–139, Warszawa.

NAWROCKA-MIKLASZEWSKA M.
1997 Minerały ciężkie z eoceńskich i czwartorzędowych osadów Lubelszczyzny, [*in:*] Kasiński J. R., Piwocki M., Saternus A., Tołkanowicz E. & Wojciechowski A., *Realizacja projektu prac geologicznych dla określenia perspektyw występowania bursztynu w utworach eocenu Lubelszczyzny*, **12**, 1–41, Centralne Archiwum Geologiczne P.I.G., Warszawa.

PAWŁOWSKI S., PAWŁOWSKA K. & KUBICA B.
1985 Geology of the Tarnobrzeg native sulphur deposit, *Prace Instytutu Geologicznego*, **114**, Warszawa.

PIWOCKI M.
1990 Utwory starsze od czwartorzędu (pre-Quaternary formations), [*in:*] *Atlas Rzeczypospolitej Polskiej, skala 1 : 1 500 000*, **21** (2), Państw. Przeds. Wyd. Kartograf., Warszawa.

PIWOCKI M. & ZIEMBIŃSKA-TWORZYDŁO M.
1995 Litostratygrafia i poziomy sporowo-pyłkowe neogenu Niżu Polskiego, *Przegląd Geologiczny*, **43** (9), 916–927, Warszawa.

1997 Neogene of the Polish Lowlands — lithostratigraphy and pollen/spore horizons, *Geological Quarternary*, **41** (1), 21–40, Warszawa.

POŻARYSKA K.
1977 Upper Eocene foraminifera of East Poland and their palaeogeographical meaning, *Acta Palaeontologica Polonica*, **22** (1), 3–54, Warszawa.

POŻARYSKA K. & LOCKER S.
1971 Les organismes planctoniques de l'Eocéne supérieur de Siemień, Pologne orientale, *Review of Micropaleontogy*, **14** (5), 57–72, Paris.

POŻARYSKA K. & ODRZYWOLSKA-BIEŃKOWA E.
1977 O górnym eocenie w Polsce, *Kwartalnik Geologiczny*, **21** (1), 59–70, Warszawa.

POŻARYSKI W.
1951 Odwapnione utwory kredowe na północno-wschodnim przedpolu Gór Świętokrzyskich, *Biuletyn Instytutu Geologicznego*, **75**, 1–70, Warszawa.

RÜHLE E.
1948 Kreda i trzeciorzęd zachodniego Polesia, *Biuletyn Państwowego Instytutu Geologicznego*, **34**, 120, Warszawa.

1955 Przegląd wiadomości o podłożu czwartorzędu północno-wschodniej części Niżu Polskiego, *Biuletyn Instytutu Geologicznego*, **70**, 159–173, Warszawa.

1985 Poszukiwanie bursztynu w utworach paleogeńskich dorzecza Prypeci i na półwyspie Sambii, *Technika Poszukiwań Geologicznych*, **24** (2), 30–33, Warszawa.

SŁODKOWSKA B.
1993 *Badania palinologiczne osadów trzeciorzędowych z arkuszy Hrebenne i Lubycza Królewska*, Centralne Archiwum Geologiczne P.I.G., Warszawa.

1996 *Wyniki badań palinologicznych próbek osadów trzeciorzędowych przeprowadzonych na arkuszu Kąkolewnica 1 : 50 000 z profili: Zosinowo 1, Rudnik 2 i Sawki 3*, Centralne Archiwum Geologiczne P.I.G., Warszawa.

[SREBRODOLSKIJ B. I.] СРЕБРОДОЛЬСКИЙ Б. И.
1980 *Янтаръ Украины*, 1–123, Наукова Думка, Киев.

1984 *Геологическое строение и закономерности размещения месторождений янтаръя СССР*, 1–167, Наукова Думка, Киев.

STRZELCZYK G.
1990 *Dokumentacja geologiczna w kategorii C_2 złoża kruszywa naturalnego (piaski budowlane) wraz z określeniem występowania bursztynu w utworach trzeciorzędowych w rejonie Górka Lubartowska, Przeds. Geol. "Polgeol"*, 1–42, Centralne Archiwum Geologiczne P.I.G., Warszawa.

TUTSKIJ W.
1997 Geologie und Entstehung der Bernsteinvorkommen in Nordwesten der Ukraine, *Metalla*, **66**, Sonderheft, 57–62, Bochum.

UBERNA J.
1981 Upper Eocene phosphate-bearing deposits in Northern and Eastern Poland. Bulletin de la Polonaise Academie des Sciences, Series Sciences de la Terre, **29** (1), 81–90, Warszawa.

UBERNA J. & ODRZYWOLSKA-BIEŃKOWA E.
1977 Nowe stanowiska osadów górnoeoceńskich na obszarze północnej Lubelszczyzny, *Kwartalnik Geologiczny*, **21** (1), 73–87, Warszawa.

VINKEN R. [ed.]
1988 The Northwest European Tertiary Basin, *Geologisches Jahrbuch, A*, **100**, 1–508.

WAŻYŃSKA H.
1997 Wyniki badań palinologicznych próbek osadów trzeciorzędowych z otworu wiertniczego Czemierniki, arkusz Radzyń Podlaski 1 : 50 000, [*in:*] Kasiński J. R., Piwocki M., Saternus A., Tołkanowicz E. & Wojciechowski A., *Realizacja projektu prac geologicznych dla określenia perspektyw występowania bursztynu w utworach eocenu Lubelszczyzny*, **11** (2), 1–12, Centralne Archiwum Geologiczne P.I.G., Warszawa.

WEITSCHAT W.
1997 Bitterfelder Bernstein — ein eozäner Bernstein auf miozäner Lagerstatte, *Metalla*, **66**, Sonderheft, 71–84, Bochum.

WOŹNY E.
1966a Eocen z Siemienia koło Parczewa, *Kwartalnik Geologiczny*, **10** (3), 843–850, Warszawa.

1966b Fosforyty i bursztyny z Siemienia koło Parczewa, *Przegląd Geologiczny*, **14** (6), 277–278, Warszawa.

1977 Pelecypods from the Upper Eocene of East Poland, *Acta Palaeontologica Polonica*, **22** (1), 99–112, Warszawa.

ZALEWSKA Z.
1974 Geneza i stratygrafia złóż bursztynu bałtyckiego, *Biuletyn Instytutu Geologicznego*, **281**, 139–173, Warszawa.

ZIEGLER P. A.
1990 *Geological Atlas of Western and Central Europe*, Shell Intern. Petr. Maatsch., Hague.

GEOLOGIE UND MINERALOGIE DES BERNSTEINS VON KLESSOW, UKRAINE

Wladimir TUTSKIJ & Ljudmila STEPANJUK

Kurzfassung

Die ersten schriftlichen Angaben über Bernsteinfunde in der Ukraine stammen aus dem Jahre 1721. Die erste nutzbare Bernstein-Lagerstätte (des oberen Eozän- unteren Oligozän Schichten) wurde in Klessow (Ukraine) in den 90ger Jahren entdeckt. In geotektonischer Hinsicht liegt die Bernsteinvorkommen von Klessow am NW-Abhang des Ukrainischen Schildes, nämlich zwischen Wolyn-Podolsker Platte und der Senke des Pripjat. Die bernsteinführende Schichten, entstanden als Resultat der Verwitterungsprozessen der kristallinen Grundgebirge, die zur oligozän Meeresküste und dort zusammen mit dem vom Festland kommenden organischen Material transportiert und eingebettet wurden. Ukrainischer Bernstein, die Varianten und organische Inklusen wurden auch beschrieben.

Historie

Die ersten schriftlichen Angaben über Bernsteinfunde in der Ukraine stammen aus dem Jahre 1721 von dem polnischen Naturwissenschaftler G. RZĄCZYŃSKI. Im Laufe der folgenden zwei Jahrhunderte berichteten zahlreiche ukrainische und ausländische Forscher über weitere Bernsteinfunde, wobei ein mineralogisches Interesse überwog.

Eine großen Beitrag zur Erforschung des ukrainischen Bernsteins leistete P. A. TUTKOWSKIJ in den Jahren 1890–1910. Er bestätigte die Bernsteinführung der paläogenen Schichtfolgen an den Flüssen Goryn und Slutsch, insbesondere im Gebiet der Station Klessow, ca. 90 km nördlich von Riwne.

Erst im Jahre 1979 begannen lagerstättenkundliche Erkundungs- und Bewertungsarbeiten in der Nähe des Ortes Klessow (Kreis Sarnenske, Bezirk Rowenske), mit dem Ergebnis, daß Areale mit abbauwürdigen Bernstein-Gehalten ermittelt wurden. Aufgrund dieser Ergebnisse wurde die Lagerstätte Klessow als erste Bernsteinlagerstätte in der Ukraine abgegrenzt. Im Jahre 1993 begann das staatliche Unternehmen „Ukrburschtyn" im Raumabschnitt Pugatsch, die Bernstein-führenden Schichten durch einen Tagebau zu erschließen.

Der geologische Rahmen

In geotektonischer Hinsicht liegt die Bernsteinvorkommen von Klessow am NW-Abhang des Ukrainischen Schildes, nämlich zwischen Wolyn-Podolsker Platte und der Senke des Pripjat. Ultrametamorphe, metasomatische und intrusive Bildungen des Unter- und Mittel-Proterozoikums bilden das kristalline Basement (Abb. 1). Es ist fast überall von Sedimenten des Mesozoikums und Känozoikums überlagert. Im südöstlichen Teil des Untersuchungs-gebietes liegt das kristalline Basement in Höhen von +130 bis +170 m NN und ist dort nur durch quartäre Sedimente überdeckt. Gegen Norden und Westen senkt sich die Oberfläche des Kristallins um ca. 10 m/km, wobei Aufragungen einzelner Schollen die Gesamttendenz der Absenkung etwas verschleiern.

Das Mesozoikum ist durch kreideartige Mergel des Turons (Obere Kreide, K_2t) und nur lückenhaft im nordwestlichen Teil des Untersuchungsgebietes vertreten. Känozoischen Alters sind die Obuchow-Folge des oberen Eozän (P_2ob), die Mezhigorsk-Folge des Unteren Oligozän ($P_3m\check{z}$) und die fluvioglazialen Bildungen des Quartärs.

Die eozäne, marine Obuchow-Folge liegt im nördlichen und westlichen Teil des Untersuchungsgebietes transgressiv auf dem Mesozoikum und auf dem kristallinen Basement. Sie besteht aus feinkörnigen, glaukonitischen Quarzsanden mit Schluff-Lagen (Aleurite: Silt-Sand Gemisch im Kornbereich 0,01–0,1 mm) von grün-grauer bis grüner Farbe, mit Glimmerführung und ist überall ziemlich gleichartig ausgebildet. Bei hohem Glaukonitanteil läßt die tiefgrüne Farbe die Folge leicht erkennen und sie als markanten Horizont hervorzuheben.

Die oligozäne Mezhigorsk-Folge ist fast überall verbreitet, ausgenommen auf den Hochlagen des kristallinen Basements. Der ungleichkörnige Quarz-

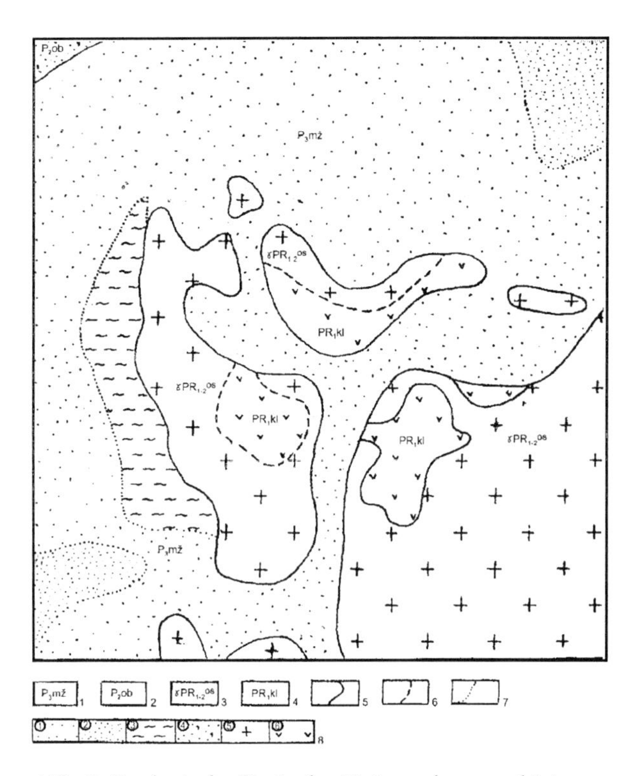

Abb. 1. Geologische Karte des Untersuchungsgebietes Klessow/Ukraine (Maßstab 1: 50 000). 1 — Ablagerungen der Mezhigorsk-Suite, Oligozän; 2 — Ablagerungen der Obuchov-Suite, Eozän; 3 — Bildungen des Osnizk-Komplex, Unteres – Mittleres Proterozoikum; 4 — Bildungen der Klessow-Serie, Unteres Proterozoikum; 5 — Glaubwürdige geologische Grenzen; 6 — Vermutliche geologische Grenzen, vermutet; 7 — Faziesgrenzen gleichalter Ablagerungen; 8 — Lithologische Zusamensetzung der Gesteine: (1) Quarzsande unterschiedlicher Körnung, (2) Aleurite (Schluffe), (3) Lehme, (4) Glaukonitischer Quarzsand, (5) Granite, (6) Gneise.

sand enthält nur einen geringen Anteil an Glaukonit (1–2%). Er ist von hellgrauer bis dunkelgrauer Farbe. Die Gerölle der Kiesfraktion bestehen aus Granit, Gang-Quarz und Kiesel. Die Mächtigkeit der Folge schwankt zwischen 2 und 8 m. Bernstein tritt vornehmlich in dieser Folge auf. In diesem stratigraphischen Niveau liegt auch die Lagerstätte von Klessow.

Das kristalline Basement und die paläogenen Schichtfolgen werden im gesamten Gebiet von quartären Sanden glazifluviatiler Bildungsweise überdeckt.

Charakteristik der bernsteinführenden Folgen

Die bernsteinführende Mezhigorsk-Folge wurde in einem Meer gebildet. Der verschieden-farbige Quarzsand enthält in Lagen und Linsen inkohlte Holzreste, die Anzeichen des nahen Festlandes sind. Das Sedimentmaterial selbst stammt aus dem Abtrag des kristallinen Basements. Die Korngrößen liegen zwischen 0,1 und 0,5 mm. Die granulometrische Zusammensetzung ist sowohl im Profil als auch regional unterschiedlich.

Der untere Teil des bernsteinführenden Sandes ist durch einen erhöhten Gehalt an Markasit gekennzeichnet. Ferner treten auf: Turmalin, Biotit, Staurolith und Granat. In der leichten Fraktion wurde ein bedeutender Anteil an Glaukonit (1–10%) sowie Spuren von Chlorit entdeckt.

Den oberen Teil des Sandes charakterisiert ein erhöhter Gehalt an Rutil, Disthen, Ilmenit, Staurolith und Granat. In der leichten Fraktion sind Quarz-Varietäten und ein geringer Gehalt an Feldspat (1–8%) vorhanden.

Geologische Entwicklung und Tektonik

Das Untersuchungsgebiet um Klessow an der NW-Flanke des Ukrainischen Massivs besitzt einen Stockwerks-Bau. Das untere Stockwerk wird durch das heterogene Kristallinmassiv vorkambrischen Alters gebildet, das obere durch den alpidisch strukturierten Komplex, zu dem die kreidezeitlichen Bildungen, die Folgen des Eozän und Oligozän und das Quartär gerechnet werden. Die beiden Stockwerke sind durch eine Schichtlücke mit Winkeldiskordanz voneinander getrennt.

Der Ukrainische Schild wurde im, jüngsten Alb bis Turon, wie man jetzt weiss, von dem Meer, das in der Unterkreidezeit im W und NW lag, weithin überflutet. Die Ursache liegt in einer epirogenen Senkung. Infolge rascher Transgression lagen die Liefergebiete fern vom Untersuchungsgebiet. Terrigene Sedimente konnten hier nicht gebildet werden.

Eine folgende Hebungsperiode bewirkte, daß vom frühen Santon bis zum späten Eozän festländische Bedingungen herrschten, wobei fluviatile Abtragung der Hauptfaktor der Erosion war.

Bei einer erneuten Überflutung im jüngsten Eozän blieben lokale Hochlagen des Basements vom Wasser unbedeckt, denn das Eozänmeer war nicht sehr tief. Die Gleichmäßigkeit der lithologische Ausbildung deutet auf eine gleichförmige Senkung des Basements hin.

An der Wende Eozän/Oligozän führte ein bedeutender struktureller Umbau am Rande der Osteuropäischen Tafel zu einer deutlichen Verkleinerung des Meeresraumes und damit zur Regression. Seither war das Untersuchungsgebiet ein Festland und unterlag vornehmlich den Prozessen der Verwitterung und Abtragung. Die quartärzeitlichen Klimaänderungen übten eine wesentlichen Einfluß auf die Reliefgestaltung aus.

Paläogeomorphologie

Für die Bildung der bernsteinführenden Folgen der Lagerstätte von Klessow waren die paläogeomorphologischen Verhältnisse des Festlandes, auf dem der Bernsteinwald wuchs, und die des marinen Sedimentationsgebietes von großer Wichtigkeit. Die umfangreichen und komplexen Untersuchungen des Gebiets haben das Paläorelief einer Küstenregion entdeckt, in der zahlreiche Inseln, die durch schmale Durchlässe getrennt sind, die Entwicklung von breiten Buchten und Lagunen vor dem Festland ermöglichten (Abb. 2). Dies ist eine ziemlich eigenständiger Typ einer abrasiv-kumulativen Küste.

Die kontrastreichsten Elemente des Paläoreliefs stellen die Inseln dar. Sie sind Erosionsreste, seltener tektonische Hochbereiche des kristallinen Basements und besitzen eine Größe von 1–2 km^2 bis zu 10–12 km^2. Solche Vorsprünge sind nur von quartärzeitlichen Sedimenten überlagert, oligozäne Bildungen fehlen. Vermutlich waren die Hochlagen nicht Sedimentationsgebiet, denn deren Oberflächen des Kristallin (heute zwischen +160 und +175 m NN) überragten im Oligozän die maximale Ablagerungshöhe in den Meerengen, die heute im Klessower Gebiet bei +155 bis +157 m NN liegt. Auch das heutige Relief orientiert sich an diesem oligozänen Paläorelief.

Die oligozän-zeitlichen „Schären" des Klessower Gebiets waren recht flach. Die maximale Höhe der Inseln, bezogen auf die Sohle des Paläogens, beträgt 30 m, die mittlere Höhe 10–20 m. Bei einer Breite der Meerengen zwischen 0,5–3,0 km und einer mittleren Höhe bis 20 m beträgt die maximale Neigung des (oligozänen) Paläoreliefs etwa 5°.

Im Niveau des oligozänen Meeresspiegels beträgt das Verhältnis der Insel-Flächen gegenüber der Fläche der sie voneinander trennenden Meerengen etwa 3 : 1. Die Zahl der Inseln und die Inselfläche verringern sich in nördliche und westliche Richtung. Umgekehrt nimmt die Meerengenfläche in südöstlicher Richtung, also in Richtung auf den Ukrainischen Schild, ab. Dies zeigt, daß der Klessower Raum vor Beginn der Bernsteinsedimentation Paläorelief aufwies, das durch Erosion erzeugt war.

Bernsteinhäufigkeit

Flüsse transportierten den Bernstein zur Küste. Deltakegel vor Küsten- oder Flußebenen waren die ersten Bernstein-Depots. Hier wurde nur ein Teil des Bernsteins akkumuliert, den Rest verbreiteten Strömungen, Wellen und Gezeiten über den ganzen Küstenraum, dessen Lage bei steigendem Meeresspiegel sich ständig veränderte. Die größte Bernstein-Konzentration ist an die Deltakegel und an Strandsedimente der Meeresbuchten, Meerengen und Golfe gebunden.

An der Küste erfolgt eine Zonierung der Sedimente nach der Korngröße: grobkörnige Sedimente liegen

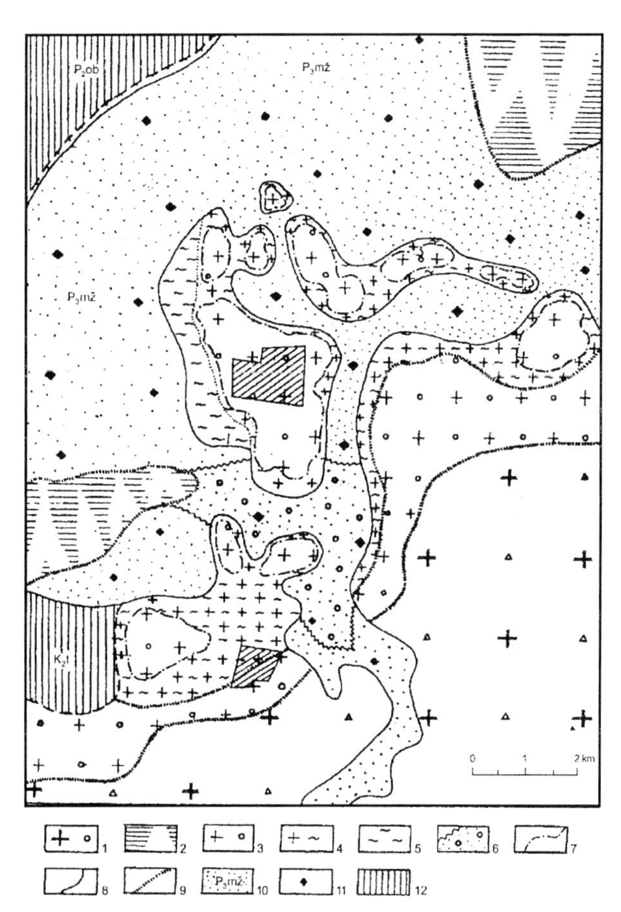

Abb. 2. Paläogeomorphologische Karte des Untersuchungsgebietes Klessow/Ukraine für das Oligozän. 1 — Schwachreliefierte Denudationsebene, Bereich des Bernsteinwaldes; 2 — Gebiet des offenen Meeres; 3 — Abrasionsrestberge oberhalb des Meeresspiegels; 4 — Abrasionsrestberge unterhalb des Meeresspiegels; 5 — Flächen der transgressiven Zehmvorkommen; 6 — Delten; 7 — Konturen der Inseln des „Schären"-Typs; 8 — Grenzen der Verbreitung des bernsteinführenden Horizontes; 9 — Formgrenzen des Paläoreliefs; 10 — Verbreitung der Oligozän-Sande; 11 — Bernstein im Sediment; 12 — Gebiete postoligozäner Erosion.

in Festlandsnähe, feinförnige Sedimente dagegen zum offenen Meere hin. Im Strandbereich werden besonders grobe Gerölle angehäuft: Gerölle aus dem Kristallinen Grundgebirge, nämlich Gerölle aus Granit und Quarz, sowie Holzreste. Sie werden von den nachfolgenden Regressionssedimenten bedeckt. Große Partikel, deren Dichte kleiner als 1g/cm^3 ist und die deshalb nicht sinken (z. B. Holz, Meerestier- und Fischkadaver) werden nicht zum offenen Meer, sondern zur Küste transportiert und dort eingebettet.

Die fossilen Harzbrocken verhalten sich unter ufernahen Bedingungen ähnlich. Sie werden längs der Küste zonar verteilt, wobei sie zumeist eine möglichst ferne Position vor der Küste im „ruhigen" Meeres einnehmen, nämlich jenseits der Sturmwellen-Basis. Wir vermuten, daß folgende Zonierung in den Oligozän-Strandsedimenten der Klessower Zone besteht:

- Zone der Anhäufung des Kohle-Detritus nahe dem Festland: verschiedenkörniger, dunkelgrauer, humusführender Sand mit einzelnen großen Bernsteinstücken;
- Zone maximaler Bernsteinführung: verschiedenkörniger, grauer Sand mit Granit-, Quarz- Kies;
- Zone passiver Wellenwirkung: fein- und mittelkörniger, grün-grauer, glaukonitischer Quarzsand mit wenig Bernstein.

Im Prinzip tritt Bernstein in allen Küstenbildungen auf, jedoch in unterschiedlicher Menge. So ist die ganze Schicht der küstennahen Meeresab-

Abb. 3. Makro Pflanzen-Fossilien: 1 — *Pinus palaeostrobus* (ETT.) HEER; 2 — Pinus parabrevis KILLPER; 3 — *Pinus thomassiana* GOEPPERT *var. kampassica* GORBUNOV; 6 — *Pinus echinostrobus* SAPPORTA; 8–10 — *Pinus* sp. Fot. Nikola DEMCENKO.

Abb. 4. Natürliche Formen des Ukrainischen Bernsteins von Klessow. Fot. Nikola DEMCENKO.

lagerungen (einschließlich der in Deltakegeln) bernsteinführend, aber es gibt im Rahmen ihrer Verbreitung unseren Angaben nach maximal bernsteinführende Zonen — nämlich Flußdelten, Strandzonen von Golfen, Meerengen und Buchten an den Küsten des „Schären"-Typs.

Paläontologie der bernsteinführenden Schichten von Klessow

Im späten Eozän und frühen Oligozän, als ein Meer das Territorium der Subparatethys bedeckte, lag das Gebiet um Klessow in einem untiefen Küstenraum. Auf dem Festland des Ukrainischen Schildes bestanden günstige Bedingungen für Nadelwälder, die Harz in großen Mengen produzierten. Ferner waren auch Vertreter des immergrünen subtropischen Pflanzenreichs verbreitet.

Im Tagebau bei Pugatsch wurden — hauptsächlich in den unteren Schichten — gemeinsam mit Bernsteinstücken auch fossiler Zapfen gefunden, die der Familie Pinaceae angehören (Abb. 3): *Pinus thomassiana* (GOEPP.), *P. palaeostrobus* (ETT.) HEER, *P. parabrevis* KILLPER, *P. echinostrobus* SAPPORTA und *P. spinosa* HERBST u. a. m.. Die Zapfen sind nicht groß, nicht entfaltet. Sie sind fossilisiert, aber im Laufe der Fossilisation nicht zusammengedrückt worden. Die Mehrheit der Fundstücke ist unverdrückt und unversehrt, denn Schildchen und Nabel, die sich inmitten des Querkiels befinden, sind deutlich erkennbar. Die keilförmigen Dornen (Stacheln) sind an fast allen Zapfen abgebrochen.

Unter den übrigen paläontologischen Resten sind große Mengen von Hai-Zähnen und Wirbel und Gräten von anderen Raub-Fischen zu erwähnen, sowie unidentifizierte organische Reste.

Eigentümlichkeiten des Bernsteins von Klessow

Der Bernstein des Klessower Vorkommens ist überwiegend von einer braunen oder dunkelbraunen Verwitterungsrinde von 1–2 mm Dicke überzogen. Die Form der Bernstein-Stücke ist äußerst verschieden (Abb. 4–6):

- kleine (1–2 cm) Stücke sind gewöhnlich rund, nußartig, schwach zusammengepresst;
- bei mittleren Stücken (3–10 cm) überwiegt die fladenartige Form, sie erscheinen zusammengepresst mit gut gerundeten Rändern; die untere Oberfläche solcher Stücke ist besser geglättet als die obere, auf der nicht selten kleine Vertiefungen, Höhlen oder Vorsprünge vorhanden sind;
- die großen Stücke (größer als 10 x 10 cm) besitzen oft ein stangenartiges, eckiges Aussehen, seltener die Form eines Brotlaibes; ihre Unterseite ist flach und völlig geglättet; die Oberseite dagegen gewölbt;
- Bernsteinbrocken von komplizierter Gestalt mit deutlichen Vorsprüngen, Eindellungen oder Sprossen, sind äußerst selten.

Ferner gibt es kugelartige Formen von 5–6 cm Durchmesser. Sie besitzen eine schalen-förmige Struktur vergleichbar der von Ooiden. Die Schalen sind leicht abzulösen. Der innere Kern von 3–4 cm Dicke besteht aus festem, nicht oxidierten Bernstein.

Bernsteinstücke von rohrartiger, linsenartiger, tropfenförmiger oder bisweilen auch eiszapfenförmiger Gestalt (externe natürliche Formen) weisen einen guten Erhaltungszustand auf. Formen des Bernsteins, die im Innern der harzproduzierenden Bäume sich anreicherten und dort die Hohlräume

Abb. 5. Natürliche Formen mit Verwitterungs-Rinde, typische für Klessow Vorkommen. Fot. Nikola DEMCENKO.

Abb. 6. Seltene Stücke des Bernsteins aus Klessow. Fot. Nikola DEMCENKO.

ausfüllte, insbesondere die internen Formen, die zwischen Stamm und Borke sich bildeten, stellen keine Seltenheit dar. Sie sind sehr flach und bilden bisweilen noch die Rindenstruktur ab.

Häufig sind auf Bernsteinstücken schmale Rinnen, Rillen oder Vertiefungen zu finden, die 0,5–2,0 cm Durchmesser besitzen und 1–3 cm tief in den Bernstein reichen. Diese Hohlformen waren uspründlich ausgefüllt durch Zweige, Rindenstücke, Zapfen oder Wurzeln, deren kohlige Substanz verschwunden ist.

Die Hauptmenge des Bernsteins, der für Schmuckstücke verwendet oder in der Schmuckindustrie verarbeitet wird, liegt zwischen 10 und 80 mm Größe. Etwa 70–80% der Fundstücke des Klessower Bernsteins wiegen zwischen 10 und 100 g. Stücke, die eine Größe von 150 mm oder das Gewicht von 1 kg überschreiten, sind äußerst selten.

Die Farbe des Klessower Bernsteins schwankt im frischen Bruch zwischen weiß porzellanartig, blaßgelb (elfenbein-farben) und gelb, rot-braun, kirschrot und braun-schwarz. Der überwiegende Teil liegt im orangegelben Farbton und dessen Schattierungen. Ein Bernsteinstück kann verschiedene Farbvarietäten aufweisen.

Die Durchsichtigkeit des Bernsteins hängt von den Hohlräumen und Blasen ab, die im Bernstein enthalten sind. Der Rauch- oder Wolkenbernstein, ein klarer Bernstein mit matten oder gelben, undurchsichtigen Streifen oder Flecken, ist recht verbreitet. Daneben tritt der Elfenbein-Bernstein auf, der wachsgelb und undurchsichtig ist und der zonar gefärbte Bernstein, der einen Wechsel von hellgelben klaren und dunkleren matten Steifen von 0,5–2,0 cm Breite zeigt. Bernsteine, die grüngelb bis hellgrün gefärbt sind, sind typisch für das Klessower Vorkommen. Graublau gefärbter Bernstein wurde vereinzelt gefunden.

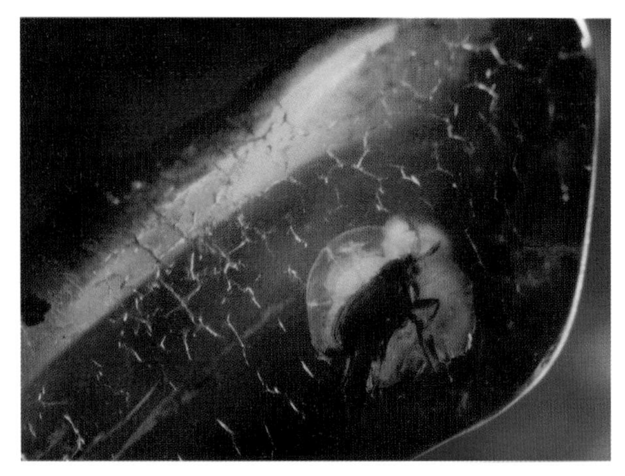

Abb. 7. Klasse Insecta, Ordnung Diptera, Familie Phoridae, *Muscidora* sp. Fot. Nikola DEMCENKO.

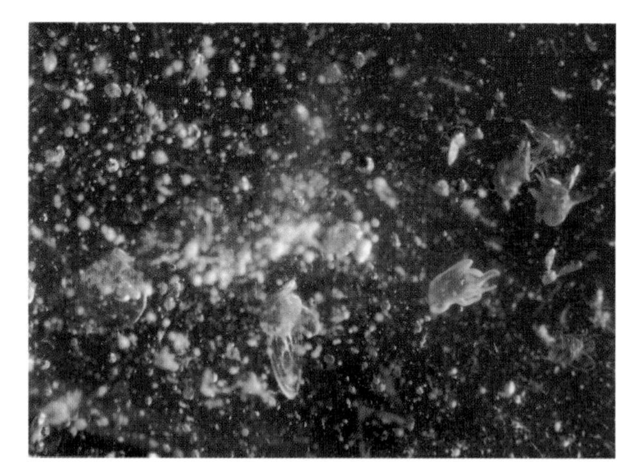

Abb. 9. Klasse Insecta, Ordnung Diptera, Familie Drosophilidae. Fot. Nikola DEMCENKO.

Abb. 8. Klasse Insecta, Ordnung Diptera, Familie Xylophagidae, *Xylophagus* sp. Fot. Nikola DEMCENKO.

Abb. 10. Klasse Insecta, Ordnung Diptera, Familie Muscidae, Made eines der Vertreters der Zweiflügler. Fot. Nikola DEMCENKO.

Einschlüsse im Bernstein

Der Klessower Bernstein enthält zahlreiche* Einschlüsse. Bei den mineralischen Einschlüssen überwiegen Kalzit und Pyrit, aber auch tonige Substanz ist enthalten. Gas- und Flüssigkeits-Einschlüsse besitzen rundliche, bisweilen ovale Formen von 0,001 bis 3 mm Durchmesser. Diese Einschlüsse erzeugen im Bernstein eigenartige, gar wunderbare Bilder und Strukturen.

Die Einschlüsse von pflanzlichen und tierischem Lebenwesen sind von besonderer Wichtigkeit (Abb. 7–13). Gegenwärtig wird eine Katalogisierung

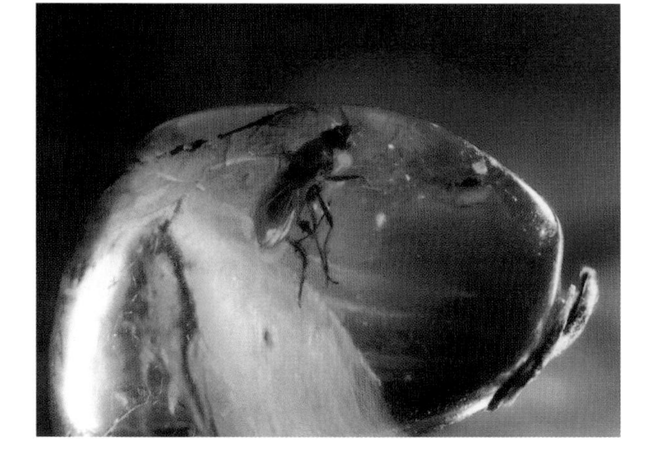

Abb. 11. Klasse Insecta, Ordnung Diptera, Familie Platypezidae. Fot. Nikola DEMCENKO.

*Tierische Inklusen in ukrainischem Bernstein sind nicht sehr zahlreich. (Ed.)

aller Insekten, die im Klessower Bernstein je gefunden wurden, durchgeführt. Hervorzuheben sind Käfer der Familie der Pselaphidae, die in verfaulenden Baustümpfen und Waldlaubstreu lebt und von Pilzen und Farnsporen lebt. Ferner von Vertretern der Dolichopodidae, Phoridae (*Musidora* sp.), Lestremiidae (*Lestremia* sp.), Lonchopteridae, Formicidae (*Lasius* cf. *fuliginsus*), Sciaridae, Drosophilidae, Pompilidae (*Agenia* sp.), Xylophagidae (*Xylophagus* sp.), Platypezidae (?*Clythia* sp.) und kleine Fliegen (Bestimmung: J. NEKRUTENKO). Ferner wurden gefunden: Cecidomyiidae (Gallier), Mycetophylidae (Pilzmücken), Psychodidae (Schmetterlings-mücken), Auchenorrhyncha (Zikaden), Helodidae (Moraster), Limoniidae (Limoniden-Mücken) und aus der Ordnung der Trichoptera (Köcherfliegen). Aus der Klasse der Spinnentiere sind vertreten Theridiidae (*Clya*), *Mizalia* cf. *pilosa*) und Acarina (Milben).

Die Mehrheit der erwähnten Familien ist auch im Baltischen Bernstein bekannt. Dies läßt auf ähnliche geographische und klimatische Bedingungen an der Wende Eozän/Oligozän in der nordwestlichen Ukraine und im Baltikum schließen. Die neuen Informationen über die Insekten, die sich im Bernstein befinden, wird zweifellos unsere Vorstellungen über die Naturbedingungen der weit zurückliegenden Epochen der Erdgeschichte verbessern, die uns einen so wunderbaren „Sonnestein“ geschenkt haben.

Wladimir TUTSKIJ & Ljudmila STEPANJUK
Ministry of Finance of the Ukraine State Enterprise
"Ukrainian Amber"
16 Lipnya st. 38
266028 Rivne, Ukraine

Abb. 12. Klasse Insecta, Ordnung Diptera, Familie Sciaridae. Fot. Nikola DEMCENKO.

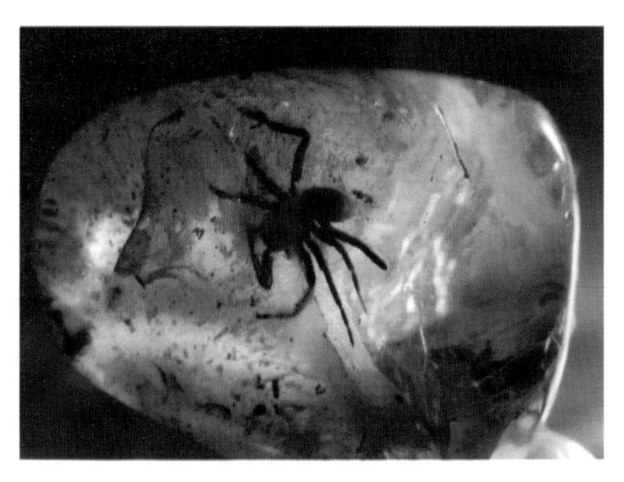

Abb. 13. Klasse Arachnida — Spinnentiere, Ordnung Araneomorphae, Familie Theridiidae, *Clya* sp. KOCH, BERENDT. Fot. Nikola DEMCENKO.

GEOLOGIA I MINERALOGICZNE WŁAŚCIWOŚCI ZŁÓŻ BURSZTYNU NA UKRAINIE

Wladimir TUTSKIJ, Ludmiła STEPANJUK

Streszczenie

Pierwsze pozytywne wyniki poszukiwań zasobów bursztynu (sukcynitu) na Ukrainie uzyskano w okolicach Klesowa (rejon Sarn, obwód Równe), gdzie ukraińscy geolodzy odkryli złoża o znaczeniu przemysłowym.

Badany obszar leży w strefie połączenia tarczy ukraińskiej, płyty wołyńsko-podolskiej i zlewni prypeckiej. Ruchy fałdowe z końca fazy laramijskiej pozostawiły obszar o złożonym i rozczłonkowanym reliefie. W późnym eocenie maksymalna transgresja morska objęła przeważającą część obszaru, z wyjątkiem wzniesień tarczy ukraińskiej.

Na przełomie epoki eoceńskiej i oligoceńskiej nastąpiła gruntowna przebudowa strukturalna zachodniej części platformy wschodnioeuropejskiej, powodująca znaczne zmniejszenie basenu morskiego. W płytkim basenie oligoceńskim, o spokojnym reżimie hydrodynamicznym, gromadziły się glaukonitowe osady piaszczysto-ilaste. Badanie rejonu klesowskiego pozwoliło ustalić istnienie bardzo specyficznego rozczłonkowanego morskiego typu abrazyjno-akumulacyjnego wybrzeża w czasie osadzania się bursztynu. Regresywne warstwy oligoceńskie (piaski bursztynonośne) reprezentują facje deltowych stożków, cieśnin, zatok i plaż.

Północno-zachodnia część tarczy ukraińskiej, z którą związane są kopalnie bursztynu, w późnym eocenie – wczesnym oligocenie stanowiła płytkowodną strefę przybrzeżną basenów morskich. Warunki klimatyczne zbliżone były do wilgotnego subtropiku.

Zespół minerałów osadów bursztynonośnych Klesowa tworzył się w wyniku sedymentacji produktów wietrzenia skał krystalicznych, które były transportowane potokami do strefy przybrzeżnej morza oligoceńskiego.

Bryłki bursztynu ze złóż Klesowa z reguły pokryte są brunatnoczarną korą zwietrzałego bursztynu o grubości 1–2 mm. Forma bryłek jest bardzo różna — od zaokrąglonych do kanciastych. Często bursztyn ma formy charakterystyczne dla współczesnych naturalnych form żywicy drzew iglastych (sople, krople, formy wewnątrzpniowe, podkorowe i inne). Barwa bursztynu z Klesowa jest bardzo zmienna — od porcelanowobiałej poprzez żółtą do czerwonobrunatnej; bardzo ciekawe są nierzadkie egzemplarze o odcieniach zielonkawych. Występują również, interesujące dla nauki, inkluzje pochodzenia organicznego i nieorganicznego. W okazach z Klesowa oznaczono kilka rodzajów owadów, pajęczaków, larw owadów, występujących w liczbie od 1 do 7 osobników w poszczególnych bryłkach bursztynu. Poszukiwania i badania inkluzji są kontynuowane.

Literatur

RZĄCZYŃSKI G.
1721 *Historia naturalis curiosa Regni Poloniae,* 176–184, Sandomiriae.

Ergänzende Literatur

[TUTKOWSKIJ P. A.] ТУТКОВСНИЙ П. А.
1911 Янтарь въ Волынской губерни, *Труды Общества Изследователей Волыни*, **6**, 21–51, Żytomir.

[DUDKIN V.] ДУДКИН В.
1984 *Отчет по групповой геологической сьемке масштаба 1 : 50 000*, Геофонд, Киев.

[MAJDANOVICH I.] МАЙДАНОВИч И.
1988 *Геология и генезис янтареносных отложений Украинского Полесья*, Наукова Думка, Киев.

[MACUJ V.] МАЦУЙ В.
1995 *Янтарь Украины*, М П „Терра", Киев.

PALAEOENTOMOLOGY IN POLAND

Jan KOTEJA

Abstract

The first fossil insect — an amber inclusion — had already been described by a Polish researcher by the end of the 19th century, but extensive palaeoentomological activity did not appear until the 1970s. Some 200 papers have been published up to date, *c.* 90% in the last two decades. The large inclusion collection at the Museum of the Earth (Warsaw) has given rise to research, but remains in other fossil resins and rocky fossils, covering a wide time span from Triassic to Quaternary, have also been studied by Polish palaeontologists. Selected groups of Arachnida, Homoptera, Heteroptera, Lepidoptera, Diptera, Coleoptera and Strepsiptera have been subjected to palaeontological research. Some forty authors have published palaeoentomological papers, but currently only about ten are active. A palaeoentomological section was established at the Polish Entomological Society in 1985, which organizes yearly meetings and issues a palaeoentomological newsletter — *Inclusion-Wrostek*.

This is only my personal view on the question, a recollection of some events, not devoid of a bit of emotion, since I have been involved in the whole adventure for more than ten years.

Studies on fossil insects and other land arthropods are carried out in many countries, currently very extensively in Germany, France and Spain; a unique palaeoentomological laboratory is in operation at the Moscow Palaeontological Institute with an enormous collection of old insect fossils; and there are many outcrops of fossil material in some countries, for instance in England. Some Americans are also very active, e.g. Dr David GRIMALDI at the American Museum of Natural History, or Dr George POINAR (who has initiated DNA studies in amber inclusions). Unique to Polish palaeoentomology is its formal organization — the Section of Fossil Insects at the Polish Entomological Society — and the newsletter — *Inclusion-Wrostek* — that it issues. Dr Edmund JARZEMBOWSKI (London) wrote in a somewhat jocose style (January 1995): "Congratulations on being ahead of Euramerica and having a palaeoentomology section in the Polish Entomological Society".

Arthropod inclusions had already been described or mentioned by Polish authors in the 17th and 18th centuries. However, the first fossil insect — a beetle embedded in Baltic amber — was described 114 years ago by A. WAGA (1883). Unfortunately, both the inclusion and the interest in palaeoentomology disappeared; only some ten papers were published up until the early 1970s. In 1973 A. SKALSKI published his first paper on butterfly inclusions, then followed the papers by P. MIERZEJEWSKI (1976) and R. KULICKA (1977). However, a real explosion of palaeoentomological activity occurred in the 1980s — twenty authors published their palaeoentomological papers and another twenty showed an active interest in various aspects of research into fossil arthropods.

I still remember the journey with Prof. Jerzy PAWŁOWSKI and assistant Prof. Dr Wiesław KRZEMIŃSKI (Zoological Institute, PAS, Cracow) to Częstochowa where we met Prof. Barbara KOSMOWSKA-CERANOWICZ and Dr Róża KULICKA (Museum of the Earth, Warsaw), Prof. Ryszard SZADZIEWSKI (University of Gdańsk), Dr Piotr WĘGIEREK (Silesian University, Katowice), Dr Henryk GARBARCZYK, Dr Stanisław GŁOGOWSKI and Dr Eligiusz NOWAKOWSKI (Zoological Institute, Warsaw), Dr Bogusław SOSZYŃSKI (Regional Museum, Piotrków Trybunalski) and the late assistant Prof. Dr Andrzej SKALSKI, head of the Regional Museum in Częstochowa. It was Spring, 19 April, 1985. Various questions were vigorously discussed at this meeting, particularly the problem of formally establishing fossil taxa, but the main question was — what to do next? There were suggestions to organize yearly palaeoentomological meetings (such as the "Fachgesprächs" now arranged by the German palaeoentomologists) or establish a formal palaeoentomological section at the Polish Entomological Society... and we left Częstochowa as members of the "Section of Fossil Insects". A few months later the Section was formally accepted by the Authorities of the PES and officially presented at the XXXIX

Congress of the Polish Entomological Society (September 1996, Tleń near Bydgoszcz).

As beginners in palaeontology we were interested in the acquisition, preparation and preservation of fossils, in geology, taphonomy, fossiliferous deposits and collections and, obviously, palaeoentomological literature. Some time later Dr KRZEMIŃSKI published a list (*c.* 400 items) of palaeoentomological papers in his library (*Inclusion-Wrostek*, **2**, 1986). A large collection of books and papers dealing with amber and inclusions was kept at the Museum of the Earth; its comprehensive annotated bibliography was published some years later (KOSMOWSKA-CERANOWICZ 1993).

I have already mentioned *Inclusion-Wrostek* — the newsletter of the Section issued since Autumn 1985. Except for the format (A5) and the editor, everything else has changed in the meantime. You must remember that computers were unavailable in Poland in the 80s and xeroxcopying was only possible under special conditions. Initially the newsletter was all in Polish, then with an increasing number of English texts, among them a palaeontological joke — *Stellate hairs — index fossils of ambers*. The original title (in Polish) of the newsletter was *Bulletin of the Section of Fossil Insects at the PES*, with a note "off record". Soon a confusion arose between the editor and the Authorities of the Society because the bulletin appeared not to be „strictly palaeoentomological". To avoid further polemic I changed the title to *Inclusion-Wrostek — a palaeoentomological workshop*, without any mention of the Section and Society (No. 7, 1988). From then on the Authorities had no reason to trouble themselves about the contents of the newsletter. Ten issues (usually 8 pp.) were produced between 1985 and 1989 with reports of meetings and symposia, paper abstracts, correspondence and news; none in 1990–1993 (see below).

Although officially registered at the Polish Entomological Society, the Section itself was an informal group: anybody who wished to be included into the mailing list of *Wrostek* became a "member" of the group, and the editor in chief was the leader and secretary of the Section at the same time. Judging by the number of names on the mailing list there were some fifty "palaeoentomologists" in the 1980s in Poland. In fact, only twenty individuals published at least one paper on fossil arthropods at that time; the others intended to study fossil insects "in the near future", or wished to know the results of palaeontological research, or simply liked amber or palaeoentomologists.

Yearly palaeoentomological spring meetings (alternating with autumn amber meetings at the Museum of the Earth) have been organized in various places since 1985, with round table discussion as the main event. Twelve to twenty-five persons participated in the symposia with more than ten paper presentations in the 80s and early 90s. This number slightly decreased in the following years. Look at the brief history below:

- I Symposium: 19.04.1985 — Regional Museum in Częstochowa; organizer: assistant Prof. Dr Andrzej SKALSKI; 12 participants. Establishment of the Palaeoentomological Section at the Polish Entomological Society.
- II Symposium: 21–22.03.1986 — Institute of Systematics and Evolution of Animals of the Polish Academy of Sciences; Cracow-Mogilany; organizer: assistant Prof. Dr Wiesław KRZEMIŃSKI; 18 participants; 11 reports (some concerning palaeobotanical questions).
- III Symposium: 23–24.04.1987 — Department of Zoology at the Silesian University, Katowice; organizers: assistant Prof. Dr Aleksander HERCZEK, Dr Jacek GORCZYCA; 21 participants (with Prof. Alexander P. RASNITSYN from the Palaeontological Institute, Moscow); 11 reports.
- IV Symposium: 19–20.04.1988 — Department of Zoology and Parasitology of the Gdańsk University, Gdańsk-Katownia; organizer: Prof. Dr Ryszard SZADZIEWSKI; 17 participants; 10 reports.
- V Symposium: 27–28.04.1989 — Institute of Applied Zoology at the Cracow Agricultural University, Cracow; organizer: Prof. Jan KOTEJA; 18 participants; 24 reports.
- VI Symposium: 10–11.05.1990 — Institute of Zoology of the Polish Academy of Sciences, Warsaw; organizers: Dr Henryk GARBARCZYK, Dr Stanisław GŁOGOWSKI; 24 participants; 12 reports.
- VII Symposium: 10.05.1991 — Department of Zoology at the Silesian University, Katowice; organizers: Dr Piotr WĘGIEREK, assisstant Prof. Dr Aleksander HERCZEK, Dr Jacek GORCZYCA; 13 participants; 10 reports.
- VIII Symposium: 9.10.1992 — Upper Silesian Museum, Natural History Division, Bytom; organizer: Dr Roland DOBOSZ; 23 participants (5 persons from the Palaeontological Institute, Moscow); 13 papers presented.
- IX Symposium: 19–20.11.93 — Museum of Natural History of the Institute of Systematics and Evolu-

tion of Animals, Cracow; organizer: assistant Prof. Dr Wiesław KRZEMIŃSKI; 25 participants; 14 reports (a joint symposium of the Palaeontological Section and the Amber Meetings of the Museum of Earth).

- X Symposium: 14.04.1994 — Department of Zoology and Ecology, Cracow Agricultural University, Cracow; organizer: Prof. Jan KOTEJA; 14 participants; 3 papers presented.
- XI Symposium: 19.05.1995 — Department of Zoology and Ecology, Cracow Agricultural University, Cracow; organizer: Prof. Jan KOTEJA; 4 participants; no paper presentations (other persons visited Cracow between May 15th and June 1st).
- XII Symposium: 18.04.1996 — Department of Zoology, Cracow Agricultural University, Cracow; organizer: Prof. Jan KOTEJA; 9 participants (3 people arrived earlier or later); 3 paper presentations.
- XIII Symposium: 2.09.1997 — Archaeological Museum, Gdańsk; 2 papers and one poster presented at the International Interdisciplinary Symposium *Baltic Amber and Other Fossil Resins. 997 Urbs Gyddanyzc — 1997 Gdańsk.*

And some meetings were very profitable and pleasurable! Look at the reports in the newsletter.

Although fossil collectors and students are enthusiastic about the object of their interest, activity and passion, palaeontological studies play a less than minor role in entomology today. Among the more than 3000 papers presented at the XX International Congress of Entomology (Florence, 1996) only seven papers slightly touched upon palaeontological questions, i.e., 0.2% of all (KOTEJA 1996c). I think this gives an accurate picture of current palaeoentomological activity, including the situation in Poland. In 1973, in the prognosis of development of palaeozoology in Poland to 1990 presented at the Second Congress of Polish Science, palaeoentomological investigations were not even mentioned, either as a fact or need (KIELAN-JAWOROWSKA 1973). Exactly twenty years later SKALSKI (1993) published a bibliography of palaeoentomological papers produced by 38 Polish authors which included approximately 150 titles between 1883 and 1993; 85% in 1973–1993. These figures support the thesis that palaeoentomology exists in Poland, but not necessarily at large meetings and in the opinion of various authorities.

Amber has given rise to Polish palaeoentomological studies: the large collection at the Museum of the Earth, Warsaw, established by Adam CHĘTNIK in the years 1951–1958, later developed (up until 1974) by Zofia Zalewska and then successively enlarged by Prof. Barbara KOSMOWSKA-CERANOWICZ to reach *c.* 20,000 specimens up to date (KOSMOWSKA-CERANOWICZ 1990); the animal inclusions, some 16,000 specimens, under the curatorship of Dr Róża KULICKA. Of special value is the collection (*c.* 8000 specimens) assembled and prepared by Tadeusz GIECEWICZ because it represents unselected material from one site — NE Gdańsk (KULICKA 1985; 1994). Both Prof. KOSMOWSKA-CERANOWICZ and Dr KULICKA encouraged entomologists to identify, classify and study the collection. No wonder the subject of arthropods embedded in Baltic amber constitutes one third of all papers published by Polish palaeoentomologists, among them a series of reviews of the insect groups housed at the Museum and published in the *Proceedings of the Museum of the Earth* (*Prace Muzeum Ziemi*), e. g., KULICKA & ŚLIPIŃSKI (1996), KULICKA & WĘGIEREK (1996), KULICKA *et al.* (1996).

In addition to materials for study, the Museum of the Earth has offered yearly amber meetings (since 1994), at which various geological, mineralogical and paleontological questions are discussed and entomological papers can presented.

Apart from the collection at the Museum of the Earth, small inclusion samples are also housed at several other institutions, e. g. at the Museum and Institute of Zoology, PAS, Warsaw; the Institute of Palaeobiology, PAS, Warsaw, at the Regional Museums in Łomża and Malbork, the Museum of Natural History at Wrocław University, the Museum of the National Wolin Reserve and the rapidly growing collection, arranged by assistant Prof. Dr Wiesław KRZEMIŃSKI, at the Natural History Museum of the Institute of Systematics and Evolution of Animals, PAS, Cracow. Unfortunately, the size and value of private arthropod inclusion collections are unknown.

Other than Baltic amber, there are very few sources of palaeoentomological material in Poland: peat bogs offering subfossil material (mainly Coleoptera) and some Tertiary and Jurassic outcrops quite recently discovered in the Holy Cross Mountains (WĘGIEREK 1995a & b). Carboniferous or Permian deposits are not currently being explored, although the oldest pterygote — *Stygne roemeri* HANDLIRSCH — was found in Upper Silesia. Thus, Polish students work on material from abroad, mainly from Russia (Palaeontological Institute, Moscow), Germany, France, England, Austria, Spain and the USA. In addition to Baltic amber, inclusions of ten other Tertiary and Cretaceous fossil resins have been studied. Investigations on other fossils (impressions and the like) covers a wide time span from Triassic to Quaternary deposits. Many papers (*c.* 20%) deal with theoretical (taphonomy, evolution, phylogeny) and technical questions.

All Polish students of fossil arthropods have a zoological education and are (or were) affiliated with various institutions, i.e. the universities of Gdańsk, Poznań, Wrocław and the Silesian Universities, zoological institutes of the Polish Academy of Sciences in Warsaw and Cracow, agricultural universities of Siedlce and Cracow, the teacher training school in Katowice, with only one person holding a full-time post as a curator of amber inclusions at the Museum of the Earth, Warsaw. This means that palaeontological research is an additional job for most of them, and that they study only those fossil groups (usually quite narrow) with which they are familiar as far as extant fauna is concerned. This also is an obstacle in palaeoentomological research — many entomologists claim that "to study fossil forms one should know the extant world fauna". Nevertheless, as many as seven persons got their PhD degrees on the basis of palaeontological research and publications. In any case, the interest of Polish palaeoentomologists is limited to a few arthropod groups, as follows (students who published at least one paper during the past ten years are listed; titles omitted):

- Aranea (selected families): Jerzy PRÓSZYŃSKI, Marek ŻABKA (investigations discontinued);
- Acarina (selected groups): Wojciech MAGOWSKI (studies temporarily discontinued);
- Blattoptera: Róża KULICKA (preliminary studies);
- Homoptera — Auchenorryncha: Cezary GĘBICKI, Jacek SZWEDO (preliminary studies, not published);
- Homoptera — Psyllinea and Aleyrodinea: Maciej S. KLIMASZEWSKI;
- Homoptera — Aphidinea: Andrzej CZYLOK, Piotr WĘGIEREK (currently only the latter);
- Homoptera — Coccinea: Barbara ŻAK-OGAZA, Jan KOTEJA (currently the latter);
- Heteroptera (Miridae and a few other groups): Aleksander HERCZEK;
- Lepidoptera (Microlepidoptera): Andrzej SKALSKI (discontinued);
- Coleoptera (selected groups and subfossil forms): Jerzy PAWŁOWSKI, Kazimierz GALEWSKI, Antoni KUŚKA, Lech BOROWIEC, Marek WANAT, Adam ŚLIPIŃSKI (currently three persons);
- Strepsiptera: Róża KULICKA;
- Diptera (Limoniidae, Ceratopogonidae and a few other groups): Ewa KRZEMIŃSKA, Wiesław KRZEMIŃSKI, Ryszard SZADZIEWSKI.

For more information consult the *Bibliography of palaeoentomological papers published by Polish authors* (SKALSKI 1993) which includes *c.* 150 items. After 1993 another 60 papers were published dealing mainly with Homoptera, Heteroptera, Diptera and Coleoptera.

It is not the aim of this paper to evaluate the scientific value of the Polish contribution to palaeoentomology (this may be made by specialists) or the activity and "productivity" of the Polish palaeoentomologists; I would only like to mention some of the papers published after 1992, i.e. not listed by SKALSKI (1993), as examples. There is a series of publications on Ceratopogonidae, including monographs on Bitterfeld and Lebanese amber (SZADZIEWSKI 1996; SZADZIEWSKI & GROGAN 1996a, b), a series of papers on Limoniidae and other primitive Diptera based on inclusions and rocky fossils, among others on the discovery of the oldest representatives of the order (KRZEMIŃSKI & LUKASCHEVITCH 1993; KRZEMIŃSKI 1993; 1996; ANSORGE & KRZEMIŃSKI 1994; KRZEMIŃSKI *et al.* 1994; KRZEMIŃSKI & ANSORGE 1995; KRZEMIŃSKI & KRZEMIŃSKA 1994a & b), a series of papers on Aphidinea, including Cretaceous fossils (WĘGIEREK 1993; 1996; WĘGIEREK & MAMONTOVA 1993), on Psyllinea (KLIMASZEWSKI & POPOV 1993; KLIMASZEWSKI 1995) and Heteroptera (HERCZEK 1993; POPOV & HERCZEK 1992; 1993), many in cooperation with the researchers of the Moscow Arthropod Laboratory; papers on Tertiary and Quaternary Coleoptera (KUŚKA 1996; PAWŁOWSKI *et al.* in press); the first Polish record on fossil Acarina (MAGOWSKI 1994); attempts at isolating DNA from Baltic amber (PAWŁOWSKI *et al.* 1996); the last available papers by the late Assistant Prof. SKALSKI (1995) on Lepidoptera; and also on some theoretical questions (KLIMASZEWSKI 1993; KOTEJA 1996a & b). In addition to original papers and reviews, several nice popular books on amber, and especially insect inclusions, have been issued (e.g., KOSMOWSKA-CERANOWICZ 1983; KOSMOWSKA-CERANOWICZ & KONART 1995; KRZEMIŃSKA *et al.* 1992, Polish edition 1993)

Many things changed at the beginning of the 1990s in eastern Europe, including research and palaeoentomology. Firstly, communication, both by post and travel, became much easier resulting in more extensive cooperation between palaeoentomologists and an increasing number of joint projects. On the other hand, the possibility (sometimes need) for activity in fields other than research meant that some paleontologists left fossils for administrative work, business, teaching or simply living insects. Dr Andrzej SKALSKI, Poland's most senior paleoentomologist, has passed away. Unfortunately (for the fossils) no new palaeoentomologists have emerged. Research institutions have a sparse budget for purchasing fossils (especially worrying is the situation at the Museum of Earth which has acquired only a few arthropod inclusions over the past years). At the same time the interest in fossils, particularly amber inclusions, has become much greater among amateurs. People search for and collect inclusions and try to identify them, sometimes asking for help or supplying materials for research. This is certainly an optimistic social phenomenon. Among those people who cooperate with palaeoentomologists I would like to mention, Mr Jacek SERAFIN (Kasparus) who selects and collects inclusions wanted by palaeoentomologists. In 1996 the "Societas Succinorum in Polonia" (Stowarzyszenie Bursztynników w Polsce) was established. One of their aims is to cooperate with palaeoentomologists.

For reasons already mentioned (the editor has been involved in administrative tasks at his university) the palaeoentomological newsletter *Inclusion-Wrostek* was not issued in 1990–1993. It was resurrected in 1994, practically as an international bulletin, computer-compiled, with contributions from many foreign authors, a current palaeoentomological bibliography and an updated mailing list containing more than 200 addresses from all over the world, and an increasing number of pages (36 in No. 26).

Palaeoentomology as a formal unit has existed in Poland for more than ten years. I would like to express my hearty thanks to all my Polish colleagues who cooperated in the organization of meetings, supplied their papers and reports, and for their assistance in all other fields. Also, I am very grateful to all foreign colleagues for their visits, correspondence, materials and friendship.

Jan KOTEJA
Department of Zoology and Ecology,
Cracow Agricultural University
Al. Mickiewicza 24,
30-059 Kraków, Poland

PALEOENTOMOLOGIA W POLSCE

Jan KOTEJA

Streszczenie

O inkluzjach stawonogów w bursztynie pisali polscy autorzy już w XVII i XVIII wieku, ale pierwszego bursztynowego chrząszcza opisał naukowo dopiero WAGA w roku 1883, czyli 114 lat temu, jednak do początków lat siedemdziesiątych opublikowano zaledwie 10 prac paleoentomologicznych. W prognozach rozwoju paleozoologii do roku 1990 przedstawionych na Drugim Kongresie Nauki Polskiej w roku 1973 badania nad kopalnymi stawonogami lądowymi w ogóle nie zostały wzięte pod uwagę (KIELAN-JAWOROWSKA 1973). Jednak już w tym roku swoją pierwszą pracę o motylach bursztynowych ogłosił A. SKALSKI, a wkrótce potem ukazały się prace P. MIERZEJEWSKIEGO (1976) i R. KULICKIEJ (1977). Prawdziwy rozkwit badań paleoentomologicznych nastąpił dopiero w latach osiemdziesiątych. W zestawieniu polskiej bibliografii paleoentomologicznej za lata 1883–1993 SKALSKI (1993) wymienia 39 autorów i 150 pozycji, z czego 85% w ostatnich dwu dekadach. Obecnie (1997) lista prac o stawonogach kopalnych przekroczyła 200, jednak liczba autorów zmieniła się nieznacznie.

Wszystkie osoby zajmujące się obecnie paleoentomologią są zoologami i badania nad stawonogami kopalnymi stanowią tylko małą część ich działalności naukowej. W instytucji powołanej do badań paleontologicznych pracuje tylko jedna osoba, niemniej jednak aż siedem osób uzyskało stopnie naukowe na podstawie dorobku paleoentomologicznego. Próby zwiększenia zainteresowania formami kopalnymi napotykają opór ze strony entomologów, którzy argumentują, że badania paleontologiczne wymagają znajomości współczesnej fauny światowej, w tym z obszarów tropikalnych. W rezultacie prace paleoentomologiczne ograniczone są obecnie (1980–1995) do niewielu taksonów, mianowicie:

Aranea (wybrane rodziny) — dwie osoby, badania przerwane. Acarina (Heterostigmata) — jedna osoba, badania chwilowo przerwane. Blattoptera — jedna osoba, badania wstępne. Homoptera: Auchenorrhyncha — dwie osoby, badania wstępne; Psyllinea i Aleyrodinea — jedna osoba; Aphidinea — jedna osoba; Coccinea — jedna osoba. Heteroptera (Miridae i kilka innych grup) — jedna osoba. Lepidoptera (Micropterygidae) — dr Andrzej SKALSKI (zmarł w 1996); Coleoptera (wybrane grupy) — trzy osoby.

Strepsiptera — jedna osoba. Diptera (głównie Limoniidae i Ceratopogonidae) — trzy osoby.

Inspirację do badań paleoentomologicznych dała bogata kolekcja inkluzji w bursztynie bałtyckim w Muzeum Ziemi PAN w Warszawie i zachęta ze strony jej kustoszy, stąd prace dotyczące inkluzji w bursztynie bałtyckim stanowią jedną trzecią wszystkich publikacji. Pozostałe obejmują inkluzje w innych żywicach kopalnych (prawie wszystkich znanych) i odciski, od triasu do form subfosylnych.

W roku 1985 zawiązała się Sekcja Owadów Kopalnych przy PTE o luźnej organizacji, zrzeszająca około 30 czynnych paleoentomologów i sympatyków. Sekcja organizuje doroczne spotkania z udziałem 15–25 uczestników (w latach dziewięćdziesiątych liczba się zmniejszała), na których wygłaszane są krótkie sprawozdania z bieżących prac.

Sekcja wydaje biuletyn informacyjny (od 1985) *Inclusion-Wrostek*, początkowo w języku polskim, od roku 1994 w języku angielskim, obecnie w nakładzie 150 egzemplarzy i o światowym zasięgu.

Bibliography

ANSORGE J. & KRZEMIŃSKI W.
1994 Oligophrynidae, a Lower Jurassic dipteran family (Diptera, Brachycera), *Acta Zoologica Cracoviensia*, **37** (2), 115–119.

HERCZEK A.
1993 *Systematic position of Isometopinae Fieb. (Miridae, Heteroptera) and their intrarelationships*, 1–88, Uniwersytet Śląski, Katowice.

INCLUSION-WROSTEK, Palaeoentomological Newsletter and Workshop, Kraków, 1994–1997 Ed. Jan Koteja, Nos. **11–26**; former: *Biuletyn Sekcji Owadów Kopalnych przy PTE*, 1985–1987, Nos **1–6**; *Inclusion, Palaeoentomological Workshop — Wrostek, małe a cieszy*, 1988–1989, Nos. **7–10**.

KIELAN-JAWOROWSKA Z.
1973 Prognoza rozwojowa paleozoologii Polskiej do roku 1990, *Przegląd Zoologiczny*, **17** (3), 296–300.

KLIMASZEWSKI S. M.
1993 The structure of hind wings in Psyllodea (Homoptera) and its possible significance in recognizing the relationships within this suborder, *Acta Biologica Silesiana*, **22** (39), 57–67, Katowice.

1995 *Succinopsylla dominicana* n. gen. n. sp., a new jumping plant louse from the Dominican amber (Insecta: Homoptera, Rhinopsyllidae), *Mitteilungen aus dem Geologisch-Paläontologischen Institut der Universität Hamburg*, **78**, 189–195.

KLIMASZEWSKI S. M. & POPOV Y. A.
1993 New fossil hemipteran insects from Southern England (Hemiptera: Psyllina + Coleorrhyncha), *Annals of the Upper Silesian Museum. Entomology. Supplement 1*, 13–35.

KOSMOWSKA-CERANOWICZ B. (ed.), [KOSMOWSKA-CERANOWICZ B., KULICKA R., LECIEJEWICZ K., MIERZEJEWSKI P. & PIETRZAK T.]
1983 *Bursztyn w przyrodzie*, 1–100, Wydawnictwa Geologiczne, Warszawa. English edition:

1984 *Amber in nature*, 1–102, Wydawnictwa Geologiczne, Warszawa.

KOSMOWSKA-CERANOWICZ B.
1990 The scientific importance of museum collections of amber and other fossil resins, *Prace Muzeum Ziemi*, **41**, 141–148.

KOSMOWSKA-CERANOWICZ B. (ed.)
1993 *Bursztyn bałtycki i inne żywice kopalne. Piśmiennictwo polskie oraz prace autorów polskich w literaturze światowej. Bibliografia komentowana 1534–1993. Część I*, Pietrzak T. & Różycka T., *Bursztyn w przyrodzie, kulturze i sztuce*, 1–164, Oficyna Wydawnicza Sadyba, Warszawa.

KOSMOWSKA-CERANOWICZ B. & KONART T.
1995 *Tajemnice bursztynu* (second edition), 1–152, Muza SA, Warszawa.

KOTEJA J.
1996a Scale insects (Homoptera: Coccinea) a day after, [*in*:] Schaefer C. W. (ed.), Studies on hemipteran phylogeny, *Proceedings of the Thomas Say Publications in Entomology*, 65–88.

1996b Syninclusions, *Inclusion-Wrostek*, **22**, 10–12.

1996c Florence viewed from Cracow, *Inclusion-Wrostek*, **23**, 10–11.

KRZEMIŃSKA E., KRZEMIŃSKI W., HAENNI J. P. & DUFOUR CH.
1992 *Les fantomes de l'ambre. Insectes fossiles dans l'ambre de la Baltique*, 1–142, Musée d'Histoire Naturelle, Neuchâtel, Polish edition:

1993 W *bursztynowej pułapce*, 1–141, Muzeum Przyrodnicze Instytutu Systematyki i Ewolucji Zwierząt PAN, Kraków.

KRZEMIŃSKI W.
1993 Fossil Tipulomorpha (Diptera, Nematocera) from Baltic amber (Upper Eocene). Revision of the genus *Helius* Lepeletier et Serville (Limoniidae). *Acta Zoologica Cracoviensia*, **35** (3), 597–601.

1996 The importance of fossil materials to investigations on higher-level phylogeny of the Diptera, *Proceedings of the XX International Congress of Entomology, Firenze, August 25–31, 1996. Abstracts*, 25.

KRZEMIŃSKI W. & ANSORGE J.
1995 New Upper Jurassic Diptera (Limoniidae, Eoptychopteridae) from the Solnhofen Lithographic Limestone (Bavaria, Germany). *Stuttgarter Beiträge zur Naturkunde, Serie B (Geologie und Paläontologie)*, **221**, 1–7.

KRZEMIŃSKI W. & KRZEMIŃSKA E.
1994a A new species of *Cheilotrichia* (Empeda) from Sakhalin amber (Diptera, Limoniidae), *Acta Zoologica Cracoviensia*, **37** (2), 91–93.

KRZEMIŃSKI W. & KRZEMIŃSKA E.

1994b *Procramptonomyia marianna*, a new species from the Upper Jurassic of Great Britain (Diptera, Anisopodomorpha, Procramptonomyiidae), *Acta Zoologica Cracoviensia*, **37** (2), 101–105.

KRZEMIŃSKI W., KRZEMIŃSKA E. & PAPIER F.

1994 *Grauvogelia arzvilleriana* sp. n. — the oldest Diptera species (Lower / Middle Triassic of France), *Acta Zoologica Cracoviensia*, **37** (2), 95–99.

KRZEMIŃSKI W. & LUKASCHEWITCH L.

1993 Ansorgiidae, a new family from the Upper Cretaceous of Kazakhstan (Diptera, Ptychopteromorpha), *Acta Zoologica Cracoviensia*, **35** (3), 593–596.

KULICKA R.

1977 *Mengea tertiaria* w bursztynie bałyckim (w zbiorach Muzeum Ziemi PAN), *Przegląd Geologiczny*, **1**, 32–33.

1985 Inkluzje zwierzęce w bursztynie bałtyckim w zbiorach Muzeum Ziemi PAN w Warszawie, *Wiadomości Entomlogiczne*, **6** (3–4), 179–186.

1994 Inkluzje zwierzęce w bursztynie bałtyckim z osadów holoceńskiej plaży Gdańsk-Stogi (kolekcja autorska T. Giecewicza), *Abstracts of the XI Amber Meeting at the Museum of the Earth, Warsaw, November 19–20, 1994.*

KULICKA R., HERCZEK A. & POPOV Y. A.

1996 Heteroptera in the Baltic amber, *Prace Muzeum Ziemi*, **44**, 19–23.

KULICKA R. & ŚLIPIŃSKI A. S.

1996 A review of the Coleoptera inclusions in the Baltic amber, *Prace Muzeum Ziemi*, **44**, 5–12.

KULICKA R. & WĘGIEREK P.

1996 Aphid species (Homoptera Aphidinea) from the collection of the Baltic amber in the Museum of the Earth, Polish Academy of Sciences, Warsaw. Part three, *Prace Muzeum Ziemi*, **44**, 41–44.

KUŚKA A.

1996 New beetle species (Coleoptera: Cantharidae, Curculionidae) from Baltic amber, *Prace Muzeum Ziemi*, **44**, 13–18.

MAGOWSKI W. Ł.

1994 Discovery of the first representative of the mite subcohort (Arachnida: Heterostigmata Acari) in the Mesozoic Siberian amber, *Acarologica*, **35** (3), 229–241.

MIERZEJEWSKI P.

1976 On application of scanning electron microscope study to organic inclusions from the Baltic amber, *Annales Societatis Geologorum Poloniae*, **46** (3), 291–295.

PAWŁOWSKI J., KMIECIAK D., SZADZIEWSKI R. & BURKIEWICZ A.

1996 Próba izolacji DNA owadów z bursztynu bałtyckiego, *Prace Muzeum Ziemi*, **44**, 45–46.

PAWŁOWSKI J., KUŚKA A., MAZUR M., STEBNICKA Z. & WARCHAŁOWSKI A.

in press Beetle remains (Coleoptera) from the Quaternary deposits in Poland, *Acta Zoologica Cracoviensia*, in press.

POPOV Y. A. & HERCZEK A.

1992 The first Isometopinae from Baltic Amber (Insecta: Heteroptera, Miridae), *Mitteilungen aus dem Geologisch-Paläontologischen Institut der Universität Hamburg*, **73**, 241–258.

1993 *Metoisops punctatus* sp. n., the second representative of the fossil genus *Metoisops* from Baltic amber (Insecta: Heteroptera, Miridae), *Annals of the Upper Silesian Museum. Entomology. Supplement 1*, 7–12.

SKALSKI A. W.

1973 Uwagi o motylach z żywic kopalnych, *Polskie Pismo Entomologiczne*, **43**, 647–654.

1993 Bibliography of palaeoentomological papers (Insecta and Arachnida) published by Polish authors in years 1893–1993, with remarks on investigations of fossil insects in Poland, *Annals of the Upper Silesian Museum. Entomology. Supplement 1*, 67–75.

1995 Study on the Lepidoptera from fossil resins. Part XI. *Baltimartyria*, a new genus for *Micropteryx proavitella* Rebel, 1936 with redescription of this species (Lepidoptera, Zeugloptera, Micropterygidae), *Amber & Fossils*, **1** (1), 26–37.

SZADZIEWSKI R.

1996 Biting midges from Lower Cretaceous amber of Lebanon and Upper Cretaceous Siberian amber of Taimyr (Diptera, Ceratopogonidae), *Studia Dipterologica*, **3** (1), 23–85.

SZADZIEWSKI R. & GROGAN W. L. JR.

1996a Biting midges (Diptera: Ceratopogonidae) from Mexican amber, *Polskie Pismo Entomologiczne*, **65**, 291–295.

1996b Biting midges from Dominican amber. II. Species of the tribes Eteromyiini and Palpomyiini (Diptera: Ceratopogonidae), *Memoir Entomol. Soc. Washington*, **18**, 254–260.

WAGA A.

1883 Note sur un Lucanide incrusté dans le succin (*Paleognatus* Leuthenr *succini* Waga), *Annales de la Société Entomologique de France*, **6** (3), 191–194.

WĘGIEREK P.

1993 Aphid remains from the Upper Cretaceous (Homoptera: Canadaphididae, Paleoaphidae), *Annals of the Upper Silesian Museum. Entomology. Supplement 1*, 57–66.

1995a Dwa nowe trzeciorzędowe stanowiska owadów kopalnych w Polsce, *Przegląd Zoologiczny*, **5**, 419–420.

1995b Jurassic (Lower Liassic) insect impressions in the Świętokrzyskie Mountains in Poland, *Inclusion-Wrostek*, **19**, 5.

1996 Aphid species (Homoptera: Aphidinea) from the collection of the Baltic amber in the Museum of the Earth, Polish Academy of Sciences, Warsaw (Part two), *Prace Muzeum Ziemi*, **44**, 25–39.

WĘGIEREK P. & MAMONTOVA V. A.

1993 A new fossil species of the genus *Stomaphis* Walk. (Aphidoidea: Lachnidae), *Annals of the Upper Silesian Museum. Entomology. Supplement 1*, 37–50.

SCALE INSECTS (HOMOPTERA, COCCINEA) IN CRETACEOUS AMBER

Jan KOTEJA

Abstract

Cretaceous fossiliferous amber is reviewed chronologically; scale insects occur in at least nine of the fourteen reviewed deposits. Thus far, two extinct species have been described from the Upper Cretaceous Canadian Cedar Lake and Siberian Taymyrian amber. Further Canadian material and a collection of New Jersey amber are currently being examined, the latter represented by as many as seven families; the remainder of existing material awaits study. The Matsucoccidae, now confined to *Pinus* and Holarctic, are evidenced throughout the Cretaceous and Tertiary, both as amber inclusions and rocky fossils. A few other archeococcid groups also occur in Cretaceous and Tertiary deposits. Two Cretaceous groups seem to be extinct. Primitive neococcids — Putoidae, Pseudococcidae and Eriococcidae — appeared no later than during the Mid-Cretaceous; advanced groups, such as Coccidae and Diaspididae, have not been observed in Cretaceous deposits up to date.

Introduction

The first fossil scale insects discovered in Baltic amber were described in the classic papers of BERENDT, GERMAR, KOCH and MENGE in the mid 19th century. These discoveries were ignored for nearly one hundred years and coccidologists believed that scale insects, because of their peculiar life history, were unavailable for palaeontological studies. FERRIS (1941) noted: "I have previously entirely ignored the fossil forms, simply because of feeling that conclusions concerning them would be merely guesses and unworthy of serious consideration ... Of the females at hand, one is a beautifully preserved specimen which can be examined with the higher powers of compound microscope and is actually as suitable for study as were most of the preparations with which students of the Coccoidea were content up to scarcely more than twenty-five years ago" — but concluded — "we would almost certainly not be able to differentiate among such species on the basis of specimens preserved in amber". Now more than 500 inclusions in Baltic amber and a number in Dominican and other Tertiary ambers have been gathered in a couple of families.

Palaeontological studies on scale insects were resurrected with the paper by BEARDSLEY (1969) who described a winged male — *Electrococcus canadensis* — from Cretaceous Cedar Lake amber. Twenty years later another Cretaceous coccid — *Inka minuta* — was described, this time from Taymyrian amber (KOTEJA 1989a). At the same time two impression fossils — *Eomatsucoccus* and *Baisococcus* — were discovered in Lower Cretaceous Siberian deposits (KOTEJA 1988b, 1889b) and another unnamed coccid in the Weald Clay, England (COOK & ROSS 1996). However, the existence of coccids in Lebanese amber, the oldest fossiliferous amber of the Cretaceous period, had already been noted by SCHLEE (1972; and by Prof. Aftim ACRA and by Mr Dany AZAR, personal communications). Since then several other Cretaceous fossiliferous ambers have been found, quite recently in the Alava Province, Spain (ARILLO 1996), and among them the New Jersey amber in which scale insects appeared to be very abundant and diverse (Dr David GRIMALDI, personal communication).

The aforementioned discovery has actually given rise to a study on Cretaceous coccids, which is planned in at least two independent parts: (a) examination and description of the New Jersey material (currently underway) and (b) gathering information on scale insects that may exist in any collection of Cretaceous amber inclusions or rock fossils. To be frank, the present article is a request for help in searching for coccid fossils and, as far as possible, making material available for examination. Obviously, first of all it is an acknowledgment of the generous assistance and friendship of many people who have already provided information, written records, their opinion and suggestions, and material; without all these contributions even a preliminary report on Cretaceous coccids could not have been sketched. My particular thanks go to:

Dr Antonio ARILLO (Universidad Complutense de Madrid), Dr John W. BEARDSLEY (University of Hawaii, Honolulu), Dr John COOPER (Booth Museum of Natural History, Brighton), Mrs Christel and Mr Hans W. HOFFEINS (Hamburg), Dr Edmund A. JARZEMBOWSKI (Maidstone Museum & Art Gallery, Maidstone), Dr Douglas J. MILLER (Systematic Entomology Laboratory, USDA, Beltsville), Dr André NEL (Muséum National d'Histoire Naturelle, Paris), Dr Vicente ORTUÑO (Madrid), Dr Edward M. PIKE (Calgary), Dr George O. POINAR (Oregon State University, Corvallis), Dr Beata M. POKRYSZKO (Museum of Natural History, Wrocław), Dr Andrew J. ROSS (Natural History Museum, London), Dr Dieter SCHLEE (Staatlichen Museum für Naturkunde, Stuttgart), Mr Jacek SERAFIN (Kasparus, Poland), Dr Bob SKIDMORE and Dr Josée POIRIER (Canadian National Collection, Ottawa), Dr Irena D. SUKACHEVA, Dr Yuri A. POPOV, Dr Alexandr P. RASNITSYN and Dr Dymitry E. SHCHERBAKOV (Palaeontological Institute, Moscow). I am especially grateful to Dr David GRIMALDI (American Museum of Natural History, New York) for his comprehensive correspondence, valuable suggestions and for lending the large collection of New Jersey amber coccids.

Cretaceous amber

Amber or fossil resin is still a mysterious mineral as regards its chemical composition, physical properties, origin, diagenesis, age and content of fossilized organisms. Among the numerous Cretaceous amber deposits only some are fossiliferous. For instance, the amber beds distributed across the Middle East (Lebanon, Israel, Jordan) are considered to be of the same age — Lower Cretaceous, presumably Hauterivian — and origin (presumably Araucarian), but arthropod inclusions have been found only in amber from the deposits at Jezzine and Dar el-Baidha in the Lebanon (SCHLEE 1972; SCHLEE & GLÖCKNER 1978; BANDEL & VÁVRA 1981; POINAR 1993). Only fossiliferous amber will be discussed further in this paper.

Obviously, palaeoentomologists are interested in the age of any fossil resin — the older the resin, the "better". With the progress of various stratigraphic and physical techniques the age of amber deposits, in terms of both geological and absolute age, can be more and more precisely determined; thus no wonder that chronological data presented by various authors, sometimes even in the same paper, differ considerably. Sometimes chronological data are lacking and the age estimates are simply guesses.

Fossiliferous amber is known from the entire Cretaceous and occurs mainly in the Northern Hemisphere: Alaska, Canada, New Jersey, Europe, Siberia, Japan and the Middle East (Tab. 1).

Identification of the resin source largely depends on the method applied: examination of plant inclusions, associated plant fossils and the chemical/physical characteristics of amber (LANGENHEIM 1990). For instance, Coniferales, Araucariaceae, Taxodiaceae, Cupressaceae and Hamamelidaceae have been suggested as the source of New Jersey resin (GRIMALDI *et al.* 1989; POINAR 1993; D. GRIMALDI personal communication). The question of whether to not assume that species of different systematic affiliation could produce large amounts of resin in the same area over the same period remains open. Resin sources have not even been suggested for most Cretaceous ambers; for the rest, the Araucariaceae (*Agathis*-like trees) are believed to have produced the resin (Tab. 1).

The occurrence of arthropod inclusions in Cretaceous amber has been mentioned in numerous papers and then repeated in various monographs and popular articles, but the identification of the fossils usually stops at order and family level. This is because entomologists generally ignore palaeontological studies and because of the problems encountered in obtaining material for examination, which task may sometimes require many years. Perhaps the most widely studied are the Taymyrian amber inclusions; only single species and selected groups from other ambers have been thoroughly examined and described. It appears that an exception will be New Jersey amber: a number of specialists have been invited by Dr David GRIMALDI (AMNH, New York) to produce a comprehensive monograph which will cover all questions relating to this particular amber and its inclusions.

Burmese amber

Burmese amber was originally believed to date from the Mid-Eocene; later, on the basis of the fauna found preserved in it, it was suggested that it is of late Cretaceous origin. RASNITSYN (1996) presented an interesting hypothesis concerning its age: "The first important impression is that this assemblage shows appreciable similarity to various assemblages from the Late Cretaceous fossil resins, and differs considerably from the Cenozoic ones [...]. This is not to say

AMBER	SITE	AGE (Myr)	ORIGIN	INC	COC	SOURCE
BURMA	Myanmar, Burma	70? Upper Cret.	?	+++	+	Rasnitsyn 1996; PC: A. Rasnitsyn, 24.08.1996
CANADA-1	Edmonton, Alberta	72 Maastrichtian	?	++	+?	Pike 1993; 1994
CANADA-2	Grand Prairie, Alberta	72 Maastrichtian	?	++	+?	Pike 1993; 1994
CANADA-3	Cedar Lake, Manitoba	79? Upper Cret.	Araucariaceae	+++	+	Beardsley 1969
CANADA-4	Grassy Lake, Alberta	79 Campanian	Araucariaceae	+++	+	Pike 1993; 1994; PC: Pike, 22.08.1997
CANADA-5	Medicine-Hat, Alberta	79 Campanian	?	++	+	PC: Beardsley, 14.04.1986
KUJI	Honshu, Japan	85 Coniacian	?	+++	?	Krumbiegel & Krumbiegel 1994
SIBERIA	Yantardakh, Taymyr	87 Santonian	?	+++	+	Zherikhin & Sukacheva 1973; Koteja 1989
NEW JERSEY	White Oaks, Middlesex Co. New Jersey	94 Turonian	Cupressaceae?	+++	+	Grimaldi *et al.* 1989; PC: D. Grimaldi 1996–97
FRANCE	Paris and Aquitan Basins	100 Alb/Cenom.	Araucariaceae	++	?	Schlüter 1978
ALASKA	Arctic Coastal Plain, Arctic Foothills	100? Lower Cret.	Araucariaceae	+	+	Usinger & Smith 1957; PC: G. Poinar, 3.09.1996
UK	UK	100? Lower Cret.	?	+	–	PC: E. Jarzembowski, 29.01.95
ALAVA	Sierra de Cantabria, Spain	112 Lower Cret.	?	+++	?	Arillo 1996; Pokryszko 1996; PC: A. Arillo, 13.08.1997
AUSTRIA	Golling, Salzburg	120 Neocomian	Araucariaceae	+	–	Schlee 1984; Vávra 1984
LEBANON	Jezzine, Dar el-Baidha, Lebanon	135 Hauterivian	Araucariaceae	+++	+	Schlee 1972; Poinar 1993; PC: A. Nel, 16.10.1996

Age (col. 3): Myr — million years, numbers refer to the oldest estimates. Arthropod inclusions (col. 5): + — some fossils present, "++" — some dozens of fossils known, "+++" — some hundred fossils reported. Coccid inclusions (col. 6): "+" — present, "-" — absent. Source (col. 7): PC — personal communication.

Table 1. Cretaceous fossiliferous amber.

that the actual age of Burmese amber is necessarily Cretaceous: the amber might well have its origin on an isolated island in the midst of the Tethys ocean, and the source fauna might be a Cretaceous relict in a Cenozoic world". There is no suggestion concerning the resin producer of this amber.

Burmese amber was brought to Europe at the beginning of the 20th century, but after 1930 its deposits became inaccessible for exploration. The collection at the Natural History Museum comprises 117 blocks which contain as many as 1200 arthropod inclusions classified within 24 groups, including 18 insect orders; Dr A. P. RASNITSYN (personal communication) recognized scale insects in 17 amber blocks, half of them being adult males (for more information see RASNITSYN 1996).

Canadian amber

Fossiliferous Upper Cretaceous amber deposits are known in at least four localities in Alberta and one in Manitoba (Tab. 1); the former are considered to lie *in situ* whereas the latter (at Cedar Lake) are thought to have been redeposition from Alberta. Araucariaceae are believed to be the resin source of Canadian amber. PIKE (1993, 1994) recorded 16 hexapod orders with at least 65 families and 130 estimated species, homopterans dominating (Aphidoidea and Phylloxeroidea). The main Canadian amber collection is housed at "Agriculture", Ottawa, the Royal Tyrrell Museum, Drumheller and Harvard University. Scale insects are represented by one established species — *Electrococcus canadensis* (BEARDSLEY, 1969) — and a couple of undescribed specimens (Dr E. PIKE, personal communication; KOTEJA, unpublished).

Siberian amber

There are numerous amber deposits in north-eastern Siberia, but those of the Taymyr peninsula are most abundant, Yantardakh being the most productive locality. Its age is estimated as Santonian (85–87 Myr) and the sediments are considered primary. There are no data concerning the resin source. Arthropod inclusions have been classified within 16 insect orders (with Diptera dominating) and several groups of Arachnida; these are housed at the Palaeontological Institute, Moscow (ZHERIKHIN & SUKACHEVA 1973). Scale insects are represented by one established species — *Inka minuta* (KOTEJA 1989) and another unidentified one.

Kuji amber

Large amber deposits of suggested Coniacian age (85 Myr) occur near Kuji on Honshu Island, Japan. The amber pieces are large, some very large (dozens of kilograms), but only a few bear arthropod inclusions; scale insects have not been reported (KRUMBIEGEL & KRUMBIEGEL 1994).

New Jersey amber

The New Jersey beds belong to the numerous Cretaceous amber deposits on the Atlantic Coastal Plain of North America (GRIMALDI *et al.* 1989); fossiliferous amber occurs mainly in White Oaks Pits, Middlesex Co., New Jersey. Its age is estimated as Turonian (90–94 Myr) and Cupressaceae are suggested as its botanic origin. Some 20 groups at order level have been recognized; housed at the American Museum of Natural History, New York (Dr David GRIMALDI, personal communication), and scale insects are represented by seven families (*c.* 70 inclusions in 50 amber nodules). As mentioned elsewhere, with respect to scale insects, the complex study of New Jersey amber will serve as a basis for investigation of other Cretaceous ambers.

French amber

The name refers to small amber samples collected in Albian-Cenomanian deposits (*c.* 95 Myr) in north-western France (Paris and Aquitan Basins); Araucariaceae are suggested as the resin source. Only some 80 arthropod inclusions in 12 orders have been found in this amber, but no scale insects so far (Schlüter 1978). This material is held at several institutions (Freies Universität Berlin, Museum of the Earth in Warsaw and others).

Alaskan amber

Cretaceous amber has been found in numerous places on the Arctic Coastal Plain and in the Arctic Foothills of Alaska (USINGER & SMITH 1957; POINAR 1993). Stratigraphical data are lacking; plant remains suggest either an Early or Late Cretaceous age. Some geologists suggest it may be the oldest amber in North America (Dr G. POINAR, personal communication). The origin of Alaskan amber is also uncertain — both Taxodiaceae and Araucariaceae are suggested as the resin producers. Less than 300 pieces have been found to contain biological inclusions (housed at the University of California, Berkeley), but among them a coccid male (POINAR, personal communication).

Amber from the United Kingdom

I can quote here only a brief note by Dr E. JARZEMBOWSKI (personal communication, January 29, 1995): "We have recently found insect-bearing Early Cretaceous amber in the UK; only Diptera inclusions so far".

Alava amber

In early 1995 Spanish newspapers reported the discovery of amber deposits near Vitoria (northern slopes of Sierra de Cantabria, Alava province). Its age is estimated as Lower Cretaceous and the inclusions are said to be very well preserved and extremely rich; housed at the Museo de Ciencias Naturales de Vitoria (ARILLO 1996; POKRYSZKO 1996). A study of the arthropod inclusions has just begun (Dr A. ARILLO, personal communication, August 13, 1997). There is hope that coccids will be found in at least some amber pieces.

Austrian amber

A small amber deposit was found near Golling, in the Salzburg province, considered to be of Neocomian (120 Myr) origin and produced by Araucariaceae. Only a very few poorly preserved inclusions have been recognized in this amber; no scale insects (SCHLEE 1984; VÁVRA 1984).

Lebanese amber

The name refers to a few amber deposits in the Middle East — Lebanon, Israel and Jordan — but arthropod inclusions have been found only in amber from Jezzine and Dar el-Baidha in the Lebanon. The scientific value of Lebanese amber lies in its age, estimated as Neocomian (130–135 Myr). Again Araucariaceae are suggested as the resin producers. At least 4000 arthropod inclusions have been discovered so far and classified within 15 orders, including scale insects. Lebanese amber collections are housed at the Staatliches Museum für Naturkunde, Stuttgart, Geologisch-Paläontologisches Institut and Museum, Universität Tübingen and in several private collections, the largest owned by Aftim Acra (SCHLEE 1972; SCHLEE & GLÖCKNER 1978; KRUMBIEGEL & KRUMBIEGEL 1994; POINAR & POINAR 1994; Dr A. NEL, Dr G. POINAR & Mr D. AZAR, personal communication).

Scale insects

Scale insects or coccids (Homoptera, Coccinea), a sister group of aphids, are represented by two superfamilies — Orthezioidea (archeococcids) and Coccoidea (neococcids) comprising some 6000 extant species in 30 families. Studies on fossil forms are scarce and deal almost exclusively with Baltic amber inclusions. Approximately 20 species (in 8 families) found in this amber have been described, and representatives of at least another 5 families are known to occur here, none being considered as extinct. New Baltic amber genera have been established for technical reasons rather than biological ones, namely because the taxonomy of extant forms is based almost exclusively on adult females (adult males and juveniles being little known), whereas fossil forms are represented mainly by adult males and juveniles.

Coccids are known to occur in almost all Mesozoic fossiliferous ambers and at two rock fossil sites which span the entire Cretaceous. Data on the occurrence of coccid fossils in pre-Cretaceous deposits must be regarded with reservation, although these insects certainly already existed as a distinct group by the beginning of Mesozoic (KOTEJA 1985).

Only a small proportion of the available or existing Cretaceous material has been studied so far, thus any general conclusions must be greatly limited.

On the basis of a single male, BEARDSLEY (1969) described *Electrococcus canadensis* from Cedar Lake amber and suggested that it is related to recent Pityococcidae. Later BEARDSLEY examined some specimens from Alberta and noted that several structures were not available in the holotype (Dr J. W. BEARDSLEY, personal communication). Dr E. PIKE (personal communication) examined inclusions in Grassy Lake amber but none was similar to *Electrococcus*.

KOTEJA (1988a) described a curious male — *Cancerococcus apterus* — found in Baltic amber and affiliated it with the Eriococcidae. However, Dr D. J. MILLER (personal communication) suggested that it may be a relative of the Pityococcidae (adult males are known but have not yet been described in detail). Quite recently I have examined several adult males and one female in Baltic and Bitterfeld amber (kindly

	Cretaceous		Tertiary		Extant
	incl.	impr.	incl.	impr.	
ORTHEZIOIDEA					
Ortheziidae	-	-	5	-	>50
Xylococcidae	-	1	?	-	5
Matsucoccidae	1	4	5	?	>30
Pityococcidae	1	-	1	-	5
Steingeliidae	+	-	+	-	1
Monophlebidae *s. l.*	+	-	+	?	>50
Margarodidae *s. l.*	?	-	+	-	>50
Phenacoleachiidae	+	-	+	-	2
New family	+	-	-	-	-
COCCOIDEA					
Putoidae	+	-	1	-	>10
Pseudococcidae	+	-	+	-	>1000
Eriococcidae	+	-	6	-	>300
Inkaidae	1	-	-	-	-
Kermesidae	-	-	1	-	>50
Coccidae	-	-	3	+	>1000
Diaspididae	-	-	+	+	>2000

incl. — inclusion, impr. — impression. Figures in each column represent numbers of described species, "+" — present, "-" — absent

Table 2. Fossil scale insects.

made available by Christle and Hans Werner HOFFEINS) which clearly show their relationship with Pityococcidae. Similar forms also occur in New Jersey amber. Thus, the contemporary Pityococcidae living on *Pinus* in North America seem to be a relict group which acquired specialized features (reduction of compound eyes, simplification of wing structure, loss of caudal extensions) and survived the Cretaceous and Tertiary periods.

Another ancient group are the Matsucoccidae (*c.* 10 fossil species), also confined to *Pinus*, but distributed throughout the whole Holarctic and represented by about 30 species. In Baltic and Bitterfeld amber they constitute about half of all inclusions (KOTEJA 1984, 1986). Upper Cretaceous examples have been found in New Jersey amber (unpublished), but have not been noted in Taymyrian amber (KOTEJA 1989a); information on their existence in other Cretaceous ambers is lacking. The earliest fossil record of this group was found in the Neocomian deposits in Siberia (KOTEJA 1988b) and England (COOK & ROSS 1996).

Close relatives of the above groups are the Xylococcidae and Steingeliidae. The former were found in Neocomian Siberian deposits (KOTEJA 1989b), the latter probably occur in both Cretaceous and Tertiary ambers (unpublished). Today they represent relict groups limited to the Holarctic and, possibly, South America.

The Ortheziidae, with many primitive features and a world-wide distribution, are known from Baltic, Bitterfeld and Dominican ambers but have not been found in Cretaceous amber so far. Presumably they lived in forest litter as do most of the extant forms.

The Monophlebidae *s. l.* and Margarodidae *s. l.* have a world-wide distribution and represent diversiform groups which have been noted in both Cretaceous and Tertiary ambers, but have not been described, because knowledge of recent forms (adult males) is too poor.

The Phenacoleachiidae is a relict group (2 species) limited to New Zealand. Fossil forms have been found in Cretaceous and Tertiary ambers, but not described (unpublished).

One group of advanced archeococcids represented by numerous specimens found in New Jersey amber seems to be extinct (unpublished).

Neococcids are represented in Cretaceous amber by four families: Putoidae, Pseudococcidae, Eriococcidae and Inkaidae. *Puto* is believed to be the ancestor of neococcids (KOTEJA 1996), the Eriococcidae and Pseudococcidae are now abundant and diversiform groups, whilst Inkaidae, a suggested relative of the Eriococcidae, seem to be extinct (KOTEJA 1989a).

Advanced neococcids, like Kermesidae, Coccidae and Diaspididae known from various Tertiary ambers have not been found among Cretaceous fossils.

Conclusions

Preliminary results of research into Cretaceous amber demonstrate that scale insects do evolve and radiate, which though rather obvious was not documented by fossil evidence.

Jan KOTEJA
Department of Zoology and Ecology,
Cracow Agricultural University
Al. Mickiewicza 24, 30-059 Kraków, Poland

CZERWCE (HOMOPTERA, COCCINEA) W BURSZTYNIE KREDOWYM

Jan KOTEJA

Streszczenie

Ze względu na specyficzną metamorfozę i biologię czerwce uważane były aż do połowy XX wieku za niedostępne do badań paleontologicznych. Obecnie zgromadzono ponad 500 inkluzji czerwców w bursztynie bałtyckim, dominikańskim i innych żywicach trzeciorzędowych. Reprezentują one kilkanaście rodzin, które sukcesywnie są badane i opisywane. Nic więc dziwnego, że zainteresowanie paleoentomologów skierowało się na czasy wcześniejsze — kredę.

Pierwszy czerwiec kredowy opisany został na podstawie jednego okazu samca z bursztynu kanadyjskiego (Cedar Lake) przez BEARDSLEY'A (1969). Dwadzieścia lat później KOTEJA (1989a) odkrył dwa gatunki (jeden został opisany) w bursztynie tajmyrskim i w tym samym czasie (KOTEJA 1988b, 1989b) odciski dwu rodzajów w dolnokredowych złożach syberyjskich nad Baisą, a COOK i ROSS (1996) jeden odcisk w Anglii. Informacje o występowaniu czerwców w różnowiekowych bursztynach kredowych były sporadycznie podawane, jednak bez opisów. Występowanie dalszych zasygnalizowane zostało korespondencyjnie przez dr. J. W. BEARDSLEY'A (bursztyn kanadyjski), dr. A. NELA (bursztyn libański), dr. E. M. PIKE'A (bursztyn kanadyjski), dr. G. POINARA (bursztyn z Alaski), dr. A. P. RASNITSYNA (bursztyn birmański) i dr. D. GRIMALDIEGO (bursztyn z New Jersey), tak że czerwce okazały się być obecne we wszystkich znanych bursztynach kredowych (Tab. 1). Specjalnie cenna jest kolekcja z New Jersey, w której rozpoznano przedstawicieli siedmiu rodzin zarówno spośród Orthezioidea, jak i Coccoidea (Tab. 2).

Wśród czerwców kopalnych najbardziej charakterystyczne są Matsucoccidae — obecnie monofagi sosny i ograniczone do Holarktyki — ich szczątki odnajdywane są w pokładach całej kredy i trzeciorzędu. Przedstawiciele kilku pokrewnych grup również występują w niektórych pokładach kredowych i trzeciorzędowych, ale opisany został tylko *Baisococcus victoriae* KOTEJA (z dolnej kredy) i *Electrococcus canadensis* BEARDSLEY z górnej kredy). Nie znaleziono w pokładach kredowych Ortheziidae, choć z całą pewnością występowały w tym czasie. Spośród Coccoidea w kredzie odnotowano tylko grupy pierwotne — Putoidae, Pseudococcidae i Eriococcidae. Młodsze grupy — Coccidae, Diaspididae — znaleziono dopiero w bursztynach trzeciorzędowych. Dwie grupy kredowe — Inkaidae zbliżone do Eriococcidae (Coccoidea) z bursztynu tajmyrskiego i nieopisana grupa spośród Orthezioidea z bursztynu New Jersey — nie są znane z trzeciorzędu ani fauny współczesnej (Tab. 2).

Przedstawiony obraz czerwców kredowych może ulec zasadniczym zmianom po zbadaniu dalszych materiałów.

Bibliography

ARILLO A.

1996 Los insectos en ámbar, *Boletin de la SEA. Vol. Monogr. PaleoEntomologia*, **16**, 47–149.

BANDEL K., VÁVRA N.

1981 Ein fossiles Harz aus der Unterkreide Jordaniens, *Neues Jahrbuch Geologie und Paläontologie, Monatshefte*, **1**, 19–33.

BEARDSLEY J. W.

1969 A new fossil scale insect (Homoptera: Coccoidea) from Canadian Amber, *Psyche*, **76** (3), 270–279.

COOK E., ROSS A. J.

1996 The stratigraphy, sedimentology and palaeontology of the Lower Weald Clay (Hauterivian) at Keymer Tileworks, West Sussex, southern England, *Proceedings of the Geologists' Association*, **107** (3), 321–239.

FERRIS, G.F.

1941 Contribution to the knowledge of the Coccoidea (Homoptera). IX. A forgotten genus of the family Margarodidae from Baltic amber. *Microentomology*, **6**, 6–10.

GRIMALDI D., BECK C. W., BOON, J. J.

1989 Occurrence, chemical characteristics and palaeontology of the fossil resins from New Jersey. *American Museum Novitates*, **2948**, 28.

KOTEJA J.

1984 The Baltic amber Matsucoccidae (Homoptera, Coccinea), *Annales Zoologici*, **37**, 437–496.

1985 Essay on the prehistory of the scale insects (Homoptera, Coccinea), *Annales Zoologici*, **38**, 461–503.

1986 *Matsucoccus saxonicus* sp. n. from Saxonian amber (Homoptera, Coccinea), *Deutsche Entomologische Zeitschrift, Neue Folge*, **33**, 55–63.

1988a Two new eriococcids (Homoptera, Coccinea) from Baltic amber, *Deutsche Entomologische Zeitschrift, Neue Folge*, **35**, 405–416.

1988b *Eomatsucoccus* gen. n. (Homoptera, Coccinea) from Siberian Lower Cretaceous deposits, *Annales Zoologici*, **42**, 141–163.

1989a *Inka minuta* gen. et sp. n. (Homoptera, Coccinea) from Upper Cretaceous Taymyrian amber, *Annales Zoologici*, **43**, 77–101.

KOTEJA J.
1989b *Baisococcus victoriae* gen. et sp. n. — a Lower Cretaceous coccid (Homoptera, Coccinea), *Acta Zoologica Cracoviensia*, **32**, 93–106.

1996 The scale insects (Homoptera: Coccinea) a day after, [in:] Schaefer, C. W. (ed.), Studies on Hemiptera phylogeny, *Proceedings of Thomas Say Publications in Entomology*, Entomological Society of America, Lanham (Maryland), 65–88.

KRUMBIEGEL G. & KRUMBIEGEL B.
1994 *Bernstein — Fossile Harze aus aller Welt*, Goldschneck-Verlag (Fossilien Sonderband 7), Weinstadt, 1–110.

LANGENHEIM J. H.
1990 Plant resins. *American Scientist*, **78**, 16–24.

PIKE E. M.
1993 Amber taphonomy and collecting biases, *Palaios* **5** (8), 411–419.

1994 Historical changes in insect community structure as indicated by hexapods of Upper Cretaceous Alberta (Grassy Lake) amber. *Canadian Entomologist*, **126**, 695–702.

POINAR G. O. JR.
1993 *Life in amber*, Stanford University Press, Stanford, 1–350.

POINAR G. & POINAR R.
1994 *The quest for life in amber*, Addison-Wesley Publishing Company, Reading Massachusetts, 1–219.

POKRYSZKO B. M.
1996 Amber from Alava, *Inclusion-Wrostek*, **23**, 13.

RASNITSYN A. P.
1996 Burmese amber at the Natural History Museum, London. *Inclusion-Wrostek*, **23**, 19–21.

SCHLEE D.
1972 Bernstein aus dem Lebanon. *Kosmos*, **68**, 460–463. Stuttgart.

1984 Notizen über einige Bernsteine und Kopale aus aller Welt. *Stuttgarter Beiträge zur Naturkunde, Serie C*, **18**, 29–38.

SCHLEE D. & GLÖCKNER W.
1978 Bersteine und Bernstein-Fossilien. *Stuttgarter Beiträge zur Naturkunde, Serie C*, **8**, 46–72.

SCHLÜTER T.
1978 Zur Systematic und Palökologie harzkonservierter Arthropoda einer Taphozönose aus Cenomanium von NW-Frankreich, *Berliner Geowissenschaftliche Abhandlungen, A*, **9**, 1–150.

USINGER R. L. & SMITH R. F.,
1957 Arctic amber. *Pacific Discovery*, **10**, 15–19.

VÁVRA N.
1984 Reich an armen Fundstellen: Übersicht über die fossilen Harze Österreichs, *Stuttgarter Beiträge zur Naturkunde, Serie C*, **18**, 9–14.

ZHERIKHIN V. V. & SUKACHEVA, I. D.
1973 On Cretaceous insect bearing ambers (retinites) of northern Siberia, *Reports of the 24th Annual Readings in Memory of N. A. Kholodkovsky* [in Russian], Leningrad, 3–48.

RARE ANIMAL INCLUSIONS IN BALTIC AMBER

Barbara KOSMOWSKA-CERANOWICZ & Róża KULICKA

Abstract

This article has been prepared as a supplement to the posters presented at the Symposium by: B. KOSMOWSKA-CERANOWICZ, R. KULICKA & G. GIERŁOWSKA — *A lizard found in Baltic amber*; and R. KULICKA & T. KUSIAK — *A new example of Heteroptera (Miridae) from Baltic amber* (cf. Annexe 1 & 2).

Alongside inclusions commonly encountered in amber, among which new species have come to light, such as the particularly noteworthy specimen of *Metoisops kusjaki* POPOV from the collections of Tadeusz KUSIAK, donated to the Museum of the Earth (cf. Annex 2), in recent years a succession of discoveries have been made of extremely rare fauna which have in one way or another left some trace in Baltic amber. The private collection belonging to the family of the late T. DROZDA includes an amber impregnated fragment of maxilla belonging to a representative of the swine family Suiformes (KOSMOWSKA-CERANOWICZ 1991/1992; KULICKA & SULIMSKI 1994; KOSMOWSKA-CERANOWICZ & KULICKA 1995). A very interesting source material for the study of small mammals is housed at the Museum of the Earth in the form of a small, yet unique collection of animal tracks captured on the surfaces of nodules of Baltic and Bitterfeld amber (cf. KULICKA & SIKORSKA-PIWOWSKA 1997; 1999). A number of successes have been noted in recent months relating to finds of reptiles in amber. Those interested in amber rarities in Poland have been waiting for such a find for over a hundred years (cf. Annex 1 and KOSMOWSKA-CERANOWICZ *et al.* 1997). Verbal notification has also been received of the discovery of a lizard in Lithuania. The last few days have yielded another rare find — that of a *Paleogammarus sp.* (Crustaceae) in Baltic amber (report in preparation). This specimen has been donated to the Museum of the Earth by A. RYBICKI from Warsaw.

The amber impregnated fragment of maxilla is of great value for research into Suiformes. We know from studies carried out by the French that these animals already existed in the Palaeogene. Unfortunately, the jaw fragment discovered is too small to enable the identification of this mammal to species. However, the process of "amberification" of bones and teeth, hence the remains of vertebrates, has never previously been recorded in literature. Various types of fossilization process, in which substances are substituted, such as silification, calcification and pyritization are excellently understood. Also well-known is the process of chemical weathering induced by carbon dioxide dissolved in rain water (carbonation), which leads to silicates being reduced to carbonates.

The jaw fragment probably became impregnated with amber by the heavily weathered bone and teeth being engulfed in liquid resin. The areas once occupied by skeletal and dental tissue (with the exclusion of enamel) were replaced with resin, without damaging the external form of the jaw fragment nor the wear pattern on the working occlusal surface of the teeth.

The first lizard to be found in Baltic amber came from the region of Sambia and was written up by KLEBS (1910). At the STANTIEN & BECKER company, where he was employed from 1872, KLEBS came across over 150,000 kg of amber stalactite and stalactite-like forms (Schlauben) — the richest source of organic inclusions — the identification of which he worked on for many years. In the period from 1884 to 1887, examining over 200 kilograms of crude amber (176 kg after polishing), KLEBS found 7,826 specimens, from a total of 13,877, to contain inclusions (making up 47% of the material examined).

Compiling his data KLEBS presented an interesting picture of the coexistence of selected groups of arthropod: Diptera 50.9%, Hymenoptera 5.1%, Phryganiden 5.6%, Microlepidoptera 0.1% (one specimen containing 22 examples of Microlepidoptera, was presented as a gift to BISMARCK by BECKER), Coleoptera 4.5%, Collembola 10.6%, Thysanura 0.1%, Rhynchoten 7.1%, Orthopteren 0.5%, spiders 4.5%, mites 8.6%, other 2.4%.

In modern-day terminology this list is as follows:
Collembola 10.6%

Insecta:

Diptera	50.9%
Hemiptera	7.1%
Trichoptera	5.6%
Hymenoptera	5.1%
Coleoptera	4.5%
Orthoptera	0.5%
Thysanura	0.1%
Microlepidoptera	0.1%

Arachnida:

Acarina	8.6%
Aranea	4.5%
other	2.4%

This data indicates that the collection investigated was selectively (?) chosen. The statistical results obtained thus far for the coexistence of insect, arachnid and myriapodan groups in Baltic amber provide a completely different picture (cf. KULICKA 1991 [*in*:] KOSMOWSKA-CERANOWICZ 1991/1992).

KLEBS also discovered a number of other very rare finds, of which he considered the lizard (*Nucras*) catalogued under inventory no. 12664 to be the most valuable specimen. The lizard remained in the BECKER Museum's collections up until their transferal to Albert University in Königsberg. After the Second World War, many thought the lizard to have been lost — a fact reflected in numerous publications. It does, however, still exist, along with the other most valuable specimens formerly housed at Königsberg's Albert University, in the collections of the Insitut und Museum für Geologie und Paläontologie der Universität in Göttingen (RITZKOWSKI 1977; 1999).

In order to make a better study of the lizard KLEBS prepared the specimen (*habe ich den Bernstein im Einschluß gesprengt*) and cleaned it, which made it easier to examine in detail the arrangement of the lizard's scales but, sadly, also damaged it quite badly. KLEBS also noted the presence of a hollow cavity (*Hohlraum*) and some slightly carbonized remains. According to the somewhat controversial opinion published by TORNQUIST (1910) the only thing to have survived of these remains was cuticle. The lizard was stored in water in several pieces in the BECKER Collection for twenty-five years. After the collection was taken over by the University, the amber fragments which had contained this specimen were stuck back together, the cavity filled with resin and the whole specimen set inside it. KLEBS wrote that the specimen became difficult to distinguish clearly. This may be the reason why numerous suggestions have been made, from the 1940s onwards, that this example of *Nucras* may have been set, either naturally or artificially, in copal rather than succinite (LOVERIDGE 1942; SCHLEE 1990 and others). This specimen was not lost (cf. RITZKOWSKI 1998), and given the limited knowledge of its history, it could well have been the subject of a variety of speculative theories.

That the lizard was naturally captured in copal (in terms of resins from the southern hemisphere) seems highly improbable. Although forgeries were already being made as long as four centuries ago, it is, nevertheless, difficult to imagine that BECKER would have mixed so-called composites (i.e. the material richest in inclusions) with specimens of uncertain origin (an inventory of vertebrate forgeries has recently been compiled by GRIMALDI *et al.* 1994).

We know from KLEBS' report (1910) that the specimen in question was examined by many researchers. Its identification by BOETTGER of Frankfurt as *Cnemidophorus* had already been verified by BOULANGER of London in 1891. BOULANGER classified the specimen as belonging to the family Lacertidae, species *Nucras tessellata* (SMITH). In 1917 the author changed the name of this species to *Nucras succinea* (after POINAR 1992). *Nucras* is present today in the tropical zone.

Since KLEBS' discovery of the first example of a lizard in Baltic amber, over one hundred years of searching have yielded little success, although eleven variously well preserved specimens of iguana and gecko have been found in Dominican amber (POINAR 1992). In his publication KATINAS (1983, 45) includes a photograph showing nodular scales on a fragment of lizard skin in Baltic amber.

The most recent discovery of a lizard in succinite was made by Gabriela GIERŁOWSKA, in June 1997, whilst sorting through a rather less impressive amount of crude amber than that available to KLEBS. The material which was being sorted had been acquired by wet-sieving Quaternary sediments from Gdańsk Stogi (cf. Annex 1).

The specimen in question is currently housed at the Museum of the Earth, Polish Academy of Sciences, Warsaw and will be on show in the amber exhibition until April '99. To avoid its damage, the specimen's owner has secured the opening to its empty section with a mixture of dammar and wax.

Barbara KOSMOWSKA-CERANOWICZ & Róża KULICKA
Muzeum Ziemi PAN
Aleja Na Skarpie 20/26
00-488 Warszawa, Poland

ZNALEZISKA RZADKICH INKLUZJI ZWIERZĘCYCH W BURSZTYNIE BAŁTYCKIM

Barbara KOSMOWSKA-CERANOWICZ, Róża KULICKA

Streszczenie

Obok inkluzji często spotykanych w bursztynie, wśród których znajduje się nowe rzadkie gatunki, jak np. zasługujący na uwagę szczególny okaz pluskwiaka ze zbiorów Tadeusza KUSIAKA, ofiarowany Muzeum Ziemi (por. Aneks 2), ostatnie lata przyniosły szereg odkryć różnego rodzaju wyjątkowo rzadkich pozostałości fauny, które w różnej formie pozostawiły swe ślady w bursztynie bałtyckim. W prywatnych zbiorach rodziny nieżyjącego już T. DROZDY przechowywany jest zbursztynizowany fragment szczęki osobnika świniowatych — Suiformes. Ostatnie miesiące były szczególnie pomyśle dla zainteresowanych bursztynowymi rarytasami. Po stu latach znaleziono drugiego gada, jaszczurkę, w bursztynie bałtyckim — pierwszy raz w Polsce (por. Aneks 1). Do zainteresowanych tematem dotarła ustna wiadomość o znalezieniu kolejnej jaszczurki na terenie Litwy. Ostatnie dni przyniosły jeszcze jeden rzadki okaz, tym razem kiełża (Crustaceae) w bursztynie bałtyckim (w opracowaniu). Okaz ten został przekazany w darze do zbiorów Muzeum Ziemi przez A. RYBICKIEGO z Warszawy.

Bibliography

GRIMALDI D. A., SHEDRINSKY A., ROSS A. & BEAR N. S.
1994 Forgeries of fossils in "Amber": history, identification and case studies, *Curator. The Museum Journal*, **37** (4), 251–274.

KATINAS V.
1983 *Baltijos gintaras*, 1–111, "Mosklas", Vilnius.

KLEBS R.
1910 Über Bernsteineinschlüsse im allgemeinen und die Coleopteren meiner Bernsteinsammlung, *Schriften der physikalisch-ökonomischen Gesellschaft zu Königsberg*, **51**, 217–242.

KOSMOWSKA-CERANOWICZ B.
1991/1992 *Spuren des Bernsteins*, 1–102, Bielefeld.

KOSMOWSKA-CERANOWICZ B. & KULICKA R.
1995 Amber Molars, *Amber & Fossils*, **1**, 38–41, Kaliningrad.

KOSMOWSKA-CERANOWICZ B., KULICKA R. & GIERŁOWSKA G.
1997 Nowe znalezisko jaszczurki w bursztynie bałtyckim, *Przegląd Geologiczny*, **10/11**, 1028–1030.

KULICKA R. & SULIMSKI A.
1994 Usual amber fossil, *Rocznik Muzeum Górnośląskiego, Przyroda*, **14**, 69–71.

KULICKA R. & SIKORSKA-PIWOWSKA Z.
1997 Traces of vertebrates in Baltic amber and Saxony amber, *Museum of the Earth / Scientific conferences / Abstracts*, **9**, 24–25, Warszawa.

1999 Mammalian ichnites in amber, *Investigations into Amber. Proceedings of the International Interdisciplinary Symposium: Baltic Amber and Other Fossil Resins. 997 Urbs Gyddanyzc — 1997 Gdańsk, Gdańsk 1997* (ed. B. Kosmowska-Ceranowicz & H. Paner), Gdańsk, ...

LOVERIDGE A.
1942 Scientific results of a fourth expedition to forested areas in East and Central Africa. IV Reptiles, *Bulletin of the Museum of Comparative Zoology*, **91**, 240–373.

POINAR G. O. JR.
1992 *Life in amber*, 1–350, Stanford, California.

SCHLEE D.
1990 Raritäten Wierbeltiere: Leguane, Geckos, Frosch im Dominikanischen Bernstein, [*in*:] Das Bernstein-Kabinet, *Stuttgarter Beiträge zur Naturkunde, S.C*, **28**, 17–21.

RITZKOWSKI S.
1997 Osobliwości w bursztynie bałtyckim z byłych królewieckich zbiorów bursztynu, obecnie w zbiorach Getyngi (Dolna Saksonia), *Muzeum Ziemi / Konferencje naukowe / Streszczenia*, **10**, 28–31, Warszawa.

1999 Die Eidechse *Nucras succinea* BOULENGER der ehem. Königsberger Bernsteinsammlung, *Investigations into Amber. Proceedings of the International Interdisciplinary Symposium: Baltic Amber and Other Fossil Resins. 997 Urbs Gyddanyzc — 1997 Gdańsk, Gdańsk 1997* (ed. B. Kosmowska-Ceranowicz & H. Paner), Gdańsk, ...

TORNQUIST A.
1910 *Geologie von Ostpreußen*, Berlin.

Annex 1

A LIZARD FOUND IN BALTIC AMBER

Barbara KOSMOWSKA-CERANOWICZ, Róża KULICKA & Gabriela GIERŁOWSKA

The first lizard to be found in amber was that identified during the examination of 200 kg of crude amber housed in the BECKER Collection. During the period from 1874 to 1900 KLEBS carried out a search for inclusions in this collection as part of his efforts to prepare it for exhibition purposes. The lizard was identified as *Nucras* and following a series of specialist analyses was recorded in the inventory under number 12664. KLEBS found over 13,000 organic inclusions during his search (in 200 kg of raw material!).

Over two hundred years of examining Baltic amber for vertebrate inclusions proved futile following KLEBS' success.

Bearing in mind this long and fruitless search, the discovery in 1997 of a **lizard** trapped in amber, which had been recovered by the wet sieving of Quaternary deposits from Gdańsk Stogi, seemed all the more unusual.

Ms G. GIERŁOWSKA, one of the authors of this article, was responsible for this discovery. She has worked in co-operation with the Museum of the Earth for many years and is well aware of the scientific significance of organic inclusions in amber. Whilst sorting amber nodules prior to working them, she noticed a curious opening together with the imprint of a small scale in one of the pieces she was handling. The material had already undergone two hours of surface cleaning to remove the weathered cortex which prevents one from being able to see inside the nodule. This process had taken place in a polishing machine using a fine-grained diamond disc operating under running water. The next stage of careful cutting and polishing revealed an incomplete, though, in some parts, well-preserved inclusion whose shape indicated that it was a **lizard** (Fam. Lacertidae; Fig. 1).

The region of Gdańsk Stogi, from which this amber had been recovered, was photographed and marked on a district map of the city. The area concerned is marshy and covered with woodland, which includes slight elevations in the form of small sand dunes. It lies to the south of the main line of dunes, approximately 1 km from the sea coast. The amber in question lay at a depth of about 5.5 m and came from the Holocene sediments of a fossil beach partly covered by sand dunes. The amber was redeposited there from the Eocene deposits of the Chłapowo-Sambian delta.

Fig. 1. The lizard (Lacertidae family) found at Gdańsk Stogi, Poland, in 1997. Photo. G. GIERŁOWSKA. Private collection.

The evidence which points to the undisputed origin of the nodule of clear yellow amber, in which the lizard was found, does not however give any indication of the type of fossil resin in which the creature must have drowned. We know that other yellow resins apart from succinite occur in the Baltic region. These include gedanite, gedano-succinite and a resin known as young amber or colophony. The piece of amber concerned was subjected to infrared spectroscopy analysis in order to identify it. The IRS curve resulting from this analysis is typical of succinite.

Barbara KOSMOWSKA-CERANOWICZ & Róża KULICKA
Muzeum Ziemi PAN
Aleja Na Skarpie 20/26
00-488 Warszawa, Poland

Gabriela GIERŁOWSKA
ul. Szara 9 m. 50
80-116 Gdańsk, Poland

Annex 2

A NEW EXAMPLE OF HETEROPTERA (MIRIDAE) FROM BALTIC AMBER

Róża KULICKA & Tadeusz KUSIAK

A new example of Heteroptera was identified in the private collection of organic inclusions in amber belonging to one of the authors of this article — Tadeusz KUSIAK. It came in the form of a member of the Miridae family. Thus far, this family has been represented in Baltic amber by insects of the following subfamilies: Cylapinae, Mirinae, Phylinae and Isometopinae. Examples of the subfamily Cylapinae appear in amber most frequently. In contemporary fauna they are among the smallest of insect groups, with the most commonly occurring insects being those belonging to the Mirinae and Phylinae subfamilies, which are very rarely found among the fossilized fauna present in amber.

Few examples of the primitive subfamily Isometopinae have been found in amber to date. Unlike all other insects of the Miridae family, they have stemmata. Their contemporary equivalents live concealed in coniferous woodland environments. They occur most commonly in tropical and subtropical zones. The specimen found in T. KUSIAK's collection was of the genus *Metoisops* and belongs to the Electromyiomma category of this subfamily. The fossil variety of *Metoisops* (POPOV & HERCZEK 1992) has until now been represented by only two species: *Metoisops kerzhneri* (POPOV & HERCZEK 1992 — from the collections of the Museum of the Earth PAN) and *Metoisops punctatus* (POPOV & HERCZEK 1993 — from the Institute of Palaeontology RAN).

The very distinctive characteristics of the head and integument structure prompted a new unit of classification to be established for this species — *Metoisops kusjaki* (Fig. 2; POPOV, in press). Even though the author of this species has not yet published a full descriptive report of his findings, he has

Fig. 2. *Metoisops kusjaki* POPOV from the collections of the Museum of the Earth, Warsaw, inv. no 23135. Photo. Y. POPOV.

allowed this data to be published as part of the *Baltic Amber and Other Fossil Resins* symposium materials.

Róża KULICKA
Muzeum Ziemi PAN
Aleja Na Skarpie 20/26
00-488 Warszawa, Poland

Tadeusz KUSIAK
Stacja Ekologiczna Uniwersytetu Wrocławskiego
ul. Leśna 9
58-540 Karpacz, Poland

Bibliography

POPOV Y. A. & HERCZEK A.

1992 The first Isometopinae from Baltic amber (Heteroptera, Miridae), *Mitteilungen aus dem Geologisch-Paläontologischen Institut der Universität Hamburg,* **73**, 241–258.

1993 *Metoisops punctatus* sp. n., the second representative of the fossil genus *Metoisops* from Baltic amber (Heteroptera: Miridae, Isometopinae), *Annals of the Upper Sillesian Museum, Entomology. Supplement*, **1**, 51–56.

MAMMALIAN ICHNITES IN AMBER

Róża KULICKA & Zofia SIKORSKA-PIWOWSKA

Abstract

The morphology of autopodial imprints preserved in Baltic amber and Bitterfeld amber has made it possible to determine the adaptive specializations of their mammalian trackmakers' locomotory apparatus, and compare them with those of other vertebrates. The proportions of autopodial segments enabled the restoration of their anatomy, resembling that of some Insectivora and Hyracoidea. Additionally, environmental and palaeoclimatic data, as well as the adaptive ranges of these taxa led to the attribution of ichnotaxa to trackmakers comparable to *Eremitalpa* (Insectivora, Chrysochloridae) and *Dendrohyrax* (Hyracoidea, Procaviidae), and to the establishment of three new ichnotaxa: *Succinoeremitalpa baltica* ichnogen. et ichnosp. n., *Succinodendrohyrax balticus* ichnogen. et ichnosp. n., *Succinodendrohyrax saxonicus* ichnogen. et ichnosp. n.

Introduction

Remains of Tertiary land mammals are known from all continents, though they are rarer from tropical areas of Africa, South America and Asia, despite the fact that most mammalian orders may have arisen there. The scarcity of well-preserved skeletons of tropical land vertebrates is due to the corrosive effect of acidic forest soil and to the prevailing climate, which encourages fast decomposition.

The size and strength of Tertiary mammals prevented their capture in the sticky resin later forming amber, but they documented their presence in the amber forests with footprints or hair inclusions. BACHOFEN-ECHT (1949) interpreted some depressions in the surface of amber as rodent footprints; he speculated, that the imprints may have been left by animals moving over not yet hardened resin. We have studied such ichnites preserved on the surface of lumps of Baltic and Bitterfeld amber.

Baltic amber (succinite) from Tertiary beds in the Sambian Peninsula (Russia) is dated to the Late Eocene. These are secondary deposits, formed on the sea bed, containing amber redeposited by rivers (KATINAS 1971). Amber-rich strata of the same age also continue into the Polish region of Gdańsk (e.g. MARZEC 1971; PIWOCKI *et al.* 1985). Along the Polish seashore amber is also found in Quaternary (both Pleistocene and Holocene) sediments (KOSMOWSKA-CERANOWICZ & PIETRZAK 1982). This amber comes from the erosion of Tertiary beds, and has been redeposited by Quaternary glacial, fluvioglacial or fluvial transport in large areas of northern and central Poland and other Baltic states.

Bitterfeld amber is also a succinite, as confirmed by physicochemical analyses (including infrared spectroscopyy). It comes from the Tertiary beds exploited by a lignite mine in Goitsche (near Halle/ Saale, Germany), and is found in black-grey argillaceous sands with muscovite. The age of these deposits is Upper Oligocene–Lower Miocene. In the heavy minerals in amber-bearing strata, the prevalent minerals are turmaline and andalusite, as in the Magdeburg sands (KOSMOWSKA-CERANOWICZ & KRUMBIEGEL 1990). The deposition of resins in the Bitterfeld strata occurred after a short transport of debris from older, Oligocene strata in changed sedimentary conditions of the littoral zone (KRUMBIEGEL & KRUMBIEGEL 1994). Thus the footprint preserved in a Bitterfeld amber lump was made by a land mammal living in a tropical succiniferous forest during the Oligocene epoch.

Material

In the amber collections of the Museum of the Earth, Polish Academy of Sciences, Warsaw (abbreviated MZ) about a dozen specimens of Baltic amber with footprints have been found, and of the few that allow a more precise interpretation, two specimens are described below (Fig. 1: 1, 2).

The footprints studied are imprinted on irregular nodules of raw, opaque amber, covered with a fairly thin, weathered but non-exfoliating cortex. The first nodule, inv. no. MZ 1692, collected in 1957, weighs 116 g (94 x 83 x 30 mm) (Fig. 1: 1). The second, inv. no. MZ 1124, collected in 1957, weighs 53 g (40 x 70 x 30 mm) (Fig. 1: 2). There are several imprints left by the same autopodium on various sides of this specimen.

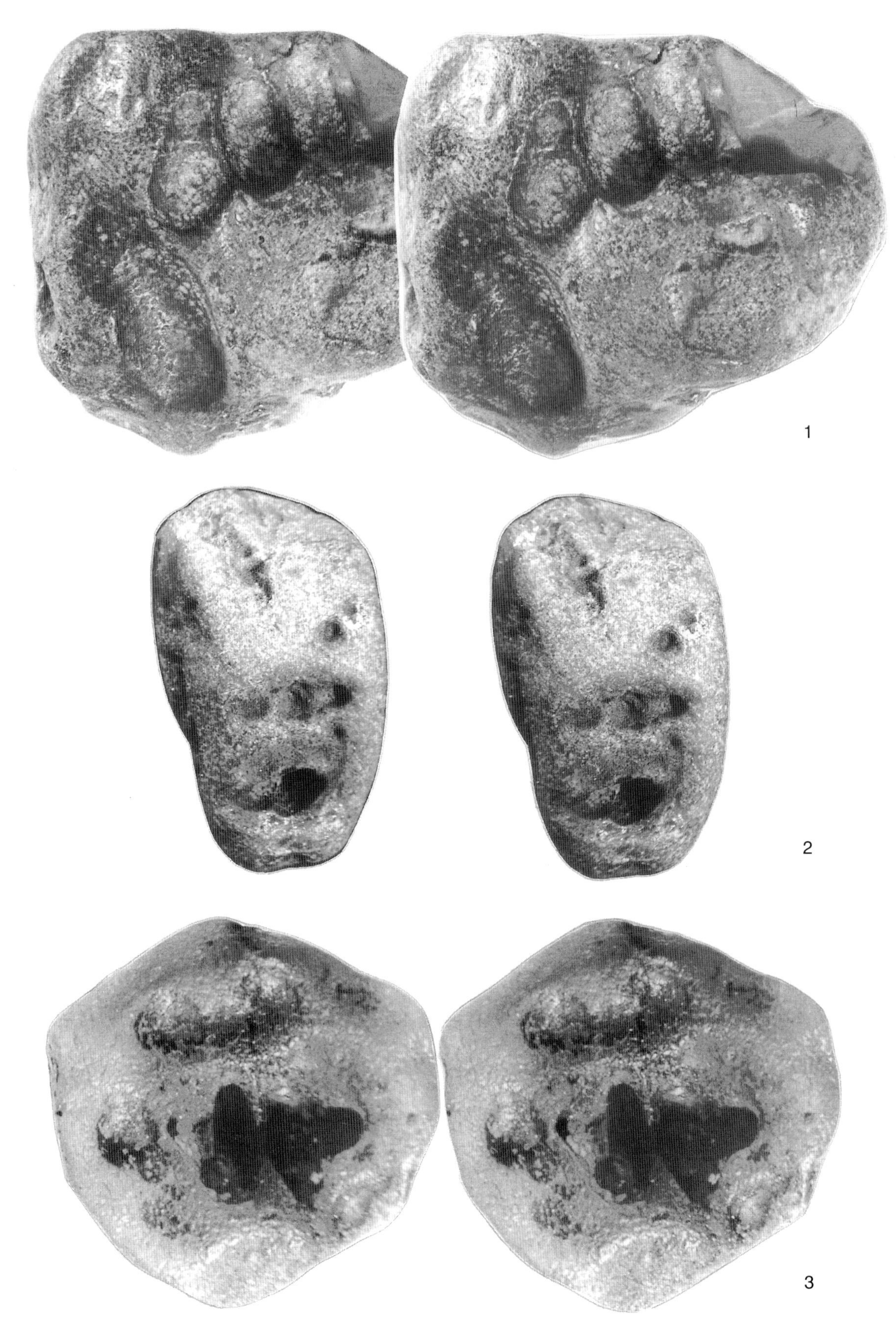

Fig. 1. See caption opposite.

The Museum of the Earth also houses a collection of Bitterfeld amber which has been compiled by B. KOSMOWSKA-CERANOWICZ and G. KRUMBIEGEL. In 1993, after amber extraction in the Goitsche mine had come to an end, the Mitteldeutsche Braukohleindustrie AG (MIBRAC) donated numerous samples of crude amber and other fossilized resins to the Museum. From that material, we have selected a specimen with a limb imprint.

The specimen in question (inv. no. MZ 21719) is a lump of translucent, honey-yellow amber, weighing 7.3 g (40 x 30 x 13 mm). It bears one complete and several incomplete impressions of the same limb, as indicated by the same size and spacing of digits (Fig. 1: 3).

Methods

The footprints were studied according to a model of tetrapod morphological development, based on the adaptive characteristics of extant and fossil vertebrates (SIKORSKA-PIWOWSKA 1984; 1993). These inherited adaptive characteristics of limbs are correlated with the whole animal structure and related to its environment.

The shape of the imprints facilitated the determination of types of adaptation and specialization of the locomotory apparatus, autopodial joint structures and basipodial specialization of the trackmakers. These types were attributed to environments (a given limb type is best for locomotion in a particular type of environment). Digit length order and position of the autopodium during locomotion were established (the foot position depends on the proportions and articulation of the more proximal parts of the limb).

Description

Order **Insectivora**

Family **Chrysochloridae**

Succinoeremitalpa baltica ichnogen. et ichnosp. n.

Derivation of the name: Latin *succinum* — amber, resembling manus of *Eremitalpa*, found in Baltic amber.

Type specimen — MZ 1692 (Fig. 1: 1; Fig. 2: 1).

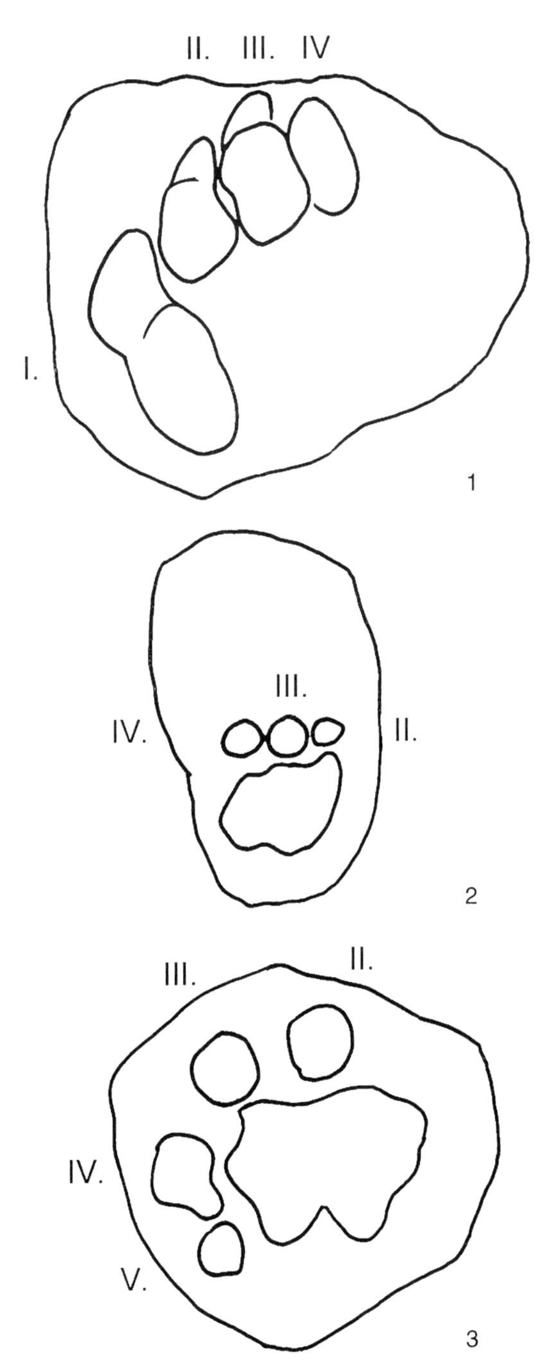

Fig. 2. Outline of autopodial mammalian imprints on nodules of amber (see Fig. 1) from the collections of the Museum of the Earth: 1 — *Succinoeremitalpa baltica* ichnogen. et ichnosp. n.; 2 — *Succinodendrohyrax balticus* ichnogen. et ichnosp. n.; 3 — *Succinodendrohyrax saxonicus* ichnogen. et ichnosp. n.

Opposite: Fig. 1. Autopodial mammalian imprints on nodules of amber (Baltic — 1 & 2, Bitterfeld — 3) from the collections of the Museum of the Earth: 1 — *Succinoeremitalpa baltica* ichnogen. et ichnosp. n., inv. no. MZ 1692, stereophotograph: 70 mm base, actual size; 2 — *Succinodendrohyrax balticus* ichnogen. et ichnosp. n., inv. no. MZ 1124, stereophotograph: 70 mm base, actual size; 3 — *Succinodendrohyrax saxonicus* ichnogen. et ichnosp. n., inv. no. MZ 21719, stereophotograph: 70 mm base, 1 : 1 scale.

Imprint of the lateral four manual digits, the first three of them ending in well developed claws. Bending of the digits indicates outward rotation of the manus. The hand is pronated. Relative digit lengths (III > II > IV > I) indicate mesaxony. Digit V is reduced.

Outward rotation with pronated position is typical for mammalian forelimbs adapted for digging (SIKORSKA-PIWOWSKA 1984): the hand is rotated and pronated to dig earth tunnels, whilst the foot is oriented straight and serves to remove the earth backwards. Digging adaptation in mammals is found among Ornithorhynchidae, Dasyuridae, Notoryctidae, Chrysochloridae, Talpidae, Geomyidae and Orycteropidae (SIKORSKA-PIWOWSKA 1984). Of these, only Chrysochloridae have four digits in an outwardly rotated manus, while the pes is five-toed and positioned straight. Of seven known genera of this family (*Amblysomus, Calcochloris, Chlorotalpa, Chrysochloris, Chrysospalax, Cryptochloris, Eremitalpa*), only *Eremitalpa* Roberts has a four-fingered manus with three claws. Extant representatives of this genus live in South Africa.

Order **Hyracoidea**

Family **Procaviidae**

Succinodendrohyrax balticus ichnogen. et ichnosp. n.

Derivation of the name: Latin *succinum* — amber, resembling foot of *Dendrohyrax* found in Baltic amber.

Type specimen — inv. no. MZ 1124 (Fig. 1: 2; Fig. 2: 2).

Imprint of a left foot with clearly imprinted heel but no other tarsal elements. This indicates superposition of astragale over calcaneum. (A similar superposition occurs in crocodiles, but the crocodilian astragal is functionally part of the crus, while the calcaneum — of the distal tarsus and metatarsus. Thus, the crocodilotarsan articulation is in mid-tarsus, and no heel imprint is possible.) The imprint under discussion can be attributed to a plantigrade mammal.

A clearly visible pad imprint can be seen in the middle of the footprint.

The anterior part of the footpint includes three deep imprints of the distal phalanges of digits II, III, IV; the middle one being the longest (mesaxonic foot). Digit IV is the shortest. A small claw is visible at the end of digit II, while the others end in nails. The imprints (especially those of the fingers, are repeated several times on various sides of the amber nodule — perhaps the animal had a branch grasped in its foot.

This set of characteristics of the autopodium studied resembles that of an arboreal mammal — tree hyrax *Dendrohyrax* (Hyracoidea) with a three-toed foot. It represents a secondary climbing adaptation type of plantigrade specialization, occurring in an arboreal environment (SIKORSKA-PIWOWSKA 1984). The genus *Dendrohyrax* comprises three extant species (*D. arboreus, D. dorsalis, D. validus*) living in African tropical forests.

Succinodendrohyrax saxonicus ichnogen. et ichnosp. n.

Derivation of the name: *saxonicus* — from the German province of Saxony (Sachsen).

Type specimen — inv. no. MZ 21719 (Fig. 1: 3; Fig. 2: 3).

Distal phalanges of digits V, IV, III, II are imprinted several times on the amber surface. Behind them are two distinct grooves corresponding to the serial arrangement of carpals. Thus it is supposedly a left manus imprint. The manus is mesaxonic, with relative digit lengths as follows: III > II > IV > V. There is no pollex imprint, so it must have been reduced. The unguals are accompanied by crescent-shaped nail imprints, resembling those of primitive hooves. Traces indicating full carpal contact with the substrate indicate plantigrade locomotion and grasping abilities of the mammal.

The following characteristics: plantigrade, mesaxonous, four-fingers and grasping hand indicate arboreality. The outlined description best fits the anatomy of recent tree hyraxes (*Dendrohyrax* sp.) and we think that *Succinodendrohyrax saxonicus* ichnogen. et ichnosp. n. has been left by their relative.

Discussion

Eremitalpa and all Chrysochloridae are limited to southern Africa. This endemic family first occurred there in the Miocene. Our finding of *Succinoeremitalpa baltica* ichnogen. et ichnosp. n. would extend their fossil record back to the Eocene and their palaeogeographic distribution to Europe. The ichnites would also indicate a larger body size of the fossil chrysochlorids than their living representatives.

Succinodendrohyrax balticus ichnogen. et ichnosp. n., and *S. saxonicus* ichnogen. et ichnosp. n. from Baltic and Bitterfeld amber are manus and pes imprints most closely resembling limbs of tree hyraxes (*Dendrohyrax*). Hyracoidea probably evolved in the Late Eocene in Northern Africa (SUDRE 1979). They rapidly evolved there and by the Early Oligocene were already fairly differentiated (MÜLLER 1989). Their first European occurrences known to-date came from Miocene deposits of southern Europe (CARRROLL 1988). Thus, it had been thought that they did not expand beyond Africa until the Miocene, when the continent was attached to Asia. Our study would suggest the presence of Hyracoidea in Europe as early as in the Eocene.

The ichnites we studied might have been left in amber by some other extinct animals, whose limb anatomical characteristics resembled those of the living mammals mentioned above. Thus, our interpretation can only be confirmed by findings of skeletal fossils of hyraxes and golden moles in the Eocene of Europe.

Róża KULICKA
Museum of the Earth, Polish Academy of Sciences
Al. Na Skarpie 20/26
00-488 Warszawa, Poland

Zofia SIKORSKA-PIWOWSKA
Agricultural and Paedagogic University,
Institute of Biology
ul. B. Prusa 12
08-110 Siedlce, Poland
(and) Dept. of Human Anatomy,
Institute of Biostructure, Faculty of Medicine
ul. Chałubińskiego 5
02-004 Warszawa, Poland

ŚLADY AUTOPODIÓW SSAKÓW ZACHOWANE NA POWIERZCHNI BURSZTYNU

Róża KULICKA, Zofia SIKORSKA-PIWOWSKA

Streszczenie

Po raz pierwszy opisano i zilustrowano trzy nowe ichnotaksony: *Succinoeremitalpa baltica* ichnogen. et ichnosp. n., *Succinodendrohyrax balticus* ichnogen. et ichnosp. n., odciścięte na powierzchni bryłek bursztynu bałtyckiego i *Succinodendrohyrax saxonicus* ichnogen. et ichnosp. n. na powierzchni bryłki bursztynu bitterfeldzkiego. Badania prowadzono na podstawie morfologii odcisków autopodiów tych ssaków, określając typ adaptacji i specjalizacji ich aparatu lokomocyjnego. Proporcje poszczególnych segmentów autopodialnych pozwoliły na określenie ich budowy anatomicznej, która jest porównywalna z cechami niektórych przedstawicieli Insectivora i Hyracoidea. Uzupełniając dane określające środowisko, klimat i możliwości adaptacyjne tych grup pozwoliły na odniesienie zbadanych śladów do ichnotaksonów odpowiadających formom *Eremitalpa* (Insectivora, Chrysochloridae) i *Dendrohyrax* (Hyracoidea, Procaviidae).

Bibliography

CARROLL R. L.
1988 *Vertebrate paleontology and evolution*, 1–698, New York.

KATINAS V.
1971 *Amber and amber-bearing deposits of Southern Baltic Land* [in Russian], 1–151, Mintis, Vilnius.

KOSMOWSKA-CERANOWICZ B. & KRUMBIEGEL G.
1990 Bursztyn bitterfeldzki (saksoński) i inne żywice kopalne z okolic Halle (NRD), *Przegląd Geologiczny*, **9**, 394–400.

KOSMOWSKA-CERANOWICZ B. & PIETRZAK T.
1982 *Znaleziska i dawne kopalnie bursztynu w Polsce*, 1–132, Warszawa.

KRUMBIEGEL G. & KRUMBIEGEL B.,
1994 Bernstein. Fossile Harze aus aller Welt, *Fossilien*, 7.

MARZEC M.
1971 Zarys budowy geologicznej utworów trzeciorzędowych i czwartorzędowych Zatoki Puckiej, *Przegląd Geologiczny*, **12**, 545–547.

MÜLLER A. H.
1989 *Lehrbuch der Paläozoologie, Vertebraten*, **3**, 1–809, Jena.

PIWOCKI M., OLKOWICZ-PAPROCKA I. & KOSMOWSKA-CERANOWICZ B.
1985 Stratygrafia trzeciorzędowych osadów bursztynonośnych okolic Chłapowa koło Pucka, *Prace Muzeum Ziemi*, **37**, 61–77.

SIKORSKA-PIWOWSKA Z.
1984 Modele biologique de l'evolution de l'appareil locomoteur des Tetrapodes, *Zoologica Poloniae*, **31** (1–4), 65–223.

1993 An estimation method of phylogenetic links of vertebrates based on characters of their locomotory apparatus, *Zoologica Poloniae*, **38** (1–4), 5–25.

SUDRE J.
1979 Nouveaux mammiféres Eocene du Sahara occidental, *Palaeovertebrata*, **9**, 83–115.

DIE EIDECHSE *NUCRAS SUCCINEA* BOULENGER DER EHEM. KÖNIGSBERGER BERNSTEINSAMMLUNG

Siegfried RITZKOWSKI

Kurzfassung

Die Bernstein-Eidechse *Nucras succinea* BOULENGER wurde von R. KLEBS etwa 1874 oder 1875 entdeckt und im Jahre 1889 erstmals veröffentlicht. Sie befand sich bis 1899 im Besitze der Bernstein-Firma STANTIEN & BECKER (Nr. 12664 des Firmen-Museums) und bis 1945 in der ehemaligen Bernsteinsammlung der Universität Königsberg. Seit 1958 wird sie im Institut und Museum für Geologie und Paläontologie der Universität Göttingen aufbewahrt. Sie ist keine Fälschung.

Identifizierung

Die Bernsteinsammlung der ehemaligen Albertus-Universität zu Königsberg i. Pr. enthielt als besondere Rarität eine Eidechse in Bernstein. Von ihr berichtet der letzte Direktor der Sammlung, K. ANDRÉE, in seinem Bernsteinbuch (1937, 60ff.). In der Annahme, daß dieses Fundstück in den Wirren des Kriegsendes verloren gegangen sei, veröffentlicht Th. KRUCKOW (1962) eine Zeichnung des Einschlusses, die er als Schüler 1935 in Königsberg angefertigt hatte (Abb. 1). Die Eidechse ging jedoch nicht verloren, sondern befand sich unter den Stücken der Bernsteinsammlung, die 1944 nach Göttingen gebracht und seit 1958 in der Obhut des Institut und Museum für Geologie und Paläontologie der Universität verwahrt werden (RITZKOWSKI 1977; 1996).

Abb. 1. *Nucras succinea* BOULENGER, Zeichnung von Th. KRUCKOW 1935 aus KRUCKOW 1962, Abb. 1 (= Eidechse Nr. 12.664 des ehem. BECKER'schen Museums).

Die erste Nachricht über den Fund gibt KLEBS im Jahre 1889 auf der Versammlung der Naturforscher und Ärzte in Heidelberg (Abb. 2; KLEBS 1910, 227). Er hatte die Eidechse bereits 1891 A. BOULENGER in London vorgelegt (KLEBS 1910, 227), der sie als einen echten Lacertiden erkannte. Da KLEBS — seit 1874 geologischer Berater der Firma STANTIEN & BECKER (KLEBS 1910, 224) — in seiner Publikation von 1910 (1910, 226) über die Inklusen berichtet, die er zwischen 1874 und 1900 hatte auffinden können, muß die Eidechse zwischen 1874 und 1889 gefunden worden bzw. in den Besitz der BECKER´schen Sammlung geraten sein. Bei KLEBS (1910, 228) findet sich ferner der Hinweis, daß er das Fundstück „ ... im BECKERSCHEN Museum durch durch 25 Jahre unter Wasser aufbewahrt hatte und so der geologischen Universitätssammlung übergab ... “. Als die BECKERSCHE Sammlung mit der Bernsteinfirma im Jahre 1899 in den Besitz des

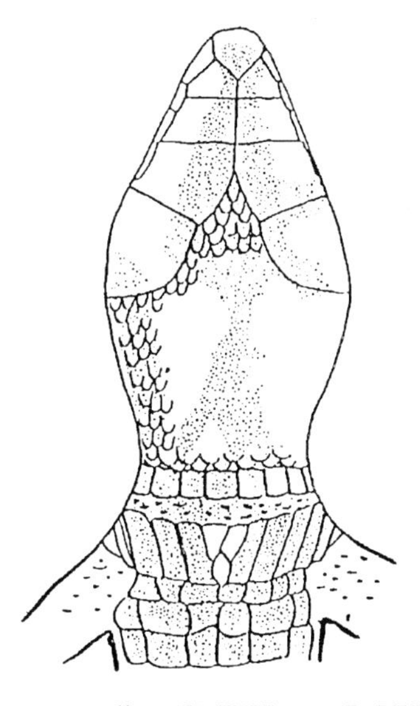

Abb. 2. *Nucras tessellata* (SMITH); aus R. KLEBS 1910, Abb. S. 228 (= Eidechse Nr. 12.664 des ehem. BECKER'schen Museums).

Preussischen Staates überging (BREKENFELD 1996, 281), entstand (im Jahre 1901) die Königliche Bernsteinsammlung der Universität Königsberg. Demzufolge muß die Eidechse um 1875 bereits in Besitz der BECKERSCHEN Sammlung gewesen sein (auch ANDRÉE 1937, 61, schreibt, daß sie um 1875 gefunden worden sei). Sie war dort unter der Nummer 12.664 registriert. Es ist allerdings zu vermerken, daß der Katalog der BECKER´schen Bernsteinsammlung von 1880 (KLEBS 1880) die Eidechse nicht aufführt und auch nur 3.504 Nummern erhält.

Die Katalog-Nummer 12.664 ist an dem Bernsteinstück, daß die Eidechse einschließt, nicht mehr zu erkennen. Doch spricht folgendes dafür, daß es sich um die Eidechse handelt, die KLEBS 1891 veröffentlicht hatte.

- TORNQUIST, der Direktor der neugeschaffenen Bernsteinsammlung der Universität, erwähnt die Eidechse in seiner „Geologie von Ostpreußen" (1910, 119f) und zitiert die Befunde von KLEBS 1910.
- ANDRÉE, der Direktor der Bernsteinsammlung von 1915 bis 1945, hat dieses Fundstück immer als die KLEBS´sche Eidechse angesehen. ANDREÉ war es auch, der die Auswahl der Stükke traf, die 1944 nach Göttingen gebracht wurden.
- Die Göttinger Eidechse weist alle Merkmale auf, die im Bericht von KRUCKOW (1962), der die Eidechse in Königsberg gezeichnet hatte, erwähnt werden.
- Die Eidechse war das einzige nahezu vollständige Exemplar der Königsberger Sammlung. Eine Verwechslung mit anderen Fundstücken ist deshalb ausgeschlossen.

Es kann deshalb davon ausgegangen werden, daß die Eidechse Nr. 12.664 der BECKER´schen Sammlung sich nunmehr in Göttingen befindet.

Orginal oder Falsifikat

Es gibt zahlreiche Berichte, daß rezente Wirbeltiere, insbesondere Eidechsen, Frösche, Fische, u. a. in Bernsteinharz eingebettet wurden. KLEBS (1910, 218) zählt eine Reihe derartiger Falsifikate auf. (Die Sammlung C. GIEBEL befindet sich im Museum für Naturkunde Coburg. Allerdings wurde die Eidechse bislang nicht wiedergefunden; frdl. mdl. Mitt. Dr. MÖNNIG, 1997). Um einer Täuschung nicht zu unterliegen, führt KLEBS Untersuchungen an dem Stück durch. (Dies könnte auch als Hinweis zu werten sein, daß das Fundstück nicht aus der Produktion der Firma stammte, sondern möglicherweise aus dritter Hand erworben wurde). KLEBS schreibt (1910, 226):

„Zum Nachweis etwa vorhandener Wirbel und um die Oberfläche des ganz undeutlichen Stückes besser sichtbar zu machen, habe ich den Bernstein im Einschluß gesprengt. Es war außer wenig kohligen Resten nur ein Hohlraum vorhanden, wie es übrigens bei allen Bernsteinstücken der Fall ist. Ob unter diesen kohligen Resten sich Muskelteile verbergen, darüber habe ich keine Untersuchungen angestellt ... " . Und (S. 227): „Nach der Sprengung und vorsichtigen Reinigung ließen sich mehr Details in der Beschuppung erkennen."

TORNQUIST, Direktor der neugeschaffenen Bernsteinsammlung der Universität, bestätigt (1910, 119f) die Ansicht von KLEBS, daß es sich nicht um eine Fälschung handele, da ein Hohlraum und nicht Skelettreste angetroffen wurden.

Während der Verweilzeit im BECKER´schen Museum, also 25 Jahre lang, wurde das Bernsteinstück unter Wasser aufbewahrt. In der Universitäts-Bernsteinsammlung erfolgte eine Präparation des Objekts. KLEBS schreibt (1910, 228):

„Durch die Präparation ist die Eidechse in der jetzigen Sammlung nicht zu ihrem Vorteil verändert worden. Die Hälften des Stückes, die ich im Beckerschen Museum durch 25 Jahre unter Wasser aufbewahrt hatte und so der geologischen Universitätssammlung übergab, sind wieder zusammengekittet und die Höhlung ist mit Harz gefüllt und das Ganze darin eingebettet. Natürlich ist das Stück dadurch undeutlicher geworden. Die feinen Abdrükke, die man nach der Teilung an der Innenseite wahrnehmen konnte, sind zum größten Teil jetzt unsichtbar geworden. Es ist kaum Aussicht vorhanden, daß das Prachtexemplar je wieder in den früheren Zustand zurückgeführt werden kann, wenn man auch die Teile wieder auseinanderschmelzen wollte. Die eingedrungene Harzmasse und namentlich ihre Entfernung durch harzlösende Mittel werden sicherlich den hauchartigen Abdruck der Beschuppung angegriffen haben."

Eine Bestimmung des fossilen Harzes steht bislang aus.

Nomenenklatur

1889 *Cnemidophorus*, Heidelberg, 62. Versammlung deutscher Naturforscher und Ärzte, 1889 (KLEBS 1910, 227).

1891 *Nucras tessellata* (SMITH) durch BOULENGER / London (KLEBS 1910, 227).

1917 *Nucras succinea* BOULENGER (teste: POINAR 1992, 238).

1962 *Nucras succinea* BOULENGER KRUCKOW 1962, 269, Fußnote.

Angaben zum Fundstück

Die Eidechse liegt in einem Bernsteinstück, das wiederum in ?Kanadabalsam in einem Glasrahmen von 10 mm Höhe eingebettet und von zwei Glasplatten bedeckt wird (Abb. 3). Das Präparat läßt das Tier auf dem Rücken liegend erscheinen.

Die Abmessungen von Tier, umhüllenden Bernstein und Präparat betragen:

Parameter	Länge (cm)	Breite (cm)	Dicke/Höhe (cm)
Eidechse	4,2		
Bernsteinstück	5,2	1,8	
Präparat (incl. 2 Deck- u. Bodenplatte)	8,1	5,1	1,9
Bodenplatte	9,9	6,3	

Die Länge der Eidechse (Fundstück) beträgt nach eigener Messung 4.2 cm. Sie wird von ANDREÉ (1937, 61) ebenfalls mit 4,2 cm angegeben. Dies wird durch die Schätzung von KRUCKOW (zwischen 4 und 4,5 cm) bestätigt (KRUCKOW 1962, 267). BACHOFEN-ECHT (1949, 180) Angabe von 10 cm Länge läßt sich weder am Tier noch am Präparat nachvollziehen (*cf.* KRUCKOW 1962, 267).

Unter der Deckplatte ist Canadabalsam hervorgeflossen und bildet auf dem Glasrahmen der linken Körperseite einen Wulst bis zu 4 mm Stärke, an den drei anderen Seiten ist er geringer. Im Dezember 1997 wurden die Seiten wieder freigelegt, um die Einsicht in das Präparat zu verbessern.

Oberfläche und Unterfläche des Bernsteins, der die Eidechse umhüllt, sind plan. Der Zeitpunkt der Anlage der Schliff-Flächen ist unbekannt, denn auch KLEBS (1910) gibt keinen Hinweis auf eine Bearbeitung. Die Fläche, an der KLEBS den Bernstein gespalten hat, lässt sich — vor der Reinigung — noch nicht genau beobachten.

Es fehlt ein Stück des Schwanzes. Das Schwanzstück der Eidechse ist abgebrochen (ANDREÉ 1937, 61). Auch BACHOFEN-ECHT (1949, 180) notiert das Fehlen der Schwanzspitze. Es wurde nicht ein Tier, dem ein Stück des Schwanzes fehlte, in Bernstein eingebettet, sondern das Schwanzstück ist nach der Einbettung mit dem bauchseitigen Bernstein abgebrochen. Angaben über den Zeitpunkt des Verlustes wurden nicht gefunden. Bei KLEBS (1910) findet sich jedoch kein Hinweis auf die Unvollständigkeit des Tieres.

Die einzige Abbildung erfolgt durch KRUCKOW 1962, 268, Abb. 1 als Zeichnung. KLEBS gibt ferner (1910, 228) eine Abbildung der Halsregion eines gleichgroßen, rezenten *Nucras tessellata*, auf den BOULENGER 1891 die Kehlbeschuppung der Bernsteineidechse eingezeichnet hatte. Diese Abbildung wurde von KRUCKOW (1962) nochmals publiziert.

Die rechten Extremitäten sind vorhanden. Das linke Hinterbein ist durch den Anschliff des Bernsteinsstückes gekappt. Das Innere weist keine Füllung durch Skelettreste oder kohlige Substanz auf. Vielmehr dürfte ein Hohlraum existiert haben, der bei der erneuten Einbettung mit Canadabalsam gefüllt wurde. Das linke Vorderbein fehlt.

In dem Bernsteinstück sind ferner ein Käfer, drei Spinnen und ein weiteres Insekt eingeschlossen.

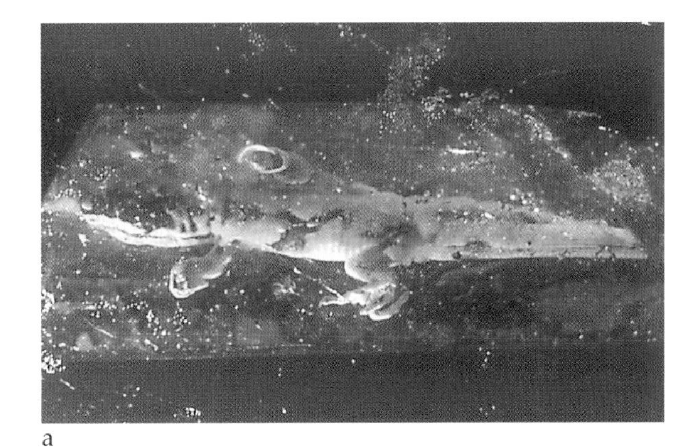

a

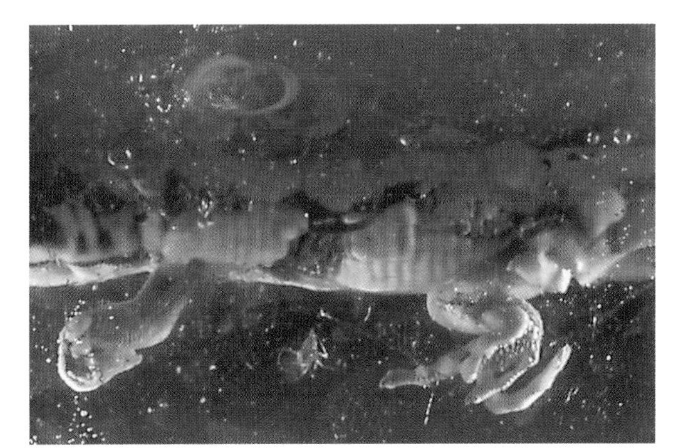

b

Abb. 3. *Nucras succinea* BOULENGER (= Eidechse Nr. 12.664 des ehem. BECKER'schen Museums), Bauchseite: a — Gesamtansicht, b — Deteil. Fot. S. RITZKOWSKI, 1998.

Bearbeitungen

Eine Bearbeitung wird durch Prof. Dr. W. BÖHME, Museum Alexander König, Bonn erfolgen.

Siegfried RITZKOWSKI
Institut und Museum für Geologie
und Paläontologie
Universität Göttingen
Goldschmidtstrasse 3
D-37077 Göttingen, Germany

JASZCZURKA *NUCRAS SUCCINEA* BOULENGER BYŁEGO ZBIORU BURSZTYNU W KÖNIGSBERGU

Siegfried RITZKOWSKI

Streszczenie

Z danych KLEBSA (1910) wynika, że „bursztynowa" jaszczurka *Nucras succinea,* ujawniona około 1874–1875 została znaleziona w bursztynowym surowcu firmy STANTIEN & BECKER. Po raz pierwszy została opisana w 1889 r. Przypisania do współczesnej afrykańskiej rodziny Lacertidae rodzaju *Nucras* dokonał BOULENGER, który w 1917 wyznaczył nowy gatunek *Nucras succinea.* Jaszczurka znajdowała się do 1899 r. u właścicieli firmy STANTIEN & BECKER, pod nr. inw. 12664.

W poszukiwaniu resztek szkieletu R. KLEBS rozłamał okaz z jaszczurką na dwie części. Niestety poza drobnymi resztkami organicznej substancji, znalazł jedynie pustą przestrzeń. W 1899 r. zbiory firmy przeszły na własność Uniwersytetu Alberta w Königsbergu. Tu obie części jaszczurki w bursztynie zostały sklejone i zatopione w (?)balsamie kanadyjskim.

R. KLEBS i dyrektorzy bursztynowych zbiorów A. TORNQUIST i K. ANDRÉE zamanifestowali wątpliwości co do prawdziwości inkluzji. Pogląd, że „bursztynowa" jaszczurka jest fałszerstwem, nie znalazł potwierdzenia w żadnych materialnych argumentach.

„Bursztynowa" jaszczurka należy do królewieckich zbiorów bursztynu, które w 1944 r. zostały przeniesione do Getyngi i złożone w kopalni soli w Volpriehausen. Od 1958 r. materiał ten znajduje się pod opieką Instytutu i Muzeum Geologii i Paleontologii Uniwerytetu w Getyndze. Obecnie jest w opracowaniu u Prof. W. BÖHMEGO w Muzeum Aleksandra Königa w Bonn.

Literatur

ANDREÉ K.
1937 *Der Bernstein und seine Bedeutung in Natur- und Geisteswissenschaften, Kunst und Kunstgewerbe, Industrie und Handel,* 1–219.

BACHOFEL-ECHT A.
1949 *Der Bernstein und seine Einschlüsse,* Wien, 1–204.

BREKENFELD A.
1996 *Die Unternehmerpersönlichkeiten Friedrich Wilhelm Stantien und Moritz Becker,* [*in*:] Ganzelewski M. & Slotta R., Bernstein — Tränen der Götter, *Veröffentlichungen aus dem Bergbau-Museum Bochum,* **64**, 277–283.

KLEBS R.
1880 *Der Bernstein. Seine Gewinnung, Geschichte und geologische Bedeutung. Erläuterung und Katalog der Bernstein-Sammlung der Firma Stantien & Becker,* Königsberg.

1910 Über Bernsteineinschlüsse im allgemeinen und die Coleopteren meiner Bernsteinsammlung, *Schriften der physikalisch-ökonomischen Gesellschaft zu Königsberg,* **51**, 217–242.

KRUCKOW TH.
1962 Eine echte Bernstein-Eidechse, *Der Aufschluss,* **13** (11), 267–270.

POINAR G. O.
1992 *Life in amber,* 1–350.

RITZKOWSKI S.
1977 Das Schicksal der Königsberger Bernsteinsammlung, *Museumskunde,* **42** (2), 87–88.

1996 *Geschichte der Bernsteinsammlung der Albertus-Universität zu Königsberg i. Pr.,* [*in*:] Ganzelewski M. & Slotta R. (Hrsg.), Bernstein — Tränen der Götter, *Veröffentlichungen aus dem Bergbau-Museum Bochum,* **64**, 293–298.

TORNQUIST A.
1910 *Geologie von Ostpreußen,* Berlin.

SOME REGULARITIES OF THE CAPTURE OF PLANT REMAINS IN NATURAL RESINS

Nikolai I. TURKIN & Elena E. EZHOVA

Abstract

Living coniferous trees of the species *Picea abies* and *Pinus sylvestris* were used to investigate the processes which govern how plants become trapped in resin. It was shown that the amount of plant inclusions is always far greater than that of animal inclusions. The taxonomic composition of the plant community influences the qualitative and quantitative characteristics of the buried plant material to a much lesser degree than species fitting of a resiniferous plant. When interpolating data on burial processes in modern resins with that of data relating to fossil resins, it is necessary to take into account the essential features of all stages of the processes of burial and fossilization.

Introduction

The processes by which animals and plants become buried in natural plant resins are very complex and influenced by numerous factors. A specific selectivity is characteristic for this kind of burial (ZHERIKHIN & SUKATSCHEVA 1989). This is determined by the following:

— the physical properties of the resin, which define the top size limit of inclusions;
— peculiarities of the landscape, the character of the adjacent plant and animal communities surrounding resiniferous plants;
— the mode of life and behaviour of the organisms which are trapped;
— the biochemical features of the initial primary resin;
— climatic and edaphic factors.

Thus, the composition of oryctocenosis partly reflects the features of the flora and fauna characteristic of the period during which a particular fossil resin was formed. Applying the actuapalaeontological approach, in which the process of an item becoming trapped in the resin of living plants is used as a model, appears very promising. The rules governing the ways in which plants are buried in resin were investigated using the living coniferous trees: *Picea abies* (L.) KARST. and *Pinus sylvestris* L.

Methods and material

Samples of fir-tree resin were collected near the village of Kosmodemyansky in the Kaliningrad region and pine resin from the Borisovsky district of Belarus. The resins were dissolved in 96% ethanol (resin to ethanol mass ratio = 1 : 1) and left to stand for 7 days. The amount of all biological inclusions measuring over 1 mm in length per unit of mass was calculated. In order to study the plant communities of the sample sites, their taxonomic composition and the abundance of different plant species were established using the Drude scale (FEDORUK 1976). Geobotanical maps were also drafted (Fig. 1).

We used crude amber recovered from the sea (near Primorsk, Kaliningrad region). We decided to use only natural amber material collected at a particular site in this study, not collection pieces, due to the fact that they had previously been selected in accordance with the aims and means of the collector. Two samples (every sample weighed 2 kg) were analysed. The overall amount of inclusions was calculated, as well as the respective number of animal and plant inclusions.

Results

Resin of modern conifers

Samples of pine resin were collected in plant community *Pinetum vaccinioso-pleuroziosum*, *Pinetum callunoso-pleuroziosum* and *Piceoso-pinetum vaccinioso-pleuroziosum*. Samples of fir-tree resin were collected *Pinoso-piceetum myrtillosum*, *Piceoso-pinetum myrtillosum*, *Pinetum pleuroziosum* and *Pinetum myrtillosum*. The frequency of occurrence of different plant species at the sample sites is shown in Tables 1 and 2.

At the sampling sites where fir resin was collected fir and pine trees dominated, oak also being present,

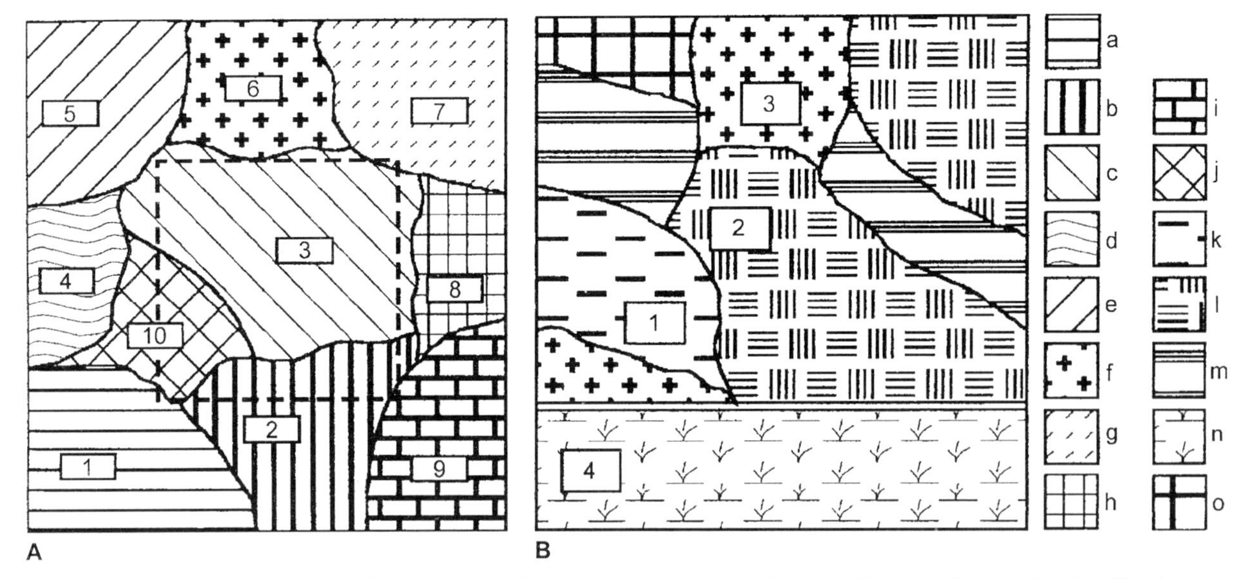

Fig. 1, A & B. Geobotanical map of the regions where pine resin (A) and fir resin (B) was collected. Dashed line in Fig. 1A shows the place where pine resin was collected. Sampling sites where plant associations were studied are shown by rectangles with a number inside. In Fig. 1B sampling sites coincide with places where fir-resin was collected. Symbols for different plant associations: a — *Pinetum vaccinioso-callunosum*; b — *Piceoso-pinetum vaccinioso-pleuroziosum*; c — *Pinetum myrtilloso-pleuroziosum*; d — *Pinetum pteridioso-pleuroziosum*; e — *Betuloso- pinetum myrtilloso-pleuroziosum*; f — *Pinetum pleuroziosum*; g — *Pinetum pteridiosum*; h — *Pinetum vacciniosum*; i — *Pinetum vaccinioso-pleuroziosum*; j — *Pinetum callunoso-pleuroziosum*; k — *Pinoso-piceetum myrtillosum*; l — *Piceoso-pinetum myrtillosum*; m — *Piceetum polytrichosum*; n — *Pinetum myrtillosum*; o — *Pinetum deschampsiosum*.

Species	Abundance (after Drude scale) Sampling			
	area 1*	area 2	area 3	area 4
Pinus sylvestris L.	cop2**	cop3	cop3	cop3
Picea abies (L.) KARST.	cop3	cop2	cop2	0
Quercus robur L.	cop2	cop2	cop2	cop1
Betula pendula L.	sp	0	0	cop2
Sorbus aucuparia L.	cop2	0	0	0
Sambucus racemosa L.	0	0	0	sp
Rubus idaeus L.	cop1	cop1	0	0
Corylus avellana L.	cop1	cop1	cop1	sp
Vaccinium myrtilus L.	cop3	cop3	cop2	cop3
Vaccinium vitis-idaea L.	cop1	0	0	cop2
Fragaria vesca L.	0	0	0	sp
Deschampsia flexuosa L.	cop2	cop2	cop2	cop2
Dryopteris filix-mas (L.) SCHOTT	cop2	0	0	sp
Polytrichum commune L.	cop2	0	0	cop1
Pleurozium schreberi (BRID) MITT.	cop2	cop2	cop3	cop2
Hypogimnia physodes L.	sp	sp	sp	sp

Table 1. Occurrence frequency of plant species at the areas of *Picea abies* L. resin sampling. *Numbers of areas according to Fig. 1B. **Legend: cop3 — > 20% of vertical projection; cop2 — 4–20% of vertical projection; cop1 — 1.5–4% of vertical projection; sp — 1–1.5% of vertical projection; sol —several exemplars; un — the only exemplar; 0 — species is absent.

rowan (*Sorbus aucuparia*) and white birch (*Betula pubescens*) occurring less often. The most common varieties of undergrowth included bilberries (*Vaccinium myrtillus*), moss (*Pleurozium schreberi*) and grass (*Deshampsia flexuosa*). Pine trees were dominant at the sites where pine resin was collected, with birch (*B. pubescens*), fir, oak (*Quercus robur*) and aspen (*Populus tremula*) occuring sporadically and one instance of rowan (*S. aucuparia*) being noted.

Undergrowth was represented primarily by bilberry and clusterberry (*V. myrtillus*, *Vaccinium vitis-idaea*), although heather (*Calluna vulgaris*) and moss (*P. schreberi*) were also plentiful. Fern (*Pteridium aquilinum*), lily-of-the-valley (*Convallaria majalis*), moss (*Polytrichum commune*), blackberry (*Rubus caesius*) and raspberry (*Rubus idaeus*) were not so abundant. In general, the flora was significantly richer due to the greater diversity in undergrowth.

Analysis of the different inclusion group ratios for fir-tree resin (Tab. 3) revealed that bark and wood remains were the most common (39.6%) together with fir-tree needles (33%), although pine-tree needles constitute only 2% of all inclusions. A share of seed scales and was 8.2%. Lichens made up 2.6% of the assemblage, which is a relatively high percentage compared to the overall presence of this group

Species	Abundance (after Drude scale)									
	1*	2	3	4	5	6	7	8	9	10
Vaccinium myrtillus L.	cop1	cop2	cop3	cop1	cop3	cop1	cop1	cop2	cop1	cop2
Vaccinium vitis-idaea L.	cop3	cop3	cop1	cop2	sp	sp	sp	cop3	cop3	cop1
Pteridium aquilinum (L.) KUHN.	0	0	0	cop3	cop1	sp	cop3	sp	0	sp
Dryopteris filix-mas (L.) SCHATT.	0	sol	sol	0	0	sp	0	sp	0	sp
Fragaria vesca L.	0	sp	sp	0	sp	0	sp	sp	0	sp
Rubus caesius L.	sp	sp	0	0	sp	sp	cop1	cop1	cop1	sp
Hamaenerium angustifolium (L.) SCOP.	0	sp	sp	sp	sp	0	sp	0	sp	0
Taraxacum officinale WEB.ex WIGG.	0	sol	0	sol	0	sol	sol	0	0	0
Vicia cracca L.	sp	0	sp	0	0	un	0	sp	sp	0
Lycopodium complanatum L.	sp	0	0	0	sp	cop1	0	sop1	sp	sp
Anemone nemorosa L.	cop1	sp	sp	0	sp	0	0	0	0	sp
Calluna vulgaris (L.) HILL.	cop3	sp	cop2	cop1	cop1	cop1	cop1	sp	sp	cop3
Platanthera bifolia (L.) L. C. RICH.	0	0	sp	0	0	0	0	0	sp	0
Viola canina L.	sp	0	sp	0	sol	sol	0	0	0	sp
Fumaria officinalis L.	0	0	sol	0	un	0	sp	sol	0	0
Hieracium pilosella L.	sp	sp	0	0	sp	0	0	cop1	cop1	cop1
Calamagrostis epigeios (L.) ROTH.	sp	sp	0	0	sp	0	0	cop1	sp	cop1
Gnaphalium sylvaticum L.	0	0	0	0	sp	cop1	0	0	sp	sp
Pyrola rotundifolia L.	0	0	0	cop1	0	sp	sp	0	cop1	cop1
Convollaria majalis L.	cop1	sp	sp	sp	sp	sp	cop2	cop1	0	sp
Melampyrum nemorosum L.	0	0	0	cop1	0	cop1	sp	0	sp	sp
Rubus idaeus L.	0	0	cop2	0	0	sp	cop1	cop1	cop1	0
Pulsatilla patens (L.) MILL.	sp	0	sp	0	sp	sp	sp	0	sp	cop1
Majanthemum bifolium (L.) Fr. SCHMIDT	0	0	0	sp	0	0	sp	0	sp	0
Paris quadrifolia L.	0	0	0	sol	0	0	sol	0	0	0
Carex rostrata STOKES	0	0	sp	0	0	sp	0	sp	0	0
Pleurozium schreberi (WILLD) MITT.	cop2	cop3	cop3	cop3	cop3	cop3	cop2	cop2	cop3	cop3
Dicranus polysetum SW.	cop2	0	0	0	sp	cop2	sp	sp	cop2	cop2
Polytrichum commune L.	cop1	cop2	cop2	cop2	cop1	cop1	0	0	0	cop2
Hypogimnia physodes L.	sol	sp	sp	sol	cop1	sol	sp	sp	sp	sp
Pinus sylvestris L.	cop3	cop3	cop3	cop3	cop3	cop3	cop3	cop3	cop3	cop3
Sorbus aucuparia L.	cop1	0	0	sol	0	sp	0	sp	sp	0
Picea abies (L.) KARST.	sp	cop3	sp	cop1	0	sp	0	0	0	0
Betula pubescens L.	sp	0	sp	cop2	cop3	0	cop2	cop1	sp	cop1
Populus tremula	sp	cop1	cop1	0	cop2	sp	cop1	sp	0	sp
Quercus robur L.	0	cop2	cop2	sp	sp	sp	0	0	0	0

Table 2. Occurrence frequency of plant species in the areas of *Pinus sylvestris* L. resin sampling. *Numbers of areas according to Fig. 1A. Other abbreviations — see legend for Table 1.

in the plant community (less than 1.5% of the grass canopy). Other plants constituted about 2.3% of the entire assemblage. In total, plant remains made up as much as 86.7% of all biological inclusions in the fir resin.

Faunal remains were much rarer — 13.3%. It is interesting that the most common animal inclusions in fir resin were puparia of the order Diptera. Other arthropods were found in 4 cases per 100 biological inclusions (4.3%). Our findings coincide well with other authors' data on fly puparia found in the resin of some conifers from Siberia and the Russian Far East (ZHERIKHIN & SUKATSCHEVA 1989). At all of the sampling sites the quantity of puparia was higher than the quantity of other animal inclusions. Animal remains consistently constituted a significantly

Object	Number of objects							
	area 1		area 2		area 3		Average	
	N/1kg	%	N/1kg	%	N/1kg	%	N/1kg	%
Fragments of lichens	9	1.5	35	3.4	10	2.3	15	2.6
Pieces of bark and wood, >1mm	231	38.4	415	39.9	183	40.6	243	39.6
Fir-tree needles	159	26.4	412	39.6	119	26.3	230	33.0
Pine needles	17	2.8	5	0.5	21	4.7	14	2.1
Seeds, seed scales	64	10.6	62	5.9	46	9.1	57	8.2
Other plant objects	9	1.5	5	0.5	10	2.3	8	2.3
Plant objects in total	**489**	**81.2**	**934**	**89.8**	**389**	**86.3**	**604**	**86.7**
Diptera, puparia	66	11.0	81	7.8	41	9.1	63	9.0
Other arthropods	47	7.8	25	2.4	20	4.6	29	4.3
Animal objects in total	**113**	**18.8**	**106**	**10.2**	**61**	**13.7**	**83**	**13.3**
Biological objects in total	**602**	**100**	**1040**	**100**	**450**	**100**	**697**	**100**

Table 3. Composition of plant and animal inclusions in *Picea abies* resin.

smaller share of all biological inclusions than did plant material, even when not taking into account wood and bark fragments, which make up the bulk of plant remains found in amber. On average, the percentage of plant remains (excluding bark and wood) was 79.5% and animal remains — 20.5%.

Rather similar results were gained from the analysis of inclusions buried in pine-tree resin. Fragments of wood and bark occurred most frequently — 43.0%. The general amount of plant remains was significantly higher than that of animal remains — 76.9% and 23.1% respectively (not including bark and wood). Fly puparia outnumbered other animal inclusions: on average, puparia constituted 10.7% of the faunal assemblage, other arthropods — only 3.4%. Large amounts of lichen (21.3%) were noted in comparison with the quantities found in fir resin (2.6%). In contrast to the fir resin samples, fir needles were entirely absent from the pine resin samples, in which pine needles constituted about 5.7%.

Thus, it is possible to identify some common features of the processes by which living matter is trapped in coniferous resins related to different species:

- Plant inclusions quantitatively dominate over animal remains;
- The majority of plant inclusions represent the remains of the resiniferous plant (excluding bark and wood).

Baltic amber

The results of qualitative analysis of two samples of crude amber, with regard to their biological inclusions, is presented in Table 4. Spores and pollen inclusions were predominant in these samples, making up 44.8% of all biological inclusions. Pieces of wood and bark of over 1 mm constituted 32.5% of the total assemblage. Oak hairs made up a significantly smaller amount of the assemblage — 17.9%. No leaves, buds, brunches, flowers, fruits, seeds or their fragments were found among the material under analysis. In total, plant remains accounted for 94.3% of all inclusions, with faunal remains amounting to only 5.7%.

Discussion

Similarly to modern resins, the proportion of plant inclusions in amber was far greater than that of animal inclusions. At the same time, inclusions such as leaves, buds, brunches, flowers, fruits, seeds or their fragments are extremely rare in Baltic amber in comparison to modern resins: they constitute 0.4–2.4% of plant assemblages found in amber (KATINAS 1971; KLEBS 1911), whilst in modern fir-resin they make up 48.2%, and in pine resin — 31.3%. Thus, there are much greater amounts of relatively large plant inclusions than in Baltic amber. Most probably, this is linked to the destruction of inclusions which are not completely covered by resin due to the influence of micro-organisms and weathering (KATINAS 1971; SAVKEVICH 1970). Consequently, data on the percentage of this kind of plant inclusions in Baltic amber does not adequately reflect the composition of its contemporary plant communities, which limits the use of this information in palaeobotanic reconstructions. From this point of view, comparative

Objects	Sample 1		Sample 2		Average	
	No./1kg	%	No./1kg	%	No./1kg	%
Pieces of bark and wood	243	36.8	210	28.1	226.5	32.5
Spores and pollen	253	38.3	383	51.3	318.0	44.8
Hairs of oak leaves	108	18.4	130	17.4	119.0	17.9
Plant inclusions in total	**604**	**91.5**	**723**	**96.8**	**663.5**	**94.3**
Animal inclusions in total	**56**	**8.5**	**24**	**3.2**	**40.0**	**5.7**
Biological objects in total	**660**	**100**	**747**	**100**	**703.5**	**100**

Table 4. Composition of plant and animal inclusions in Baltic succinite.

studies of small plant inclusions (spores, pollen, microalgae etc.) in fossil and modern resins look far more promising.

As we have already seen, the content of own leaves (needles) in the resin of resiniferous trees is relatively high — up to 33% of all inclusions for fir and about 5.7% for pine. The higher proportion of own needles in the fir resin can be explained by the biological characteristics of these two species: the leaf index (area of leaf surface per unit of area covered by krone projection) for fir is twice as high as that of pine (GRODZINSKY & GRODZINSKY 1973) and, furthermore, the surface area of fir needles is several times less than that of pine needles. Consequently, the amount of needles per unit of area covered by crown projection is several times greater for fir. The low frequency of small objects, such as as fir needles, in Baltic amber probably indirectly attests to the fact that fir was not an amber-producing tree. As regards the rather low incidence of pine needles in Baltic amber, this might be explained by the generally poor preservation of large (long) objects during the processes of immersion and fossilisation.

The high percentage of lichen inclusions in pine resin compared with that in fir resin is linked to the positive consortive relationship between some lichen species and pine. Otherwise, it is not characteristic for fir trees. The low incidence of lichen inclusions is most likely explained by the fact that talloma are often destroyed during fossilisation, because they are seldom totally covered by resin.

Other plants which were present, often in abundance, in the plant communities under consideration are only very rarely found in modern resins and their amount was not comparable with the remains of the resiniferous plant. The taxonomic composition of the plant community influences the qualitative and quantitative characteristics of buried plant material much less so than species fitting of a resiniferous plant. This conclusion, as seems, can be useful for understanding of burial process in fossil resin also.

Conclusion

The process of living matter becoming trapped in the resin of some conifers is characterized by a number of common features:

- plant inclusions outnumber faunal inclusions;
- the majority of plant inclusions consist of the remains of the resiniferous plant (excluding bark and wood).
- the taxonomic composition of the plant community influences the qualitative and quantitative characteristics of buried plant material to a much lesser degree than species fitting of a resiniferous plant.

It seems, that when interpolating these results to a burial process regarding to fossil resins, Baltic succinite in particular, it is necessary to take into account an essential difference between the resin-egestion physiology of the investigated conifers and the physiology of a fossil resin source-plant. The processes taking place during subsequent stages of burial are important also in this case.

Nikolai I. TURKIN
Kaliningrad State University
Al. Nevskogo 14
Kaliningrad, Russia

Elena E. EZHOVA
P. P. Shirshov Institute of Oceanology, Atl. Branch
Pr. Mira 1
236000 Kaliningrad, Russia
lena@ioran.kern.ru

NIEKTÓRE PRAWIDŁOWOŚCI ZATAPIANIA SZCZĄTKÓW ROŚLINNYCH W NATURALNYCH ŻYWICACH

Nikolai I. TURKIN, Elena E. EZHOVA

Streszczenie

Procesy zatapiania obiektów zwierzęcych i roślinnych w naturalnych żywicach roślinnych są bardzo złożone i wpływa na nie wiele czynników. Specyficzna selekcja jest charakterystyczna dla tego rodzaju zatapiania (ZHERIKHIN & SUKATSCHEVA 1989). Jest ona określona przez:

— fizyczne właściwości żywicy, które warunkują górną granicę rozmiaru inkluzji (selekcja wg wielkości),

— cechy krajobrazu, charakter otaczającego środowiska roślinnego i zestawu zwierząt związanych z żywicującą rośliną (selekcja ekologiczna),

— sposób życia i zachowywania się zatapianych organizmów (selekcja behawioralna),

— biochemiczne właściwości pierwotnej żywicy (selekcja fizjologiczna),

— czynniki klimatyczne i edaficzne (glebowe).

Skład oryktocenozy[1] odzwierciedla zatem częściowo właściwości rzeczywistej fauny i flory charakterystycznej dla okresu tworzenia się kopalnej żywicy.

Obiecujące wydaje się zastosowanie zasady aktualizmu paleontologicznego, gdy proces zatapiania w żywicy żyjących drzew jest użyty jako model procesów zachodzących w przeszłości.

Prawidłowości zatapiania obiektów roślinnych w żywicach były badane na współczesnych drzewach *Picea abies* i *Pinus sylvestris*. Próbki żywicy świerka i sosny zebrano i dokonano analizy składu taksonomicznego i stosunku liczbowego różnych grup inkluzji roślinnych. Opisano zbiorowiska roślinne w miejscu pobrania próbek.

Wykazano, że liczba inkluzji roślinnych jest zawsze większa niż zwierzęcych. Maksymalny procent inkluzji roślinnych w żywicy danych gatunków stanowią szczątki i fragmenty tych właśnie żywicyjących gatunków roślin. Skład taksonomiczny zbiorowiska roślinnego w znacznie mniejszym stopniu wpływa na jakościową i ilościową charakterystykę zatopionego materiału roślinnego niż przynależność gatunkowa żywicodajnej rośliny.

Przy interpolowaniu wyników na procesy zatapiania odnoszące się do żywic kopalnych, a w szczególności bursztynu bałtyckiego, należy koniecznie brać pod uwagę zasadniczą różnicę pomiędzy fizjologią wydzielania żywicy badanych iglastych i fizjologią rośliny dostarczającej kopalnej żywicy. Ważne są także procesy, którym podlegały obiekty już zatopione, w kolejnych etapach przekształcania się świeżej żywicy w bursztyn.

[1] [red.] Oryktocenoza — etap zachowania się skamieniałości (zespół wszystkich sfosylizowanych szczątków występujących w osadzie).

Bibliography

GRODZINSKY A. M. & GRODZINSKY D. M.
1973 *Kratky spravochnik po physiologii rastenii*, 1–592, Naukova dumka, Kiev.

FEDORUK A. G.
1976 *Botanicheskaya geographia. Polevaya praktika*, 1–226, Minsk.

KATINAS V. I.
1971 Yantar i jantarenosnyje otlozhenija juzhnoj Pribaltiki, *Trudy Inst. Geol. Vilnius*, **20**, 1–156.

KLEBS R.
1911 Über Bernsteineinschlüsse im allgemein und die Coleopteren meiner Bernsteinsammlung, *Schr. phys.-ökon. Ges. zu Konigsberg i. Pr., Konigsberg*, **51**, 217–242.

SAVKEVICH S. S.
1970 *Amber*, 1–190, Leningrad, Nedra.

ZHERIKHIN V. V. & SUKATSCHEVA I. D.
1989 Zakonomernosti zakhoroneniya nasekomykh v smolakh, *Doklady sovetskikh geologov na 28 sessii mezhdunar. geologocheskogo congressa, Washington, July 1989*, Nauka, Moskwa, 84–92.

LATE PALAEOLITHIC AMBER IN NORTHERN EUROPE

Jan Michał BURDUKIEWICZ

Abstract

This paper deals with the beginnings of amber-working. The oldest amber ornaments in northern Europe date from the Late Palaeolithic (13,000–10,000 BP). The Hamburgian and Creswellian groups were the first to fashion ornaments from this fossil resin. The earliest amber artefact to have been found thus far is a plaque recovered from dead-ice lake Meiendorf (Schleswig-Holstein). Other amber artefacts dating from the same period were also discovered at Siedlnica 17a (SW Poland). These consisted of two ornaments (in the shape of a reversed icicle and a parallelogram respectively) and six small amber fragments. Other amber finds are known from Deimern 41 (Lower Saxony), Ureterp I in Holland and Gough's Cave (Wales).

The fragments of an amber plaque have recently been found at the Federmesser site of Weitsche in, Lower Saxony. The plaque was carved into the shape of a horse figurine similar to that found at Dobiegniew in Western Pomerania. It is possibly a figurine of a large herbivore, but this supposition requires better verification.

Introduction

Amber nodules are known from Middle Palaeolithic sites in Europe. Such finds are rather rare and do not bear any traces of working. It is believed that amber was first used as an incense by Neanderthal man (ROTTLÄNDER 1973).

The oldest known amber artefacts in Europe come from Late Upper Palaeolithic sites in the Ukraine (Fig. 1). An anthropomorphic figurine and beads made of amber were found at Dobranichevka in the Dnieper river basin (SHOVKOPLAS 1972). Similar amber beads are also known from Mezin and Mezhirich (PIDOPLICHKO 1976; ROGACHEV & ANIKOVICH 1984, 201). Natural amber plaques are quite numerous in the Dnieper river basin near Kiev (SREBRODOLSKIJ 1980).

All of the aforementioned sites belong to the so-called Mezinian culture. Numerous decorative items made mostly from bone and ivory typify their material culture. Extensive assemblages of mammoth bones were found at these sites in the Dnieper and Desna river basins. These were usually arranged in circular formations and bear evidence of having comprised intentional constructions. Several of them have been interpreted as the remains of "huts" or "ritual objects" (SHOVKOPLAS 1972; PIDOPLICHKO 1976; BOSINSKI 1987).

The mammoth hunters left numerous *objets d'art* such as painted mammoth bones, an ivory bracelet and other bone artefacts decorated with various geometric and linear designs (details — see BOSINSKI 1987). Amber objects are rather rare in the ivory and bone artefact assemblages of the Ukrainian Palaeolithic. The Mezinian culture is usually compared to the older Magdalenian in Western Europe and dated to 17,000–15,000 years BP (PIDOPLICHKO 1976). According to several sets of data, the Mezinian culture should be younger and parallel to the middle Magdalenian and Hamburgian in Western Europe (13,500–11,800 years BP; GAMBLE 1986; BOSINSKI 1987).

The Late Palaeolithic in Northern Europe

The oldest objects fashioned from amber by man in Northern Europe are known from the Late Palaeolithic, conventionally dated by ^{14}C to 15,000–10,000 years ago. In recent research on Late Glacial chronology conventional ^{14}C dates have been calibrated and therefore proved to be earlier: 11,800–17,000 years BP (details see STREET *et al.* 1994; VEIL 1997). A settlement gap (20,000–15,000 BP) in Europe, north of its main mountain chains, was caused by the last Scandinavian glaciation (Fig. 2). Northern Europe was resettled by two culturally distinct groups. According to recent research the Highland zone (Mittelgebirge in German) was colonised during the Epe (Pre-Bølling) Interstadial (*c.* 15,000–14,000 BP) by Magdalenian groups from southern France

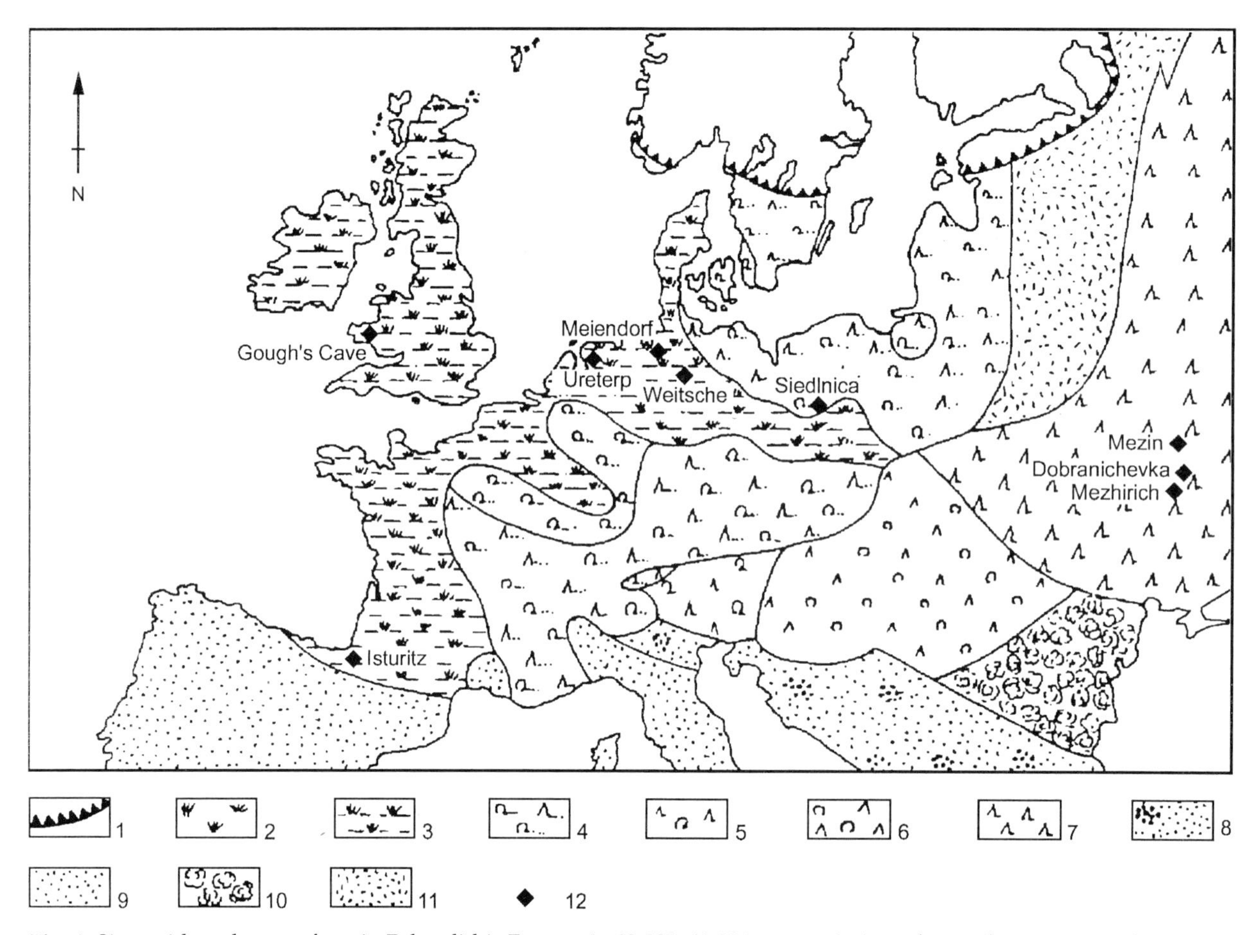

Fig. 1. Sites with amber artefacts in Palaeolithic Europe (*c.* 13,000–12,000 years BP): 1 — glacier sheet, 2 — tundra, 3 — tundra "xeric" variant, 4 — birch-pine forest, 5 — mixed forest, 6 — northern mixed conifer-deciduous forest, 7 — spruce dominated forest, 8 — steppe with Gramineae, 9 — steppe, 10 — mixed-deciduous forest, 11 — mixed forest, 12 — sites with amber artefacts (after HUNTLEY & BIRKS 1983, simplified).

(BURDUKIEWICZ 1993). The European Plain was settled by a unit known as the Shouldered Point Technocomplex (SPT — Hamburgian and Creswellian) 13,000–12,000 years BP (BURDUKIEWICZ 1986; 1993). According to A. RUST (1937; 1962) the Hamburgian culture originated in Eastern Europe and was already present in northern Germany by 20,000 BP, as an hunting community living near the glacier. Later, numerous radiocarbon dates as well as palinological and stratigraphic data from Hamburgian and Creswellian sites clearly indicated that they cannot be older than the Bølling Interstadial, i.e. 13,000–12,000 years BP (BURDUKIEWICZ 1986, 50). The Scandinavian glacier was already in southern Sweden (Figs. 1 & 2). The European Plain was covered by park-tundra in the so-called old moraine zone and by shrub-tundra in the young moraine zone (USINGER 1975). Reindeers migrating in large herds were the main species of game (BURDUKIEWICZ 1986).

Recently, it has become clear that the SPT originated from the Magdalenian (Fig. 2). There are numerous similarities between both taxonomic units, especially phases IV–VI of the classic Magdalenian). The most important of these similarities is the concurrence of flint industries with such diagnostic tools as shouldered points and Zinken perforators. The bone and antler industry of the SPT, however, was not rich and was very similar to the Magdalenian one. Demonstrable proof of genetic ties between the SPT and the comparable Magdalenian (phases IV–VI) has long been impossible due to the apparent chronological discrepancy just mentioned. Comparing radiocarbon dates, it is evident that Magdalenian IV–IV sites are older or parallel to SPT sites (details see BURDUKIEWICZ 1986, 185; 1989). Taking all arguments into account there is no good reason to seek the origin of the SPT in Eastern Europe.

The next taxonomic unit was the Arch-backed Point Technocomplex (APT — Federmesser) which appeared in the subsequent, warmer period — the Allerød Interstadial (12,000–11,000 years BP). At this time the climate of northern Europe changed to a subarctic one (Fig. 2). Plant cover was already much denser and developed into pine-birch woodland (Fig. 1). APT lithic assemblages are twice as numerous as SPT. They are characterised by a general simplification of lithic technology and new diagnostic tools such as arch-backed points and short end-scrapers. Woodland animal species were the main game (elk, red deer, etc.). Evidence for amber usage by APT societies is not as abundant as in the SPT.

In conclusion, we can surmise that the amber-working tradition of north-western Europe developed as an independent process from the east European one. The main feature is a lack of bone and ivory art — so rich in south-eastern Europe. Amber working in north-western Europe began slightly later or parallel in time to the eastern European tradition. It is necessary to stress that amber nodules are known from several Magdalenian sites between the Pyrenees (Auersan, Isturitz — ROTTLÄNDER 1973) and Moravia (Zitného and Kůlna caves — SKUTIL 1930).

Meiendorf-pond, Schleswig-Holstein

The first amber artefact was found in 1934 by A. RUST (1937) during excavations of the Meiendorf-pond site in the Ahrensburg Valley near Hamburg, northern Germany (Figs 3 & 4). The discovery of the Meiendorf site with its thick organogenic layers deposited in a dead-ice lake presented an opportunity to recover numerous bone, antler and exceptional amber objects (RUST 1937, 112).

The top deposit was meadow peat *c.* 2 m thick. Below were thinner layers of organic silt, peat with remains of macrofauna and poorly sorted sand containing displaced prehistoric artefacts. The bottom sediment (4–6 m from the top) was calcareous gyttja containing Hamburgian artefacts (Fig. 3).

The bottom of the dead-ice lake yielded 284 flint artefacts. Faunal remains included bones of: reindeer (71 specimens), horse (1), hare (2 or 3), polar fox (1 or 2), wolverine (1) and some polar birds. Of particular significance is the domination of reindeer bones in the faunal assemblage (*c.* 99 %). This assemblage does not, of course, portray a representative sample of the animal population. The remains constitute only part of the food waste of Hamburgian hunters, and of material that found its way into the lake and was preserved in its sediments. Nevertheless, the remains are obviously of arctic species typical of a tundra environment. Such an interpretation of the ecological setting is also supported by pollen analysis carried out by R. SCHÜTRUMPF (1937) and later revised by H. USINGER (1975). The site can be dated to the Bølling Interstadial (Fig. 2; see also USINGER 1975; BURDUKIEWICZ 1986).

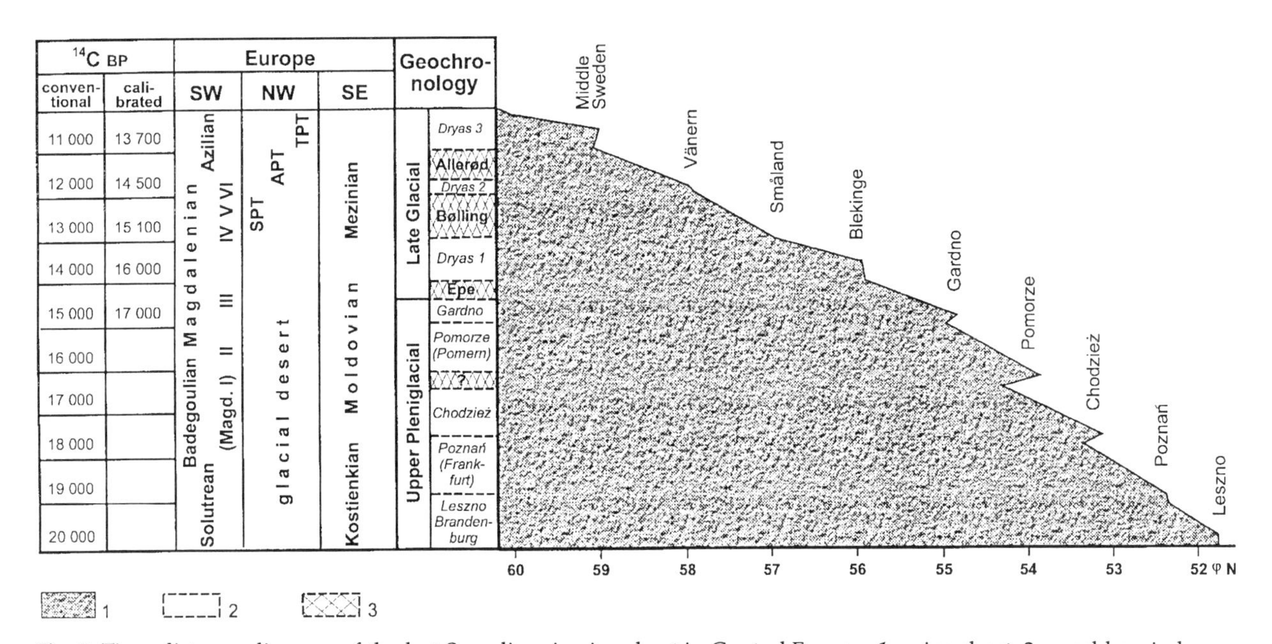

Fig. 2. Time-distance diagram of the last Scandinavian ice-sheet in Central Europe: 1 — ice-sheet, 2 — cold periods, c — warmer periods.

An important find in the field of prehistoric amber artefact research was that of a well-preserved amber plaque from dead-ice lake Meiendorf. This item was found together with two other amber nodules, which were not described in detail (RUST 1937, 112). The plaque is of opaque yellow amber. It may originally

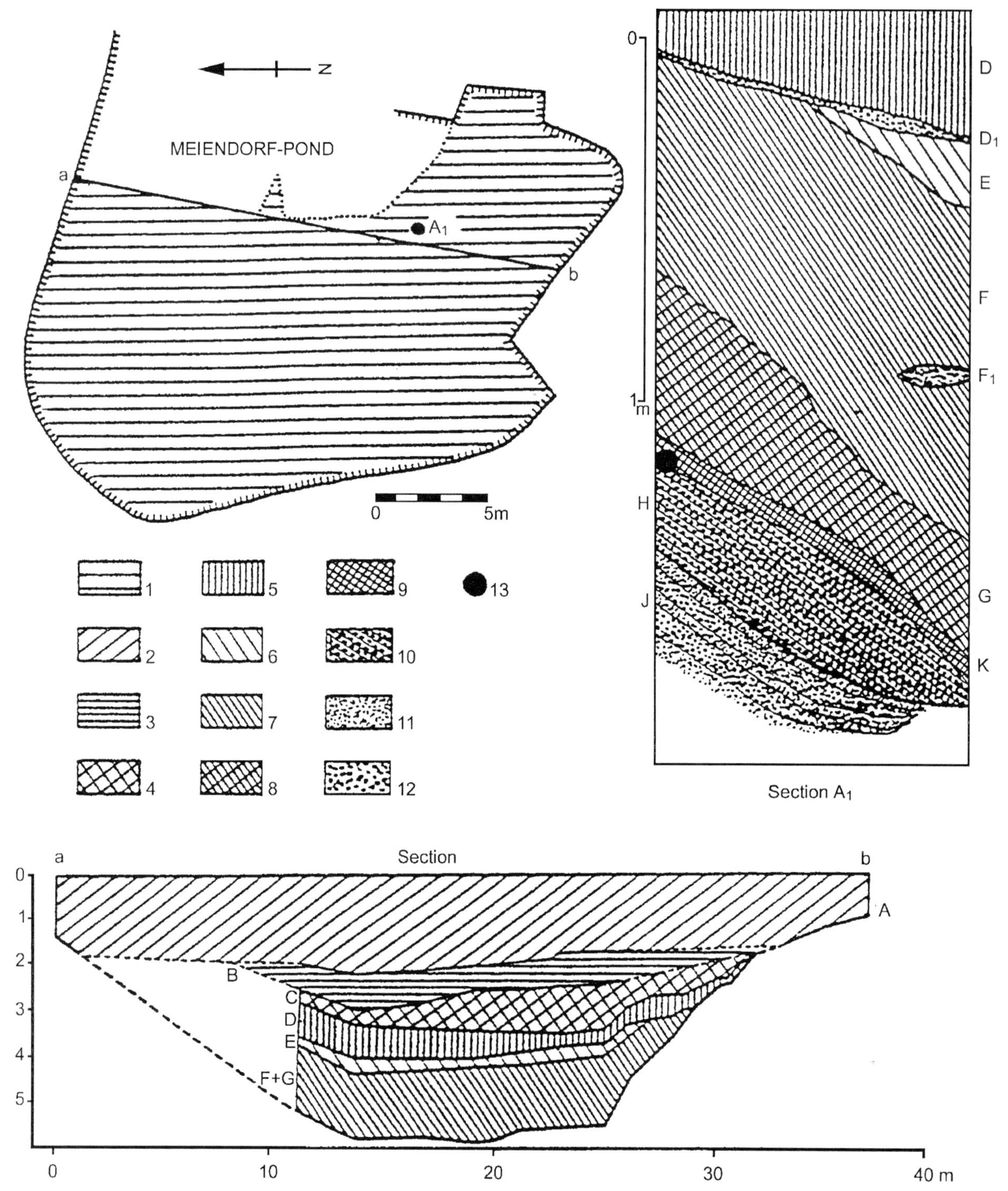

Fig. 3. Stratigraphy of Meiendorf-pond near Hamburg, NW Germany: 1 — excavated area; 2 — meadow peat; 3 — low *Phragmites* peat; 4 — organic silt; 5 — *Carex* peat with macroremains of plants; 6 — gyttja; 7 — sandy gyttja; 8 — "cultural layer"; 10 — sandy gyttja with Hamburgian artefacts; 11 — fine sand; 12 — coarse sand; 13 — Reindeer bone (after RUST 1937).

have been approximately trapezium-shaped (55 x 43 mm) and had a hole drilled in its centre. The edges of the plaque bear traces of processing. A series of intersecting X-shaped crosses has been engraved on one edge. The concave edge has been damaged. Other edges are also slightly damaged but their primary shape has been retained. Both flat surfaces of the plaque are engraved with irregularly arranged lines (Fig. 4).

A. RUST (1937, 112) believed that he was able to identify a schematic outline of a horse's head on one side of the plaque. The drilled hole represents, in his opinion, the eye of the horse. On the reverse side RUST (1937, 112) thought it possible to discern the faint outline of a reindeer. Such an interpretation is rather difficult to accept because the engraved lines are so numerous and irregular. The hole was simply used to suspend the plaque as an ornament.

Fig. 4. Amber plaque from Meiendorf-pond near Hamburg, NW Germany (after RUST 1937).

Siedlnica 17a, SW Poland

For many years the amber plaque from Meiendorf was the only find attesting to so early a date for the use of amber as a raw material for ornament production. Other finds have now been excavated from the Hamburgian site of Siedlnica 17a, Leszno voivodeship, in south-western Poland (BURDUKIEWICZ 1980). Siedlnica 17a is located in the valley of the River Kopanica microregion in the middle section of the Oder river basin. Numerous archaeological sites with SPT (Hamburgian), APT (Federmesser) and Tanged Point Technocomplex (TPT — Ahrensburgian and Swiderian) have been discovered in the Kopanica valley (BRODZIKOWSKI & van LOON 1987; BURDUKIEWICZ 1987).

Siedlnica 17a lies on a sandy terrace, *c.* 60 cm below the APT artefact concentrations of Siedlnica 17 (Fig. 5). This difference in depth may be explained by the accumulation of local deposits at Siedlnica 17 during the colder period of Dryas 2 (BRODZIKOWSKI, & van LOON 1987). The excavated area of 247 m^2 yielded 6141 flint artefacts, 170 fragments of stone slabs and roundstones and 6 pieces of unidentified animal bones. Almost all finds were located in a single large concentration, 6 m in diameter (Fig. 6). The preliminary refitting analysis shows a single refitting network. It is necessary to stress the absence of a structured hearth, such as that found at Olbrachcice 8 — another rich Hamburgian site, lying 1 km to the north (BURDUKIEWICZ 1986, 118).

A considerable number of the flint artefacts from Siedlnica, made mostly of brown silex, were covered with patinas of various colours. This came about as a result of local geochemical features. The stratigraphy of the site has been slightly disturbed by deep ploughing. In general, Hamburgian artefacts were located in kGcn kA/E/gg soil horizons developed in the river-bed facies (KOWALKOWSKI & MYCIELSKA-DOWGIAŁŁO 1983). Soil horizons were disturbed by small frost fissures up to 70 cm deep.

Amber objects and a limestone bead were found in a rich Hamburgian flint artefact concentration, prompting us to associate them with the Hamburgian culture (Fig. 6). These finds were recovered from the top part of the fossil soil horizon (kA/E/gg), 52 cm below the present-day ground surface (Fig. 7).

This assemblage contained two amber ornaments and 6 small amber fragments — possibly waste material from ornament production. The larger of the two ornaments was a reversed icicle-shaped piece of bone-coloured white amber measuring 41 x 14 x

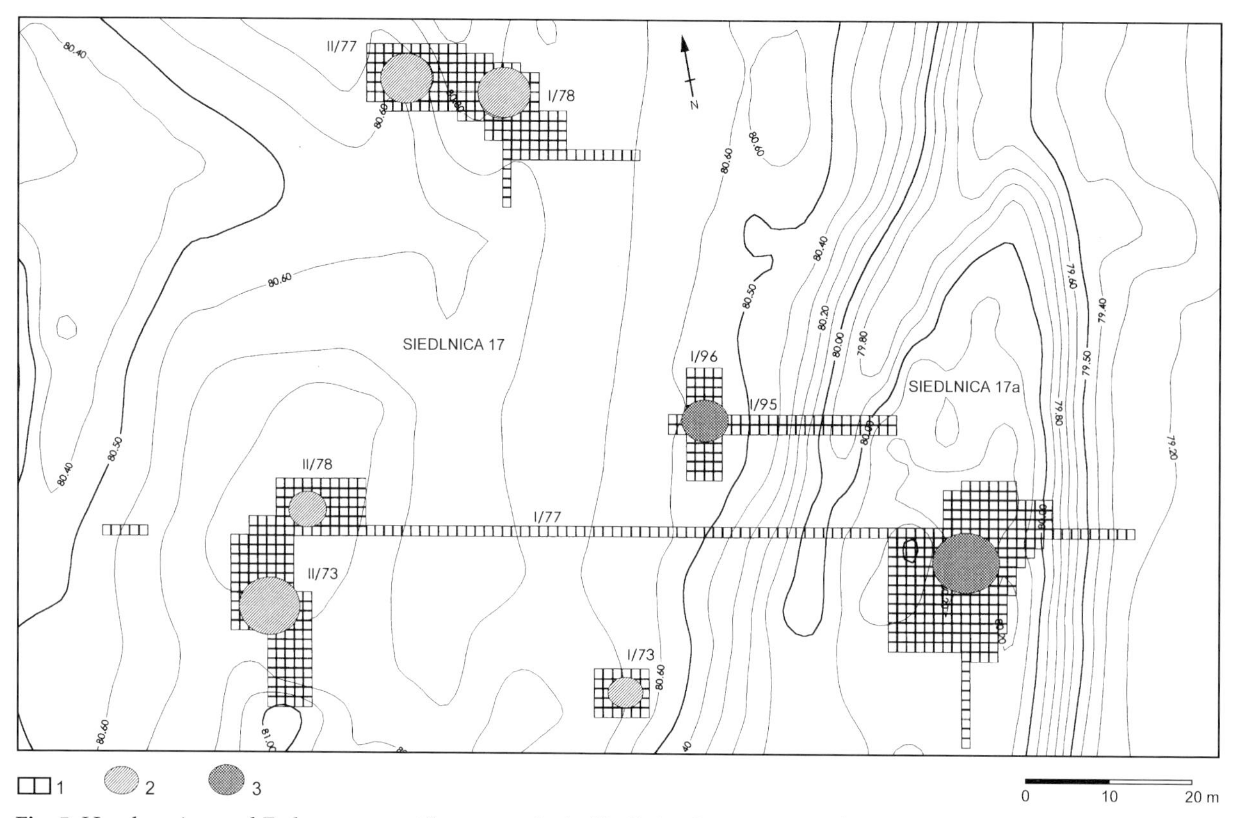

Fig. 5. Hamburgian and Federmesser settlement units in Siedlnica, Leszno voivodeship, SW Poland.

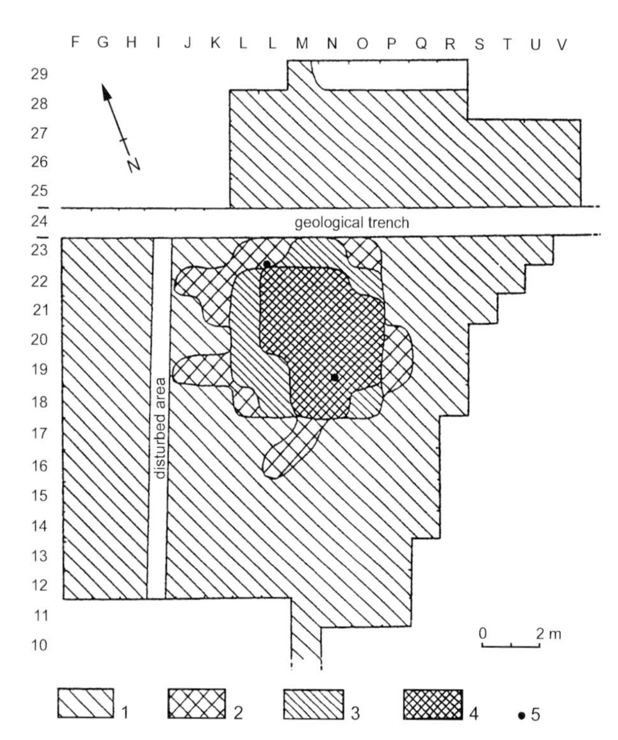

Fig. 6. Hamburgian concentration with amber objects Siedlnica 17a, Leszno voivodeship, SW Poland. Frequency distribution of the flint artefacts per one square meter: 1 — 1–24; 2 — 25–49; 3 — 50–99; 4 — < 99; 5 — amber objects.

Fig. 7. Amber ornaments in situ at Siedlnica 17a, Leszno voivodeship, SW Poland. Photo. R. SIERKA.

6.5 mm (Fig. 8). Its surface was covered by a light brown external cortex, with a red, strongly oxidised internal cortex. The nucleus of the object was not oxidised. The colour of the amber objects was so similar to the colour of ironpan that it was very difficult to recognise them. One of them was slightly damaged during excavation, making it distinguishable as amber.

The edges and the surface of the ornament carry traces of processing in the form of grinding and polishing. Only a part of the edge in the proximal fragment still shows the natural impressions of wood. According to R. MAZUROWSKI from the Institute of Archaeology, Warsaw University, a so-called subcortical slab was used in the production of this ornament (MAZUROWSKI 1983).

Later, natural post-depositional processes caused the piece to crack and split into five separate fragments, which refit well. This cracking was probably caused by frost action. The post-depositional nature of the cracking is indicated by the sharp edges of the cracked surfaces, where the cortex is much thinner than on the original surface of the object (Fig. 8). The artefact was probably a pendant, attached by some other material to a leather strap.

Fig. 9. Amber ornament, roughly shaped in the form of a parallelogram from Siedlnica 17a, Leszno voivodeship, SW Poland. Photo. R. SIERKA.

Fig. 8. Icicle-shaped amber ornament from Siedlnica 17, Leszno voivodeship, SW Poland. Photo. R. SIERKA.

The second ornament from Siedlnica 17a was found in juxtaposition to the first one (Fig. 7). It consisted of a roughly parallelogram-shaped piece, measuring 21.5 x 14.8 x 8.6 mm (Fig. 9). The original colour of this transparent amber artefact cannot be established as it is very badly weathered. An internal red cortex is present beneath the 2 mm-thick, light-brown, external cortex. The nucleus of the object is not visible, but is probably not oxidised.

The edges of the object display traces of grinding and polishing. One edge was slightly damaged during the Hamburgian period, as indicated by the thickness of the cortex in the flaw, which is the same as elsewhere (Fig. 9). Similarly to the other ornament from Siedlnica, this one may have been a pendant, although neither had had holes bored in them. Their shapes, however, suggest such a function.

Six small, sharp-edged and completely patinated amber fragments, found 4 m to the south, were associated with these ornaments. They are so oxidised that the variety of the amber is impossible to ascertain. These pieces might be production waste or fragments of natural nodules. Poor preservation and the small dimensions of these pieces preclude a more detailed interpretation of amber processing at the Hamburgian site of Siedlnica 17a.

The amber artefacts from Siedlnica 17a were associated with a small limestone bead — the only known example of such a find in the Hamburgian culture. The bead comprises a 1 mm-thick, oval-shaped, unifacially convex limestone disc (specific gravity 2.43) with a relatively thin edge and a maximum diameter of 7.6 mm (Fig. 10). The material was analysed by specialists from the University of Wrocław: T. CZYŻEWSKA from the Department of Palaeontology and A. GRODZICKI from the Department of Petrography. The bead was shaped by careful grinding and polishing of its flat surface. The convex side, with a depression in the middle, is a natural surface (Fig. 10). The 0.7 mm hole may have been made either by drilling or, more likely, by grinding the flat surface until the depression became visible from the other side. It is also possible, however, that the depression has a natural origin, as suggested by traces of unspecified circular structures of similar diameter to that of the depression (Fig. 10).

Beads very similar to this one, both in size and technology, are known from younger Magdalenian sites (LEROI-GOURHAN & BRÉZILLON 1966; BOSINSKI 1969). A comparable bead made from amber was found at the Magdalenian site of Isturitz in the western Pyrenees (DELPORTE 1981).

Amber from other SPT sites

Other finds of Baltic amber at SPT sites are known from Deimern 41 near Soltau in Lower Saxony (TROMNAU 1975) and Ureterp I in Holland (BOHMERS 1947). One amber nodule from Ureterp I bears traces of cutting. The westernmost discovery of an amber nodule was made at the Creswellian site of Gough's Cave (Fig. 1; BECK 1965). This nodule was accompanied by two pierced teeth of *Vulpes vulpes* which were most probably used as pendants (CAMPBELL 1977).

The Federmesser site Weitsche in Lower Saxony

The most recent amber finds are known from the APT (Federmesser) site of Weitsche near Lüchow-Dannenberg in Lower Saxony, Germany (Fig. 1; VEIL & BREEST 1995; VEIL 1997). During field prospection of a sandy valley in Weitsche archaeologists found a collection of APT artefacts and a single oval, amber bead, the chronology and cultural affinity of which were, however, not verified (VEIL & BREEST 1995, 29). In 1994 a fragment of worked amber was found at the Weitsche 16 site. Further amber pieces were recovered during excavation. Together, these made up an single artefact which had, unfortunately, been crushed and scattered by ploughing (Fig. 11).

After collecting the broken pieces (*c.* 20 specimens) it was possible to reassemble part of an animal figurine with rhomboid engravings (Fig. 12). It was made from amber slab, now covered with a 1.5 mm-thick, reddish-brown cortex. Fresh amber is only visible in recently broken edges (VEIL & BREEST 1995, 35).

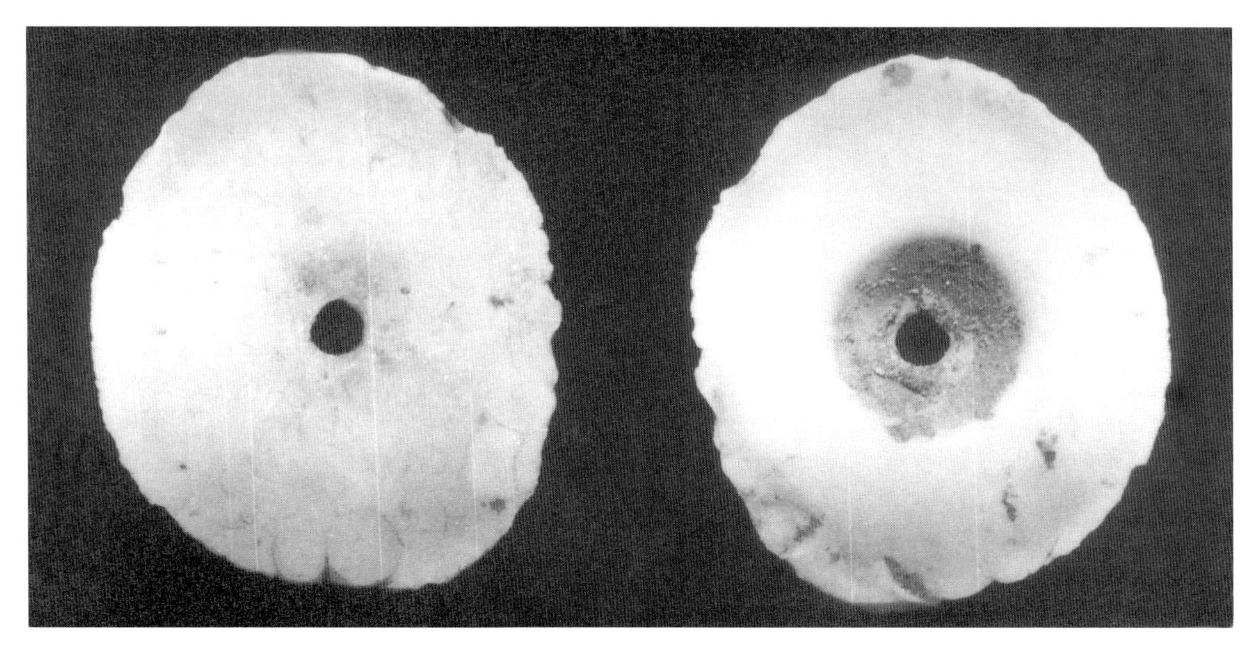

Fig. 10. Limestone bead from Siedlnica 17a, Leszno voivodeship, SW Poland. Photo. R. SIERKA.

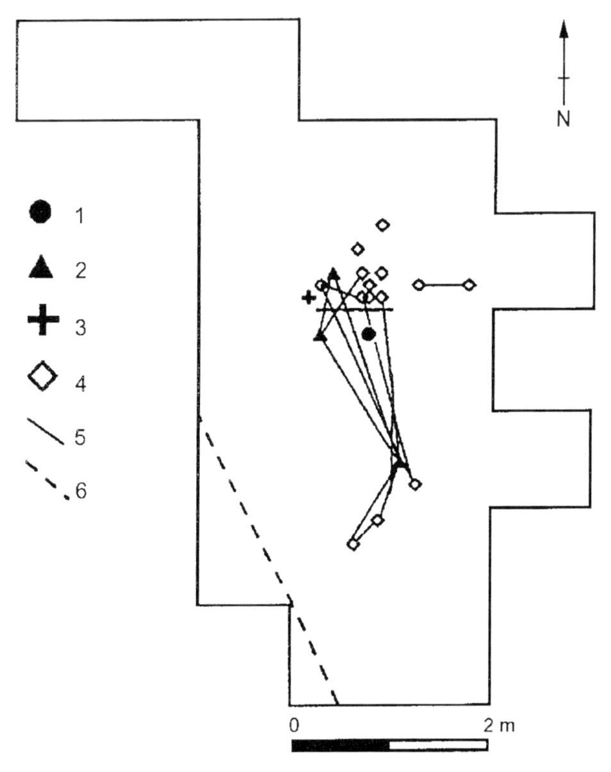

Fig. 11. Trench 1994 with amber finds at Weitsche 16, Lower Saxony (after VEIL & BREEST 1995): 1 — oval fragment (leg's connection); 2 — body of animal figurine; 3 — decorated fragment; 4 — amber chips; 5 — replacement direction; 6 — plough direction.

The figurine fragments (83 x 56 x 15 mm) represent the body and hind legs of an animal (Fig. 12, 1; VEIL 1997). The forelegs (still missing — their possible position is indicated by a dotted line on Fig. 12, 1) and hind legs were connected by a pointed, oval piece of amber as in the famous example of an amber horse figurine (115 x 85 x 30 mm) from Dobiegniew in Western Pomerania (Fig. 12, 2; MATTHES & SCHULENBURG 1881; VIRCHOW 1984). In preliminary interpretations it was seen as a figurine of a large herbivore without any detailed identification (VEIL & BREEST 1995; VEIL 1997). Both figurines are shaped in very similar fashion,however they represent different animals. Their ornamentation is also different (Fig. 12).

An important question is the chronology of the amber figurine and amber bead from Weitsche. The amber fragments were found in a ploughing horizon of the site. Are they really connected with Federmesser artefacts or are they much younger loose finds? The area of the site was carefully examined and excavated by S. VEIL and K. BREEST (1995) which established that the only Federmesser artefacts were present, calcified bone fragments and a hearth (VEIL 1997). The authors believe that they all belong to a Federmesser occupation.

If the Weitsche figurine is associated with the Late Palaeolithic then this possibility cannot be excluded for the Dobiegniew figurine, usually interpreted as being of Mesolithic or Neolithic date (KOZŁOWSKI 1972, 190; MAZUROWSKI 1883). Numerous Late Palaeolithic sites have recently been found in the Dobiegniew Lake District, and thus the Palaeolithic age of the Dobiegniew amber figurine deserves consideration (BAGNIEWSKI 1994).

Geometric motifs are common at Ukrainian Palaeolithic sites. However, similar motifs also appear on Maglemosian amber animal figurines (horse, bear and wild boar), which are mostly known from the northern European Mesolithic (see KOZŁOWSKI 1972, 190–192; VEIL & BREEST 1995, 40; VEIL 1997). Late Palaeolithic sites in the European Plain were also frequently visited by Mesolithic groups. Thus, the Mesolithic origin of the Weitsche amber figurine cannot be excluded.

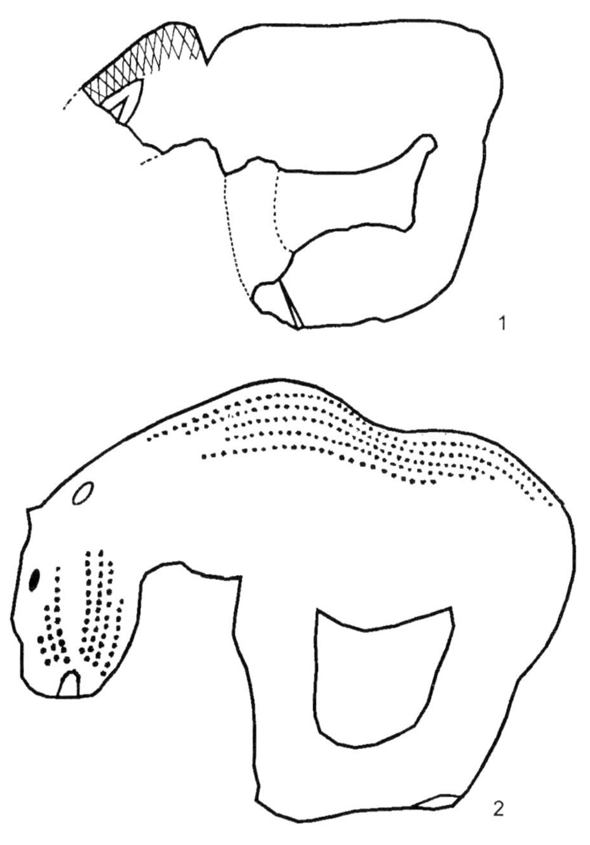

Fig. 12. Fragment of animal figurine in amber from Weitsche 16, Lower Saxony (after VEIL 1997) and horse figurine from Dobiegniew, Western Pomerania (after KOZŁOWSKI 1972). Scale 1:1.5.

Natural amber nodules have been found at two other ATP (Federmesser) sites. The first of these was Hainholz-Esinger Moor A (5 specimens) in Schleswig-Holstein, northern Germany (BOKELMANN 1983, 205), the other being Witów (several pieces up to 20 mm in diameter), located near Łęczyca in central Poland. CHMIELEWSKA (1978, 167) believes that it is a local amber.

Final remarks

Current knowledge on amber usage and processing during the Palaeolithic period is still very limited. Only isolated and very dispersed finds, which can be described as individual objects, have been found thus far. Amber was probably used much more widely but the preservation of such artefacts is generally poor or they are very difficult to recognise. Crude amber is available in large areas of Europe (ROTTLÄNDER 1973, map; SREBRODOLSKIJ 1980) and may have been exploited on the spot. Amber occurs in numerous locations in the north European moraines (MAZUROWSKI 1980). One cannot exclude, however, that the amber was brought to sites lying in the southern part of the European Plain (like Siedlnica) from the shores of the Baltic Ice lake by migrating Late Palaeolithic hunters and gatherers (Fig. 1).

Research into identifying the origin of amber has only been carried out on a rare number of occasions. For example, G. W. BECK (1965) proved the Baltic origin of an amber nodule recovered from Gough's Cave in Wales. Other finds have not undergone the same type of analysis. Another dilemma is how to distinguish amber from the Baltic shore from local amber from moraine outcrops if they are both of similar origin.

The function of Late Palaeolithic amber artefacts has already been briefly discussed. There is no doubt that they were used as decorations. However, our knowledge about the Late Palaeolithic in Northern Europe is very limited in this field. They were probably also used as amulets or other objects of religious significance. Even in this case, however, an unambiguous interpretation is impossible because such artefacts are very rare and not usually found in their original context.

Jan Michał BURDUKIEWICZ
Uniwersytet Wrocławski
Katedra Archeologii
ul. Szewska 48
50-139 Wrocław, Poland

BURSZTYN W PÓŹNYM PALEOLICIE EUROPY PÓŁNOCNEJ

Jan Michał BURDUKIEWICZ

Streszczenie

Początki obróbki bursztynu w Europie Północnej sięgają około 13 000 lat temu i wiążą się z technokompleksem z jednozadziorcami (kultura hamburska i creswellska). Są one nieco młodsze lub równoległe chronologicznie do wyrobów bursztynowych kultury mezyńskiej na Ukrainie.

Do najwcześniejszych znalezisk należy płytka bursztynowa z wytopiska martwego lodu w Meiendorf (Szlezwik-Holsztyn), której towarzyszyły dwie naturalne bryłki bursztynu. Płytka z dobrze zachowanego sukcynitu o kształcie zbliżonym do trapezu (56 x 46 mm) miała przewiercony otwór w środkowej części i ślady obróbki krawędzi. Na płaskich powierzchniach występują ślady licznych linii rytych.

Dalsze wyroby bursztynowe zostały odkryte w bogatym skupisku artefaktów kultury hamburskiej w Siedlnicy 17a, na pograniczu Śląska i Wielkopolski. Wśród nich wyróżnić można dwie ozdoby w postaci płytek bursztynowych oraz 6 drobnych ułamków bursztynu, prawdopodobnie odpadów produkcyjnych. Większa z płytek miała kształt soplowaty (41 x 14 x 6,5 mm) i wykonana została z sukcynitu odmiany kościanej o barwie białej. Druga ozdoba z Siedlnicy 17a jest płytką o kształcie równoległoboku (21,5 x 14,8 x 5,6 mm). Została ona wykonana z sukcynitu odmiany przezroczystej o nieokreślonej barwie. Powierzchnie krawędzi płytek noszą ślady szlifowania i gładzenia. Zapewne były używane jako zawieszki, mimo że nie miały one otworów.

Kolejne znaleziska bursztynu pochodzą ze stanowisk hamburskich z Deimern 41 (Dolna Saksonia), Ureterp I (Holandia) i stanowiska creswellskiego Gough's Cave (Walia).

Ostatnio paciorek i fragment płytki bursztynowej (83 x 56 x 15 mm) zostały znalezione na stanowisku technokompleksu z tylczakami łukowymi (Federmesser) w Weitsche (Dolna Saksonia). Płytka przedstawia duże zwierzę roślinożerne, trudne do bliższej identyfikacji, które wykonane zostało w podobny sposób jak bursztynowa figurka konia z Dobiegniewa na Pomorzu. Pochodzi ona z powierzchni i wymaga lepszego powiązania z artefaktami paleolitycznymi.

Znaleziska dowodzą, że od czasu schyłkowego paleolitu bursztyn stał się surowcem do wyrobu przedmiotów pełniących zapewne funkcje estetyczne i religijne (np. amulety) itp.

Bibliography

BAGNIEWSKI Z.
1994 O paleolicie schyłkowym Pomorza w świetle badań na Pojezierzu Dobiegniewskim, *Śląskie Sprawozdania Archeologiczne*, **35**, 445–470.

BECK G. W.
1965 The Origin of the Amber found at Gough's Cave, *Proceedings of the University of Bristol Speleological Society*, **10**, 272–276.

BOHMERS A.
1947 *Jong-Palaeolithicum en Vroeg-Mesolithicum*, [*in*:] *Gedenkboek A. E. van Giffen*, J. A. Bloom & Zoon, Meppel.

BOKELMANN K.
1983 Fundplätze des Spätglazials am Hainholz-Esinger Moor, Kreis Pinneberg, *Offa*, **40**, 199–239.

BOSINSKI G.
1969 Der Magdalénien-Fundplatz Feldkirchen-Gönnersdorf, Kr. Neuwied, *Germania*, **47**, 1–38.

1987 Die große Zeit der Eiszeitjäger, *Jahrbuch des Römisch-Germanischen Zentralmuseums Meinz*, **34**, 3–139.

BRODZIKOWSKI K. & VAN LOON A. J.
1987 Palaeogeographic Development of the Kopanica Valley (Southern Great-Polish Lowland) during the Late Pleistocene and the Holocene, [*in*:] Burdukiewicz J. M. & Kobusiewicz M. (Ed.): Late Glacial in Central Europe. Culture and Environment, *Prace Komisji Archeologicznej PAN – Oddział we Wrocławiu*, **5**, Ossolineum, Wrocław, 215–239.

BURDUKIEWICZ J.M.
1980 Stanowisko kultury hamburskiej Siedlnica 17a, gm. Wschowa, *Śląskie Sprawozdania Archeologiczne*, **22**, 5–11.

1986 *The Late Pleistocene Shouldered Point Assemblages in Western Europe*, E.J. Brill Publishing House, Leiden.

1987 Late Palaeolithic Settlements in the Kopanica Valley, [*in*:] Burdukiewicz J. M. & Kobusiewicz M. (Ed.) *Late Glacial in Central Europe. Culture and Environment, Prace Komisji Archeologicznej PAN – Oddział we Wrocławiu*, **5**, Ossolineum, Wrocław, 183–214.

1989 Le Hambourgien: origine, évolution dans un contexte stratigraphique, paléoclimatique et paléogéographique, *L'Anthropologie* (Paris), **93**, 189–218.

1993 The beginning of settlement of northern Europe after the last glaciation, [*in*:] Pavuk J. (Ed.), *Actes du XIIe Congrès International des Sciences Préhistoriques et Protohistoriques, Bratislava, 1-7 septembre 1991*, **2**, 32–35, Bratislava.

CAMPBELL J. B.
1977 *The Upper Palaeolithic of Britain. A study of Man and nature in the Late Ice Age*, Clarendon Press, Oxford.

CHMIELEWSKA M.
1978 *Późny paleolit pradoliny warszawsko-berlińskiej*, Ossolineum, Wrocław.

DELPORTE H.
1980–1981 La collection Saint-Périr et le Paléolithique d'Isturitz: une aquisition pristigieuse, *Antiquités nationales*, **12/13**, 20–26.

GAMBLE C.
1986 *The Palaeolithic settlement of Europe*, Cambridge University Press, Cambridge.

HUNTLEY B & BIRKS H. J. B.
1983 *An atlas of past and present pollen maps for Europe: 0 – 13.000 years ago*, Cambridge University Press, Cambridge.

KOWALKOWSKI A. & MYCIELSKA-DOWGIAŁŁO E.
1983 The Stratigraphy of Fluvial and Eolian Deposits in the Kopanica Valley based on Sedimentological and Pedological Investigations, *Geologisches Jahrbuch*, **71**, 119–148.

KOZŁOWSKI S. K.
1979 *Pradzieje ziem polskich od IX do V tysiąclecia p.n.e.*, PWN, Warszawa.

LEROI-GOURHA & BRÉZILLON
1966 L'Habitation no 1 de Pincevent près Montereau (Seine-et-Marne), *Galia Préhistoire*, **9**, 263–385.

MATTHES F. & VON SCHULENBURG W.
1881 Geschnitzte Tierfigur aus Bernstein, *Zeitschrift für Ethnologie*, **13**, 297–298.

MAZUROWSKI R.F.
1983 Bursztyn w epoce kamienia na ziemiach polskich, *Materiały starożytne i wczesnośredniowieczne*, **1**, 7–134.

ROTTLÄNDER R. C. A.
1973 Der Bernstein und seine Bedeutung in der Ur- und Frühgeschichte, *Acta Praehistoria et Archaeologica*, **4**, 11–32.

RUST A.
1937 *Das altsteinzeitliche Rentierjäger Lager Meiendorf*, Karl Wachholz Verlag, Neumünster.

1962 *Vor 20 000 Jahren, Rentierjäger der Eiszeit*, Karl Wachholz Verlag, Neumünster.

SCHÜTRUMPF R.
1937 Die paläobotanisch-pollenanalytische Untersuchung, [*in*:] Rust A., *Das altsteinzeitliche Rentierjäger Lager Meiendorf*, Karl Wachholz Verlag, Neumünste, 11–47.

SKUTIL J.
1930 Les trouvailles d'obsidienne et d'ambre dans les stations paléolithiques, *L'Homme préhistorique*, **17**, 100–112.

STREET M., M. BAALES & WENINGER B.
1994 Absolute Chronologie des späten Paläolithikum und Frühmesolithikum im nörslichen Rheinland, *Archäologisches Korrespondenzblatt*, **24**, 1–28.

TROMNAU G.
1975 Die Fundplätze der Hamburger Kultur von Heber und Deimern, Kreis Soltau, *Materialhefte zur Ur- und Frühgeschichte*, **9**.

USINGER H.
1975 Pollenanalytische und stratigraphische Untersuchungen an zwei Spätglazial-Vorkommen in Schleswig-Holstein, *Mitteilungen der Arbeitgemeinschaft Geobotanik in Schleswig-Holstein und Hamburg*, **25**.

VEIL S.
1997 Archäologie als Fortsetzungsroman: Die Entdeckung des Bernsteintieres von Weitsche 1994–1996, *Berichte zur Denkmalpflege in Niedersachsen*, **1**/97, 70–72.

VEIL S. & BREEST K.
1995 Figurinfragmente aus Bernstein vom Federmesser-fundplatz Weitsche bei, Lkrs. Lüchow-Dannenberg (Niedersachsen), *Archäologisches Korrespondenzblatt*, **25**, 29–47.

VIRCHOW R.
1884 Alte Tierfigur aus Bernstein, *Zeitschrift für Ethnologie*, **16**, 566–569.

[PIDOPLICHKO I. G.] ПИДОПЛИЧКО И. Г.
1976 *Межирчские жилища из костей мамонта*, Наукова думка, Киев.

[ROGACHEV A. N. & ANIKOVICH M. W.] РОГАЧЕВ А. Н. & АНИКОВИЧ М. В.
1984 *Поздний палеолит Русской равнины и Крым*, [*in*:] Борисковский П.И. (ed.), *Палеолит СССР*, 162–271, "Наука", Москва.

[SREBRODOLSKIJ B. I.] СРЕБРОДОЛЬСКИЙ Б. И.
1980 *Янтарь Украины*, Наукова думка, Киев.

[SHOVKOPLAS P. G.] ШОВКОПЛЯС П. Г.
1972 Добраничевская стоянка на Киевщине (некоторые итоги исследования), *Палеолит и неолит СССР*, 7.

BERNSTEINFUNDE IM KARPATENBECKEN

Klára MARKOVÁ

Kurzfassung

Fossile Harze waren im Karpatenbecken schon seit Jungpaläolithikum bekannt, aber erst in der Bronzezeit nimmt die Zahl der Bernsteinfunde wesentlich zu. Bernsteinschmuck kommt besonders zahlreich in Gräbern vor. Zwei Hauptverbreitungsrichtungen des Bernsteins wurden ausgesondert: die Donaustrassse und die Theißstrasse.

Die Beziehungen der Kulturträger untereinander sind Bewegungen artigen Arten ihren in qualitativ verschieden an relativ starren archäologischen Quellen erkennbar. Die Bewegungen zeigen sich als direkte Verschiebung von Nomaden-, Hirten- und militärischen Einheiten, als unmittelbaren oder allmählichen Austausch von Erzeugnissen oder Informationen. Wir erkennen außen einem Idealfall diese Bewegungen an der anthropologischen Unterschiedlichkeit der wandernden Bevölkerung, durch Vermittlung von Rohstoff- sowie Fertigerzeugnisimporten.

Im archäologischen Zusammenhang sind Importgegenstände bedeutsam: vom Gesichtspunkt der Parallelisierung und der relativen Chronologie der Kultureinheiten und von der Beurteilung des Charakters der gegenseitigen Beziehungen der Kulturen. Bedeutung können sie auch vom Gesichtspunkt der Bestimmung der gesellschaftlichen und wirtschaftlichen Charakteristik der untersuchten Gesellschaft haben. Die Feststellung der Anwesenheit von Importen erfordet außer der Typologie, in gewissen Fällen, auch eine Bestimmung der Rohstoffquellen des Materials. Dazu gehören auch die Funde fossiler Harze.

Die Region des Karpatenbeckens hielt man im Vergleich zu den nordwesteuropäischen Gebieten lange Zeit für ein Gebiet in dem fossile Harze im archäologischen Milieu gar nicht bzw. nur in geringer Menge vorkommen. Die Ausgrabungen im Gelände und die Begehungen ergaben jedoch zahlreiche Belege und ermöglichten es, sich ein anderes Bild ihrer Verbreitung zu verschaffen.

Die Zahl der Fundstellen mit aus Bernstein angefertigten Gegenständen nimmt im Karpatenbecken zu. Im archäologischen Zusammenhang kommt fossiles Harz im nördlichen Teil des Karpatenbeckens schon im Jungpaläolithikum vor. Die Herkunft der Rohstoffe Nordwestungarischer Funde ist uneinheitlich, obwohl es in einem Falle nicht ausgeschlossen ist, daß sie aus den Baltikum stammen (FÖLDVÁRI 1983; 1992). Unveröffentlicht sind bisher zwei Einzelfunde aus der Südwestslowakei, die auf einen baltischen Ursprung hinweisen.

In den neolithischen Kulturen bilden fossile Harze ebenfalls eine Ausnahme. Der Fund aus dem Brandgrab eines Erwachsenen von Zalužany-Lažňany im Theißtal in der Ostslowakei, das in die Lažňany-Gruppe des Polgár-Bereiches des älteren eneolithikum gehört, wurde bisher nicht analysiert (ŠIŠKA 1972). Die Rohstoffzusammensetzung für die Spaltindustrie verweist auf Fernkontakte (Hornstein aus dem Gebiet Wolhyniens und Bug; Quarzit aus dem Gebiet von Świeciechów und Feuerstein aus Südpolen) und Kontakte mit der Kultur Tripolje IIC/II, bzw. mit der Lublin-Wolhynien-Gruppe der Lengeyel-Kultur.

Zum regelmäßigen Zustrom fossiler Harze in das Karpatenbecken kam es erst während der älteren Bronzezeit. Das hat tiefere Ursachen, die mit den wirtschaftlichen und gesellschaftlichen Bedingungen ihrer Verbreitung zusammenhängen. Darauf verweist auch das Fehlen von Bernstein im expandierenden Kreis der schnurkeramischen Kulturen des Spätäneolithikums im Norden des Karpatenbeckens (Kultur der ostslowakischen Hügelgräber) oder in der Südwestslowakei mit dem Kulturkomplex Kosihy-Čaka, der Glockenbecherkultur, wie auch in der beginnenden Entwicklung der älteren Bronzezeit in der Košťany-Kultur, der Kultur Periamos-Szöreg, Maros, in der älteren Entwicklungsstufen der Vatya-Kultur, Košťany und Nitra-Kultur.

Zusammen mit der Verbreitung der Aunjetitzer-Kultur trifft man auf erste Bernsteinvorkommen. Dafür sprechen die Funde aus dem Umfeld des Hurbanovo-Typs (Nesvady) oder aus dem Aunjetitz-Maďarovce-Horizont (Matúškovo, Branč — MARKOVÁ 1993). Dutzende chemischer Analysen von fossilem Harz in der Slowakei (von Prof. C. W. BECK — Amber Laboratory in Vassar College, Poughkeepsie, NY), Ungarn, Jugoslawien und in Rumänien ergaben, daß es sich um Succinit mit dem typischen Charakter seines IRS (TODD *et al.* 1976; SPRINCZ & BECK 1981; BECK & SPRINCZ 1981; COLTOŞ 1981; BECK & MARKOVÁ 1996) handelt. Auch die typologische Gliederung in Fundgruppen der Bernsteinperlen weist auf einen einheitlichen Ursprung der Funde im Raum des Karpatenbeckens hin. Im Vergleich zu den Ausgangsrohstoffgebieten fehlen im Karpatenbecken die im nördlichen Gebiet vorkommenden Formen, z.B. die Perlen mit „V" Bohrung, viereckige und dreieckige Anhänger (KOŚKO 1979; MAKAROWICZ & CZEBRESZUK 1995; MAZUROWSKI 1983). Unter den Funden aus der Bronzezeit befindet sich eine amorphe Form, etwa ein Rohstoffstück, das gegenwärtig nur durch ein einziges Exemplar vertreten ist (Tiszafüred — KOVÁCS 1975).

Für Polen ist erwähnenswert, daß mit der Verbreitung des Bernsteins in das Karpatenbecken während der älteren Bronzezeit vor allem die Aunjetitzer Kultur nordwärts vordringt. Der Beginn wurde schon in den Proto- und Frühaunjetitzer Entwicklungsstufen festgestellt, aber deutlich erkennbar ist sie erst in den weiterentwickelten Phasen nicht nur in Schlesien. Sie ist auch an der Bildung örtlicher Kulturen in Richtung zu den Rohstoffquellen hin beteiligt.

Eine deutliche Zunahme von Bernsteinerzeugnissen im Karpatenbecken ist an der Wende der älteren und mittleren Bronzezeit (z.B. Majcichov, Sládkovičovo, Jászdózsa, Köttegyán-Gyepespart, Battonya, Szöreg C) zu verzeichnen. Damit hängt offenbar auch der Fund im Grab der Monteoru-Kultur in der Phase Ia hinter dem äußeren Karpatenbecken zusammen. Das Fundmilieu war damals verschiedenartiger als bisher. Weiterhin, wie auch später, kam Bernstein häufiger auf Gräberfeldern vor. Hier lässt es sich am besten feststellen wofür der Bernstein in der zeitgenössischen Tracht verwendet wurde. Perlen nutzte man im Karpatenbecken während der Bronzezeit als Halsketten, weniger als Anhänger oder Haarschmuck. Nur vereinzelt, bisher lediglich im Norden des Karpatenbeckens, wurde Bernstein zur Anfertigung von Schiebern mehrreihiger Halsketten gebraucht (Nesvady, Nižná Myšla).

Im Koszider-Horizont fand man Bernstein am häufigsten in Hortfunden, später sind diese Vorkommen in Depots seltener. Das Depot in der Höhle von Remete deutet A. MOZSOLICS (1988) als eine Verbindung zu dem Kult der Sonnenbewegung. Eine andere Gruppe von Bronzehortfunden dieses Zeithorizontes befindet sich zusammen mit Bernstein in Horten aus Siedlungen. Man fand sie auf den Fußböden oder bei den Hütten, in denen auch weiterer Zierrat untergebracht war (Nitriansky Hrádok, Spišský Štvrtok). In den Siedlungen befand sich Bernstein auch in den Siedlungsschichten (Nižná Myšla, Nitriansky Hrádok) und in den Gruben (Nižná Myšla, etwa Kosziderpadlás III). Eine besondere Art von Bernsteinfunden stellt eine Grube dar, wohl eine Opfergrube, die speziell bezeichnet ist und in der der Tote in ungebräuchlicher Lage zusammen mit weiteren Funden untergebracht war: mit einem Armband, einer Nadel, mit Keramik und Skelettresten von Tieren (Nitriansky Hrádok, Grube 328 — TOČÍK 1978).

Die Kartierung der Funde weist auf zwei Hauptverbreitungsrichtungen des Bernsteins im Karpatenbecken während der älteren und an der Wende der mittleren, daß vor allem die Lagerstätten auf dem Wege über Prosna und mittlere Warta, der Oder entlang benützt wurden und Bronzezeit hin. Ihre Achse bildeten die wichtigsten karpatischen Wasseradern Donau und Theiß. Die Funde verbreiteten sich von Norden bis nach Siebenbürgen und im Süden, bis zum Marosfluß. Wie eine Perle aus baltischen Bernstein aus Sarata Monteoru zeigt, erreichte ein Teil der Funde auch die Gebiete hinter dem östlichen, äusseren Karpatenbogen. In Richtung zu den nördlichen Ursprungslagerstätten läßt die Fundstellendichte vermuten, daß vor allem die Lagerstätte auf dem Wege über mittlere Warta, der Oder entlang benützt wurden und ein Zusammenhang mit der Verbreitung der Aunjetitzer Kultur besteht. Ein Teil der Funde der älteren Bronzezeit reichte bis nach Südmähren und gelangte durch das Burgenland bis nach Transdanubien. Hier erfolgte der Anschluß an die Hauptrichtung — die Donaustrasse (Franzhausen I; Franzhausen II; Gemeinlebarn F; Unterhautzenthal). An die Lagerstätten schloß sich auch die östliche Theißstrasse an. Der Anschluß erfolgte teilweise hinter dem äußeren Karpatenbogen im Norden, teilweise im hügeligen Teil innerhalb des Karpatenbogens, um zur Waag und durch die Pässe der Westkarpaten nach Mähren zu gelangen. Von einem derartigen Verlauf deutet auch der typologisch identische Anhänger aus dem Hortfund bei der Hütte der Mad'arovce-Kultur in Nitriansky Hrádok und aus dem Grab 595 auf dem Gräberfeld Franzhausen I an (NEUGEBÄUER & NEUGEBÄUER 1997).

In der mittleren Bronzezeit weißen die Bernsteinfunde im Karpatenbecken hin, daß es trotz der im Vergleich zu den wenigen Funden im Hügelgräber-Milieu in Westeuropa, zu keiner Unterbrechung der Einfuhr kam. Funde gab es weiterhin auf Gräberfeldern, während in Hortfunden Bernstein fehlt. In Brand- wie auch in Körpergräbern der mitteldanubischen Hügelgräberkultur (Smolenice), der karpatischen Hügelgräber- Kultur (Malá n. Hronom, Detek, Jánoshida) und der Pilinyer Kultur (Radzovce, Zagyvapálva) gab es Bernsteinfunde. Im Südteil des Karpatenbeckens überschreiten die Funde, zum Unterschied von der älteren Bronzezeit, die Flüsse Drau und Save und weiterhin nach Süden zu den Innenhängen des Adria-Gebirges (Jezero, Vrčin). Die Verbreitungsrichtung des Bernsteins in diese Gegend ist in der jüngeren bis späten Bronzezeit noch stärker (Lika-Beznajdač,Predgrad-Debeli Vrh).

In der Urnenfelderzeit blieben die Bernsteinfunde in Richtung nach Siebenbürgen hin seltener (Igriţa — EMÖDI 1980). Bernstein kam abermals, wenn auch in geringerem Umfang als in der älteren Bronzezeit, auch in Hortfunden von Bronzgegenständen vor (Kurd, Vršac-Majdan, Cioclovina). Zu einer wesentlichen Verschiebung in der Verbreitungsrichtung des Bernsteins kam es erst in der Hallstattzeit (HC, HD).

Die Frage der Wechselseitigkeit, bzw. weiterer Austauschobjekte in der angedeuteten Richtung ist von großer Komplexität. Die Beantwortung ist mit weiteren umfangreichen Grabungen verknüpft. Dazu gehört ebenfalls die Möglichkeit eines Zusammenhanges mit der Verbreitung des Goldes und der metallurgischen Quellen.

Bekannterweise verbreitete sich im Karpatenbecken Gold fremden Ursprungs der Gruppe B schon im Äneolithikum. Die Rohstoffquelle A3 — die siebenbürgische, wurde erst in der älteren Bronzezeit genutzt (Nesvady). Wie bereits erwähnt wurde, verbindet man auch die Ausbreitung des Bernsteins im Karpatenbecken erst mit der Verbreitungszeit der Aunjetitzer Kultur ostwärts. In der älteren Bronzezeit fand sich Bernstein südwärts bis zum Maros-Fluß. Die Kompliziertheit ist jedoch durch die Ausnutzungsmöglichkeit auch örtlicher Goldlagerstätten angedeutet, auf welche im Falle der Goldfunde aus Nižná Myšl'a (OLEXA 1996) hingewiesen wurde.

Die Bernsteinfunde im Karpatenbecken aus der Bronzezeit sind Importe, die nachweisbar baltischen Ursprungs sind. Die Verbreitung des Bernsteins in diesem Gebiet, mit allmählich sich ändernden Austauschrichtungen ist eine ausgeprägte und einzigartige Äußerung des komplizierten und vielseitig sich entfaltenden Prozesses der Kulturbeziehungen des ausgedehnten geographischen Raumes und der inneren Welt seiner Nutzer.

Klára MARKOVÁ
Slovenská Akadémia Vied, Archeologický ústav
Akademická 2
949 21 Nitra, Slowakei

ZNALEZISKA BURSZTYNU W KOTLINIE KARPACKIEJ

Klára MARKOVÁ

Streszczenie

Żywice kopalne były znane w kotlinie karpackiej już w kulturach młodopaleolitycznych i starszego eneolitu. Regularny, znaczący dopływ bursztynu bałtyckiego rozpoczął się już we wczesnym okresie epoki brązu jednoczesnym z klasyczną fazą kultury Unietyckiej. Na podstawie źródeł i jednolitej typologii można mówić o dwóch głównych szlakach — rzekami Cisą i Dunajem.

Bursztynowe perły wykorzystywano szczególnie jako ozdobę — były częścią naszyjników albo ozdób włosów, rzadziej służyły jako zawieszki. Tylko w północnej części kotliny karpackiej były wykorzystywane jako rozdzielniki.

Bursztyn znajdowany jest przede wszystkim na terenach pochówków, ale także w skarbach i w obiektach mieszkalnych. Niektóre znaleziska potwierdzają jego zastosowanie i związek z kultem. Do zmian w rozprzestrzenieniu bursztynu doszło w młodszej fazie epoki brązu, a szczególnie w okresie halsztackim.

Literatur

BECK C. W. & MARKOVÁ K.

1996 Finds of Amber in the Carpathian Basin in the Bronze Age, *[in:] Acta of the XIII. International Congress UISPP*, im Druck.

BECK C. W. & SPRINCZ E.

1981 A szegedi Móra Ferencz múzeum bronzkori borostyánykő gyöngyeinek eredete, *Archeológiai Értesítő*, **108**, 206–210.

COLŢOŞ C.
1981 Étude d'échantillons archéologiques d'ambre, *Dacia*, **25**, 193–195.

EMÖDI I.
1980 Necropola de la sfîrşitul epocii bronzului din Peştera Igriţa, *Studii si cercetari de Istorie Veche si archeologie*, **31** (2), 229–273.

FÖLDVÁRI M.
1983 Analysis of an amber bead from Pilismarót-Pálrét, *Folia Archaeologica*, **36**, 39–41.

1992 Analysis of the amber from Mogyorósbánya, *Communicationes Archaeologicae Hungariae*, 16–17.

KOŚKO A.
1979 *Rozwój kulturowy społeczeństw Kujaw w okresach schyłkowego neolitu i wczesnej epoki brązu*, Poznań, 62–63, 169–173.

KOVÁCS T.
1975 *Tumulus Culture Cemeteries of Tiszafüred*, Budapest.

MAKAROWICZ P. & CZEBRESZUK J.
1995 Ze studiów nad tradycją pucharów dzwonowatych w zachodniej strefie niżu Polski, *Folia Praehistorica Polski*, **7**, 100–131.

MARKOVÁ K.
1993 *Bernsteinfunde in der Slowakei während der Bronzezeit*, [*in*:] Beck C. W. & Bouzek J. (eds), *Amber in Archaeology*, Praha, 71–173.

MAZUROWSKI R. F.
1983 Bursztyn w epoce kamienia na ziemiach polskich, *Materiały Starożytne i Wczesnośredniowieczne*, **5**, 7–134.

MOZOLICS A.
1988 Der Bronzefund aus der oberen Remete Höhle, *Acta Archaeologica Academiae Scientiarum Hungariae*, **40**, 27–63.

NEUGEBÄUER C. & NEUGEBÄUER J. W.
1990 *Franzhausen* I, Wien.

OLEXA L.
1996 *Ein paar Bemerkungen zu den Ausgrabungen in Nižná Myšla*, 1–10. Unpublizierter schrifttlichen Text des Vortrages an Kolloquium „Chronologia otomanskej kultury", 28-29.06.1996, Nižná Myšla.

SPRINCZ E. & BECK C. W.
1981 Classifications of the Amber Beads of the Hungarian Bronze Age, *Journal of Field Archaeology*, **8**, 469–485.

ŠIŠKA S.
1972 Gräberfelder der Lažňany-Gruppe in der Slowakei, *Slovenská Archeologia*, **20** (1), 107–175.

TOČÍK A.
1978 *Nitriansky Hrádok*, *AU* Nitra.

TODD J. M., EICHEL M. H. & BECK C. W.
1976 Bronze and Iron Age Amber Artifacts in Croatia and Bosnia-Hercegovina, *Journal of Field Archaeology*, **3**, 107–120.

AMBER IN ANCIENT EGYPT

Andrzej NIWIŃSKI

Abstract

Following the discovery of the tomb of Tutankhamun in 1922, which brought to light some objects containing elements made of a brown brittle material, as well as a quantity of powdered red substance of resinous character, it was postulated that the ancient Egyptians knew real amber. However, chemical analysis has shown that only semi-fossil copal can be considered, being a very rare and imported material.

In November 1922 Howard CARTER unearthed the tomb of TUTANKHAMUN[1], which was probably the most spectacular discovery ever made in Egyptian archaeology. Among the several thousand objects[2] recovered were a number of pieces of jewellery made of gold, lapis lazuli, cornelian and coloured glass paste. A few objects, however, also included other, atypical materials. Most famous in this respect remains the iron dagger, the handle of which is made partly of rock crystal. Several other items of jewellery: a necklace of 60 (actually of only 55) beads, another necklace, one hair-ring and a pair of ear-adornments[3] contained some elements made of a dark brown, brittle amber-like material[4].

These objects were discovered in various parts of the tomb. The ear-rings (Fig. 1) were found in the so-called Treasury: a room situated behind the Sarcophagus Chamber, deposited in a wooden case in the form of the royal cartouche[5], beside some other pieces of jewellery. Another chest (ornamented with ivory and ebony) found in the "Treasury" contained "...one crude string of alternate dark resin and lapis lazuli beads"[6]. Sixty amber-like beads, once belonging to the necklace mentioned earlier, came from the famous casket painted with hunting scenes and depictions of the Pharaoh's triumph over his enemies. This box was found in the Antechamber[7]. "[...] part of large cylindrical bead in dark resin (?)"[8] was also found here, as well as "[...] a ring of resin, or possibly amber"[9]. In addition to all these objects (to which we can further add "two knuckle bones, the knob of a box", and "ear studs"[10]) resin in its crude form was found in several parts of the tomb. These were mentioned in CARTER's handwritten catalogue as "fragments of resin", "fragments of resin (one showing inlay)", "lumps of resin", "scraps of red resin", and "five pieces of red resin tied up in linen"[11]. Red (or amber-like) resin should not be confused with the black variety. Therefore, a finger-ring and the "black resin scarab suspended on gold wire"[12], which were found on the King's mummy, are not considered here.

Thirty-two years after the discovery of the tomb, in 1954, the German scholar Heinrich QUIRING

---

1 Howard CARTER, *Tut-ench-Amun. Ein ägyptisches Königsgrab*, I–III, Leipzig 1924–1934.

2 Helen MURRAY, Mary NUTTALL, *A Handlist to Howard Carter's Catalogue of objects in Tutankhamun's Tomb* (= *Tutankhamun's Tomb Series* I), Oxford 1963.

3 Sinclair HOOD, "Amber in Egypt", in: *Amber in Archaeology*, Prague 1993, 230.

4 Illustrated in publications are only: 1 — the ear-rings (Carter's catalogue no. 269 A.2): Milada VILIMKOVÁ, *Altägyptische Goldschmiedekunst*, Prague 1969, no. 59; Alix WILKINSON, *Ancient Egyptian Jewellery*, London 1971, pl. XLV, and 2 — the necklace of 60 (55) beads (Carter's catalogue no. 21dd): S. HOOD, *op. cit.*, pl. XVI.

5 Carter's catalogue no. 269a.

6 Howard CARTER, *The Tomb of Tutankhamen*, London 1972 (edition by Sphere Books Ltd.),182; probably Carter's catalogue no. 267m (2). Notes from the excavation records kept at the Griffith Institute, Oxford, state that the necklace was composed of sixteen beads of resin and sixteen of lapis lazuli, with one large bead of gold in the centre: cf. S. HOOD, *op. cit.*, 230.

7 Carter's catalogue no. 21dd.

8 Carter's catalogue no. 43q (2); H. MURRAY, M. NUTTALL, *op. cit.*, 2.

9 Carter's catalogue no. 54gg (*idem*, 5).

10 Cf. S. HOOD, *op. cit.*, 230.

11 H. MURRAY, M. NUTTALL, *op. cit* , 1 (no. 1 d; no. 12h), 2 (no. 32s; no. 43q (6), 10 (no. 261 f (8).

12 *Op. cit.*, 9 (no 256q). The Catalogue does not enumerate, evidently by a slip of the pen, another black resin scarab decorated with a Bennubird figure (CARTER, *Tut-ench-Amun* ..., II, pl. 26C) also found on the King's mummy. It would, therefore, be reasonable to propose the "free" number 256ii for this second scarab. In the exhibition opened in the Cairo Egyptian Museum in 1997 these artefacts are well lit and appear very dark brown in colour. S. HOOD (*op. cit.*, 231) designates them as "amber scarabs".

a

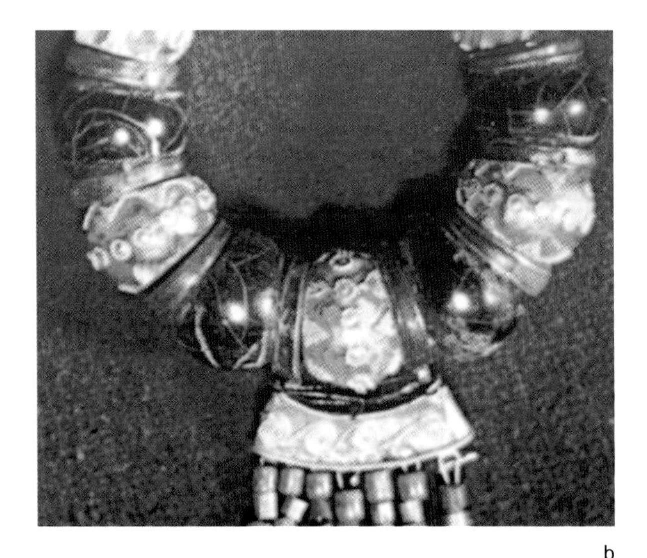

b

Fig. 1. Ear-ring from the tomb of TUTANKHAMUN (a); b — detail. Reproduction from M. VILIMKOVÁ, *Altägyptische Goldschmiedekunst*, Prague 1969, pl. 59.

published an article entitled *Die Herkunft des Bernsteins im Grabe des Tutanchamon*[13], in which, for the first time, the hypothesis that the ancient Egyptians knew amber was put forward *expressis verbis*[14]. Since then the word "amber" has appeared from time to time in publications in "an Egyptian context"[15], and the entry: "Bernstein" exists in the *Lexikon der Ägyptologie*[16]. However — as is stated in the *Lexikon* — this issue seems far from being a proven certainty, at least as regards fossil resin (real amber).

The first objection concerns the aesthetic value of the "burning stone" — this is, after all, the etymological explanation of the word "amber" in some languages (Bernstein, bursztyn). This aesthetic value lies in the colour of amber, varying from yellow, through orange to red or brown, and also in its surface transparency and beautiful shine which may evoke that of the sun (amber is sometimes designated in German by the use of poetic term: Sonnenstein). However, it was cornelian that played the role of a "burning stone" in ancient Egypt, often being used in pieces of jewellery to render the idea of a solar disc (or solar sphere).

Secondly, some amber-like material from TUTANKHAMUN'S tomb was subjected to chemical analysis, and was subsequently declared to be resin[17]. One should, however, repeat after HOOD, that none of the material under discussion has been tested by modern methods[18], and "...all speculation about amber in Egypt

[13] *Forschungen und Fortschritte* 28 (1954), 276–278.

[14] Earlier some individual objects in collections could, however, be described as made of amber: E. VERNIER, *La Bijouterie et la Joaillerie*, Cairo 1907, 26 pl. XVII no. 2; H. R. HALL, *Catalogue of Egyptian scarabs, etc., in the British Museum, I: Royal scarabs*, London 1913, pl. XXIX.

[15] Françoois DAUMAS, "Quelques notes sur l'ambre jaune dans l'ancienne Egypte", *Chronique d'Egypte* 46 (1971), 50–58 („Il s'agit ici de la résine fossile appelée ambre jaune"); Elżbieta MIERZWIŃSKA, *Dzieje bursztynu*, Malbork 1989, 12 (however with a correct observation that resin, and not amber is involved); Sydney AUFRÈRE, *L'univers minéral dans la pensée égyptienne*, Cairo 1991, 591–592; S. HOOD, "Amber in Egypt", [*in*:] *Amber in Archaeology*, Prague 1993, 230–235, pl. XVI ("in this paper I am using amber in the sense of Baltic amber").

[16] Wolfgang HELCK, in *Lexikon der Ägyptologie*, (= *LÄ*), Wiesbaden 1975, vol. I, 710–711.

[17] Alfred LUCAS, "Die Chemie in dem Grabe", in: H. CARTER, *Tut-ench-Amun* vol. II, Leipzig 1927, 206–207, 218; Alexander SCOTT, "Bemerkungen über einige im Grab Tut-ench-Amuns gefundene Stoffe", in: H. CARTER, *op. cit.*, vol. II, 232–233; H. I. PLENDERLEITH, "Bericht über die Untersuchung einiger besonderer Stoffe aus dem Grab des Königs Tut-ench-Amun", in: H. CARTER, *op. cit.*, vol. II, 240.

[18] S. HOOD, *op. cit.*, 230.

must be tentative until adequate tests can be made on the relevant material"[19]. In this situation, the results of investigations carried out by Curt W. BECK at the Amber Research Laboratory of Vassar College on a pendant from Eshnunna (Tell Asmar) seem to be of importance. This object, dated to *c.* 2500–2400 BC, was long believed to be amber. However, testing by infrared spectroscopy has recently determined that the material is copal from a region of East Africa[20].

In one of the corners of the Sarcophagus Chamber, thus just beside the mummy ensemble of TUTANKHAMUN, a double wooden naos was discovered, containing two different powdered substances, enclosed in stone vessels[21]. One of these substances appeared to be natron — the well-known natural soda used in the mummification process — the other being resin of a reddish colour. From the archaelogical context we can assume that this material might also have been used for the same purpose, i.e. for mummification. Chemical analysis has shown that this substance dissolved very easily in alcohol and also, to some extent, in water. When heated up, the powder emitted a smell similar to that of turpentine. This and other observations resulted in the conclusion that the substance was copal which came from Angola or the Sudan[22].

In addition to those objects from the tomb of TUTANKHAMUN, red amber-like material (probably copal) was used in the production of a big scarab inserted in the centre of the wooden, gilded pectoral of a scribe and prince named Ibay (Fig. 2)[23]. A further three small scarabs — one in the British Museum and two others from the collections of University College, London — are mentioned in publications as having been made of amber[24].

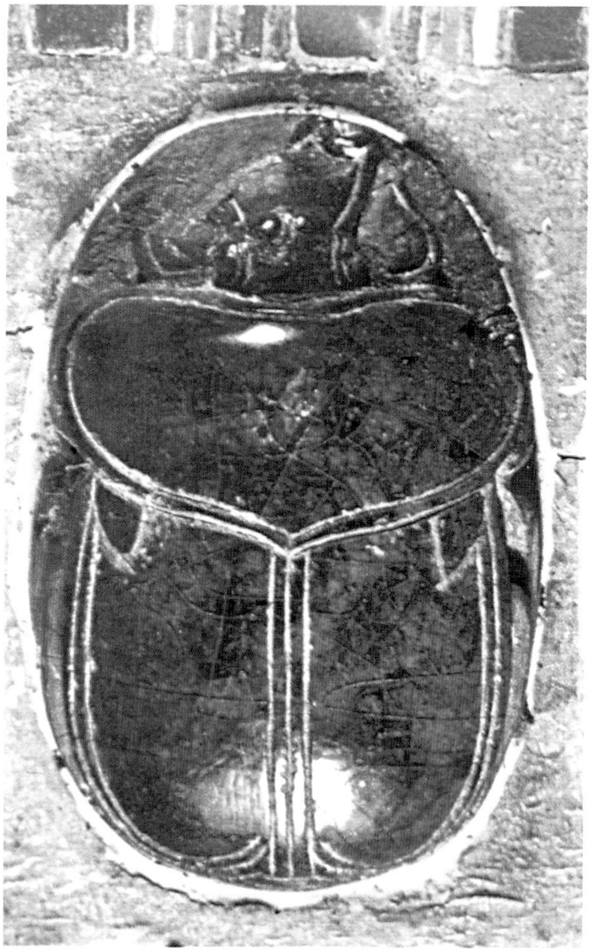

Fig. 2. Scarab in the pectoral of IBAY, Cairo Egyptian Museum, no. CG 53201. Reproduction from M. VILIMKOVÁ, *Altägyptische Goldschmiedekunst*, Prague 1969, pl. 72.

Apart from the problem of identifying the material from which the objects in question are made, another debate concerns textual evidence of the possible occurrence of amber in ancient Egypt. Here, a stimulus was furnished by Pliny's mention of the substance called sacal; from the context it appears to be identical to amber[25].

Undoubtedly, the word sacal can be understood as a derivative of the Egyptian *š/3/kr//š/3/kl*[26] or *s/3/hrt//s/3/hlt*[27], which is, however, the designation of a

[19] *Idem*, 233.

[20] Carol MEYER, Joan MARKLEY TODD and Curt W. BECK, "From Zanzibar to Zagros: a copal pendant from Eshnunna", in: *Journal of Near Eastern Studies* 50 (1991), 289. I am greatly indebted for this information to Dr. Klára MARKOVÁ of Nitra (Slovakia).

[21] Carter's catalogue no. 193: H. CARTER, *op. cit.*, II, 87, pl. 5 and 53B.

[22] H. I. PLENDERLEITH, *op. cit.*, 240.

[23] Cairo Egyptian Museum no. CG 53201, discovered in a tomb at Gurna in 1896. Apart from the discussion of the material, of which the scarab was made ("amber" is voted by E. VERNIER, *op. cit.*, 26; id. *Bijoux et Orfèveries, Catalogue Général des Antiquités Egyptiennes du Musée du Caire*, Cairo 1927, 397–398, and F. DAUMAS, *op. cit.*, 57, while "resin" by M. VILIMKOVÁ, *op. cit.*, 72–73), different opinions are given concerning the date of the object (New Kingdom versus 21st Dynasty, this advocated by DAUMAS), as well as the name of the owner (Hatiay versus Ibay). The present writer shares VILIMKOVÁ's view both of the date (New Kingdom) and the name (Ibay); the word *h3ty-a* seems to be title rather than name.

[24] S. HOOD, *op. cit.*, 230, with notes 3, 4 and 5.

[25] Pliny, *Hist. Nat.* XXXVII, 36.

[26] *Wörterbuch der Aegyptischen Sprache* (= *WB*) vol. IV, 550; about this identification, cf. F. DAUMAS, *op. cit.*, 51–53.

[27] *WB IV*, 208.

kind of resin[28]. This resin was undoubtedly an imported substance. From some sources among texts in Egyptian temples of very late (Greek and Roman) periods we learn that this resin was used in some rituals:

A — In the temple at Dendera a figure of the god SOKAR — patron of the dead — was made of the *s3hl*-resin during the feast of OSIRIS[29]; thus the funerary context of this substance is confirmed here.

B — In Edfu a text mentions a number of objects of some liturgical meaning, also made of the same resin, which is described as "coming from Baharia-oasis"[30].

C — In Kom Ombo the name *s3hrt* is linked with that of the lioness-headed goddess SAKHMET[31], who is associated in various myths with fire, blood and — geographically — with Nubia, from whence, after her mythical damaging activity, the tempered, calmed lioness came back to Egypt.

D — In Dendera another liturgical text says: "I am filling the *udjat*-eye with *s3hr*-resin and am calming this what is inside"[32]. A short explanation is necessary here, as not everybody is familiar with ancient Egyptian mythology: the *udjat*-eye plays a similar symbolic role to the lioness. During the mythical fight between OSIRIS's son HORUS and his terrible uncle SETH, one of HORUS's eyes was wounded, which led to its bleeding and becoming inflamed. However, it was subsequently healed in a magical way, and thus calmed. One should stress that the same word (*s3hri*) also means "to calm"[33]. Any ritual act of filling a mould in the form of an *udjat*-eye with *s3hl*-resin was equal to a symbolic act of victory over (and thus, a calming of) any evil.

Similarly, the heart of the dead subjected to the last judgement had to be calmed in order not to rebel against its owner — informs Chapter 29 of the *Book of the Dead*. In one of the *Books of the Dead*-papyri (known as Busca-papyrus, now in Milan) this text is entitled: "Spell over a heart made of stone *s3hrt*"[34]. The heart was, in the ancient Egyptian art, usually symbolized by a scarab. This hieroglyphic sign meaning: "to become, to come into existence", symbolized the rising sun and all types of revival, the most important of which was, of course, the resurrection of the dead. It is interesting to note that the same text which is found in the Busca-papyrus in connection with the "stone *s3hrt*", is also engraved on the reverse of the scarab inserted in the pectoral of IBAY, discussed above.

All of the earlier quoted texts confirm that the copal *s3hrltl* was probably used for excactly the same purposes in the Egyptian liturgy as was PLINY's "sacal", however, at the same time these texts also hint at its rather rare use[35] and foreign origin. A substance of a similar name — *škr/škl*[36] — is, for instance, listed among the tributes brought to Egypt by foreign countries in the 37th year of the reign of King TUTHMOSIS III (*c.* 1443 BC)[37]. The key question is: where did this mineral, most probably a sub-fossil resin, originate from? There are three possibilities, which do not, however, necessarily exclude each other:

A — Import from the West. The aforementioned inscription at Edfu which refers to the Baharia-oasis can be regarded as supporting this possibility. Baharia might have been only a stop on the way of the *s3hrt* or *škl*-mineral from West Africa[38]. The characteristics of the powdered substance from the tomb of TUTANKHAMUN are similar to those of a copal from Angola.

B — Import from the South. The mythical allusions to the lioness-headed goddess SAKHMET, coming from Nubia and called "mistress of *s3hrt*"[39], may find some support in PLINY's *Historia Naturalis*. This author mentions a theory, according to which "Egyptian amber" would have originated from Ethiopia, where — in accordance with a version of the famous myth — HELIOS's son FAETON crashed his chariot. From then on, FAETON's sisters, who had been turned into trees,

[28] S. AUFRÈRE, "Résine — *s3hrt* et ambre", in: *L'univers minéral dans la pensée égyptienne*, Cairo 1991, 591–592, with notes 5–12. The present writer does not share AUFRÈRE's opinion that the word *shrr* from a late text could have meant amber, in contrast with *s3hrt*-resin; most probably we have only to do with a writing variant of one and the same designation of copal.

[29] S. AUFRÈRE, *op. cit.*, 592: "En effet, il ne fait aucun doute que *s3hrt* est soit une résine d'acacia, soit du copal — en l'absence d'analyse, il est difficile de juger".

[30] *Op. cit.*, 592 and note 7.

[31] *Op. cit.*, 592 and note 8.

[32] *Op. cit.*, 268–269.

[33] *WB IV*, 207.

[34] M. HEERMA VAN VOSS, "Texten over amuletten in de Papyrus Busca", in: *Schrijvend verleden*, Leiden 1983, 290–293; S. AUFRÈRE, *op. cit.*, 592.

[35] The rarity of this amber-like copal must not, however, be indicative of a great value of this material in the eyes of the ancient Egyptians (as is suggested by S. HOOD, *op. cit.*, 230–231). Both the ear-rings and the necklace of 60 beads found in the tomb of Tutankhamun had evidently been rejected by thieves who had considered these as less valuable than other pieces of jewellery deposited in the same chests.

[36] *WB IV*, 550.

[37] Kurt SETHE, *Urkunden des Ägyptischen Altertums. IV. Urkunden der 18. Dynastie*, Berlin 1907, 715 (translation by Karl-Heinz PRIESE, Berlin 1984, 214).

[38] S. AUFRÈRE, *op. cit.*, 592.

[39] *Op. cit.*, 592 and note 8.

wept tears in the form of copal pieces[40]. The Egyptian expeditions, attested to in many sources, to the land of Punt, localised somewhere on the East-African coast (Somalia?) provide a possible historical basis for the hypothesis that copal was an import of southern origin. The regions of Somalia, Tanzania and nearby Zanzibar abound in copal[41]. A number of exotic products were imported to Egypt from Punt, among them incense-gum and myrrh, which are also varieties of resin[42].

C — Import from the North-East. Again, on the linguistic level, one should point to a similarity between the Egyptian words *škl* or *s3hr* and the Mesopotamian word ŠAG.KAL used in a cuneiform letter sent by Tušratta, king of the realm of Mitanni, to his son-in-law Amenhotep IV; some elements of necklaces sent from Mitanni to Egypt were made of a stone: ŠAG.KAL[43]. TUŠRATTA also sent other gifts to his daughter TADUHEPA, among them two daggers, the description of which corresponds to the actual objects found on the mummy of TUTANKHAMUN[44]. Heinrich QUIRING, who may be right in supposing that TADUHEPA was TUTANKHAMUN's mother[45], suggests that some other atypical objects, such as ear-rings containing some elements made of brown resin, might also have originated from Mitanni. The raw material used for producing these elements, according to QUIRING, had earlier been imported to Mitanni from a northern region on the Black Sea and River Dnieper[46].

Whatever can be supposed about the origin of the resin found in Egypt, one should conclude that the semi-fossil copal, which can be identified with the substance called *s3hrt* or *škl* in some Egyptian texts, was a rare and exotic material that never played any essential role in Egyptian imports, when compared for instance with lapis lazuli (imported from Afghanistan).

As to the title of the present paper — *Amber in Ancient Egypt* — we should perhaps regret that Egyptian mythology does not include any allusions to amber, which was, after all, so popular a subject in legends. It is suffcient to mention the one about the tears of FAETON's sisters, or another, Lithuanian legend of an amber palace broken into small pieces by the furious father of enamoured JURATA. Unfortunately, the only legend linking the real (fossil) amber with Egypt seems to be the modern myth which claims that amber was known by the ancient Egyptians and that amber objects have been found in Egypt.

Andrzej NIWIŃSKI
ul. Zagłoby 35 m. 1
02-495 Warszawa, Poland

BURSZTYN W STAROŻYTNYM EGIPCIE

Andrzej NIWIŃSKI

Streszczenie

Po odkryciu w roku 1922 grobowca TUTANKHAMONA, w którym znaleziono kilka przedmiotów zawierających elementy wykonane z brązowego kruchego surowca, a także pewną ilość sproszkowanej czerwonej substancji o cechach żywicy, pojawiła się hipoteza, że starożytni Egipcjanie znali bursztyn.

Analiza chemiczna wykazała jednak, że może chodzić jedynie o kopal, będący rzadkim produktem importu. Owa substancja prawdopodobnie była nazywana w Egipcie *s3hr/t/* (czyt. "saher/et/" lub "sahel/et/") albo *škr* (czyt. "szeker/szekel") i te słowa zawierają podobieństwo zarówno do babilońskiego ŠAG.KAL jak wspomnianego przez Pliniusza w *Historii naturalnej* słowa "sacal".

[40] PLINY, *Hist. Nat.* XXXVII, § 31–33.

[41] David A. GRIMALDI, *Amber. Window to the Past.*, New York 1996.

[42] W. HELCK, "Harze", [*in:*] *LÄ II*, 1022–23; S. AUFRÈRE, *op. cit.*, 591.

[43] H. QUIRING, *op. cit.*, 277; *LÄ I*, 710–711.

[44] H. QUIRING, *op. cit.*, 278.

[45] H. QUIRING, *op. cit.*, 278.

[46] H. QUIRING, *op. cit.*, 278; S. AUFRERE, *op. cit.*, 591.

EXPLOITATION AND WORKING OF AMBER DURING THE LATE NEOLITHIC PERIOD IN THE ŻUŁAWY REGION

Ryszard F. MAZUROWSKI

Abstract

In this article the author deals with the issue of amber exploitation, working and the development of exchange networks by groups of people considered to belong to the Rzucewo culture, who inhabited the coastal areas of the Bay of Gdańsk, the Vistula Bay and Vistula Estuary during the second half of the 3rd millennium BC.

At that time, a sudden increase arose in the demand for succinite among the Late Neolithic communities of the Globular Amphora culture, the Funnel Beaker culture and, in particular, the Złota culture in southern Poland. This led to the intensive exploitation of this raw material, not only on the Baltic coast, but also in the lowland territories of Żuławy, which had recently dried out after the waters of the Vistula Bay had receded in a north-easterly direction.

In this area — during the summertime — amber was collected from the surface or extracted, then worked and taken to permanent settlements. Numerous traces of Neolithic amber-workers' seasonal settlements have survived to this day in the areas of Niedźwiedziówka, Wybicko and Stare Babki (Elbląg voivodeship). Large numbers of amber-processing workshops as well as archaeological evidence of their inhabitants' everyday life have been discovered there.

The Institute of Archaeology at Warsaw University, the State Sites & Monuments Protection Office in Elbląg, and even Malbork Castle Museum have been conducting intensive archaeological research on settlement transformation in the Younger Stone Age, i.e. during the Neolithic Period, within the areas of the Elbląg Upland and the mouth of the River Vistula since the beginning of the 1980s (JAGODZIŃSKI 1987; MAZUROWSKI 1980; 1983a; 1983b; 1984; 1987a; 1987b; 1996a; 1996b). During the Late Neolithic (2500–1700 years BC), the coastal areas of the Bay of Gdańsk and the Vistula Estuary were inhabited by groups of people considered to belong to the Rzucewo culture (a name derived from the village of Rzucewo, near Puck, where one of the largest settlements of its type was discovered). They raised their settlements close to the seashore, which was of particular strategic significance, as both in the past as nowadays, the largest amounts of amber (succinite) and other fossil resins (gedano-succinite, gedanite, beckerite, stantienite and, very rarely, glessite) could be found there. These resins were washed up by the sea during heavy storms. The author believes that the Rzucewo culture developed as a result of conflict between various groups of people from inland territories who attempted to gain control of the strategically and economically important region encompassing the whole "amber-bearing" coastal zone of the Bay of Gdańsk, the Vistula Estuary and Kurski Gulf. The specific nature of this settlement pattern helped command the coastline. Extensive settlements, which remained in use for several hundred years, were located on the edges of the seashore in such a way that the surrounding coastline was visible over a stretch of several kilometres. In between the large permanent settlement areas, temporary settlements or camps were added, mainly for the purpose of realizing a variety of specific tasks. The most famous permanent settlements which have been discovered to date are located in Rzucewo near Puck, Gdynia, Modrzewina near Elbląg, Suchacz, Tolkmicko, Święty Kamień, Przylesie near Tolkmicko, Narusa and Garbina near Frombork.

The remains of numerous rectangular houses, mostly of post-built construction with wattle-and-daub infilled walls, were discovered at these permanent settlements. The houses were about 8–9 m long and not more than 5 m wide. They were divided into several rooms, each of which served different purposes. Small underground stores were situated below the houses, where large pots containing a variety of foodstuffs were kept. The site at Suchacz is a typical example of one of these settlements. Excavations had already taken place here during the inter-war period. A huge number of the features discovered indicated that the basic subsistence economy of this settlement's inhabitants consisted of fishing, hunting and gathering. Crop cultivation and animal husbandry appeared in the late phases of this culture's development, although both forms of farming had already been introduced in the remaining areas of Poland by the mid 5th millennium BC and constituted basic food procurement strategies. Thus, the extraordinary prosperity of the large and stable settlements of the Rzucewo culture, whose economic model continued as a valid tradition among hunter-gatherers, is rather surprising in this context. The reasons for the wealth and distribution of Rzucewo settlements (restricted, as they were, to a narrow belt along the Bay of Gdańsk) should be sought in the huge importance of amber. This extremely popular raw material was mainly acquired by collecting it from beaches after heavy storms. Some indications also point to the possibility that succinite may have been mined with the help of simple tools, especially in open parts of cliffs and on dunes and meadows.

Evidence of amber-working was found at all permanent settlements in the form of workshops, semi-finished and complete products, waste material and crude amber rejected during the selection process. Hoards of decorative amber goods, containing from several dozen up to more than one hundred items, were found both in and around the houses at Suchacz. Furthermore, the few burials discovered at Rzucewo culture settlements, both of adults and children, were furnished with very lavish necklaces which included amber ornaments. Burying amber hoards in an area where this fossil resin was a readily available natural resource, indicates that it was a very highly valued and prestigious material, which could be bartered for products which were difficult to obtain in the coastal region. Evidence that some permanent settlements (e.g. Suchacz) were completely destroyed by fire clearly suggests that visiting groups of people who lived further inland did not always carry out peaceful trade negotiations, or perhaps did not supply a sufficient quantity of products to constitute the equivalent value of the amber they wished to acquire. Thus, attacks and robberies would have taken place, which further explains the presence of the aforementioned hoards of amber ornaments at Suchacz (MAZUROWSKI, 1987b).

Many groups of Rzucewo peoples also collected amber from the dune areas of the Kurski Bar and probably the Vistula Bar. This is evidenced by the remains of numerous seasonal settlements, especially at the Kurski Bar (KILIAN 1955; MAZUROWSKI 1983b), and by the famous collection of finds from Juodkrante as well as the TYSZKIEWICZ collection from Palanga (KLEBS 1882; ŠTURMS 1953; 1956; LOZE 1975; MAZUROWSKI 1983b). This raw material was collected from the seashore. The people who put together these assemblages provided for themselves by fishing as well as hunting for seals and other animals. After a certain time they would have returned to their permanent settlements situated, for instance, in the Elbląg and Gdańsk Uplands.

During the second half of the 3rd millennium BC a sudden increase arose in the demand for amber in southern Poland and in the areas to the south of the Carpathian Mountains, as well as across the whole of the European Plain, where it was impossible to obtain this material in large amounts from local sources. Rzucewo culture groups, maintaining reciprocal trade relations with people from these territories, started to seek out new regions for amber extraction. Thus, they undertook exploration expeditions into the area of the Vistula Lowlands, which had become accessible due to a rapid increase in the prevalent temperature, which led to a drop in the existing water table. According to all available data, this took place during the second phase of the Subboreal period (SB_2) and in parallel with the littoral transgression along the whole Baltic coast.

The western coastline of the Vistula Bay receded considerably east. The area which it covers at present — near Ostaszewo, Niedźwiedziówka, Wybicko, Stare Babki, and Wiśniówka near Nowy Dwór Gdański — constituted a zone of coastal swamps and waterlogged meadows with numerous marginal lakes in lower-lying points, and small, sandy mounds measuring 0.5–1.0 m high. Seasonal settlements used for amber collection and processing were generally located on these mounds. The largest number of these camps was located near the mouth of the Vistula, on the banks of one of its tributaries, which used to run from south to north across the fields near Stare Babki and Niedźwiedziówka, turning east towards Wybicko and Szkarpawa (a tributary of the Vistula extant to this day) (Fig. 1).

Intensive archaeological work has been conducted in this region since 1980. The results obtained to date indicate a world-wide sensation. Along the high, sandy banks of this tributary of the Vistula, a large number of seasonal settlements were discovered with evidence of amber-processing sites

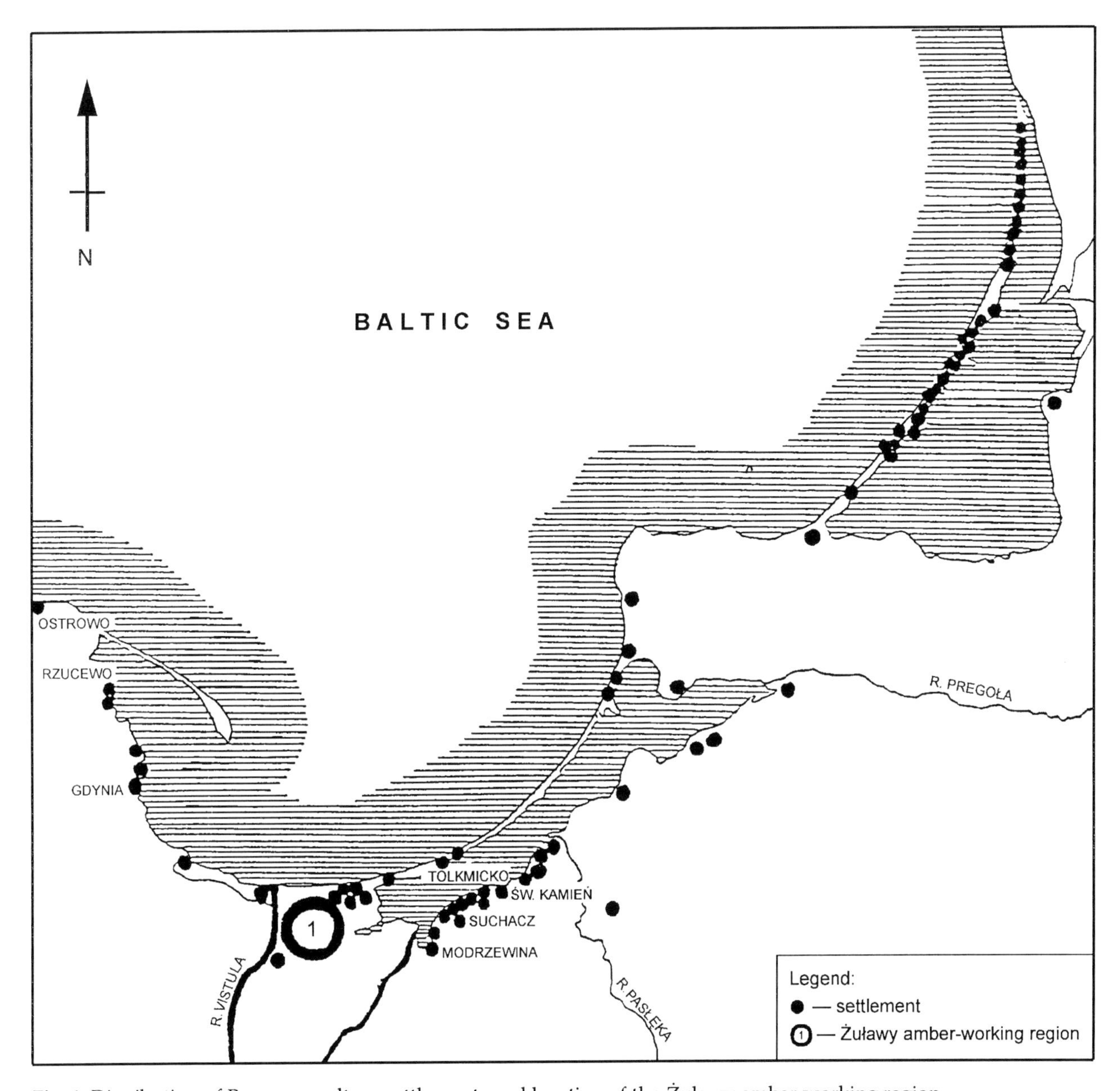

Fig. 1. Distribution of Rzucewo culture settlements and location of the Żuławy amber-working region.

in the form of damaged and semi-finished amber goods, production waste, raw material for processing as well as material rejected during the initial selection process, and a few completed ornaments. These finds were accompanied by stone, flint, bone and antler tools which were used in the amber-working process. The coexistence of amber finds with the remains of hastily produced flint and stone tools which were damaged during production is a frequently recorded circumstance. This is why raw materials such as flint, silica rock and other types of stone are often found at amber-working sites, together with pieces of cortex, chips, flakes, cores (usually decorticated), damaged borers, awls, scrapers, retouched tools, and bone and antler fittings. It is very clear that other forms of production were secondary to the main occupation, i.e. the working of succinite.

Items of everyday use were also recovered from these seasonal camps. These consisted primarily of numerous thick-walled pots ("kitchen utensils") and carefully made forms of ceramic tableware (dishes, beakers, vases, miniature forms), as well as tub-like vessels probably used for lighting rooms. The overwhelming majority of pottery was made at local settlements situated on loamy base sediments from the Vistula Bay, enriched with clay brought to the

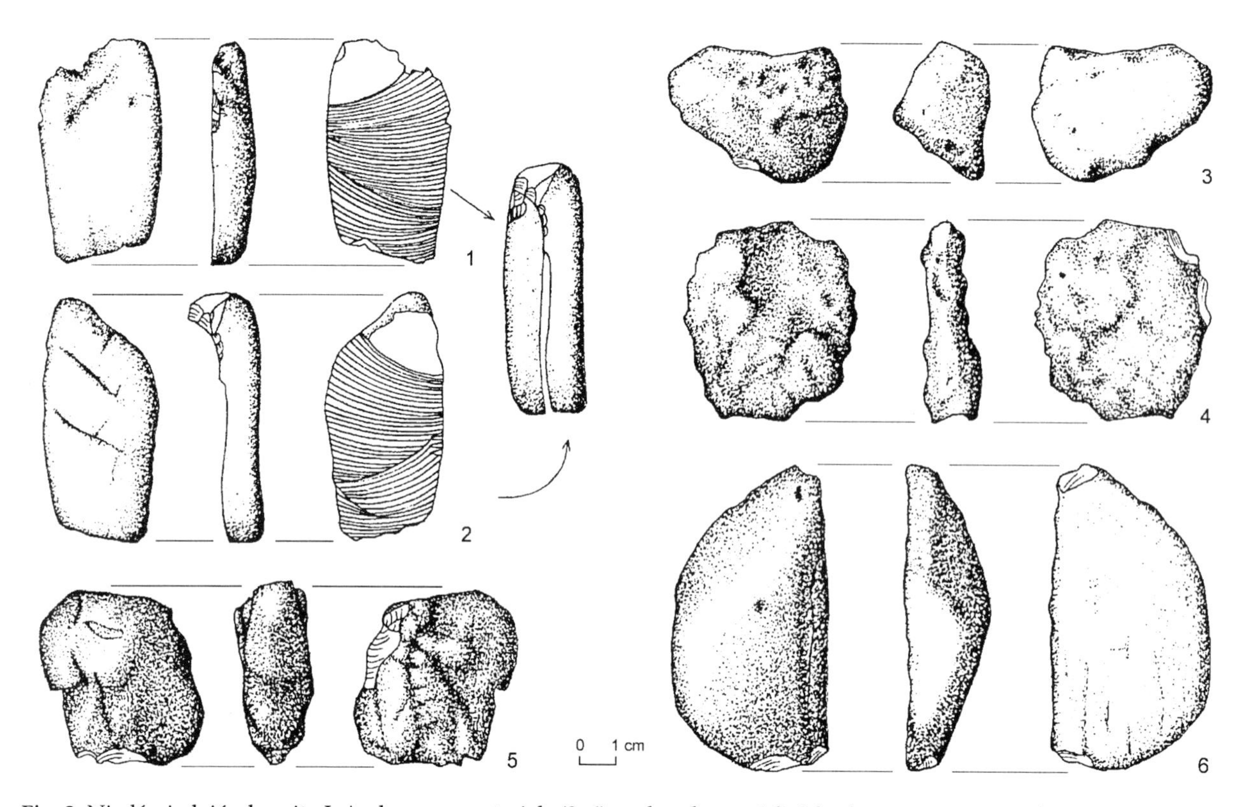

Fig. 2. Niedźwiedziówka, site I. Amber raw materials (3–6) and early semi-finished ornaments (1–2) from deposit.

settlements and medium to large stones. Stone axes and tools for woodworking were also common finds. These were used to make lightweight shelters, which when excavated appear as round spaces in between a huge number of amber, flint, stone, and pottery finds. This may signify that people only used these shelters at night. They spent the days outside, where they produced not only amber ornaments, flint and stone tools, but also prepared their food. This is indicated by the remains of ovens on stone grates or open ovens built directly on the ground and by the large amounts of charcoal in the occupation levels of the settlements. People ate mainly game (deer, roe-deer, rabbit and young seals), and freshwater fish (roach, bream, sander, salmon and pike). Pike were fished in the old river-bed, which also supplied potable water. The remains of several simple foot-bridges leading from the bank to the river-bed further emphasized the important role of the old river-bed in the functioning of these seasonal settlements. Evidence of these bridges came in the form of pairs of black alder wood posts skilfully honed to a sharp end using a stone axe.

Varying concentrations of amber-working sites were located in different zones of those settlements excavated. They frequently overlapped, which may signify that at different periods of time, the same or different groups of people returned to the same places. The author's calculations, which are only an estimate (and most probably too modest an estimate), indicate that, for instance, at Niedźwiedziówka (Stegna parish) alone, there were over 900 amber-working sites in operation within an area of 1 km^2. Over 150,000 amber objects — including 17,000 semi-finished products and complete ornaments — were recovered from excavations conducted at Niedźwiedziówka and Stare Babki. The author believes that several times as many prehistoric amber artefacts have been discovered in this region by local inhabitants and amber-hunters from the 1960s to this day.

The amber-working sites and settlements found so far date from approximately 2500–2200 BC and belonged exclusively to groups of peoples representing the Rzucewo culture. A lack of a differentiation in the pottery and flint products, as well as the very uniform style, morphology and technological execution of the amber artefacts allows us to assume that the occupation period in question was, in fact, much shorter (100–150 years). Conventional radiocarbon (^{14}C) dates obtained so far do not exclude this possibility.

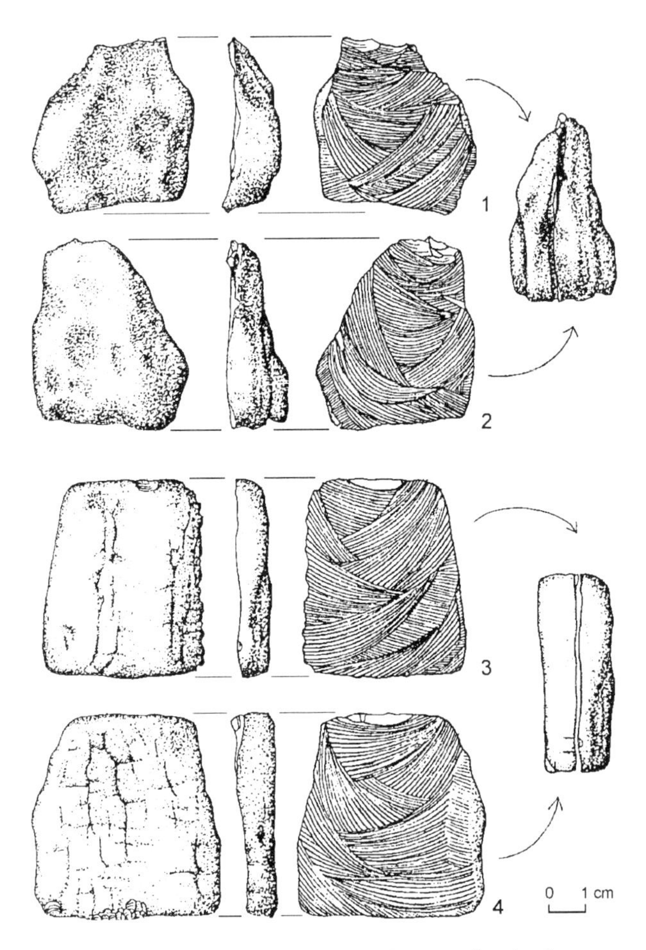

Fig. 3. Niedźwiedziówka, site I. Early semi-finished amber ornaments from deposit.

In the case of site I at Niedźwiedziówka, the author has obtained the following dates: Gd-2767 — 2590 ± 70 BC; Gd-2776 — 2320 ± 90 BC; Gd-5238 — 2190 ± 40 BC; and Gd-5253 — 2620 ± 50 BC.

Generally speaking, the amber-working sites can be divided into two basic types:

a) single use — i.e. those which were made as a result of one person processing amber in a given place on only one occasion;

b) repeated use — i.e. those which were formed as a result of overlapping single use work areas during the functioning of several settlements founded in the same place at different periods of time.

The surfaces of the amber artefacts, from various stages of processing, were well preserved. This, together with the size and concentration of the scatter created at each amber-working site and the associated finds of flint, stone, bone and antler tools, presented us with a unique opportunity to fully reconstruct the process of amber goods production during this period, and assess its efficiency and scale, as well as determine the skills and predispositions of individual craftsmen (MAZUROWSKI 1984; 1987a).

Among the numerous forms of ornaments made at various amber-working sites, by far the most common were single cylindrical beads, both tubular and oval, as well as round, nodular beads each with a V-shaped hole. Less common finds included

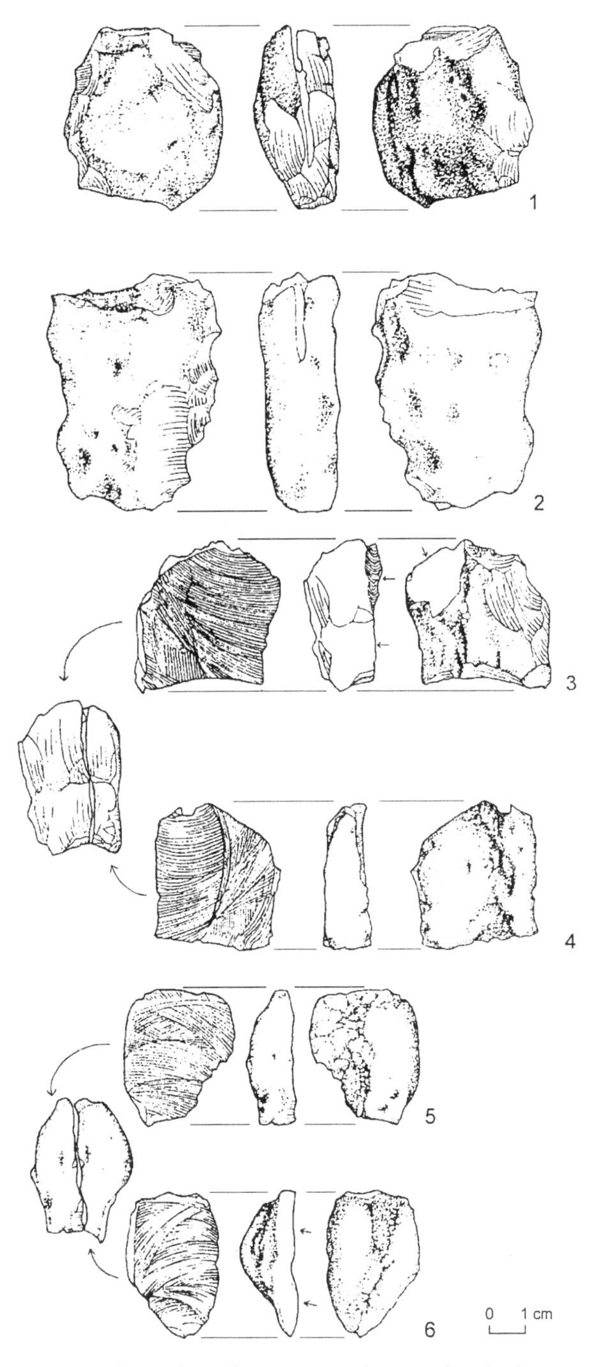

Fig. 4. Niedźwiedziówka, site I. Early semi-finished amber ornaments from deposit.

miniature copies of double-headed axes (laborise, Amazonenaxte), oval or rectangular in shape. Rarer still were trapezoidal, oval, rectangular and asymmetrical ornaments, plain discs and rings, squarish and triangular beads with V-shaped holes, and boat-shaped beads. Extremely rare finds included ornaments with surface decoration consisting of a linear series of hollow dots. These items were mostly so-called "badges" which would have been sewn onto clothes.

The fact that we are dealing with areas where amber was worked is confirmed by the predominance of semi-finished articles representing various stages of the working process which led to the production of finished goods. These do not constitute more than 0.1–0.2% of finds in relation to semi-finished products. The presence of large quantities of production waste and raw materials indicates that ready-made products were taken from the camps to permanent settlements located in the coastal zones of the Gdańsk and Vistula Bays. Amber goods produced at the settlements were mainly made to satisfy the needs of their inhabitants and for local exchange with neighbouring communities within easy reach of Rzucewo sites. The aesthetic tastes of these groups were similar, therefore the products intended for this local market were uniform in design. The same cannot be said of amber goods destined for long-distance exchange with social groups from central and, especially, southern Poland, as well as those living south of the Carpathian mountains, whose cultural traditions were often totally different and led them to prefer different forms of composition and ornamentation. This is why the author believes that large amounts of first-class crude amber and pieces worked into flat, regular shapes which could later be used for the production of very different forms of ornaments, were exported from Rzucewo settlements. This fact is attested to by hoards of amber objects found at Niedźwiedziówka. One of the hoards contained a used necklace consisting of seventeen long, cylindrical, tubular beads. Thus, even within their own community, someone hid their amber possessions under a black alder tree-trunk because they were worried about losing them. This also gives an indication of the value of amber products within the communities which had easiest access to this raw material. Hoards found at Suchacz further confirm these assumptions. In relation to the aforementioned aspects of inter-group relations, three other hoards are of great significance. These consisted of very fine quality crude amber and pieces on which only preliminary working had been carried out. In 1991 a hoard weighing a total of 660 grams was discovered in Niedźwiedziówka. It included the following objects:

1) fine nodules of raw material, particularly subcortical plates (Fig. 2, 3–6);
2) nodules of fine raw material cut with a thread or thong into two or three flat pieces (Fig. 4, 1 & 2);
3) flat rectangular pieces with cut marks (made using thread or thong) on one surface, which can be refitted to form complete initial blocks (Figs. 2, 1 & 2; 3, 1–4; 4, 3–6).

A very important feature of all the objects deposited in the hoard was the presence of one to three negatives impressions showing the interior structure and colour of the raw material on their surfaces. It seems that these early forms of plates were designated for barter. Their shape and size offered the potential recipient the possibility to produce every form of ornament known in the Late Neolithic period in Europe. Unlike ready products, the universal character of early initial forms did not result in a limitation of the range of recipients within the exchange process.

People belonging to the Rzucewo culture acquired mainly agricultural products, meat, and sometimes other raw materials for their amber. The Vistula river played an important role in the development of exchange. It provided a connection between those inhabiting the amber-rich northern region and their potential "clients" based mainly in the Sandomierz Uplands. Amber later became widespread throughout the Złota culture, the Globular Amphora culture and Bell Beaker culture, reaching other regions of southern Poland as well as south of the Carpathian Mountains. The River Vistula was the easiest communication route for relatively safe and quick access to the Bay of Gdańsk by boat or raft. After exchanging various goods for amber, people came back home (MAZUROWSKI 1983b).

Geological and archaeological data indicates that frequent spring and autumn floods took place during the season when Neolithic amber-diggers came to the region of Żuławy. Water from the Vistula Bay damaged settlements. Taking into consideration climatic conditions as well as social and cultural factors, it can be assumed that ancient amber-diggers could stay in this region from late spring till autumn. Some of them returned there several times. They were forced to build new seasonal settlements every time, because their previous camps were damaged by floods. This is why such large quantities of amber and other objects are found during excavations conducted in the old river-bed, especially in its bottom. Further motivation for coming to the Żuławy region during the aforementioned seasons of the year was the possibility of finding large amounts of amber when it was normally thrown out by the sea onto beaches in small amounts.

Amber lay in the lowermost sediments of the retracting Vistula Bay. Large amounts of this material could also be found in areas flooded by water, where amber formed surface sediments which could then be collected in summer. It is still found nowadays in many places in Żuławy during field works, melioration and building works, as well as during the washing of amber by specialized brigades of legally functioning companies and all sorts of unofficial groups. In the case of contemporary washing, this takes place along the whole coast of the Bay of Gdańsk from Port Północny (the Northern Port) in Gdańsk up to Krynica Morska. Information obtained by the author and direct observations indicate that the stratigraphy beneath contemporary dunes, at least between Sobieszewo and Stegna, is the same as that found at the sites excavated near Żuławy Wiślane. Thus, it becomes clear why beads made in the Neolithic and Bronze Age, the Roman and Early Medieval Periods could be washed out together with natural amber. Moreover, these finds are also sometimes accompanied by fragments of pottery, flint and stone products from the Neolithic up to the Early Medieval Period. Identical objects are also thrown up on the beaches of the Bay of Gdańsk. This surely means that in the Late Neolithic, the Bay of Gdańsk must have had a different coastline than at present, and that its dunes are relatively young. This conclusion is of paramount importance to archaeology. It explains the lack of a similar density of settlements in the Vistula Bar to that encountered in the Kurski Bar during the same period of the Neolithic. The author believes that Neolithic amber workshops, analogous to those discovered in the Żuławy region, may lie beneath the present-day dunes. Certain late 19th-century historic sources relating to the village of Stegna confirm this supposition (KLEBS 1882; LA BAUME 1935; KILIAN 1955). On the other hand, discoveries made during the same century of finds associated with the Rzucewo recovered from layers of peat in Wikrowo Wielkie near Elbląg (KLEBS 1882, 46, table XI: 4), as well as the information being collected at present about other neighbouring areas of the Żuławy region, seem to indicate very clearly that the spatial extent of amber exploitation by Late Neolithic communities in Żuławy was vast. Excavations in the amber mesoregion near Niedźwiedziówka suggest that this may have been the largest extraction area, but not the only one.

The economic penetration of the Żuławy region by Rzucewo culture communities took place during a period when amber became hugely popular, not only among Neolithic communities inhabiting Polish territories, but also among those living to the south of the Carpathian Mountains. This was the apogee of amber's popularity in the European Neolithic period. The author has already supported his claims for the existence of long-distance exchange, mainly of crude amber and early initial forms, i.e. nodules cut into flat plates. Such an opinion is also confirmed by comparative analysis of the morphological and technological features of ornaments discovered in settlements, graves, and hoards of the Rzucewo culture in Gdańsk and the Vistula Bay, with finds of amber products in the southern areas mentioned above. Major differences between these finished products exclude the possibility of ready-made ornaments having been exchanged. They suggest that at the end of the Neolithic period, groups of individual cultures worked out styles typical for them, or even canons (composition set-up) of ornamenting the body and clothes with amber. Furthermore, they also worked out a number of secondary and tertiary features of its processing. According to the author, these constitute a very important criterion differentiating the ornaments produced by various cultures — ornaments which have, to date, been considered pan-European (intercultural).

It is doubtless that people of the Rzucewo culture took excellent advantage of the great economic chance offered to them by the huge popularity of amber during the development period of the European Bronze Age, including also the Mediterranean region. At the very end of the Neolithic and the beginning of the Bronze Age, the fact of possessing amber was more important than the type of amber ornament itself. This was, of course, connected with the rapidly increasing real value of succinite. Its high value is evidenced by the following circumstances:

1) A clear attempt to standardize forms of ornaments was made by late Neolithic communities inhabiting the middle and upper Vistula river region and the areas to the south of the Carpathian Mountains. This is visible in the dominance of beads and other ornaments of square, rectangular or round

shapes (see Złota culture, Corded Ware culture, Bell Beaker culture). Such a phenomenon facilitated the most productive use of this precious raw material acquired by means of exchange. It was frequently accompanied by a reduction in the number of categories of mass-produced ornaments.

2) A lack of production waste and raw material which cannot be used for production purposes at the settlements of potential purchasers of succinite in the south. This was connected with the aforementioned endeavour to get the maximum potential use out of this precious raw material. The standardization of forms of ornaments limited its losses to a minimum. Amber dust was frequently the only production waste left after cutting the succinite plates into squares and rectangles. Further processing was, in fact, limited to polishing lateral edges and surfaces. In the case of round and oval shapes, only the corners were polished. Moreover, any small waste pieces and amber powder were probably used for other purposes: ritual (in the form of incense), medicinal, etc.

3) The appearance of natural or drilled amber beads together with bronze and gold products in Early Bronze Age hoards.

4) The presence of gold rings with amber in the Wessex culture in England and the sporadic connection of amber with gold in other regions of Europe during the Early Bronze Age. These circumstances directly indicate the value of amber both in the Early Bronze Age and probably earlier, at the end of the Neolithic.

Cultural transformations at the beginning of the Bronze Age resulted in an increase in the interest in amber in Jutland, *via* which it starts to reach central and southern Europe in large quantities. Conversely, the more humid climate and higher water-table in the Żuławy region at the beginning of the Bronze Age resulted in the necessity to put an end to amber extraction, not only near Niedźwiedziówka, but also in the whole depression zone at the mouth of the River Vistula. This led to a clear decline in the importance of amber to the economies of extant Rzucewo culture communities. It continued, however, to be used as an exchange commodity with peripheral uniethic civilizations.

Ryszard F. MAZUROWSKI
Warsaw University
Institute of Archaeology
ul. Żwirki i Wigury 97/99
02-089 Warszawa, Poland

EKSPLOATACJA I OBRÓBKA BURSZTYNU W OKRESIE PÓŹNEGO NEOLITU NA ŻUŁAWACH

Ryszard F. MAZUROWSKI

Streszczenie

Ogromne rozpowszechnienie bursztynu w drugiej połowie III tysiąclecia p.n.e. na rozległych obszarach Niżu Europejskiego doprowadziło do wzrostu zainteresowania grup ludności różnych kultur późnoneolitycznych na ziemiach polskich strefą pobrzeża Zatoki Gdańskiej, jak również Zalewu Wiślanego. Jest to obszar bardzo bogatego występowania naturalnych zasobów bursztynu (sukcynitu). Zdaniem autora, efektem dążeń do opanowania wspomnianego wybrzeża przez różnorodne kulturowo grupy jest wykształcenie się kultury rzucewskiej. Jej ludność niewątpliwie sprawowała kontrolę nad „bursztynodajnym wybrzeżem" przez okres ponad pięciuset lat. Długotrwałe osady odznaczają się dużą zamożnością mieszkańców, której źródłem był bursztyn.

Zdaniem autora, ludność kultury rzucewskiej osiągnęła bardzo wysoki stopień specjalizacji grupowej w pozyskiwaniu, obróbce, a także wymianie sukcynitu. Początkowo wykorzystywano przede wszystkim surowiec bursztynowy wyrzucany przez fale w okresach jesiennych i wiosennych sztormów. Spadek poziomu wód gruntowych oraz Bałtyku w drugiej fazie klimatu subborealnego (SB_2) stworzył wyjątkową możliwość poszerzenia obszarów bursztynonośnych o tereny obecnych Żuław Wiślanych. Po cofnięciu się Zalewu Wiślanego w kierunku północno-wschodnim stały się one obszarem podmokłych łąk z licznymi odnogami ujścia ówczesnej Wisły. Sytuacja taka istniała przede wszystkim w zachodniej i południowo-zachodniej części strefy przybrzeżnej Zalewu Wiślanego przed jego cofnięciem się .

W odsłoniętych osadach dennych Zalewu Wiślanego, jak również na powierzchni niszczonych wodami powodziowymi powstających łąk występował surowiec bursztynowy, który mógł być łatwo pozyskiwany w okresie letnim. A zatem w czasie, kiedy sukcynit był w bardzo małych ilościach wyrzucany przez wody Bałtyku, powstała możliwość zebrania jego znacznych ilości w osuszonych rejonach Żuław. Okoliczność ta stała się powodem masowych wypraw ludności kultury rzucewskiej, których celem było pozyskanie bursztynu oraz jego obróbka. Śladami tych działań w okresie od 2500 do 2200 p.n.e. są pozostałości przynajmniej kilkuset obozowisk sezonowych, które dokładnie zlokalizowano w Niedźwiedziówce (stan. IV), Starych Babkach (stan. I) i Ostaszewie, jak również w Wybicku. Wyznaczają one wyraźny mikroregion osadniczy, który autor nazwał „niedźwiedziówieckim mikroregionem bursztyniarskim". Jego powierzchnia wynosi około 30 km^2.

Wszystkie obozowiska sezonowe w niedźwiedziówieckim mikroregionie bursztyniarskim usytuowane są na piaszczystych łachach po obydwu stronach koryta jednej z odnóg ujścia Wisły, które w drugiej fazie klimatu subborealnego (SB_2) stało się zastoiskiem. Lokowano tu lekkie szałasowe konstrukcje mieszkalne, po których zachowały się pojedyncze jamy posłupowe, paleniska lub koliste wolne przestrzenie pomiędzy pozostałościami osadniczymi. Liczne pracownie bursztyniarskie zaznaczały się na poziomie ówczesnego gruntu jako koncentracje półwytworów ozdób, ich form finalnych, odpadów produkcyjnych, grudek surowca, jak również okazów o licznych wadach, a także narzędzi do obróbki. Wyraźnie przy tym widać, że procesowi poszukiwania bursztynu i jego obróbki podporządkowane były inne formy aktywności mieszkańców obozowisk. O wielkości obróbki bursztynu świadczy fakt, że w trakcie badań wykopaliskowych od 1981 r. odkryto łącznie ponad 150 000 zabytków bursztynowych, w tym ponad 17 000 półwytworów i form finalnych ozdób. W rejonie Niedźwiedziówki na powierzchni 1 km^2 funkcjonowało ponad 900 pracowni bursztyniarskich.

Pozyskany na Żuławach bursztyn i wykonane z niego ozdoby przenoszone były do osad stałych usytuowanych głównie na Wysoczyźnie Elbląskiej. Stąd zaś drogą wymiany surowiec i wyroby rozchodziły się na obszar środkowej i południowej Polski, jak również na tereny położone na południe od Karpat. Niezwykle istotną rolę w rozwoju ówczesnych kontaktów wymiennych pomiędzy ludnością rzucewską i nabywcami sukcynitu z południa odegrała Wisła. Szereg danych zebranych przez autora wskazuje, że mikroregion niedźwiedziówiecki był tylko jednym z kilku ośrodków na obszarze Żuław Wiślanych, w których pozyskiwano i obrabiano bursztyn w późnym neolicie.

Bibliography

JAGODZIŃSKI M.
1987 Results of tests on amber workshops of the Rzucewo culture in sites No. 1 and 2 in Wybicko, Stegna commune, Elbląg voivodeship, [*in:*] *Archaeological Research in Elbląg Voivodeship in the years 1980–1983*, 121–128, Malbork.

KILIAN L.
1955 *Haffküstenkultur und Ursprung der Balten*, Bonn.

KLEBS R.
1882 *Der Bernsteinschmuck der Steinzeit von der Baggerei bei Schwarzort und anderen Lokalitäten Preussens aus den Sammlungen der Firma Stantien und Becker und der physik.-ökonom. Gesellschaft*, 1–75, Königsberg.

LA BAUME W.
1935 Zur Naturkunde und Kulturgeschichte des Bernsteins, *Schriften der Naturforschenden Gesellschaft in Danzig*, **20** (1), 5–48, Danzig.

LOZE I.
1975 Neolithic amber ornaments in the eastern part of Latvia, *Przegląd Archeologiczny*, **23**, 49–82.

MAZUROWSKI R. F.
1980 History, state and further directions of research on amber working in the Stone and Early Bronze Ages in Europe, *Światowit*, **XXXVI**, 7–32.

1983a Bursztyn, [*in:*] Kozłowski S. K. & Kozłowski J. K. [eds] *Człowiek i środowisko w pradziejach*, 177–188, Warsaw.

1983b Bursztyn w epoce kamienia na ziemiach polskich, *Materiały Starożytne i Wczesnośredniowieczne*, **V**, 7–130.

1984 Amber treatment workshops of the Rzucewo culture in Żuławy, *Przegląd Archeologiczny*, **32**, 5–60.

1987a Badania żuławskiego regionu bursztyniarskiego ludności kultury rzucewskiej. Niedźwiedziówka, stanowiska 1–3, [*in:*] *Badania archeologiczne w woj. elbląskim w latach 1980–1983*, 79–119, Malbork.

1987b Nowe badania nad osadnictwem ludności kultury rzucewskiej w Suchaczu, woj. elbląskie w latach 1980–1983, [*in:*] *Badania archeologiczne w woj. elbląskim w latach 1980–1983*, 141–163, Malbork.

1996a Żuławski region bursztyniarski 4500 lat temu, [*in:*] *AMBERIF 1996*, 13–17, Gdańsk.

1996b Prywatne zbiory zabytków bursztynowych z obozowisk kultury rzucewskiej w Niedźwiedziówce, woj. elbląskie, Elbląg, [*in:*] Nowakowski W. [ed.] *Concordia. Studia ofiarowane Jerzemu Okuliczowi-Kozarynowi w sześćdziesiątą piątą rocznicę urodzin*, 171–182, Warsaw.

ŠTURMS E.
1953 Der Ostbaltische Bernsteinhandel in der vorchristlichen Zeit, *Commentationes Balticae*, **I**, 167ff.

1956 Der Bernsteinschmuck der östlinchen Amphorenkultur, *Rheinische Forschungen zur Vorgeschichte*, **5**, *Documenta Archaeologica Wolfgang La Baume Dedicata*, 13–20.

THE PROCESSING OF AMBER DURING THE MIDDLE NEOLITHIC IN LATVIA

Ilze B. LOZE

Abstract

The processing of amber during the Middle Neolithic in Latvia was investigated through a study of amber ornaments and their precursors in two regions: the Sārnates wetland settlement on the Baltic Sea Coast and the Lake Lubāna Basin in eastern-central Latvia. Previously, this material had been studied without attention being paid to the importance of the processing of amber and production of amber ornaments in the overall contacts between eastern Baltic and other Eastern European Forest Zone Middle Neolithic cultures.

Amber processing workshops

In the Sārnate settlement, studied between 1940 and 1960 by Lūcija VANKINA, an amber workshop in dwelling 2 was described, where amber was processed by a Comb- and Pit-marked Pottery culture. In dwelling 3 of the same culture, large amounts of amber were processed (VANKINA 1970, 112, 114). Attention was also paid to Sārnate pottery settlements, especially dwelling E, where numerous sites used for processing amber were found (VANKINA 1970, 112).

Between 1960 and 1990 many amber processing sites were found in the Lake Lubāna Basin (Fig. 1), including workshops in the Nainiekstes, Sulkas and Zvidzes settlements, where amber was processed by peoples of the Comb- and Pit-marked Pottery and Post-Narva Pottery cultures (LOZE 1988, 43–47, 84–85, 95–96). In the amber workshop in the Zvidzes settlement in dwelling B, a high concentration of amber items was found, including finished and semi-finished products and their fragments, production waste and excess, as well as raw material and amber off-cuts.

Types of amber items

The amber materials found in the Lake Lubāna Basin and Sārnates settlements were classified into types, indicating the production methods used. Amber

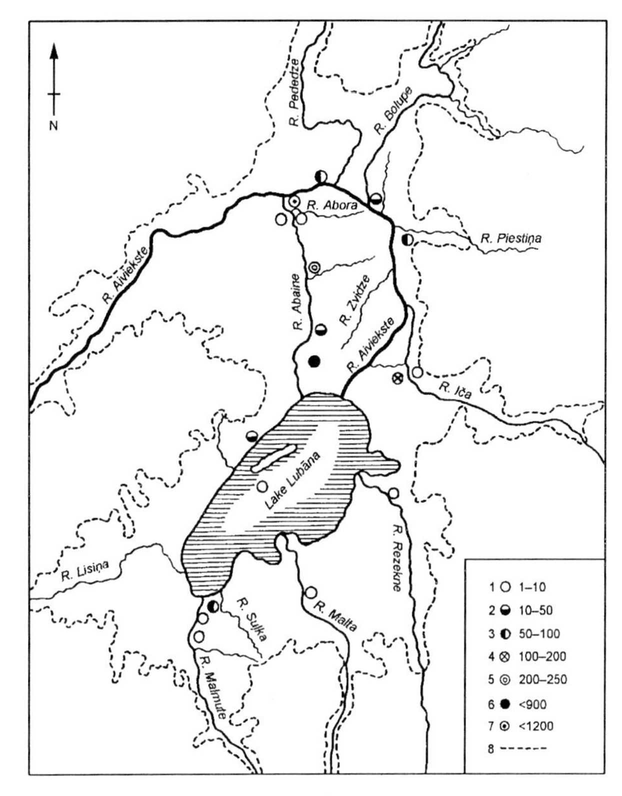

Fig. 1. Number of amber artefacts at Stone Age settlements in the Lake Lubāna Depression: 8 — border of the Lake Lubāna Depression.

ornaments were classified as follows: necklaces, rings, discs and pendants.

Necklaces

Necklace ornaments were subdivided into tubular (Group A), barrel-shaped (Group B), disc-shaped (Group C), globular (Group D) and button-shaped (Group E).

Tubular bead necklaces

Tubular bead necklaces (Group A) dominated in all of the amber production workshops, including dwelling B in the Zvidzes settlement (Fig. 2). They can be divided into long-, medium- and short-length necklaces. The cross-sections of tubular beads are:

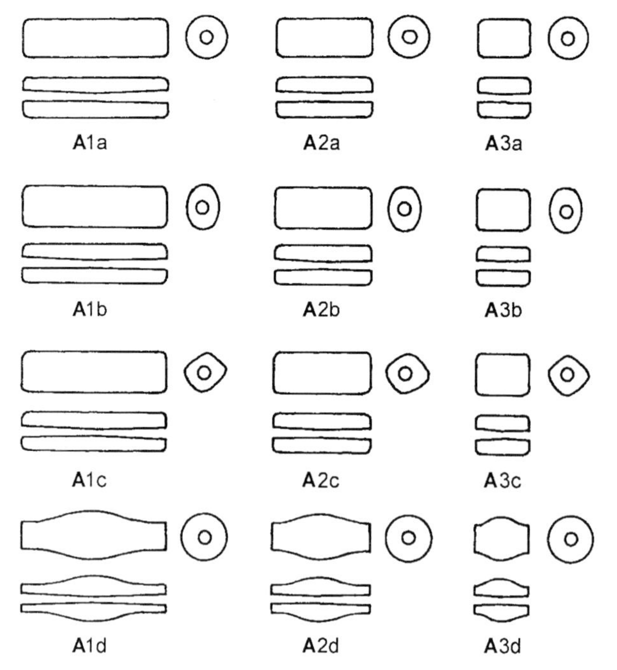

Fig. 2. Classification of amber tubular beads.

(a) cylindrical or round; (b) oval, less common; (c) rounded, square or irregular; and (d) with a broader middle part. The latter are characteristic of the Zvidzes and Nainiekstes amber processing workshops (LOZE 1988, Table XLIII: 2).

There did not seem to be a uniform method for the manufacture of tubular bead necklaces. The Zvidzes material indicated that the most suitable procedures for boring holes in these beads was as follows:

1) holes were bored from both polished ends, before polishing the remaining surface;
2) a hole was bored from a roughly- or finely-cut end, prior to processing the other end;
3) a hole was bored after polishing one end;
4) a hole was bored after polishing the entire piece.

The highest percentage of longitudinal fractures were found in the first three of the above procedures.

Barrel-shaped bead necklaces

Barrel-shaped bead necklaces (Group B) were not common. Two types were found: long, narrow and short, thickened. The best long, narrow pieces were made from transparent and high-quality amber ranging from 2.3 to 2.65 cm in length. Short, thickened beads were made from transparent or white amber.

Disc-shaped bead necklaces

The disc-shaped beads (Group C) produced were flat and thin with rounded edges. The ends were cut straight or, very ocassionally, had rounded ends. Their thickness was 0.2–0.6 cm. The following procedures were used in manufacture:

1) prior to boring a hole, the disc form was produced by coarse- and fine-cutting and polishing;
2) the hole was bored before polishing;
3) naturally-shaped amber pieces not requiring further processing were used.

Globular bead necklaces

Globular bead necklaces (Group D), which were rarely found, had a precise geometric form, with diameters reaching 1.35–1.49 cm. Single finds were recorded from the Sulkas and Zvidzes settlements.

Button-shaped bead necklaces

Button-shaped bead necklaces (Group E) were the second most common necklace type in the Middle Neolithic, including those manufactured at the dwelling B workshop at the Zvidzes settlement (Fig. 3). These beads are round, or ocassionally oval with lens-shaped (E1a, 2a), segmented (E1b, 2b) or conical (E1c, 2c) cross-sections.

The diameters of round, button-shaped beads range from 0.99 to 2.5 cm, with smaller beads being more numerous. The standard thickness is 0.45 cm, less commonly 0.25–0.35 cm.

Partly processed button-shaped beads provide evidence of the manufacturing process. After knapping a piece from an amber core, the surface was finely cut and polished before a hole was bored through the bead. Quite often, the sides were not finely finished. Fairly thin amber pieces were also used without surface finishing, the hole being bored as part of the surface processing.

Some necklaces were found which had oval beads with conical cross-sections (E2c type) and 4 facets dividing the beads into 4 even sectors (Fig. 3).

Discs and rings

Amber disc ornaments (Group F) were large (up to 3–4.5 cm) or small (0.25–2 cm), with triangular, segmented or square cross-sections (Fig. 4).

Ring ornaments (Group G) were found at all of the amber workshops investigated. These were large (3–6.7 cm) or small (1.81–2.7 cm) with widths of 0.2–0.6 cm. The ring ornaments were somewhat rounded, triangular, segmented, segmented with a sharp corner on the inner side, or square.

The disc and ring ornaments may have been manufactured simultaneously. This is suggested by

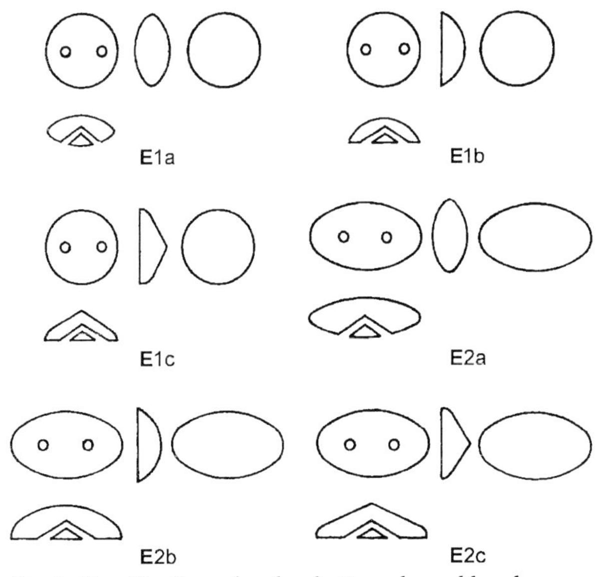

Fig. 3. Classification of amber button-shaped beads.

a partly processed artefact, from which a disc of small diameter has been removed from a ring of larger diameter by boring.

Pendants

Pendant ornaments (Group I) were the second most common amber ornament type in the Middle Neolithic in Latvia. Some of the artefacts found in each workshop were natural amber pieces (one sixth of the pendant ornaments in the Zvidzes settlement).

The predominant form of pendants in the Middle Neolithic was regular trapezoid tablets, rarely with right-angled corners. The pendant ornaments, including those from the Lake Lubāna Basin, have lens or right-angled cross-sections. Their surface forms can be divided into 6 types: (1) rectangles; (2) short and wide trapezoids with curved bases; (3) long and wide trapezoids with somewhat curved bases; (4) trapezoids with straight bases and finely engraved sides; (5) trapezoids with regularly decorated sides and (6) rounded with a broadened base with wide curves (Fig. 5).

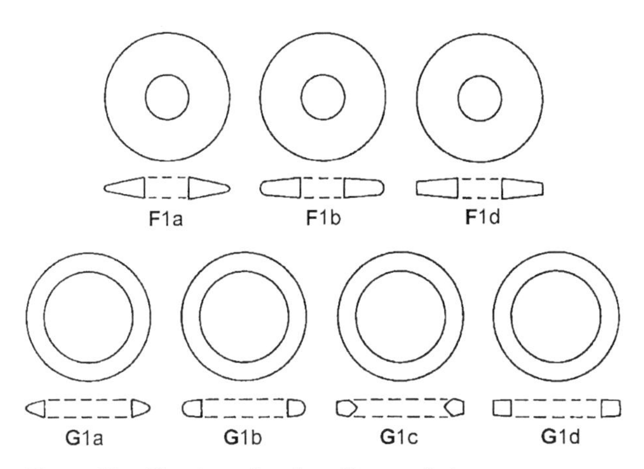

Fig. 4. Classification of amber discs and rings.

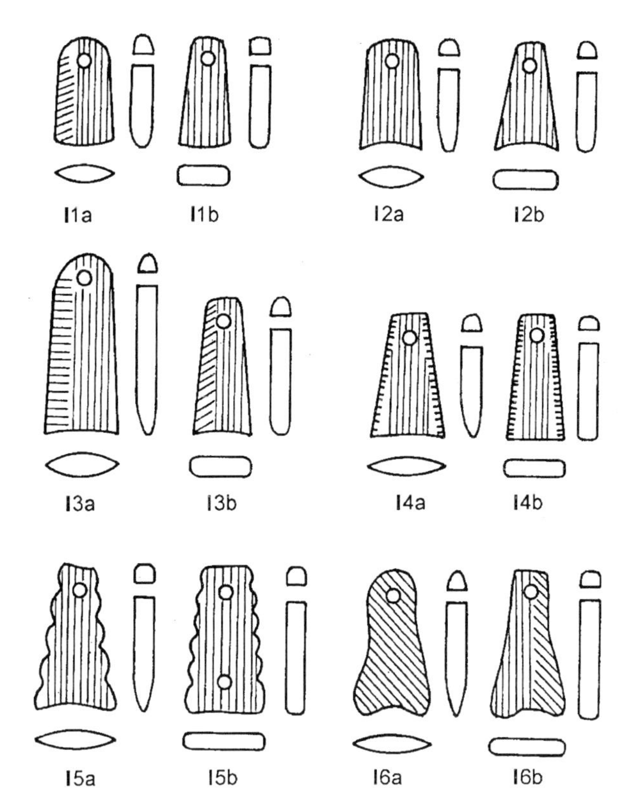

Fig. 5. Classification of amber trapezoid pendants.

The semi-finished trapezoid pendant artefacts mostly had carefully cut sides with finely cut surfaces, sometimes polished. Pendant surfaces and sides were polished vertically, but traces of horizontal polishing are also evident. Holes were bored from one or both sides, depending on the tablet width, after polishing.

Trapezoid pendants outnumber tubular or button-shaped bead necklaces, including at dwelling B of the Zvidzes settlement (Fig. 6) and Nainiekste settlement (Fig. 7). The other type of pendants is represented by oval ones (Fig. 8).

Amber figurines

The amber figurines produced were tablet-shaped or of complex manufacture. The tablet figurines found depict: birds (4 artefacts), including water fowl (Fig. 9); bear and elk (one artefact of each); and some are fragments of anthropomorphic figurines (LOZE 1983, Fig. 9). A relatively realistic figurine of an elk head was found in dwelling 23 of the Sārnates settlement, and a depiction of a human head was recovered from the Babite Lagoon Coast near Riga.

Trading value of amber in the Middle Neolithic

Amber was important in the Middle Neolithic in Latvia and in the remaining eastern Baltic region. It was used as a standard value in transactions between tribes, being exchanged for high-quality raw materi-

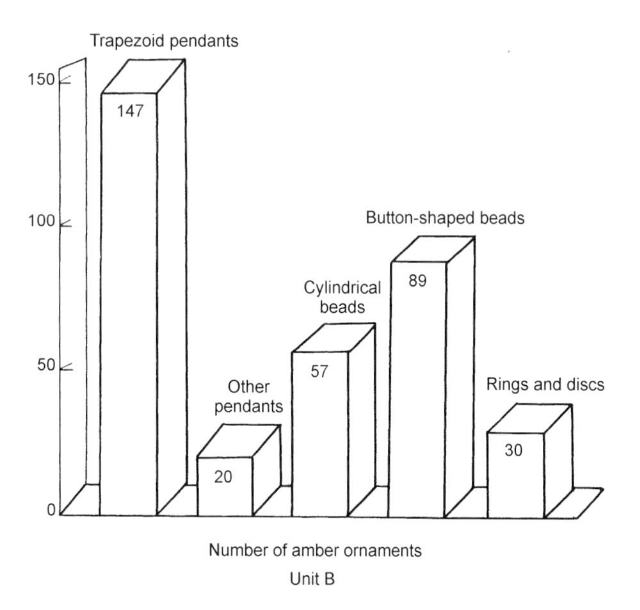

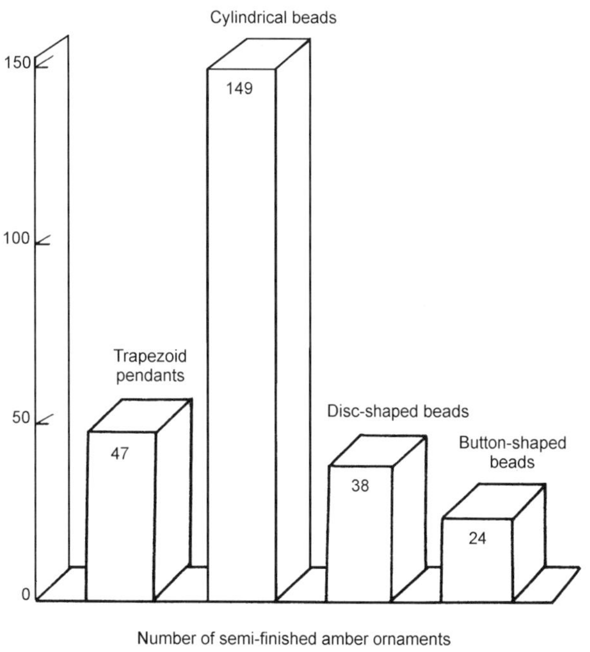

Fig. 6. Amber ornaments found at the Zvidze settlement (Unit B = dwelling B).

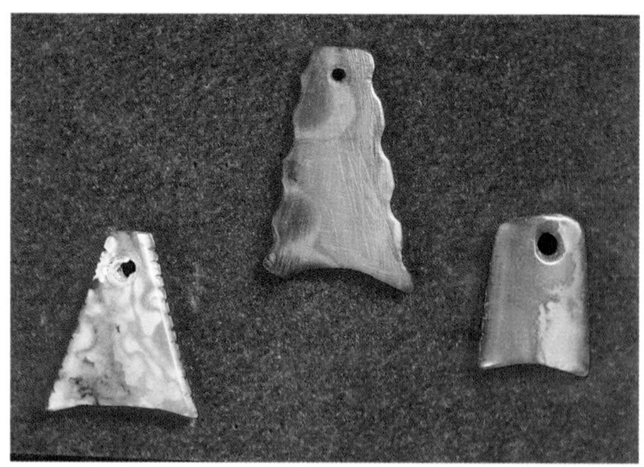

Fig. 7. Trapezoid amber pendants, Nainiekste settlement.

als, such as flint. The amber manufacturing centres in Latvia (Lake Lubāna Basin and the Coastal Sārnates wetland), and in Lithuania (Šventoji Lagoon and Kuršu Dunes) supplied the Eastern European Forest Zone peoples with amber ornaments.

In the Eastern European Forest Zone, amber production is known only from grave good, indicating the importance of Latvian and other Eastern Baltic amber production centres. The presence of amber in graves in the Eastern European Forest Zone shows that the buried individuals had worn these ornaments, but were unable to manufacture the artefacts themselves due to lack of raw material.

The Neolithic amber processing centres of Latvia, especially that situated in the Lake Lubāna Basin, were ideally located near the transport arteries of the Daugava and Volga Rivers and their tributaries, which enabled easy access to large regions. This is verified by maps showing the presence of amber grave goods

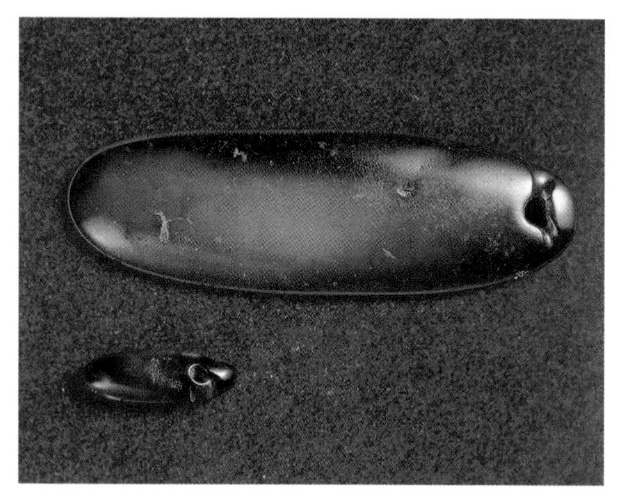

Fig. 8. Oval amber pendants, Dzedziekste settlement.

in 70 burial sites in the Eastern European Forest Zone, including the following regions: the Upper and Middle Volga, the Lower Oka, north-western Russia, and Karelia, the region between the Upper Daugava and the White Sea. In the Novgorod region, 10,000 amber artefacts have been found at the Končanskas burial site in the Msta River Basin (ZIMINA 1993, 222). Also, in the Zalavruga II settlement (Karelian region) on the Lower Viga River, 68 button-shaped bead necklaces were found in a stone-enclosed grave, which were similar to those produced at the Zvidze settlement (SAVVATEYEV 1977, 188–189: Figs 89 & 90). Rich amber artefact assemblages were also found during excavations at the Sahtisas settlement in the Klazmas River Basin on a tributary entering the right bank of the Volga River (KRAINOV 1973, Fig. 4; 1992, Figs 146 & 149).

Fig. 9. Amber bird figurine, Dzedziekste settlement.

In the Middle Neolithic, amber manufacturing centres existed in Latvia, especially in the Lake Lubāna Basin, which supplied amber goods to a wide region. The excellent water transport system ensured both a steady supply of amber from the Baltic Coast, and also enabled easy access for trading the manufactured items. The amber producers at Sārnate and Lubāna developed ornament forms as decorative elements of clothing, and also made zoomorphic and anthropomorphic figurines which were characteristic of the existing ideology of Middle Neolithic peoples.

Ilze B. LOZE
Institute of History
Academican sq. 1
LV-1003 Riga, Latvia

OBRÓBKA BURSZTYNU W ŚRODKOWYM NEOLICIE NA ŁOTWIE

Ilze B. LOZE

Streszczenie

Obróbkę bursztynu w okresie środkowego neolitu na Łotwie dokumentują ozdoby i półwytwory z Sārnate i z osad w basenie jeziora Lubāna. Warsztaty znaleziono w obiekcie mieszkalnym 3 w Sarnate i w obiekcie B w osadzie Zvidze, te ostatnie zawierały najwięcej gotowych i częściowo obrobionych wyrobów.

Bursztynowe ozdoby zostały zakwalifikowane do czterech typów: paciorki, krążki, pierścienie i zawieszki. Paciorki w kształcie rurek i guzków stanowiły typy dominujące, używane do modnych naszyjników. Serie półwytworów i ich fragmenty pozwoliły opisać metody produkcji.

Centra bursztynowej manufaktury na Łotwie, szczególnie w basenie jeziora Lubāna, oraz te z wybrzeża litewskiego dostarczały bursztynowych wyrobów dla zespołów leśnej strefy kulturowej Europy Wschodniej. Wśród znalezisk środkowoneolitycznych z leśnej strefy Europy Wschodniej bursztyn jest znajdowany tylko w wyposażeniu grobów, a nie w postaci surowca. Fakt ten określa znaczenie wschodnich centrów obróbki bursztynu bałtyckiego w handlowej komunikacji międzykulturowej.

Bibliography

KRAINOV D. A.
1973 Settlement and cemetery Sahtish VIII, *Caucasus and Eastern Europe in Ancient times* [in Russian], Moscow, 46–54.
1992 Neolithic and Eneolithic Art of the Russian Plain, *Stone Age Art* [in Russian], Moscow, 68–111.

LOZE I.
1983 *Akmens laikmeta maksla Austrumbaltija* (Stone Age art in the Eastern Baltic), Riga, 1–135.
1988 *Stone Age Settlements in the Lubāna Lowland: Mesolithic, Early and Middle Neolithic* [in Russian], Riga, 1–209.

SAVVATEYEV J. A.
1977 *Zalavruga. The Archaeological Monuments of the Lower Vyig River, Part II* [in Russian], Leningrad, 1–324.

VANKINA L. V.
1970 *The Peat Bog Settlement Sarnate* [in Russian], Riga, 1–176 & 88 Tables.

ZIMINA M. P.
1993 *The Stone Age in the Msta River Basin* [in Russian], Moscow, 1–268.

DIE NEOLITHISCHEN BERNSTEINARTEFAKTE DER BERNSTEIN-SAMMLUNG DER EHEMALIGEN ALBERTUS-UNIVERSITÄT ZU KÖNIGSBERG I. PR.

Siegfried RITZKOWSKI & Gerd WEISGERBER

Kurzfassung

Diese Darstellung gibt eine Übersicht aller 18 Bernsteinartefakte,

- die neolithischer Entstehung sind,
- die der ehemaligen Bernsteinsammlung der Universität Königsberg i. Pr. angehörten,
- die vor den Kriegswirren gerettet und nunmehr im Institut und Museum für Geologie und Paläontologie der Universität Göttingen aufbewahrt werden.

Das Verzeichnis hält die Daten fest, die bislang zu diesen 18 Fundstücken bekannt sind. Es werden die Beschreibungen von KLEBS (1882) für die Objekte mitgeteilt, die ihm seinerzeit vorlagen. Erstmals werden die Objekte mit Schnittzeichnungen abgebildet, so daß eine dreidimensionale Vorstellung ermöglicht wird. Die Zeichnungen geben de Stücke in Originalgröße wieder.

Geschichte der Fundstücke

Als im Jahre 1858 die Firma STANTIEN & BECKER begann, am Grund des Kurischen Haffes bei Schwarzort in Ostpreußen systematisch nach Bernstein zu baggern, fielen Bernsteinstücke an, die bearbeitet waren. KLEBS (1882, 4) berichtet, daß zunächst nur die auffälligen, großen Stücke von den Arbeitern separiert wurden.

Die älteste Sammlung bearbeiteter Stücke hatte der Ingenieur Schmitz in Schwarzort angelegt und diese im Jahre 1867 — so verzeichnet es KLEBS (1882, 4) — seinem späteren Nachfolger Glaubitz gezeigt. Da die Stücke als Merkwürdigkeiten angesehen wurden, wurden sie an bevorzugte Besucher oder an scheidende Beamte zur Erinnerung verschenkt. Auf diese Weise gelangte wohl auch die Nachbildung einer menschlichen Figur in den Besitz eines Herrn REGALL in New York (KLEBS 1882, 4).

Die Bedeutung der Funde erkannte G. BERENDT, der als erster Direktor die Provinzial-Sammlungen der Physikalisch-ökonomischen Gesellschaft zu Königsberg i. Pr. leitete. Er veranlaßte die Firma STANTIEN & BECKER, einzelne Artefakte der Gesellschaft zu schenken. Dennoch, so vermutet KLEBS (1882, 4) „... ging die Mehrzahl der gefundenen Gegenstände verloren". TISCHLER, der die steinzeitlichen Artefakte der kurischen Nehrung untersuchte, konnte später weitere Stücke für das Provinzial-Museum gewinnen.

R. KLEBS, der 1874 als Berater der Firma STANTIEN & BECKER tätig wurde, schreibt (1882, 4–5), daß im Jahre 1879 die Leitung der Firma beschloß, „... alle irgend wie wissenschaftlich interessanten Bernsteinstücke von dem Handel auszuschliessen und sich selber eine Sammlung anzulegen, welcher Alles, was von Bernstein-Einschlüssen, Farbvarietäten und Alterthümern im Geschäftsbetrieb gefunden würde, einverleibt werden sollte." Er bringt zum Ausdruck, daß er seit jener Zeit alles, was in Schwarzort anfiel, zu Gesicht bekommen habe.

KLEBS verfaßt auch 1880 einen Katalog dieser Bernsteinsammlung der Firma STANTIEN & BECKER für eine Ausstellung in Berlin. Er verzeichnet 3.504 Nummern. Unter den Positionen 2.001 bis 2.072 und 2.076 bis 2.078 werden die Bernstein-Artefakte aus dem Alt-Alluvium von Schwarzort aufgeführt, darunter „Altheidnische Götzenbilder aus Bernstein, von der Firma dem Provinzial-Museum zu Königsberg geschenkt" (Pos. 2.076–2.078). Diese fortlaufende Numerierung des Katalogs der Firma STANTIEN & BECKER von 1880 ist später offensichtlich nicht mehr benutzt worden. Die Artefakte tragen jetzt zumeist die Katalog-Nummern des Provinzial-Museums.

Zwei Jahre (1882) später veröffentlicht KLEBS alle Fundstücke, die ihm bislang in die Hände gekommen waren. Diese Monographie erfaßt auf 75 Seiten und 12 Tafeln vornehmlich Fundstücke Ostpreußens, berücksichtigt aber auch Funde anderer Regionen. Bis zu diesem Zeitpunkt lagen KLEBS insgesamt 434 Artefakte vor. Die Tafelerläuterungen verzeichnen 206 Stücke.

Die Darstellung der Artefakte durch KLEBS (1882) ist die einzige ihrer Art geblieben. Spätere Autoren griffen auf Informationen und bisweilen auch auf die Abbildungen dieser Veröffentlichung zurück.

In EBERT´s *Reallexikon der Vorgeschichte, Bd. 1* (1924) führt W. La BAUME unter dem Stichwort „Bernstein" die Schwarzorter Funde auf und bildet auf Tafel 133 drei Artefakte aus Schwarzort ab (Fig. c, d, i). In *Band 11* des *Reallexikons* (1927, Stichwort „Bernstein") werden von E. STURM die Tafeln 119 und 120 aus Nachzeichnungen der KLEBS´schen Abbildungen zusammengestellt.

K. ANDRÉE, der von 1915 bis 1945 Direktor der Universitäts-Bernsteinsammlung in Königsberg war (RITZKOWSKI 1995), gibt in seinem Bernstein-Buch von 1937 auf Taf. 15 insgesamt 9 Artefakte der Universitäts-Bernsteinsammlung wieder. Dasselbe Photo wird von ANDRÉE 1951 als Abb. 22 erneut verwendet. Verzeichnisse oder Aufzeichnungen, aus denen hervorginge, wie viele Artefakte letztlich der Königsberger Universitäts-Bernsteinsammlung angehörten, existieren nicht.

Ein Teil der Bernsteinsammlung der Albertus-Universität Königsberg (Pr.) — so lautete die offizielle Bezeichnung der Sammlung (ANDREÉ 1937) — wurde gegen Kriegsende (1944) in einem Kalibergwerk bei Göttingen eingelagert und 1958 — nach Umwegen über Kunstgutlager der Alliierten — dem damaligen Geologisch-Paläontologischen Institut der Universität Göttingen übergeben (RITZKOWSKI 1977; RITZKOWSKI *in* GANZELEWSKI & SLOTTA 1996a). Die übrige Sammlung verblieb in Königsberg und wurde offensichtlich bei den dortigen Kämpfen im April 1945 vernichtet.

In der Sammlung des Instituts und Museums für Geologie und Paläontologie der Universität Göttingen befinden sich insgesamt 18 Artefakte der ehemaligen Königsberger Bernstein-Sammlung, die früher der jungsteinzeitlichen „Schwarzorter Kultur" zugerechnet wurden.

Sechs der bei ANDRÉE 1937, Taf. 15 (b. S. 128) abgebildeteten Stücke sind vorhanden; die übrigen drei Objekte des Photos (rechtes, unteres Viertel) fehlen. Nach Göttingen ist auch die Phallus-Perle (58-004) gelangt, die bei KLEBS 1882, Taf. 8, Fig. 13 abgebildet ist. Ferner existieren acht Bernstein-Anhänger (58-054 bis 58-061), sowie ein Bernstein-Ring (58-070), eine linsenförmige Perle (58-071) und ein Doppel-Knopf (58-005), die in der Literatur vor 1945 nicht erwähnt sind.

Eine Darstellung von 13 der 18 Stücke, die in der Bernsteinausstellung im Deutschen Bergbau-Museum Bochum 1996 gezeigt wurden, wurde von RITZKOWSKI im Katalogteil des Buches von GANZELEWSKI & SLOTTA (Hrsg.) *Bernstein — Tränen der Götter* (1996b, 548–552, *Katalog*, Nr. 278–289a mit 5 Photos) gegeben, ein weiteres Photo findet sich bei WEISGERBER 1996 (ebd. S. 416). Eine kritische Bearbeitung des Materials steht bislang aus.

Zu den Bernsteinfunden

In der Archäologie gibt es seit längerem die Vorstellung, daß der baltische Bernstein seit der Jungsteinzeit in immer weiter reichenden Wellen verhandelt wurde. Bis 2.500 v.Chr. soll er hauptsächlich bei den Vorkommen, also den Küsten von Jütland und Baltikum und ihrer unmittelbaren Nachbarschaft benutzt worden sein. Danach erreichte er während der Kupfer- und frühen Bronzezeit bis ca. 1900 v.Chr. West-, Mittel- und Osteuropa von der Normandie bis zur Ostgrenze des heutigen Polens, um während der Bronzezeit um 1300 v.Chr. schließlich bis England, Frankreich, zu den Alpenländern und nach Rußland und erst später darüber hinaus zu gelangen. Dies ist aber eine sehr allgemeine Vorstellung, denn es sind gerade die Ausreißer aus diesem Schema, welche belegen, daß man Bernstein bereits sehr früh als etwas Besonderes ansah, das sich aufzuheben, zu erstehen oder weithin mitzunehmen lohnte.

Zu Beginn des Neolithikums war die Nutzung von Bernstein in Mitteleuropa, etwa der Bandkeramik, extrem selten. Vor einigen Jahren wurde aber in Erkelenz/Kückelshoven ein 13 m tiefer bandkeramischer Brunnen entdeckt, der wegen der guten Erhaltung seiner Hölzer dendrochronologisch auf die Zeit um 5.070 v.Chr. datiert werden konnte. Auf seiner Sohle wurde ein einzelner Bernsteinanhänger aufgefunden, bei dem das erste Bohrloch ausgebrochen und die Bruchstelle überschliffen worden war (WEISGERBER 1996). Er zeigt, daß Bernstein gelegentlich seinen Weg auch nach Südwesten fand, auch wenn Schmuck daraus damals noch nicht groß in Mode war.

Aber bereits vorher, im Mesolithikum Nordeuropas, spielt Bernstein erstmals eine größere Rolle, besonders in Dänemark. Spektakulär sind die jetzt auftauchenden Anhänger. Diese können geometrische Formen haben oder als Teildarstellungen von Tieren auftreten. Einige Tierfigürchen weisen keine Ösen zum Hängen auf. Allen ist eine weitgehende Abstraktion der Formen auf die Grundzüge eigen, so daß sie fast modern anmuten. Oft sind die Oberflächen durch Strich- oder Punktmuster gestaltet. Eine Farbauswahl des Bernsteins hat nicht stattgefunden. Die kleinen Kunstwerke gehören zur Maglemose-Kultur und datieren zwischen 8.000 und 6.600 v.Chr. Man kann davon ausgehen, daß der hier verwendete Bernstein an den dänischen Küsten aufgesammelt worden war und damit lokalen Ursprungs ist.

Das gilt erst recht für die größte Ansammlung neolithischer Bernsteinobjekte im ehemaligen Ostpreußen. Hierbei handelt es sich um eine sekundäre Fundstelle bei dem kleinen Fischerdorf Schwarzort (heute Juodkrante in W-Litauen) auf der Haffseite der Kurischen Nehrung. Dort wurde in der zweiten Hälfte des vorigen Jahrhunderts im Haffboden systematisch nach Bernstein gebaggert (SLOTTA 1996; BREKENFELD 1996).

Das Fundgebiet der Artefakte war eine flache Sandbank nördlich von Schwarzort, ca. 650 m vom Haffufer entfernt. Sie lagen 2–4 m unter dem Haffboden oft nestweise zusammen mit Rohbernstein. Sie waren zweifellos in sekundärer Lage zusammengeschwemmt worden, nachdem sie vorher an anderer Stelle weggespült worden waren. Bei dem weitaus größten Teil handelt es sich um fertige Perlen, es kommen aber auch nur teilweise oder fehlgebohrte Halbfabrikate vor. Viele Perlen weisen ein so großes Loch in der Mitte auf, daß sie eigentlich als Ringe angesprochen werden müssen. Kennzeichnend sind rechteckige, quadratische, runde und langovale konische Knöpfe mit V-förmigen Bohrungen auf der Unterseite. Die Oberflächen sind gelegentlich mit Mustern aus Punktreihen verziert. Zahlreich sind auch längliche Anhänger, die meistens nur ein Ösenloch aufweisen. Seltener sind diskusartige Scheiben mit einer umlaufenden Rille und Anhänger von Miniatur-Bernstein-Äxten.

Die alles überragenden Artefakte aus Schwarzort aber sind die aus Bernstein gefertigten flachen Idole in menschlicher Gestalt. Einige davon wurden nach Göttingen gerettet, andere sind verloren. Die Darstellungen weisen eine große Variationsbreite auf, die von totaler Abstraktion fast nach Art ägäischer Violin-Idole (THIMME 1976) reicht bis zu flachen Figürchen mit angedeuteten Gliedern und Gesichtern.

Bei aller Schwierigkeit mit den vagen Fundumständen der Schwarzorter Artefakte seien doch Gedanken zur Chronologie erlaubt. Der „Komplex" von Schwarzort muß keinesfalls einer einzigen Zeitperiode angehören, und er muß keineswegs auf einem einzigen Ursprungsort für die Artefakte basieren. Betrachtet man die Fundgruppen also differenziert, etwa die ovalen, rechteckigen, trapez- oder kissenförmigen Anhänger, so finden sich Parallelen in den mesolithischen Bernsteinfunden Dänemarks. Auch der hier aufgeführte 6,6 cm lange Tierkopf (Abb. 8) kann in gewisse Beziehung zu einem mesolithischen Elchkopf aus Egemarke/Dänemark (7,5 cm lang) gesetzt werden (Ertebolle Kultur), der allerdings Kopf und Hals aufweist (WEISGERBER 1996). Auch zu der Perle (Abb. 4) sind auffallende Parallelen vorhanden. Sie sind aus Knochen gefertigt und in großer Zahl im Orient gefunden (DUBIN 1987). Dort datieren sie aber in die Natufian Kultur (10.000–8.000 v.Chr.). Ohne einen Zusammenhang herstellen zu wollen, kann vielleicht doch so viel aus diesen Funden abgeleitet werden, daß die Perle ein höheres Alter haben kann.

Das könnte insgesamt bedeuten, daß ein Teil der Funde älter ist als die überwiegende Mehrzahl, für die ein endneolithisches Alter aufgrund der anderenorts beobachteten Vergesellschaftung mit einer bestimmten Keramik naheliegt.

Intensive Bernsteinnutzung ist auch sonst für die Kultur der Kamm- und Grübchenkeramik nachgewiesen. Diese datiert von ca. 3.000 bis 1.600 v.Chr.. Viele der Schwarzorter Funde gehören ihr an und können etwa um 2.000 v.Chr. datiert werden. Dazu gehören auch die Idole in Menschengestalt. Deren stilistische Vielfalt fällt auf. Es gibt grob naturalistische Exemplare, deren Gesicht jeweils bis auf die Nase und die Augen reduziert ist, bis zu weitgehender geometrischer Abstraktion. Allein diese stilistischen Unterschiede legen eine chronologische Differenzierung nahe. Zum Unterschied von vielen Idolen ähnlicher Zeitstellung und ähnlicher Abstraktion anderenorts (THIMME 1976) weisen die Idole aus dem Haff keine Geschlechtsmerkmale auf. Sie sind auch nicht nackt, jedenfalls ist bei einigen Kleidung angedeutet.

Es bleibt die Frage nach Herkunft und ursprünglicher Funktion der Fundstücke. Bei den vielen halbfertigen Perlen liegt es nahe anzunehmen, daß sie in Siedlungen in den Boden gerieten. Perlenhalbfabrikate, deren Bohrungen sich nicht trafen oder die beim Bohren zu Bruch gingen, wurden verworfen und gelangten in den Boden. Die Bohrungen sind meistens von zwei Seiten angesetzt worden, wobei doppelkonische Lochquerschnitte entstanden. Scharfe Riefen an den Bohrlöchern zeigen, daß dazu meist Feuersteinbohrer mit ziemlich stumpfer Spitze benutzt wurden. Anhänger können verloren gegangen sein, denn oft genug sind die Bohrungen ausgerissen. Man kann allerdings nicht ausschließen, daß Anhänger auch in Gräbern deponiert waren. Für die Idole erscheint dies als die plausibelste Erklärung. Nimmt man an, daß sie Toten mit in die Gräber gegeben wurden (wie es z.B. in der Ägäis üblich war, THIMME 1976) und daß diese Gräber später erodiert wurden, so könnte man sich ihre Konzentration im Haff erklären. In prähistorischen Siedlungen kommen sie jedenfalls kaum vor. Allerdings kann die Möglichkeit, daß sie (etwa im Sinne vom Flußfunden) als Opfer ins Wasser gerieten nicht ausgeschlossen werden. Es sollte nicht übersehen werden, daß nicht alle der sich heute in Göttingen befindlichen Funde in Schwarzort geborgen wurden.

Bei einem der Idole ist eine Durchbohrung so angebracht, daß das Stück (umgekehrt hängend) hätte als Anhänger getragen werden können (Abb. 3). Das modern anmutende Stück (Abb. 7) hat eine Längsbohrung und konnte wie eine Anhänger getragen werden. Das spatelförmige Stück (Abb. 6) könnte sowohl als Anhänger getragen worden sein als auch, wegen der vielen Bohrlöcher, als Applikation gedient haben. Bei den anderen sitzen die Bohrungen so, daß es offensichtlich ist, daß sie auf einem anderen Träger angebracht waren. Dazu käme Kleidung in Frage.

Verzeichnis der Artefakte

Das folgende Verzeichnis hält die Daten fest, die bislang zu diesen 18 Fundstücken bekannt sind. Es werden ferner die Beschreibungen von KLEBS (1882) mitgeteilt. Sowohl in der Tabelle als auch im gesamten Text werden gelegentlich Abkürzungen verwendet. Es bedeuten:

P. O. G. — Physikalisch-ökonomische Gesellschaft zu Königsberg i. Pr.;

IMGPGö — Institut u. Museum für Geologie u. Paläontologie der Universität Göttingen;

Orig.-Nr. — Original-Nummer des wissenschaftlichen Katalogs des IMGPGö;

[ ] — Einfügungen der Autoren;

L., B., H., D. — Abmessungen in (in cm): Länge, Breite, Höhe, Dicke.

Die Zeichnungen wurden dankenswerterweise von Frau Irene STEUER-SIEGMUND, Göttingen 1996 und 1997 angefertigt. Bei den Abbildungen werden die verschiedenen Ansichten mit A, B, C, D, E gekennzeichnet. Sie sind in natürlicher Größe und dadurch auch in demselben Maßstab gehalten.

Lfd.-Nr.	Orig. Nr. IMGPGö	Bezeichnung	Abbildung
1	58-001	Idol	Abb. 1, A–D
2	58-002	Idol	Abb. 2, A–D
3	58-003	Idol	Abb. 3, A–D
4	58-004	Phallus-Perle	Abb. 4, A–D
5	58-005	Doppel-Knopf	Abb. 5, A–C
6	58-050	Idol	Abb. 6, A–D
7	58-051	Idol	Abb. 7, A–D
8	58-052	Tierkopf	Abb. 8, A–D
9	58-054	Anhänger, länglich, trapezförmig, 1 Bohrung	Abb. 9, A–C
10	58-055	Anhänger, trapezförmiges Bruchstück, ursprgl. vielleicht dreieckig, 13 Grübchen, 1 Bohrung	Abb. 10, A–B
11	58-056	Anhänger, viereckig, kissenförmig, 2 Bohrungen	Abb. 11, A–D
12	58-057	Anhänger, keilförmig, 1 Bohrung	Abb. 12, A–E
13	58-058	Anhänger, länglich, Fragment,1 zentrale Bohrung	Abb. 13, A–C
14	58-059	Anhänger, viereckig, kissenförmig	Abb. 14, A–D
15	58-060	Anhänger, trapezförmig	Abb. 15, A–D
16	58-061	Anhänger, oval, 3 Bohrungen	Abb. 16, A–C
17	58-070	Ring	Abb. 17, A–B
18	58-071	Linse	Abb. 18, A–C

Tabelle 1. Übersicht der Fundstücke.

Lfd.-Nr.	Orig. Nr. IMGPGö	Bezeichnung	Fundort
1	58-001	Idol	Kurisches Haff b.Schwarzort(Juodkrante)
2	58-002	Idol	Kurisches Haff b.Schwarzort(Juodkrante)
3	58-003	Idol	Kurisches Haff b.Schwarzort(Juodkrante)
4	58-004	Phallus-Perle	Kurisches Haff b.Schwarzort(Juodkrante)
5	58-005	Doppel-Knopf	unbekannt
6	58-050	Idol	bei Neidenburg (Nidzica)
7	58-051	Idol	Krucklinnen (Kruklanki) Kr. Lötzen (Giżycko)
8–16			unbekannt
17	58-070	Ring	Samland (Sambia)
18	58-071	Linse	unbekannt

Tabelle 2. Übersicht der Fundstücke nach Fundorten.

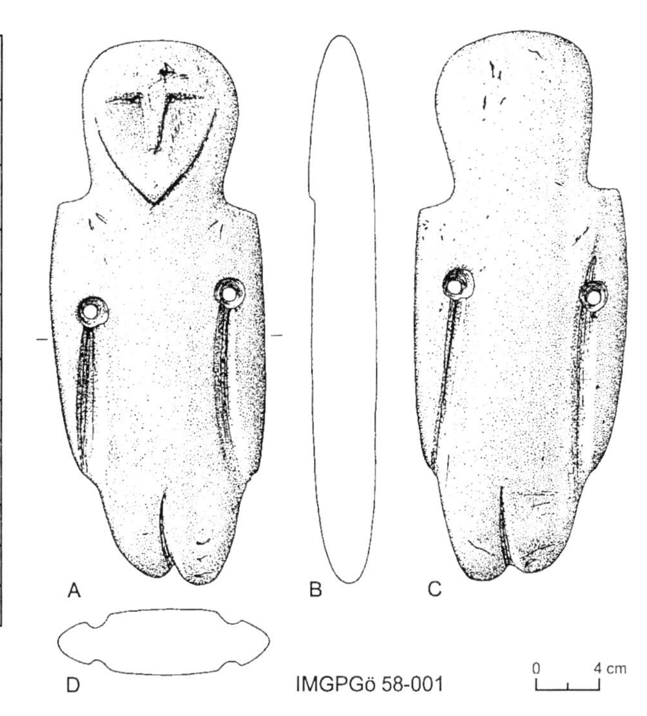

Abb. 1.

Katalog der Fundstücke

Orig.-Nr.	IMGPGö 58-001
Bezeichnung	**Idol in Menschengestalt**
Abbildung Nr.	1, A–D
Ehem. Nr.	P.O.G. 1.049
Fundort	Kurisches Haff bei Schwarzort/Ostpr. (Juodkrante, Litauen)
Original zu	KLEBS 1880, Nr. 1.076–2.079 (pt), KLEBS 1882, Taf. 9 Fig. 2
Weitere Erwähnungen	STURM 1927, S. 435 & Taf. 133 Fig. c RITZKOWSKI 1996b, S. 548, Nr 278
Abmessungen	L.: 14,3 cm; B.: 5,70 cm; D.: 1,67 cm
Gewicht	80,30 g
Kopie	Bernstein-Museum Palanga, Litauen

Beschreibung, KLEBS 1882, 29:

„Eine grosse sauber gearbeitete Figur, bei welcher der Kopf sich hoch aus den Schultern erhebt, der breite Hals aber bis an die Stirne verlängert erscheint, das Gesicht einrahmend. Das Gesicht tritt hervor und endet oben stärker gewölbt. Neben der Nase ist der Bernstein ausgeschabt, so dass dieselbe hervortritt. Augen und Mund sind als tiefere Gruben konisch eingedreht. Die Grube auf der Stirne ist ein natürlicher Fehler des Stückes. Die Arme liegen wieder an und sind beiderseits durch tiefe stark geschrammte Furchen vom Leibe getrennt. Es befinden sich nur am oberen Ende, in der Achselgegend, zwei doppelt konische gereifte Löcher, die Beine sind zwei kurze, nur durch eine Furche getrennte Stumpfe. Das Stück ist beim Herausnehmen oberhalb der Beine etwas geschrammt (durch die Baggereimer), so dass die helle Kumstfarbe unter der dunklen braunen Rinde hervortritt ... . Die ganze Oberfläche ist gut polirt bis auf die Furchen an den Armen und Beinen, bei denen die Schrammen sich nicht mehr wegnehmen liessen. Das Stück muss entweder in der Mitte einer Schnur oder an einer Oese hängend getragen worden sein."

Anmerkung:

„Kumstfarben" von der ostpreußischen Bezeichnung „Kumst" für „Kohl" abgeleitet, stellt einen sog. Bastard, d.i. eine getrübte Variante eines gelben Bernsteins dar (cf. KLEBS 1880, 13).

Orig.-Nr.	IMGPGö 58-002
Bezeichnung	**Idol in Menschengestalt**
Abbildung Nr.	2, A–D
Ehem. Nr.	P.O.G. 1.015
Fundort	Kurisches Haff bei Schwarzort/Ostpr. (Juodkrante, Litauen)
Finder	STANTIEN & BECKER, vor 1882
Original zu	KLEBS 1882, S. 28f., Taf. 9 Fig. 1a, 1b. (Front- u. Rückenansicht)
Weitere Erwähnungen	KLEBS 1880, Pos. 1.076–2.079 (pt) STURM 1927, S. 376, Taf. 120 Fig. a;

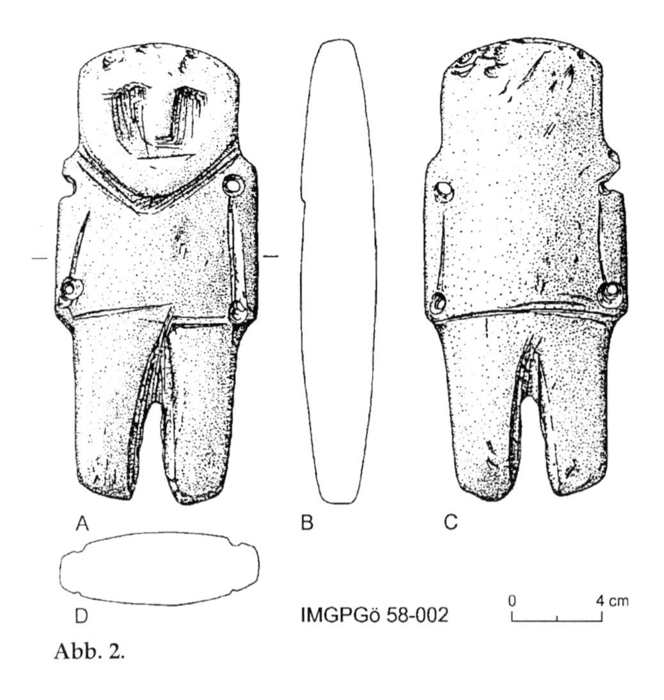

Abb. 2.

	RITZKOWSKI 1989, S. 36, Farbtaf. S. 37; WEISGERBER 1996, S. 416
Abmessungen	L: 9,57 cm; B.: 4,35 cm; D.: 1,54 cm
Gewicht	36,28 g

Beschreibung, KLEBS 1882, 28–29:

Die Figur „hat einen breiten niedrigen Kopf, dessen stumpfes Kinn unter die Schultern herabreicht. Die Stirn schneidet oben convex ab, die Nase liegt in demselben Niveau wie die Umrandung und ist ihre Umgebung ausgeschabt, ohne dass die Augen besonders charakterisiert sind. Die Stelle kennzeichnet die Anwendung eines Feuersteinschabers, die Linien sind sehr grob und unsicher eingeritzt, wie es bei Stahl- oder Eisengeräthen nie möglich wäre.

Unter der Nase ist der Mund noch besonders durch zwei tiefe Furchen getrennt, welche dieselbe unregelmässige Streifung zeigen. Die Arme liegen am Leibe an und sind nur durch zwei Furchen davon getrennt. Die Beine werden durch zwei Stumpfe vorgestellt. Als Andeutung eines jackenartigen Kleidungsstückes hat man jedenfalls die mässig tief eingravierte Linie zu sehen, die oberhalb der Beine um den ganzen Körper herumläuft, hinten in einem Zuge (d.h. durch wiederholtes Ritzen hergestellt), vorne in zwei Zügen, auf der einen Seite doppelt.

Auf der Rückseite (Abb.2, B) wird die Trennung der Arme nur durch eine ähnliche Furche dargestellt. Das ganze Stück ist gut polirt, mit Ausnahme der ausgravierten Stellen, doch sind die feinen Ritzen vom ersten Schaben meist noch durchzuerkennen.

Die Figur hat vier Löcher, je eins am oberen und unteren Ende jedes Armes, ist also wohl als Mittelstück einer doppelten Perlschnur getragen worden. Die Löcher sind doppelt-konisch. Nur das obere links ist einfach-konisch von aussen ganz hindurchgebohrt."

Orig.-Nr.	IMGPGö 58-003
Bezeichnung	**Idol in Menschengestalt,** stark abstrahiert
Abbildung Nr.	3, A–D
Ehem. Nr.	P.O.G. 1016
Fundort	Kurisches Haff b. Schwarzort/Ostpr. (Juodkrante, Litauen)
Original zu	KLEBS 1882, S. 29, Taf. 9 Fig. 3
Weitere Erwähnungen	KLEBS 1880, Pos. 2076–2079 (pt) STURM 1927, S. 376, Taf. 120 Fig. d
Abmessungen	L.: 7,74 cm; B.: 3,10 cm; D.: 0,59 cm
Gewicht	7,78 g

Beschreibung, KLEBS 1882, 29:

„Die Figur ist die roheste. Das Gesicht ist fünfeckig mit concaver Oberkante, und es tritt nur die Nase hervor, um welche sowohl die Gegend der Augen, als auch die ganze untere Hälfte ohne besondere Charakterisirung des Mundes ausgeschabt ist. Die Arme, von denen einer ausgebrochen ist, sind vom Körper wirklich getrennt, indem die Ausschabung das dünne Stück durchbrach, vereinigen sich aber wieder unten mit ihm. Die Beine sind nicht angedeutet. Ein Loch sitzt

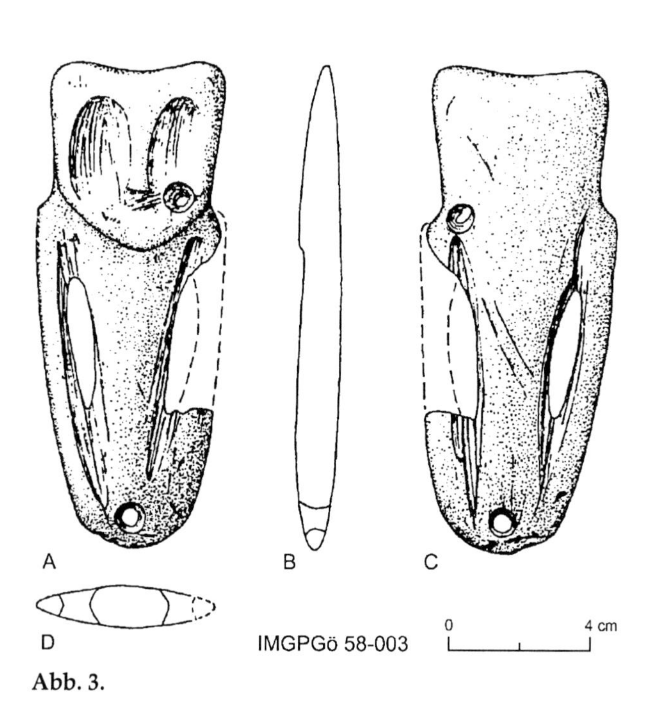

Abb. 3.

unten, das andere seitwärts unter der Nase. Wenn man die concave scharfe Kante, welche den Kopf begrenzt, betrachtet und das Stück umdreht, so macht es vollständig den Eindruck, als sei es aus einem axtförmigen Hängestück gearbeitet, welchem man später durch Schaben diese Form gab und das man dann noch einmal bohrte, um es aufrecht tragen zu können. Es wird die Verlängerung der Arme annähernd die Ecken der concaven Kante treffen. Diese ziemlich wahrscheinliche Hypothese nähert die Figur noch mehr den Hängestücken, denen sie sich ohnedies durch ihre Technik anschliesst."

Orig.- Nr.	IMGPGö 58-004
Bezeichnung	**Phallus-Perle**
Abbildung Nr.	4, A–D
Ehem. Nr.	P.O.G. 1012
Fundort	Kurisches Haff bei Schwarzort/Ostpr. (Juodkrante, Litauen)
Original zu	KLEBS 1882, S. 28, Taf. 8 Fig. 13a & 13b
Weitere Erwähnungen	STURM 1927, S. 376, Taf. 119 Fig. t
Abmessungen	H.: 1,72 cm; B.: 1,80 cm; D.: 0,90 cm
Gewicht	0,92 g

Beschreibung, KLEBS 1882, 28:

„Eines der merkwürdigsten Stücke ist ein kleiner Phallus aus Schwarzfirnis. Derselbe hat an einer Seite eine tiefgehende Furche und schneidet oben viereckig ab. Die Bohrung zeigt sehr schön die beiden konischen, stark gereiften Löcher ... und beweist, daß auch dieses auffallende Stückchen durchaus mit allen übrigen bisher beschriebenen in eine Klasse gehört und dass man mithin berechtigt ist, ihm denselben Ursprung zuzuschreiben. Es kann dasselbe daher durchaus nicht für römische Arbeit gehalten werden, obwohl gerade dort solche Symbole mit Vorliebe als

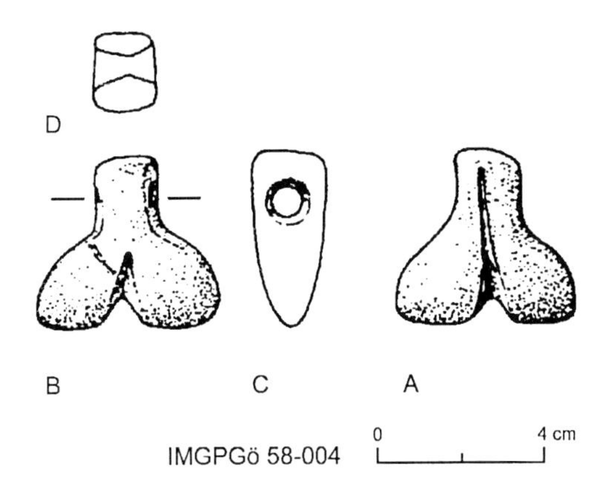

Abb. 4.

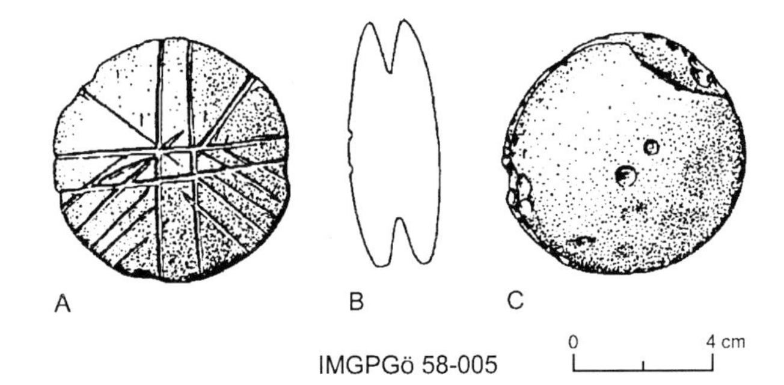

Abb. 5.

Amulette getragen wurden. Die Technik der Bohrung wäre eine ganz andere und vollkommenere gewesen. Ebenso kann man sich wohl davon enthalten, weitere religiös-ethnographische Speculationen hieran zu knüpfen."

Orig.-Nr.	IMGPGö 58-005
Bezeichnung	**Doppel-Knopf**
Abbildung Nr.	5, A–C
Fundort	unbekannt
Erwähnungen	cf. STURM 1927, S. 376, Taf. 119 Fig. e
Abmessungen	∅ außen: 3,26 cm; D.: 1,26 cm; Schlitztiefe: 0,52 cm
Gewicht	7,53 g

Über die Doppelknöpfe schreibt KLEBS (1882, 12–13):

„Gegenstände von höchst befremdlicher, Anfangs zum Verdacht reizender Form sind die Doppelknöpfe, zwei Scheiben oder Knöpfe, welche durch einen dünnen Stiel verbunden sind. Die Scheiben sind meist linsenförmig, nicht sehr dick, so dass ihre Kanten ziemlich scharf oder etwas abgerundet erscheinen ... . Die Dimensionen gehen von ... 10 mm Breite 8 mm Dicke bis 26 mm Breite und 28 mm Dicke herauf."

Das Stück zeigt auf einer Seite ein Kreuzmuster aus Doppellinien, wobei das entstandene Viertel noch einmal mittels einer Linie geteilt wurde. Im unteren Bereich treten weitere Schraffen hinzu. Die Rückseite ist bis auf einige Abplatzungen glatt.

Orig.-Nr.	IMGPGö 58-050
Bezeichnung	**Idol** in stark abstrahierter Menschengestalt
Abbildung	6, A–D
Ehem. Nr.	1922 (P.O.G.)
Fundort	Gegend um Neidenburg (Nidzica, Polen)
Finder	Donator: Sanitätrath MARSCHALL, Marienburg (Malbork, Polen)

Original zu	KLEBS 1882, S. 44, Taf. 10 Fig. 1
Weitere Erwähnungen	ANDRÉE 1937, Taf. 15 unten; ANDRÉE 1951, Abb. 22 unten; RITZKOWSKI 1989, S. 34, Farbtaf. S. 25
Abmessungen	L.: 10,6 cm; B.: 5,42 cm; D.: 0,90 cm
Gewicht	25,52 g

Beschreibung, KLEBS 1882, 44:

„Die Form ist rätselhaft, doch kann man es wohl als einen rohen Versuch zu einer Menschen- oder figürlichen Darstellung ansehen. Die Abbildung kennzeichnet es besser als jede Beschreibung.

Hervorzuheben ist besonders der lange Hals. Das Stück trägt an den Seiten vier Löcher. zwei oben, zwei in der Mitte, ähnlich wie ... [IMGPGö 58-002]; würde also die Rolle eines Mittelstückes gespielt haben. Auffallend ist das große Loch zwischen den beiden oberen, doch etwas tiefer, welches der Figur wieder den Charakter eines Hängestückes verleiht. Vielleicht sollen die drei Oeffnungen Augen und Mund andeuten. Sie sind alle doppelt-konisch gebohrt, und besonders bei dem oberen treffen die beiderseitigen Bohrungen nicht sehr genau aufeinander. Unter den beiden unteren Löchern geht eine Furche querüber und zwei längs der Seitenkanten hinab.

Auf der Rückseite läuft eine doppelte Furche querüber oberhalb beider Löcher. Ueber die schulterartigen Absätze zieht sich je eine Kerbe herum. Die Furchen sind ganz im Charakter der Steinzeit unsicher eingeritzt. Das Stück ist kumstfarbig mit stark nachgedunkelter Rinde und wird im Torf gelegen haben."

Orig.-Nr.	IMGPGö 58-051
Bezeichnung	**Idol** in stark abstrahierter Menschengestalt, zum Hängen
Abbildung	7, A–D
Ehem. Nr.	1923 (P.O.G.)
Fundort	Krucklinnen (Kruklanki), Krs. Lötzen (Giżycko, Polen)
Finder	Donator: Gutsbesitzer SKRZECKA

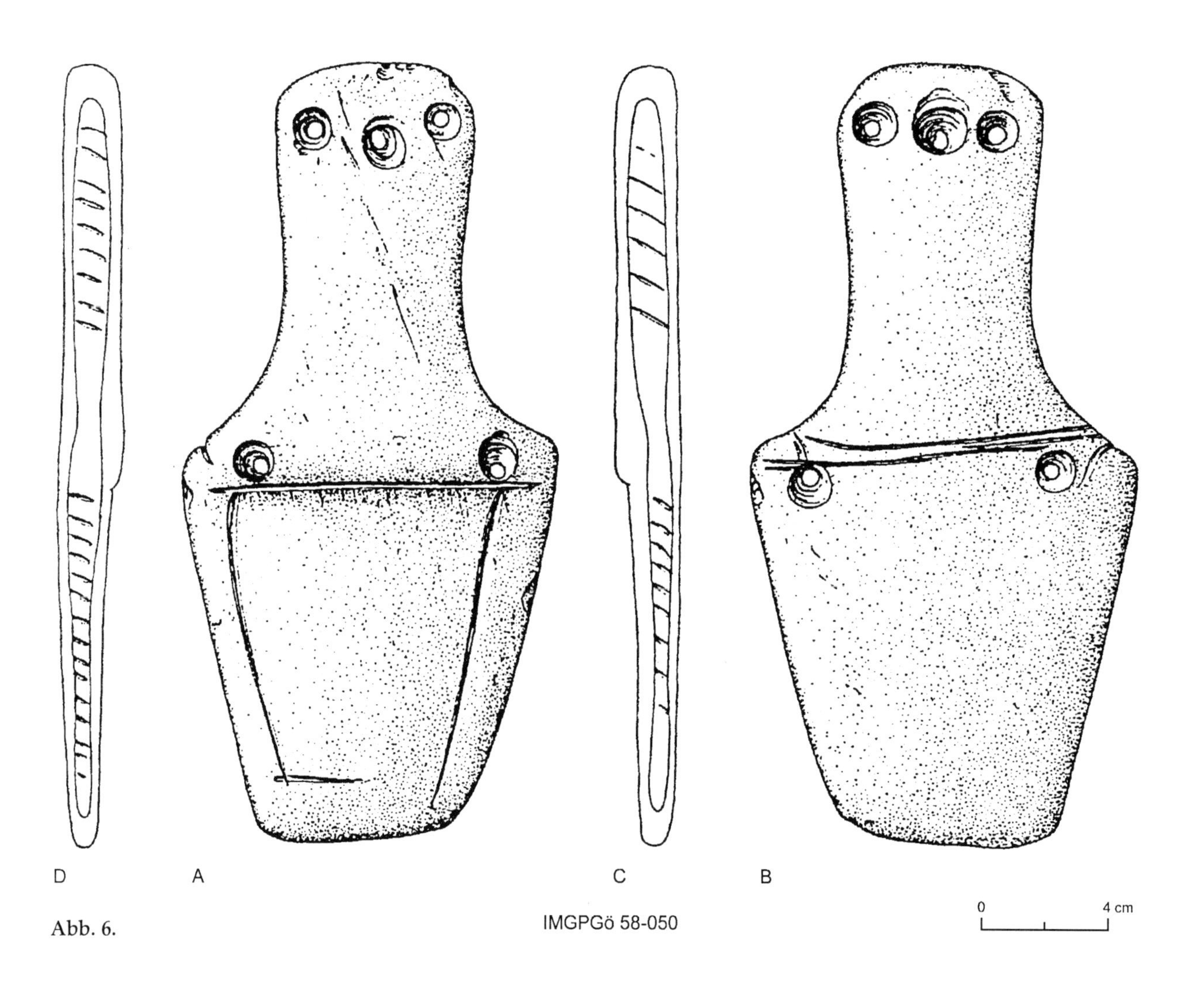

Abb. 6. IMGPGö 58-050

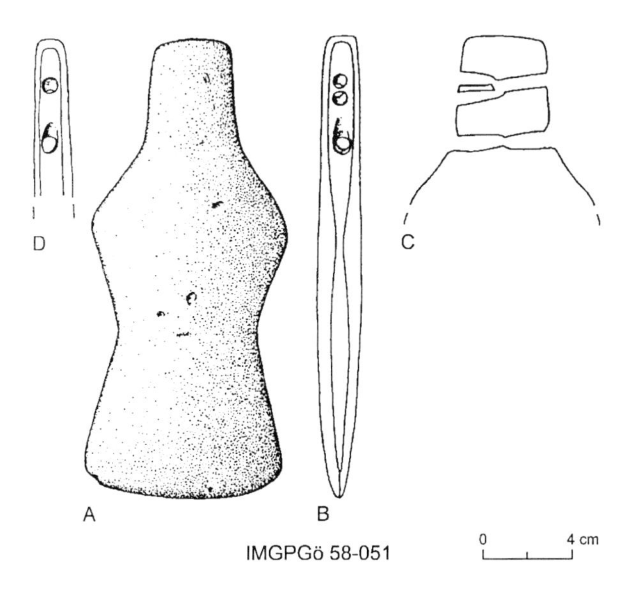

Abb. 7.

Original zu	KLEBS 1882, S. 44, Taf. 10 Fig. 3a–3c
Weitere Erwähnungen	ANDRÉE 1937, Taf. 15 unten, Fig. unten mittig; ANDRÉE 1951, Abb. 22 unten
Abmessungen	H.: 8,92 cm; B. oben - mittig - unten: 3,90 cm - 2,76 cm - 3,97 cm; D.: 0,80 cm
Gewicht	16 g (mit Wachsresten auf der Rückseite)

Beschreibung, KLEBS 1882, 44:

„Die Figur ... [IMGPGö 58-051] ... ist zwar noch weniger im Detail ausgeführt als ... [IMGPGö 58-050], aber auch hier kann man wohl die schwache Andeutung einer menschlichen Figur erkennen, bei der die Schultern und die Taille stark ausgedrückt sind, die Arme und Beine aber fehlen und der Kopf auch nur durch ein quadratisches Stück dargestellt ist. Die Vermuthung gewinnt an Wahrscheinlichkeit, wenn man dies Stück mit den primitiven Darstellungen menschlicher Gestalten in Thon, wie sie der Pfahlbau des Laibacher Moores geliefert hat, vergleicht (Mitth. der Wiener anthrop. Ges. VIII), wo bes. Fig. 12 mit unserer Fig[ur] ... in den Contouren eine gewisse Ähnlichkeit besitzt, obwohl die Stücke sonst in Technik und Decoration ganz verschieden sind. Das platte Stück, kumstfarbig mit stark nachgedunkelter Rinde, ist vorzüglich gearbeitet und polirt, so dass die vom Schaben herrührenden Furchen fast ganz verschwunden sind.

Das Kopfstück ist zweimal in der Richtung der Platte durchbohrt, und wird die Figur als herabhängendes Mittelstück einer doppelten Perlschnur gedient haben. Die beiden langen Löcher sind jedenfalls mit Knochennadeln gebohrt, sie sind von beiden Seiten begonnen, so dass die Bohrungen etwas schief aufeinander treffen. Beim obern Bohrloch ist rechts zweimal angesetzt, weil das eine Loch zu schief geraten war, so dass hier zwei Röhren in die gegenüberliegende münden."

Orig.-Nr.	IMGPGö 58-052
Bezeichnung	**Tierkopf, Elch**
Abbildung	8, A–D
Fundort	unbekannt
Original zu	ANDRÉE 1937, Taf.15, Mitte links; ANDRÉE 1951, S. 44, Abb. 22, Mitte links
Abmessungen	L.: 6,61 cm; H.: 2,35 cm; D.: 1,23 cm
Gewicht	10,98 g

Nur die Rille über Nase und der stark ausgeprägte Kinnbacken legen eine Deutung dieses Stücks als Tierkopf nahe. Die Augen zeigen starke Spuren vom Drehen des Flintbohrers und dienten wohl zur Aufnahme der Trageschnur. Wegen der hervorgehobenen Kinnlade kann das Tier als Elch angesprochen werden. Bei Pferdeköpfen in Bernstein sitzt sie im Halsbereich.

D

A

C

IMGPGö 58-052

0 4 cm

Abb. 8.

Diese Figur ist weder bei KLEBS (1882), noch im *Reallexikon der Vorgeschichte, Bd. 11* (1927) aufgeführt. ANDRÉE bildet sie gemeinsam mit den anderen Schwarzorter Figuren auf einem Photo ab, das er sowohl 1937 (Taf. 15 bei S. 128) als auch 1951 verwendet.

Der andere „Pferdekopf" des Bildes (ANDRÉE 1937, Taf. 15, Mitte rechts) entspricht der Beschreibung und Abbildung bei KLEBS 1882 (S. 28 und Taf. VIII Fig. 21), das die Katalog-Nummer des Provinzial-Museums P.O.G. 1013 trug. Dieses Stück befindet sich nicht unter den Stücken, die in Göttingen liegen. Es ist zu vermuten, daß es in den Kriegswirren verloren gegangen ist.

Ferner führt KLEBS 1882 auf S. 28 und Taf. VIII Fig. 1 (Katalog Nr. St. B. 237 = STANTIEN & BECKER) ein weiteres Artefakt an, das er als „Thierkopf ?" bezeichnet. Es geht aus den Veröffentlichungen ANDRÉES nicht hervor, ob das Stück überhaupt in der Bernstein-Sammlung der Universität Königsberg sich befunden hat. Der Verbleib ist unbekannt.

Orig.-Nr.	IMGPGö 58-054
Abbildung	9, A–C
Bezeichnung	**Anhänger**, länglich, trapezförmig
Fundort	unbekannt
Abmessungen	L.: 6,20 cm; B.: 2,90 cm; D.: 0,90 cm
Gewicht	9,47 g

Das Stück besitzt eine geglättete Oberfläche. Das doppelt-kegelförmige Bohrloch weist eine Glättung des oberen Bohrrandes auf. Dies ist vermutlich durch die schleifende Wirkung des Tragebandes entstanden.

Das Stück ist weder von KLEBS (1882) noch von ANDRÉE (1937) aufgeführt. Es ist zu den „axtförmigen Hängestücken" KLEBS (1882, 20) zu zählen, die „... von dreieckiger oder trapezförmiger Gestalt ..." sind. KLEBS nennt „... unabhängig von jeder Hypothese, die untere Seite des Trapezes oder Dreieckes die ‚Schneide' ..., auch wenn sie in eine breite Kantenfläche übergegangen ist (breite Schneide), die durchbohrte Seite das ‚Bahnende', die Kante die ‚Bahn'. ... Die Seiten und die Schneide sind oft scharf, manchmal auch abgeschliffen, so dass eine schmale Kante senkrecht zur Platte entsteht."

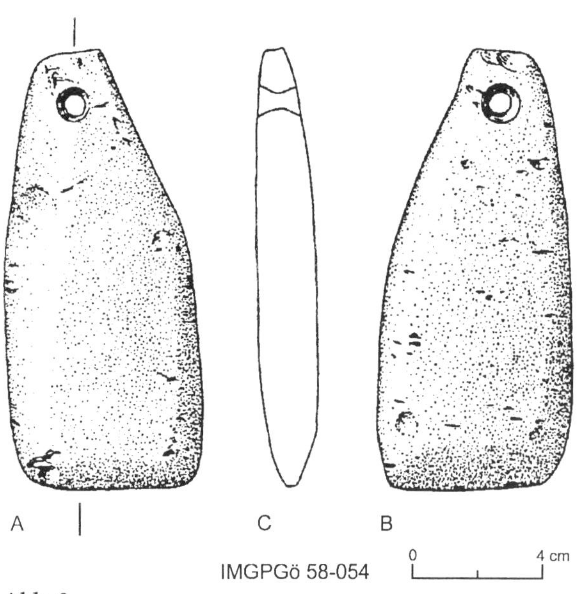

Abb. 9.

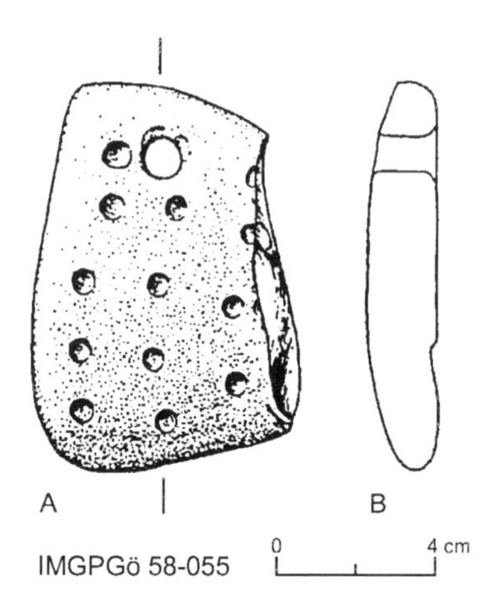

Abb. 10.

Orig.-Nr.	IMGPGö 58-055
Bezeichnung	**Anhänger**, Bruchstück
Abbildung	10, A–B
Fundort	unbekannt
Erwähnungen	Keine Erwähnung bei KLEBS 1822 und ANDRÉE 1937
Abmessungen	L.: 4,23 cm; B.: 3,93 cm; D.: 0,90 cm
Gewicht	5,90 g

Das schildförmige Hängestück besitzt eine ebene Rückseite und eine asymmetrisch zugeschärfte Schneide. Die Vorderseite ist mit Grübchen (ca. 3 mm ∅) verziert. Diese 13 schalenförmigen Vertiefungen sind in drei Reihen angeordnet. Der rechte Seitenrand hat zwei Grübchen angeschnitten; der Rand ist daher eine Bruchfläche, die nach der Verzierung erfolgt ist. Die Bruchfläche ist angewittert. Die einzige Bohrung von 4 mm ∅ besitzt eine zylindrische Form. Die Glättung des Randes des Bohrloches liegt auf beiden Seiten außerhalb der Verbindungslinie von Schwerpunkt und Bohrung. Daraus folgt, daß der Hänger getragen wurde, bevor er zerbrach. Es ist auch zu vermuten, daß er dreieckig war, denn die Glättung des Randes der Bohrung läßt auf eine randliche Lage des Schwerpunktes schließen.

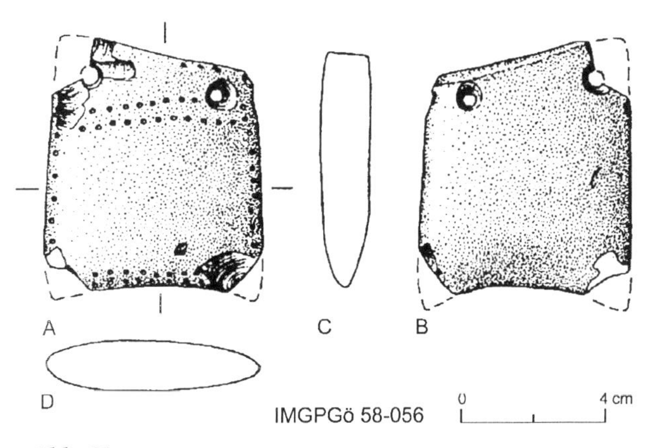

Abb. 11.

Orig.-Nr.	IMGPGö 58-056
Bezeichnung	**Anhänger**, kissenförmig
Abbildung	11, A–D
Fundort	unbekannt
Abmessungen	H.: 3,30 cm; Br.: 2,93 cm; D.: 0,68 cm
Gewicht	3,49 g

Das Stück hat am oberen Ende zwei doppelkonische, feine Bohrungen, von denen eine ausgebrochen ist. Die Vorderseite war links, rechts und oben mit einer Reihe einfacher Punkte verziert, während entlang der Unterkante und unterhalb der Bohrlöcher Doppelpunktreihen auftreten.

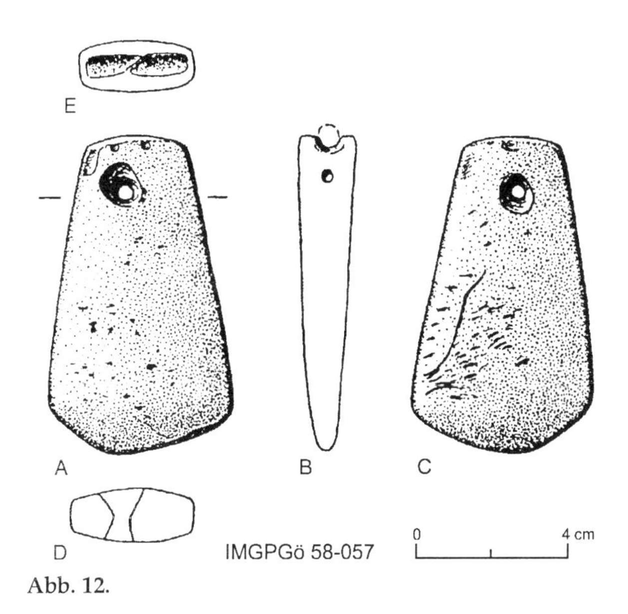

Abb. 12.

Orig.-Nr.	IMGPGö 58-057
Bezeichnung	**Anhänger**, keilförmig
Abbildung	12, A–E
Fundort	unbekannt
Abmessungen	H.: 3,92 cm; Br.: 2,35 cm; D.: 1,72 cm
Gewicht	4,06 g

Dieser grob trapezförmige Anhänger besaß am oberen Schmalende eine Längsbohrung, die sich auf der Bruchfläche hälftig abzeichnet. Eine versuchte, neue Längsbohrung wurde nur ansatzweise begonnen. Eine doppelkonische Querbohrung erwies sich als dauerhafter. Die Oberfläche ist glatt, die Schneide unregelmäßig rundlich ausgebaucht.

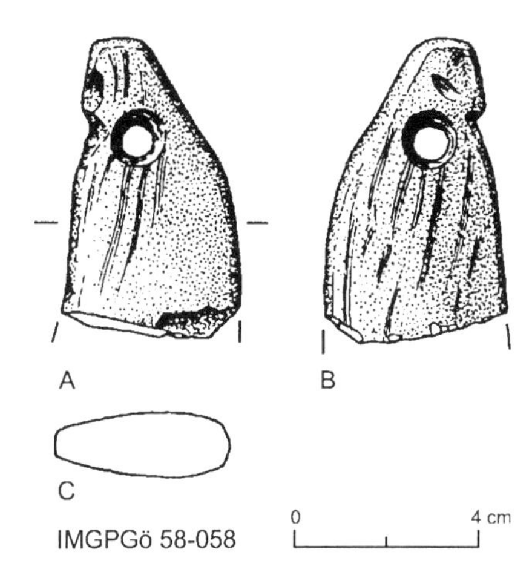

Abb. 13.

Orig.-Nr.	IMGPGö 58-058
Bezeichnung	**Anhänger**, fragmentarisch
Abbildung	13, A–C
Fundort	unbekannt
Abmessungen	H.: 5,14 cm; Br.: 2,44 cm; D.: 0,58 cm
Gewicht	5,30 g

Bei dem Stück handelt es sich um das obere Ende eines länglichen Anhängers, das unverziert ist. Außer der Bruchkante sind alle Kanten gerundet. Feine Querbohrung.

Orig.-Nr.	IMGPGö 58-059
Bezeichnung	**Anhänger**, rechteckig
Abbildung	14, A–D
Fundort	unbekannt
Abmessungen	H.: 3,28 cm; Br.: 1,91 cm; D.: 0,72 cm
Gewicht	2,33 g

Rechteckiger Anhänger mit gezipfelten Ecken. Die Bohrung ist fast zylindrisch, ist jedenfalls nicht so doppelkonisch wie üblich, sie sitzt außerhalb des Zentrums. Die Oberflächen sind nicht sehr geglättet

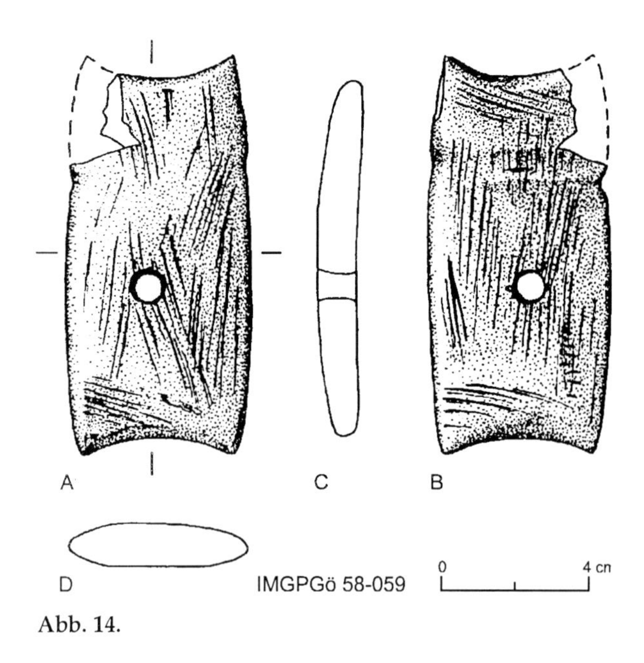

Abb. 14.

und weisen noch die Kratzspuren der Feuersteinwerkzeuge auf.

Orig.-Nr.	IMGPGö 58-060
Bezeichnung	**Anhänger**, trapezförmig
Abbildung	15, A–D
Fundort	unbekannt
Abmessungen	H.: 4,68 cm; Br.: 3,86 cm; D.: 0,64 cm
Gewicht	6,68 g

Der ungefähr trapezförmige Anhänger besitzt eine doppelkonische Bohrung unterhalb der Schmalseite, nicht mittig. Eine Oberfläche ist weitgehend geglättet, die andere weist entlang den Kanten noch deutliche Kratzspuren auf.

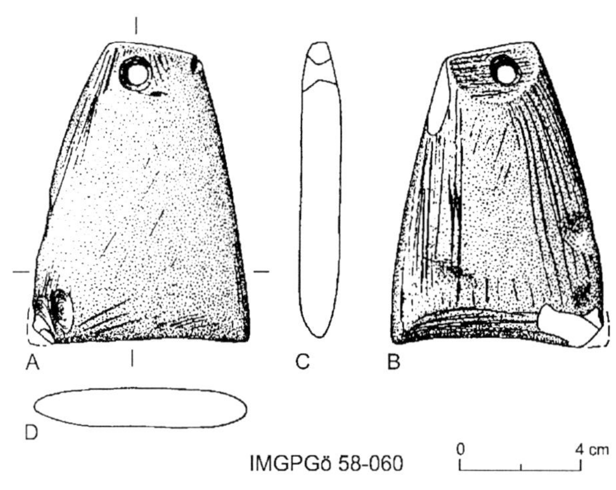

Abb. 15.

Orig.-Nr.	IMGPGö 58-061
Bezeichnung	**Anhänger**, schildförmig
Abbildung	16, A–C
Fundort	unbekannt
Erwähnungen	Weder KLEBS (1882) noch ANDRÉE (1937) erwähnen dieses Stück
Abmessungen	L.: 4,75 cm; B. 3,49 cm; D.: 0,70 cm
Gewicht	6,42 g

Das Stück weist heute drei doppelkonische Bohrungen auf, von denen eine angebrochen ist. Während eine Oberfläche fast glatt erscheint, weist die andere drei ähnliche Ziermotive auf, eines allerdings nur noch in Resten. Es handelt sich um lanzettliche oder spitzovale Flächen, die mit sich kreuzenden Schraffen gefüllt sind. Sie erwecken den Eindruck von Fischdarstellungen. Der Längsschnitt durch das Stück legt nahe, daß die Zeichnungen sich auf der Rückseite des Anhängers befanden. Das würde auch die Abnutzung der Muster erklären.

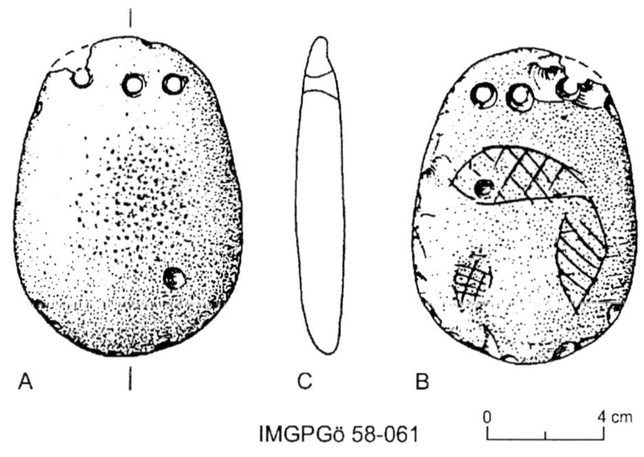

Abb. 16.

Orig.-Nr.	IMGPGö 58-070
Bezeichnung	**Ring**
Abbildung	17, A–B
Ehem. Nr.	P.O.G. 3370
Fundort	unbekannt, höchstwahrscheinlich Ostpreußen, Samland (KLEBS 1882, 45)
Finder	„... vom Bernsteinarbeiter GERBER in Königsberg gekauft, der ihn wiederum von einem Händler erstanden hat, welcher meist im Samland reist, er dürfte also wohl von daher stammen." (KLEBS 1882, 45)
Original zu	KLEBS 1882, S. 45, Textfig. 3

Abmessungen	∅ außen: 11,4 cm; ∅ innen: 4,56 cm; D.: 2,2 cm
Gewicht	133,49 g

Beschreibung, KLEBS 1882, 45:

„Der Ring ist hellkumstfarbig mit vereinzelten klaren Stellen, mit dicker rothbrauner, zersprungener, vielfach abgesprungener Verwitterungsrinde die schon auf ein sehr langes Verweilen in der Erde hindeutet.

Der äußere Durchmesser ist 115 mm, der innere ca. 50 mm und nimmt nach der Mitte zu noch ab. Die Dicke ist 23 mm, die Oberfläche wölbt sich beiderseits nach dem ziemlich scharfem Rande, ist ersichtlich nicht abgedreht, und ist die grosse Regelmässigkeit nur scheinbar, indem sich besonders innen Abweichungen von der Kreislinie finden. Der innere Cylindermantel zieht sich nach der Mitte zu etwas zusammen und somit steht dieser Ring in seiner Form den Schwarzorter Ringen, besonders Taf. IV Fig. 8 [bei KLEBS 1882; dort nicht für die Bernstein-Sammlung der Universität Königsberg erwähnt, nicht in Göttingen], ganz nahe, die er nur durch seine riesige Grösse weit übertrifft.

Die Technik läßt sich bei der zerbröselten Oberfläche nicht weiter studiren, dass er nicht modern ist und sehr lange in der Ede gelegen hat, zeigt aber die Verwitterungsrinde und der Mangel an Abdrehung. In Gräbern der Metallzeit ist hier nichts entfernt Aehnliches vorgekommen. Daher kann man ihn unbedenklich zusammen mit den Schwarzorter Ringen der ostpreussischen Steinzeit überweisen."

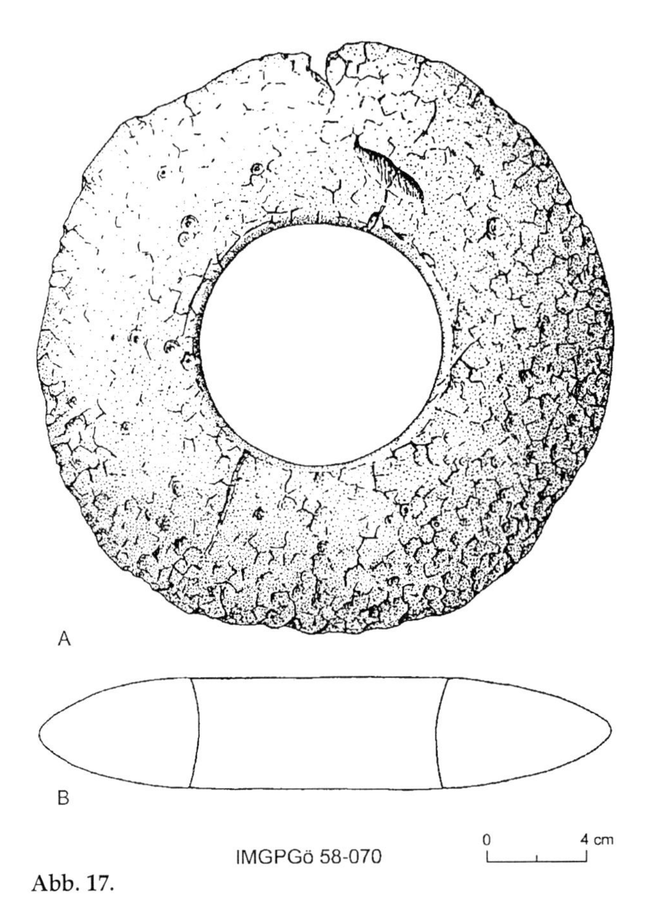

Abb. 17.

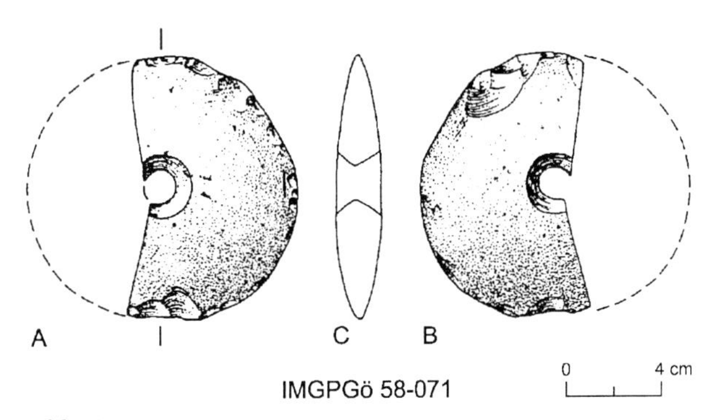

Abb. 18.

Orig.-Nr.	IMGPGö 58-071
Bezeichnung	**Perle**, linsenförmig, Bruchstück
Abbildung	18, A–C
Fundort	unbekannt
Abmessungen	∅ außen: 5,14 cm; ∅ innen: 0,61 cm; D.: 0,92 cm
Gewicht	8,11 g

Das Stück ist auf beiden Seiten gut geglättet. Entlang der Außenkanten sind Beschädigungen durch Absplisse zu verzeichnen. Die doppelkonische Bohrung liegt ziemlich in der Mitte.

Dieses Stück wurde weder von KLEBS (1882) noch von ANDRÉE (1937) erwähnt oder abgebildet. Die Form wird von KLEBS (1882, 15) als bikonvexe „Linse" bezeichnet. „Scheiben" dagegen werden durch annähernd ebene, parallele Flächen begrenzt. Die Größe (Durchmesser) der Linsen liegt nach KLEBS (1882, 15) zwischen 17 mm und 60 mm. Das Bohrloch ist, wie bei den Bochlöchern anderen Linsen, „... von beiden Seiten konisch eingedreht, stark gereift und ... annähernd central ... ."

Siegfried RITZKOWSKI
Institut und Museum für Geologie und Paläontologie
Universität Göttingen
Goldschmidtstrasse 3
D-37077 Göttingen, Germany

Gerd WEISGERBER
Deutsches Bergbau-Museum
Am Bergbaumuseum 28
D-44791 Bochum, Germany

NEOLITYCZNE WYROBY WYKOPALISKOWE Z DAWNEGO ZBIORU BURSZTYNU UNIWERSYTETU ALBERTA W KÖNIGSBERGU

Siegfried RITZKOWSKI,
Gerd WEISGERBER

Streszczenie

Autorzy przedstawili opis 18 neolitycznych wyrobów, które należały do dawnych zbiorów bursztynu Uniwersytetu w Königsbergu. Zabytki te zostały uratowane przed zniszczeniami wojennymi i ostatecznie przekazane Instytutowi i Muzeum Geologii i Paleontologii Uniwersytetu w Getyndze.

Opracowano katalog, który zawiera wszystkie dane o 18 ocalałych obiektach, łącznie z literaturą i wskazaniem gdzie znajduje się kopia danego okazu. W katalogu umieszczono również pierwsze opisy okazów, sporządzone przez KLEBSA (1882), które w owym czasie uważano za wzorcowe.

Opisane w pracy zabytki zostały po raz pierwszy pokazane na rysunkach w rzucie poziomym i pionowym, co umożliwia ich trójwymiarowe odwzorowanie. Rysunki prezentują wszystkie obiekty w naturalnej wielkości.

Literatur

ANDRÉE K.

1937 *Der Bernstein und seine Bedeutung in Natur- und Geisteswissenschaften, Kunst und Kunstgewerbe, Technik, Industrie und Handel. Nebst einem kurzen Führer durch die Bernsteinsammlung der Albertus-Universität*, 1–219 & 51 Abb., Gräfe & Unzer, Königsberg.

1951 *Der Bernstein. Das Bernsteinland und sein Leben*, 1–96 & 24 Abb., 3 Ktn., *Kosmos*, Stuttgart.

BREKENFELD A.

1996 *Die Unternehmerpersönlichkeiten Friedrich Wilhelm Stantien und Moritz Becker*, [*in*:] Ganzelewski M. & Slotta R. (Hrsg.), Bernstein — Tränen der Götter, *Veröffentlichungen aus dem Bergbau-Museum Bochum*, **64**, 277–283.

DUBIN L. S.

1987 *The history of beads, from 30,000 B.C. to the present*, New York.

KLEBS R.

1880 *Der Bernstein. Seine Gewinnung, Geschichte und geologische Bedeutung. Erläuterung und Catalog der Bernstein-Sammlung der Firma Stantien & Becker*, 1–32, Königsberg.

1882 *Der Bernsteinschmuck der Steinzeit von der Baggerei bei Schwarzort und anderen Lokalitäten Preußens aus den Sammlungen der Firma Stantien & Becker und der Physik.-ökonom. Gesellschaft*, 1–75 & 12 Taf., W. Koch, Königsberg i. Pr..

LA BAUME W.

1924 *Bernstein*, [*in*:] Ebert (Hrsg.), *Reallexikon der Vorgeschichte*, **1**, 431–441 & Taf. 133–135.

RITZKOWSKI S.

1977 Das Schicksal der Königsberger Bernsteinsammlung, *Museumskunde*, **42** (2), 87–88.

1989 *Bernstein — ein fossiles Harz*, [*in*:] Barfod J., Jacobs F. & Ritzkowski S., *Bernsteinschätze in Niedersachsen*, 10–39, Knorr & Hirt Verlag, Seelze.

1995 *Geowissenschaftler der Albertus-Universität in Königsberg*, [*in*:] Rauschning D. & v. Neree D. (Hrsg.), Die Albertus-Universität zu Königsberg und ihre Professoren, *Jahrbuch der Albertus-Universität Königsberg/Pr.*, **29**, 743–754, Duncker & Humblot, Berlin.

1996a *Die Geschichte der Bernsteinsammlung der Albertus Universität zu Königsberg i. Pr.*, [*in*:] Ganzelewski M. & Slotta R. (Hrsg.), Bernstein — Tränen der Götter, *Veröffentlichungen aus dem Bergbau-Museum Bochum*, **64**, 293–298, Bochum.

1996b *Die jungsteinzeitliche Bernstein Artefakte von Schwarzort in der Sammlung des Institut und Museum für Geologie und Paläontologie der Universität Göttingen*, [*in*:] Ganzelewski M. & Slotta R. (Hrsg.), Bernstein — Tränen der Götter, *Veröffentlichungen aus dem Bergbau-Museum Bochum*, **64**, 548–552, Bochum.

SLOTTA R.

1996 *Die Bernsteingewinnung im Samland (Ostpreußen) bis 1945*, [*in*:] Ganzelewski M. & Slotta R. (Hrsg.), Bernstein — Tränen der Götter, *Veröffentlichungen aus dem Bergbau-Museum Bochum*, **64**, 277–214, Bochum.

STURM E.

1927 *Schwarzort*, [*in*:] Ebert, *Reallexikon der Vorgeschichte*, **11**, 373–379 & Taf. 119–120.

THIMME J.

1976 *Kunst und Kultur der Kykladeninseln im 3. Jahrtausend v. Chr*, Badisches Landesmuseum, Karlsruhe.

WEISGERBER G.

1996 *Vor- und frügeschichtliche Nutzung des Bernsteins*, [*in*:] Ganzelewski, M. & Slotta, R. (Hrsg.), Bernstein — Tränen der Götter, *Veröffentlichungen aus dem Bergbau-Museum Bochum*, **64**, 413–426, Bochum.

LATE BRONZE AGE AND EARLY IRON AGE AMBER FINDS FROM THE CATCHMENT AREAS OF THE RIVERS ODER AND VISTULA

Zbigniew BUKOWSKI

Abstract

Until recently, archaeological research pointed to the existence of two main centres in Europe dealing with the acquisition and distribution of amber: Jutland in the west and Sambia in the east. It is believed that two long-distance reciprocal exchange zones (or routes) developed from the Late Neolithic through the Bronze Age and up to the Early Iron Age (HaD_3) — one in the west (from the eastern Alps to Jutland, incorporating Elbe), the other in the east (from the Carpathian Basin across the Moravian Gate towards the Vistula estuary and Sambia). These were used in particular during the Early Bronze Age and the Hallstatt B/C to D_3 periods. The latest results of geological and archaeological research in Central Europe has thrown fresh light on the significance of amber for the societies of the period in question.

Recent archaeological and geological research has led to a reappraisal of existing theories concerning the significance of amber in antiquity. Both the extent of the amber-bearing zone in the southern Baltic region has been reassessed as well as the scale on which amber was exploited by Bronze Age and Early Iron Age communities, in particular by peoples of the Lusatian culture and related societies, inhabiting the catchment areas of the Rivers Elbe, Oder, Warta and Vistula.

The most commonly published hypothesis as to the role of this raw material in long-distance exchange and cultural contacts between Northern and Southern Europe during this period of prehistory was based on direct analogies with the situation encountered during the Roman period (e.g. KOSSACK 1983, 95 ff.). This was one of the precepts which led J. de NAVARRO (1925, 496 ff. and map) to postulate the existence of a long-distance reciprocal exchange mechanism between the Baltic and Balkan regions and the northern Mediterranean coast in the Early Iron Age, which he dubbed the "amber route".

Until recently Jutland and the Sambian Peninsula were thought to have been the only two amber-bearing zones in the Southern Baltic region (e.g. JENSEN 1965, *passim*; SHENNAN 1982, 13 ff.). The aim of key works on the cultural and economic links between these far-removed areas of Europe was to establish the hypothetical courses of these routes of trade and exchange in amber during the subsequent phases of the Bronze Age and Early Iron Age. The Elbe and Saale rivers and the Alpine Pass were believed to have been of strategic significance during the first of these phases, whilst during the second phase the Oder and Vistula, the Moravian Gate, the Eastern Alps and the western part of the Carpathian Basin became of crucial importance to this route (BUKOWSKI 1990, 82 ff.).

Current research into amber deposits in Central Europe indicates that they covered a broad stretch of both the aforementioned regions, not only along the Baltic coast, but also over a wide area further inland (KOSMOWSKA-CERANOWICZ 1988, 173 ff., 174–175, Fig. 2 — map). The amber resource in the eastern part of these territories covered virtually the entire eastern Baltic zone (SAVKEVICH & SOKOLOVA 1993, 49, Fig. 1 — map).

Finds from Jutland dating from *c.* 3500–3200/3000 BC, associated with the Globular Amphorae culture and later the Corded Ware culture (e.g. EBESSEN 1993, 123 ff.), provide vital evidence of the fact that specialist workshops producing amber beads, chiefly for use as exchange goods, appeared at a very early stage. Large hoards of this type of artefact offer further support for this theory. This form of exchange was, no doubt, already well-developed on a local or maybe even inter-regional scale. The zone in question, encompassing Denmark and southern Sweden, was not only attractive by reason of its amber resources. During both the early and late phases of the Bronze Age it played a crucial intermediary role, providing a link with hunter-gatherer societies throughout Scandinavia. We can assume that this must have been an important region, supplying territories to the south with natural produce.

New evidence for the existence of sub-surface amber deposits in the border area between Brandenburg and Mecklenburg has led to a reconsideration of existing theories on the significance of amber in the cultural development of southern Baltic societies. Furthermore, it has forced researchers to take into account the possibility that this raw material may also have been easily accessible to prehistoric communities in north-eastern Germany (SCHULZ 1993, 32 ff. & 40, Fig. 4 — map; for a review of this work see also BUKOWSKI 1994, 138 ff.). R. SCHULZ has put forward a hypothesis, based on his analysis of available archaeological materials, that the development of Bronze Age societies may have been shaped to a considerable degree by local exploitation of amber and its long-distance distribution. In northern Brandenburg and south-west Mecklenburg in particular a concentration of distinctive bronze hoards and richly furnished burials dating from the earlier phases of the Bronze Age has been noted, although older finds associated with the Unětice culture have also been recorded here. Schulz also presents a distribution map of settlements of this period in the Eberswald Palaeovalley between the Lower Oder and the catchment area of the Havel (1993, 40, Fig. 4). Regional communities — the Seddin and Prignitz societies — emerged in this zone during phases IV and V of the Bronze Age. The former (WÜSTEMANN 1974, 67 ff., 94, Fig. 9 — map), which has been divided into a series of hypothetical tribal groups, is characterised by a large number of richly furnished burials dating from all three late phases of the Bronze Age (IV–VI).

Geological data for the area between the Oder and Sambia indicates that amber-bearing deposits covered large areas of Pomerania, the valley of the Middle and Lower Noteć and practically the whole of the Masurian-Ermland Lake District (KOSMOWSKA-CERANOWICZ & PIETRZAK 1982; MAZUROWSKI 1983, 177 ff., 178, Fig. 1, 179, Fig. 2 — map showing amber-bearing regions in Poland). Amber-bearing territories in the east included the Tuchola Forests, the Bay of Gdańsk and the Vistula estuary. In the last of these areas, the region of Żuławy is particularly noteworthy for the large number of Late Neolithic/ Early Bronze Age (mostly Corded Ware culture) amber workshops discovered there (MAZUROWSKI 1983, 189 ff.; 1985, 5 ff.).

Amber was not merely a sporadically occurring raw material in these territories, but was widespread as a result of the deep erosion of Tertiary amber-bearing sediments and glacial action which carried this fossil resin in boulder-clay from its primary deposition area forming Quaternary deposits here. This explains the presence of amber in such large quantities, especially along the edges of river valleys, lakes and the southern Baltic coast. This fact must be taken into account in any further consideration of this raw material's significance on a regional or broader scale to the prehistoric and early historic settlement of this zone.

In the Late Neolithic and Early Bronze Age amber was widely used by the communities inhabiting the catchment areas of the Oder and Vistula (mainly by the Corded Ware culture and related groups, such as the Rzucewo culture), who, during the Early Bronze Age, also participated in the exchange of amber goods with societies in the Carpathian Basin. However, by the time of the Unětice culture and the first stages of the development of 'tomb cultures', amber had almost entirely disappeared from the everyday items of material culture of these peoples. This is particularly noticeable throughout the Lusatian culture. The theory that this phenomenon can be explained by the increasingly widespread practice of cremation as the main form of burial is far from accurate. Although burial inventories were undoubtedly destroyed during the cremation process, in some cases small grave goods, including amber beads, were placed inside cinerary urns after the cremation ritual.

The large series of cemeteries in Upper Silesia and western Little Poland, dating from the HaB_3 to HaD_2 period, at which two burial customs were practised, internment being the more common of the two (BUKOWSKI 1995, 25 ff.), can be considered as representative of most Lusatian culture burials. If we assume that the type of burial custom practised by the communities in question was typical of this entire cultural complex, the sporadic appearance of single amber beads in one or two graves seems to be an accurate indication of how widespread the use of this material was and of what significance it was to these societies.

This problem is somewhat complicated when viewed in the context of the Hallstatt period "amber route" (see BUKOWSKI 1992, 39 ff.; 1993, 73 ff.). It is widely believed that amber played a major role in long-distance exchange with Southern Europe. Amber artefacts are almost entirely absent from the archaeological record of Czech, Moravian and Slovak territories. This holds true both for Lusatian culture groups (Platěnic) and for Bylany, Horákov and northern Kalenderberg culture communities (HRALA & PLESL 1989, 219 ff.). The recent discovery at a Horákov culture settlement in Kuřím near Brno of an amber workshop dating from the HaD_1 period (ČIŽMÁR 1996, 231 & 235) has, however, led to a reconsideration of the role played by these societies in the distribution of amber to areas lying further south. Another find of particular interest related to this culture comes from Býčí Skála cave near Brno. This constituted either a local cult site or the richly furnished burial of a prominent member of society who was associated with metalworking. Among the inventory of items recovered here were 1310 amber beads, including two necklaces, one comprising 656, the other 512 amber beads (ANGELI 1970, 147).

Considerable quantities of amber beads have also been noted in many of the burials at Scythian cemeteries of the Vekerzug culture in southern Slovakia and northern Hungary. Such instances were recorded at the Chotín cemetery, in the district of Komarno (Slovakia), dating from the HaC/D or HaD_1 period (BECK & DUŠEK 1969, 274 ff.), and at the HaD_1 cemetery in Szentes-Vekerzug (northern Hungary). A total of 1152 amber beads were recovered from four female burials at the latter of these two cemeteries, whilst a further grave yielded a necklace consisting of 279 such beads (GALÁNTHA 1993, 181 & footnotes). A large necklace made up of a variety of elements, including numerous amber beads, was found in an HaD_2 grave at the Kalenderberg culture cemetery at Sopron-Krautacker, on the border between Austria and Hungary (JEREM 1981, 114 & 115, Fig. 7:1–5).

Similar discoveries in northern Yugoslavia also deserve mention, particularly those from Slovenia. Recent excavations have provided yet more evidence of the presence of this type of burial, stretching across a far greater area of the western Balkans than had previously been assumed (see PALAVESTRA 1993, 182 — distribution map). The majority of these were richly furnished burials referred to as princes' or aristocratic graves. To give an example, two barrow graves at Atenica near Čačak (Slovenia) contained over 2400 various amber ornaments, whilst another cemetery at Novi Pazar yielded over 1000 amber beads (PALAVESTRA 1984, 14, Map 1 & Pl. 4–8; see also SPRINCZ 1993, 179 & 183, footnote 1 — list of graves containing amber finds discovered in this part of the Balkans).

There is no doubt that the items in question were made of Baltic amber. The ornaments found in graves, dating mostly from the 6th to 4th centuries BC, indicate that amber was a highly prized commodity amongst the higher classes of Ilyrian societies. Some of the artefacts under consideration consisted of figurines and pendants clearly linked to the indigenous culture and customs of this region, thus confirming the fact that they were produced locally from imported raw material. Further research is required to establish the time-span and territorial extent of long-distance trade in amber during this period and to verify the role played in this exchange by societies inhabiting the catchment areas of the Lower Vistula, Warta and Oder. Other than those communities living in the eastern Alps and northern Italy it is becoming increasingly clear that during the HaD period the population of the Little Hungarian Lowlands (Kiss-Alföld) and northern Yugoslavia was also of significance to the operation of the amber route. The early Celtic presence in the Upper Danubian river basin should also be taken into consideration in this respect.

It is difficult to gain a precise picture of the mechanics involved in this long-distance exchange of amber, particularly given that amber ornaments appear only very sporadically in the archaeological record of the Oder and Warta catchment areas, with hoards of this raw material being virtually non-existent. The one exception to this rule is the widespread use of amber beads in Gdańsk-Pomerania during the HaD_{1-2} period among societies belonging to the East Pomeranian culture. These beads constituted decorative elements of the bronze jewellery used for adorning face urns.

To date most published theories on the amber route, particularly those of B. STJERNQUIST (1967, 7 ff. — see also diagrams depicting exchange mechanisms) and K. W. STRUVE (1979, diagram shown in Pl. 72), support the idea that transactions took place in a number of regional phases, rather than having been carried out directly in the form of long-distance journeys organised by traders from the south. Bearing in mind the nature and extent of imports from the Eastern Alps and Northern Italy, exchange in local stages must have been the fundamental way of distributing more commonplace and less valuable commodities from the south in this region. In contrast, it is not unreasonable to assume that luxury goods destined for use only by elite members of society (items such as large bronze bowls, chain-link elements of horse harness fittings or distinctive ornaments) may have been traded by means of well-organised, long-distance expeditions to the north, although these were probably infrequent. Support for this theory can be found in the emergence of what was possibly a trading post at Gorszewice-Komorowo, Poznań province. Amber finds have been recovered both in the form of raw material and numerous beads from this settlement complex, which lies in close proximity to one of the most accessible crossings of the Noteć valley, leading on towards central Pomerania (BUKOWSKI 1989, 197 & 201, Fig. 1 — map).

There are two feasible explanations for the evident lack of interest in amber displayed by Lusatian culture societies. The first possibility is that amber may have been amassed primarily for the purpose of using it as an exchange commodity, although virtually no hoards of this raw material have ever been found at any contemporary settlements. The few examples discovered in the Königsberg region and in Lower Silesia have not been dated with any great accuracy. The alternative explanation is that amber may simply have not been in fashion, or its use may have been limited by the beliefs and customs of the of the time. This, however, must remain a tentative hypothesis.

Zbigniew BUKOWSKI
Ośrodek Ratowniczych Badań Archeologicznych
ul. Czackiego 15/17
00-043 Warszawa, Poland

ZNALEZISKA BURSZTYNOWE Z MŁODSZYCH FAZ EPOKI BRĄZU I WCZESNEJ EPOKI ŻELAZA W DORZECZU ODRY I WISŁY

Zbigniew BUKOWSKI

Streszczenie

W literaturze archeologicznej panował dotąd pogląd o wykształceniu się w Europie dwóch ośrodków wydobywania i dystrybucji bursztynu: Jutlandii na zachodzie i Sambii na wschodzie. W odniesieniu do sytuacji kulturowej i osadniczej od schyłku neolitu, poprzez epokę brązu, do wczesnej epoki żelaza (HaD_3), przyjmowano wykształcenie się dwóch stref (tzw. szlaków) dalekosiężnych kontaktów wymiennych: zachodniej (od wschodnich Alp do Jutlandii z wykorzystaniem Łaby) oraz wschodniej (z obrębu Kotliny Karpackiej poprzez Bramę Morawską ku ujściu Wisły i Sambii), użytkowanych zwłaszcza we wczesnych fazach epoki brązu oraz w okresie HaB/C – HaD_3. Aktualny stan badań geologicznych oraz archeologicznych w Europie Środkowej stanowi podstawę do wysunięcia nowych spostrzeżeń i propozycji w odniesieniu do miejsca i roli bursztynu dla tamtejszych społeczności w omawianym czasie.

W oparciu o rezultaty badań geologicznych, a także informacje historyczne, potwierdzono występowanie łatwych do wykorzystania złóż bursztynu w całym pasie południowobałtyckim, od Jutlandii, poprzez Meklemburgię, północną Brandenburgię, Pomorze Zaodrzańskie (*Vorpommern*), Pomorze (włączając tu dolinę Noteci i Zatokę Gdańską) oraz Pojezierze Mazursko-Warmińskie. Do strefy tej należy włączyć też pas wschodniobałtycki, zwłaszcza nadmorską część Litwy i Łotwy. Stwierdzenie to zezwala na wysunięcie przypuszczenia, że obok obu wspomnaianych rejonów bursztynonośnych na surowiec ten można było natrafić przaktycznie na rozległym obszarze południowobałtyckim. Pociąga to za sobą daleko idące konsekwencje w odniesieniu do przyjmowanych dotąd poglądów tak dotyczących hipotetycznych szlaków dalekosiężnej wymiany, jak i roli bursztynu dla rozwoju tamtejszych ugrupowań osadniczych.

W odniesieniu do ziem w dorzeczu Odry i Wisły aktualny stan badań zezwala na przedstawienie następującej sytuacji. Dla tamtejszych społeczności z eneolitu i początku epoki brązu, zwłaszcza kultur sznurowych, kultury amfor kulistych oraz kultury unietyckiej, bursztyn należał do powszechnie wykorzystywanych surowców zarówno do produkcji ozdób, jak i stanowiących ważny ekwiwalent w dalekosiężnej wymianie, głównie ze społecznościami Kotliny Karpackiej.

Od fazy rozwoju ugrupowań tzw. „mogiłowych" oraz niemal przez cały okres istnienia kultury łużyckiej, tj. od BB_2 do HaD, uderza niemal zupełny brak wytworów bursztynowych oraz surowca w grobach, osiedlach i skarbach tych ludów. Analiza znalezisk zaprezentowana przez autora tych uwag potwierdza sporadyczną obecność w grobach pojedynczych paciorków, a wyjątkowo tylko wieloelementowych naszyjników bursztynowych. Wysuwane poglądy o zniszczeniu inwentarzy grobowych wskutek powszechnie panującego zwyczaju kremacji zmarłych mogą tylko częściowo uzasadnić powyższy stan rzeczy. Obszerna seria birytualnych cmentarzysk z Górnego Śląska i zachodniej Małopolski z przewagą pochówków szkieletowych, a więc zachowujących inwentarz, w tym bursztyn, potwierdza brak zainteresowania nim od schyłku epoki brązu (HaB_3), poprzez okres Hallstatt (C–D) przez tamtejsze społeczności.

Dla tej fazy od HaB_3 do HaD_3 przyjmuje się wykształcenie tzw. „szlaku (lub szlaków) bursztynowego", w wyniku czego w obręb Śląska, zachodniej Małopolski oraz Wielkopolski i Kujaw sprowadzono liczne importy, głównie metalowe, ze wschodnich Alp i północnej Italii, w obrębie których bursztyn tzw. bałtycki był wówczas powszechnie wykorzystywany, głównie jednak przez warstwy uprzywilejowane. Z kilku skarbów surowca bursztynowego, znanych z rejonu Królewca, ujścia Wisły oraz Dolnego Śląska, tylko znalezisko z Juszkowa (woj. gdańskie) datowane jest na początki epoki żelaza, przy czym mamy tu zapewne do czynienia z surowcem na użytek własny. Z obrębu stref bursztynonośnych brak jest niemal zupełnie wspomnianych importów, ujawnionych na obszarach położonych od nich na południe.

Rozpowszechnienie ozdób bursztynowych w HaD w dorzeczu Odry i Wisły dotyczy tylko społeczności związanych z kulturą wschodniopomorską (kulturą urn twarzowych).

Na aktualnym etapie badań nie jest możliwe wyjaśnienie mechaniki eksportu bursztynu z północy w obręb dorzecza Dunaju, wschodnich Alp i północnej Italii. Jednym z założeń badawczych może być przyjęcie możliwości, że bursztyn gromadzony był głównie dla dalekosiężnej wymiany (w tym przypadku etapowej), a tylko sporadycznie wykorzystywany w charakterze ozdób (moda, zwyczaje, tabu?). Zaś jego wykorzystywanie przez społeczności w dorzeczu Odry i Wisły mogło się łączyć głównie z obrzędami kultowymi, które w materiale archeologicznym nie pozostawiły niemal żadnych śladów.

Bibliography

ANGELI W.
1970 Zur Deutung der Funde aus der Býčískála-Höhle, [*in:*] *Krieger und Salzherren. Hallstattkultur im Ostalpenraum, Mainz*, 139–150.

BECK C. W. & DUŠEK M.
1969 Die Herkunft des Bernsteins vom thrakischen Gräberfeld von Chotín, *Slovenská archeológia*, **17**, 247–258.

BUKOWSKI Z.
1989 Szlaki handlowe z Południa na Pomorze w młodszej epoce brązu i we wczesnej epoce żelaza, [*in:*] *Problemy kultury łużyckiej na Pomorzu*, Słupsk, 185–208.

1990 Critically about the so-called amber route in the Odra and Vistula river basins in the Early Iron Age, *Archaeologia Polona*, **28**:1988 (1990), 71–122.

1992 Tzw. szlak bursztynowy z wczesnej epoki żelaza na obszarach na południe od Sudetów i Karpat w świetle importów pochodzenia południowego, [*in:*] *Ziemie polskie we wczesnej epoce żelaza i ich powiązania z innymi terenami*, Rzeszów, 39–54.

1993 Tzw. szlak bursztynowy z wczesnej epoki żelaza z międzyrzecza Łaby i Wisły w świetle skarbów z importami i bursztynem, [*in:*] *Miscellanea archaeologica Thaddaeo Malinowski dedicata ...*, Słupsk - Poznań, 73–97.

1994 Review of *Veröffentlichungen des Brandenburgischen Landesmuseums für Ur- und Frühgeschichte*, **27**:1993, [*in:*] *Archeologia Polski*, **39**, 136–144.

1995 Niektóre dyskusyjne problemv związane z rozwojem grupy górnośląsko-małopolskiej, [*in:*] *Dziedzictwo kulturowe epoki brązu i wczesnej epoki żelaza na Górnym Śląsku i w Małopolsce. Tzw. grupa górnośląsko-małopolska kultury łużyckiej. Śląskie Prace Prehistoryczne*, **4** (Bytom), 25–46.

ČIŽMÁŘ M.
1996 Das hallstattzeitliche Gehöft in Kuřím, *Pravěk* N.Ř., **5**:1995 (1996), 217–254.

EBESSEN K.
1993 Sacrificies to the powers of nature, [*in:*] Digging to the past, Larhus, 121–130.

GALÁNTHA M.
1986 The Scythian Age cemetery of Csanytelek-Ujhalastó, [*in:*] *Hallstatt Kolloquium Veszprém 1984*, Budapest, 69–77.

HRALA J. & PLESL E.
1969 Nálezy jantaru v kulturách mladši doby bronzové, a časné doby železné v Čechách, [*in:*] *Problemy kultury łużyckiej na Pomorzu*, Słupsk, 209–226.

JENSEN J.
1965 Bernsteinfunde und Bernsteinhandel der jüngeren Bronzezeit Dänemarks, *Acta Archaeologica* (Kopenhagen), **36**, 43–86.

JEREM E.
1981 Zur Späthallstatt- und Frühlaténezeit in Transdanubien, [*in:*] *Die Hallstattkultur. Symposium Steyr 1980*, Linz, 105–136.

KOSMOWSKA-CERANOWICZ B.
1988 Niektóre złoża bursztynu i próby klasyfikacji żywic kopalnych, [*in:*] *Surowce mineralne w pradziejach i we wczesnym średniowieczu Europy Środkowej*, Wrocław, 173–188.

KOSMOWSKA-CERANOWICZ B. & PIETRZAK T .
1982 *Znaleziska i dawne kopalnie bursztynu w Polsce*, Warszawa.

KOSSACK G.
1983 Früheisenzeitlicher Gütertausch, *Savaria*, **16**, 95–112.

MAZUROWSKI R.F.
1983 Bursztyn, [*in:*] *Człowiek i środowisko w pradziejach*, J.K. Kozłowski & S.K. Kozłowski (eds), Warszawa, 177–188.

MAZUROWSKI R.F.
1985 Amber treatment workshops of the Rzucewo culture in Żuławy, *Przegląd Archeologiczny*, **32**, 5–50.

De NAVARRO J. M.
1925 Prehistoric routes between Northern Europe and Italy defined by the amber trade, *The Geographical Journal*, **56**, 496–521.

PALAVESTRA A.
1984 *Kneževski grobi starijeg gvozdenog doba na Centralnom Balkanu*, Beograd.

1993 *Prehistoric amber in Central and Western Balkans*, Beograde.

SAVKEVICH S.S. & SOKOLOVA T. N.
1993 Amber-like fossil resins of Asia and the problem of their identification in archaeological context, [*in:*] *Amber in archaeology*, Proceedings of the Second International Conference on Amber in Archaeology, Liblice 1990, C.W. Beck & J. Bouzek (eds), Praha, 48–50.

SCHULZ R.
1993 Die natürlichen Vorkommen von Bernstein in Nordbrandenburg und die Besiedlung in der Bronzezeit, *Veröffentlichungen des Brandenburgischen Landesmuseums für Ur- und Frühgeschichte*, **27**, 32–46.

SHENNAN E. H.
1982 Exchange and ranking: the role of amber in Earlier Bronze Age in Europe, [*in:*] *Ranking resource and exchange*, C. Renfrew & S. Shennan (eds), Cambridge, 13–25.

SPRINCZ E .
1993 Veneter – Skythen – Bernsteinhandel, [*in:*] *Amber in archaeology*, Proceedings of the Second International Conference on Amber in Archaeology, Liblice 1990, C.W. Beck & J. Bouzek (eds), Praha, 179–186.

STJERNQUIST B.
1967 Models of commercial diffusion in prehistoric times, *Scripta Minora 1965–1966*, **2**, Lund.

STRUVE K. W.
1979 Die jüngere Bronzezeit, [*in:*] *Geschichte Schleswig-Holsteins*, **2**: K.W. Struve, H. Hingst, & H. Jankuhn, *Von der Bronzezeit bis zur Völkerwanderungszeit*, Neumünster, 97–144.

WÜSTEMANN H.
1974 Zur Sozialstruktur im Seddiner Kulturgebiet, *Zeitschrift für Archäologie*, **8**, 67–107.

AMBER IN THE MATERIAL CULTURE OF THE COMMUNITIES OF THE REGION OF KUIAVIA DURING THE ROMAN PERIOD

Aleksandra COFTA-BRONIEWSKA

Abstract

Excavations carried out over the past thirty years by the Kuiavia Research Group have shown that amber played a vital role in the culture of prehistoric societies inhabiting this region. A thriving amber-working industry developed here, despite the fact that Kuiavia lies a considerable distance from those areas richest in this natural resource. The link between amber-working in Kuiavia during the Roman period and the presence of a major trade route passing through this region is examined and an assessment made of the impact which becoming part of a long-distance exchange mechanism had on Kuiavian communities.

Introduction

It is generally known that the various communities of the Stone Age were not only aware of the existence of amber, but also used it as a material for their artefacts. Initially, this fossil resin was employed only in the regions where its rich deposits were found; fairly soon, however, it began to be purchased both in neighbouring and in distant areas.

In view of its physical characteristics and decorative value, amber was employed mainly for the fashioning of jewellery and became a significant factor of the cultural symbolism of prehistoric humanity; at the same time, it found no technological application. Wealthy societies that represented a higher level of civilizational development, appreciated its pleasing appearance and imported it either in the form of finished products or as raw material to be cut in their workshops.

Among the most important sources of amber during that epoch, were its deposits on the Baltic sea coast. It was this amber which reached the peoples of southern Europe. Due to its geographical location, the present-day territory of Poland served as a transit area for this long-distance exchange, although the actual routes of transportation varied from one period of history to another. Nevertheless, apparently no communities of the lowlands and uplands of Poland attempted to process this material on a significant scale during the early days of its exportation. This situation changed only in the first centuries AD, when amber workshops appeared throughout the region of Kuiavia (Kujawy), which had not previously participated in the trading relations between the Baltic coast and southern Europe, except for brief episodes in the Neolithic period and early Bronze Age.

The aforementioned change took place after the Celts had started using a route which crossed Kuiavia. The communities of that region had earlier been involved in the long-distance trading of bronze jewellery before participating in the transcontinental exchange of crude amber and amber goods. A breakthrough discovery revealing this fact was made during archaeological exploration at the village of Jacewo in 1968. Excavation results not only proved that this fossil resin had been worked in a region far removed from the natural deposits of this mineral, but also revealed that local amber-working techniques had been significantly advanced — at the time of the discovery, the latter fact was not yet generally accepted with reference to the Roman period. Although initially archaeologists were quite sceptical about such an interpretation of the source material discovered at Jacewo, subsequent studies of this region provided further data from the realm of material culture, confirming and expanding the original conclusion.

The discovery at Jacewo initiated investigations into the importance of amber to the prehistoric population of Kuiavia. This research was carried out as part of a long-term project by the Team for the Study of Kuiavia (TSK) of the Adam MICKIEWICZ University in Poznań, whose work to date has yielded abundant source material relating to the issue under discussion. In virtually every excavated settlement dated to between the 1st and 5th centuries AD, either specialist amber workshops or at least indications of the practice of amber-working were discovered.

Beside Jacewo, amber workshops were discovered in: Łojewo and Konary (both in 1975), Gąski (sites 18, 24 & 25) and Inowrocław (site 100) (all four in the late 1980s), and in Kuczkowo (in 1996). Moreover, source material identified in the settlements at Inowrocław (site 95) and Krusza Zamkowa (site 3) (Fig. 1) suggests that amber-working was also practised there. The available source material was collected from various locations throughout the region in question and accordingly warrants certain general conclusions.

The aim of this paper is to discuss the extent of amber-working in Kuiavia, to offer specific deductions, and to emphasise the importance of this activity in the overall development of local societies during the Roman period.

The geographical scope of this discussion is the Inowrocław Plain in Kuiavia, i.e. the territory where the Adam Mickiewicz University in Poznań conducts archaeological exploration, this being the only area for which we possess a body of source material relevant for the subject under consideration in this paper.

Although the available sources are abundant, comprising ten settlements where both dwellings associated with amber-working and large amounts of production waste were discovered, we must admit that our data are not fully comprehensive and that the finds were only fragmentary. This was due to the nature of our exploration, which in Łojewo, Konary and Kuczkowo consisted of rescue excavations, in Gąski 24, of reconnaissance, and in Gąski 25, of detailed surface survey. The majority of the most relevant data was recovered during excavations at the settlements of Jacewo (site 4b) and Gąski (site 18). Some specimens of amber were, however, poorly preserved and disintegrated.

On the whole, it must be assumed — in respect of the rescue work and especially in the case of chance finds reported by local residents — that many small fragments of amber might have been dispersed or overlooked. Consequently, as amber is located at every site which is being investigated now, one may ask how reliable the conclusion was that this material did not appear in similar settlements which were discovered during earlier stages of research.

It is now thirty years since the first amber workshop was discovered in Kuiavia and since studies on the significance of this fossil resin were initiated — studies addressing themselves not only to the Roman period, but to the entire prehistory of this region.

The available source material and data relating to material culture proves that this issue evolved continuously throughout history. At each stage of their development, the societies of Kuiavia used amber in a different manner, on a different scale and for different reasons. The results and progress of our research have been presented regularly in various forms, such as short communiqués[1], reports summarising the discoveries of individual collections of source material[2], more extensive articles brought out

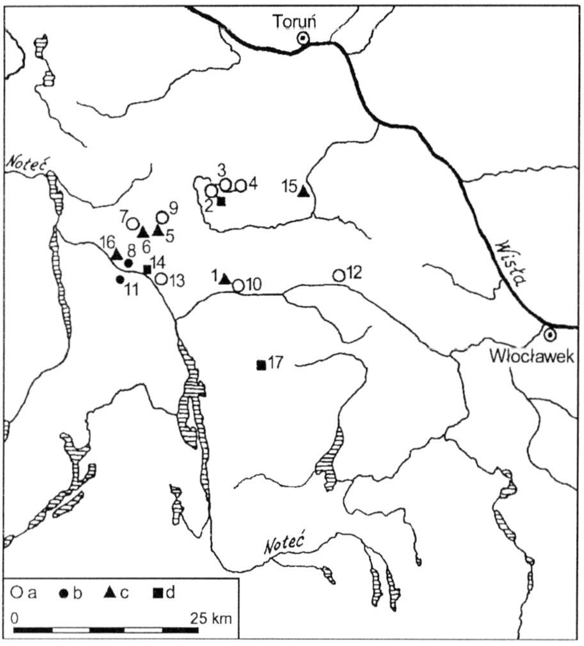

Fig. 1. Locations of amber finds : a — amber workshops; b — home-production sites; c — isolated lumps of raw material; d — burial grounds.

[1] "Krusza Zamkowa gm. Inowrocław st. 3", *Informator Archeologiczny* (research of 1973, published in 1974), 146; (research of 1974, published in 1975), 37. "Konary gm. Dąbrowa Biskupia st. 28", *Informator Archeologiczny* (research of 1975, published in 1976), 126–127. "Łojewo gm. Inowrocław", *Informator Archeologiczny* (research of 1975, published in 1976), 138–139. "Inowrocław st. 95", *Informator Archeologiczny* (research of 1979, published in 1980), 125–126. "Gąski gm. Gniewkowo st. 18," *Informator Archeologiczny* (research of 1987, published in 1988), 114–115; (research of 1988, published in 1992), 71.

[2] Aleksandra COFTA-BRONIEWSKA, "Badania archeologiczne w Jacewie pow. Inowrocław w 1968 roku" ["Archeological Research in Jacewo, District of Inowrocław, in 1968"], *Komunikaty Archeologiczne*, Bydgoszcz, 1972, 44–54. Idem, "Warsztaty produkcyjne z okresu wpływów rzymskich w Jacewie pow. Inowrocław" ["Manufacturing Workshops from the Roman Period in Jacewo, District of Inowrocław"], *Sprawozdania PTPN*, 1970, **1** (84), published in 1972, 93–94. Idem, "Badania archeologiczne w Inowrocławiu 1968 r." ["Archeological Research in Inowrocław in 1968"], *Sprawozdania archeologiczne*, 1972, **24**, issue 9, 163–164. Idem, "Badania archeologiczne w Inowrocławiu i powiecie inowrocławskim w latach 1967–1973 r." ["Archaeological Research in Inowrocław and the District of Inowrocław in 1967–1973"], *Wiadomości Archeologiczne*, **40**, 1975, issue 3, 411–417. Idem, "Rozwiane wątpliwości" ["Resolved Doubts"], *Z otchłani wieków*, **42**, 1976, issue 2, 119–123. Idem, "Osada z okresu wpływów rzymskich na st. 4b w Jacewie pow. Inowrocław" ["A Settlement from the Roman Period at Site 4b in Jacewo, District of Inowrocław"], *Komunikaty*

in research journals and collections of studies[3] and in treatises offering a broader view of this topic[4].

Although it is an established policy of the TSK to have complete specialist anlyses of the collected material carried out after each season of excavation, this turned out to be impossible in the case of Kuiavian amber, as we were not able to find a laboratory willing to undertake this work. We only managed to commission three analyses of the amber from the workshop at Jacewo, which were conducted soon after its discovery and whose results were published in 1971[5].

Finally, for clarity's sake, let us explain that as monographs of individual settlements are not yet ready, the collections of source material identified in this paper will be dated only approximately, either to the early or to the late Roman period.

Description of the Source Material

Jacewo, Site 4b

The discovery in Jacewo, which — as we have mentioned — was the first one in a long series, still remains the most significant collection of information on amber-working in the Roman period. It provided the richest and most exhaustive corpus of source material, even though the identified amber was confined to a relatively small area, appearing in two marked clusters, one approximately 10 m away from the other.

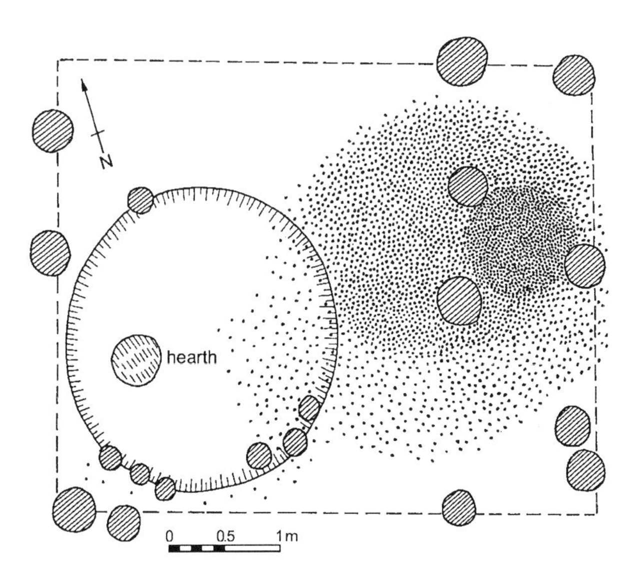

Fig. 2. Jacewo, site 4b: amber workshop No. 1.

The Jacewo 4b site is located along the eastern boundary of the city of Inowrocław. It was catalogued in 1967, underwent detailed reconnaissance and fieldwalking in 1968 and was excavated in the 1969 and 1970 seasons. A total area of approximately 500 m^2 has been uncovered[6].

Cluster No. 1 which was very large and typical of an amber-working site, deserves particular attention. On the other hand, apparently only a part of Cluster No. 2 was unearthed.

Cluster No. 1

This consisted of 4,940 pieces of amber. These were found inside a rectangular structure measuring 4 x 5 m, whose area was outlined by fourteen post-holes of 30–50 cm in diameter. The interior of the structure was divided into a main room of 16 m^2 and a smaller one of 4 m^2 (Fig. 2). At the time when this building was inhabited, its floor was a layer of very black humus; within the structure, the humus was very hard and covered with a thin layer of highly viscous oily clay. In the western part of the structure, a shallow, oval hollow measuring 2.40 x 2.70 m was discovered,

---

Archeologiczne, 1978, 143–147. Idem, "Z dziejów badań nad późnym okresem lateńskim i okresem wpływów rzymskich na Kujawach" ["From the History of the Research into the Late La Téne Period and the Roman Period in Kuiavia"], *Komunikaty Archeologiczne*, 1990, 63–79. Aleksander Kośko, "Badania w strefie nadnoteckiej Inowrocławia w 1972 r." ["Research in the Area of Inowrocław on the Noteć in 1972"], *Komunikaty Archeologiczne*, 1978, 17–21.

3 A. COFTA-BRONIEWSKA, "Amber Craft in Kuiavia in the Era of Przeworsk Culture," Archeologia Polona, **23**, 1984, 149–165. Idem, "Badania stanowiska 18 w Gąskach gm. Gniewkowo, woj. bydgoskie" ["Research at Site 18 in Gąski, Commune of Gniewkowo, Province of Bydgoszcz"], *Ziemia Kujawska*, **9**, 1993, 201–224. A. KOŚKO, "Specyfika rozwoju kulturowego społeczeństwa Niżu Polski w dobie schyłkowego neolitu i wczesnej epoki brązu. Zarys problematyki" ["The Distinctive Features of the Cultural Development of the Society of the Plains of Poland toward the Close of the Neolithic Period and in the Early Bronze Age. A Summary of the Subject"], [*in:*] *Lubelskie materiały archeologiczne* [*The Lublin Archeological Studies*], **6**, Lublin, 1991, 23–37. A. COFTA-BRONIEWSKA, "Metodyka badań regionalnych na Kujawach" ["The Methodology of Regional Research in Kuiavia"], [*in:*] *Kontakty pradziejowych społeczeństw Kujaw z innymi ludami Europy* [*The Relations of the Prehistoric Communities of Kuiavia with Other European Peoples*], Inowrocław, 1988, 15–27.

4 A. COFTA-BRONIEWSKA, *Grupa Kruszańska kultury przeworskiej* [*The Krusza Group of the Przeworsk Culture*], Poznań, 1979. A. KOŚKO, *Rozwój kulturowy społeczeństw Kujaw w okresach schyłkowego neolitu i wczesnej epoki brązu* [*The Cultural Development of the Communities of Kuiavia toward the Close of the Neolithic Period and in the Early Bronze Age*], Poznań, 1979. A. COFTA-BRONIEWSKA & A. KOŚKO, *Historia pierwotna społeczeństw Kujaw* [*The Earliest History of the Communities of Kuiavia*], Warszawa–Poznań, 1982.

5 Stanisław BERNAT, "Bursztyn w podczerwieni" ["Amber in the Infrared Light"], *Problemy*, **27**, 1971, No. 12, 36–37.

6 Cf. footnote 2, A. COFTA-BRONIEWSKA.

containing a hearth, i.e. a circle of burnt-out red pug, 50 cm in diameter and 20 cm thick, covered with a layer of ash and charcoal.

More than 4,500 pieces of amber were scattered across the eastern part of the room over an area of 7–8 m^2, 3,500 of these (including 100 damaged beads and 10 lumps of crude amber) appearing within an area of 4 m^2. Most amber was found in small hollows (5–10 cm in diameter and 5–20 cm deep) which had been made all over the floor in this part of the structure. Since the structure contained many potsherds, animal bones and household objects, it was most certainly a dwelling. The breakdown of the identified amber proves, in turn, that this fossil resin was worked there on an extensive, professional scale. Out of the 4,940 pieces of amber, we discovered: 13 nodules of raw material; 13 unfinished products on which work had barely begun; 238 beads broken as holes had been drilled in them; 333 fragments measuring between 0.8 and 1 cm; 4,337 shavings from 2 to 8 mm in size, and only 6 finished products (Figs. 3–5; 6: 1–2). The total weight of this assemblage,

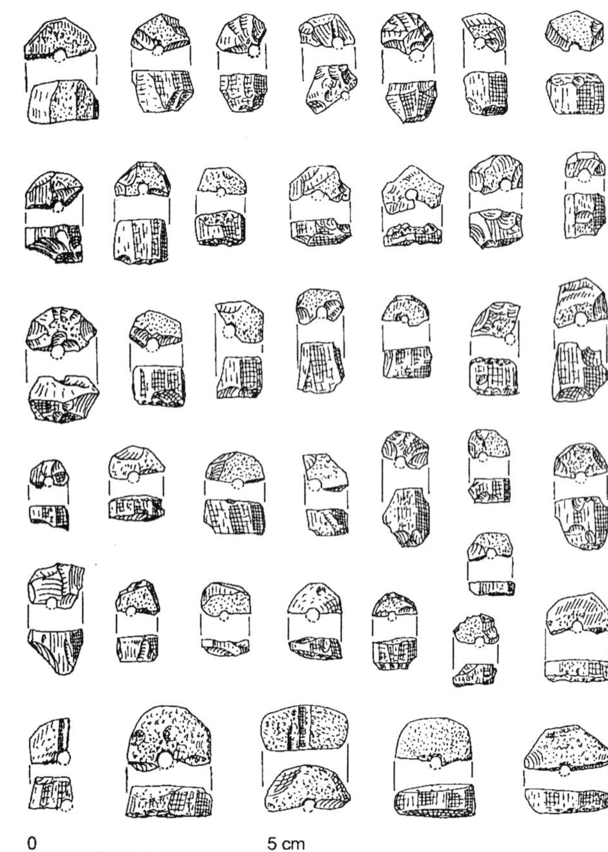

Fig. 4. Jacewo, site 4b: pieces of amber from workshop No. 1.

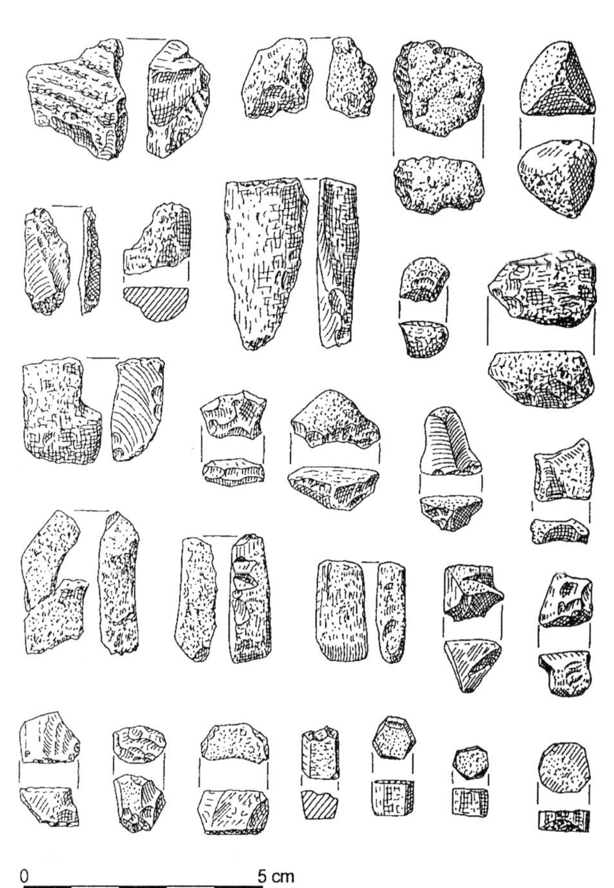

Fig. 3. Jacewo, site 4b: pieces of amber from workshop No. 1.

including many pieces measuring less than 2 mm, was 390 g[7]. As this large amount of source material was concentrated in a small area near the north-eastern corner of the room, it suggests that this was the location of the amber workshop. Moreover, if we assume that the dozens of small hollows seen on the floor in this part of the room were traces of the legs of a portable piece of equipment, it may be supposed that the artisan often changed his work place.

Finally, eight imported objects were found in the amber workshop: fragments of four glass vessels, three glass beads and a cowrie shell (Fig. 6: 19–25). These may suggest the extent of the importance of this particular craft.

Cluster No. 2

Approximately 10 m away from the first scatter, another much smaller cluster of amber pieces was identified. Comprising only about 80 specimens, it is neverthe-

[7] A. COFTA-BRONIEWSKA, "Amber Craft in Kuiavia ..."

less notable for its contents: 3 lumps of raw material, 15 beads broken as holes were drilled in them, and approximately 60 shavings (Fig. 6: 3–17). Thus, it may suggest that a second amber workshop had operated nearby, in a structure that was situated outside the excavated area. There was, however, a specialist bone and antler workshop close to the cluster, which might have undertaken amber-working as well. A fragment of a *terra sigillata* vessel (Fig. 6: 18), found in the same location as the second cluster of amber, may be considered another significant piece of evidence.

Łojewo, Site 4

This settlement is located approximately 5 km away from the north-eastern perimeter of Inowrocław. It was founded on a small sandy hill, on the eastern slope of the valley of the River Noteć. Due to the scarcity of sand in Kuiavia, a region of chernozem soil, this spot — like many other hills — was eventually turned into a sand quarry. In 1975, a rescue excavation was conducted at this site in the extant part of the scarp of the deep pit. Out of the 41 structures discovered, one (No. 2)[8] is of interest to our topic. Its shape was reconstructed as having been that of a rectangle measuring 3 x 4 m. The material found inside suggests that it was both a dwelling and an amber workshop (Fig. 7). The latter function is evidenced by the relatively large number of amber pieces recovered from this feature and by their breakdown: the 400 recorded pieces included an unfinished bead, 43 items broken as holes were drilled in them, 34 pieces of crude amber (measuring approximately 1 cm) and 320 smaller pieces (2–8 mm in size) (Fig. 8). The total weight of this assemblage, inclusive of small grains, was about 54 g[9]. The amber was found at the bottom of the structure, and was noticeably concentrated in an area of approximately 2.5 m², situated in its centre, which might be

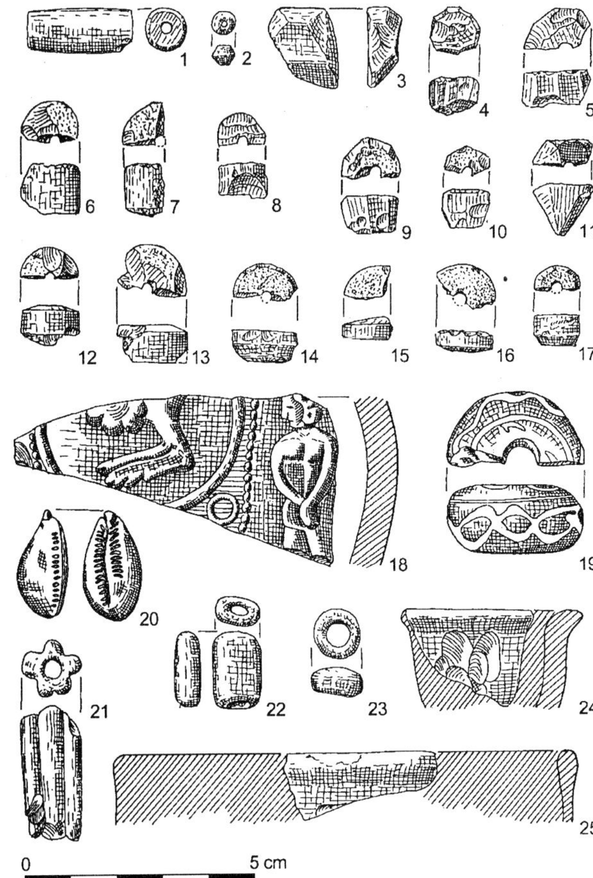

Fig. 6. Jacewo, site 4b: 1–2 — two beads from workshop No. 1; 3–17 — pieces of amber from workshop No. 2; 18–25 — imported objects from workshop No. 1.

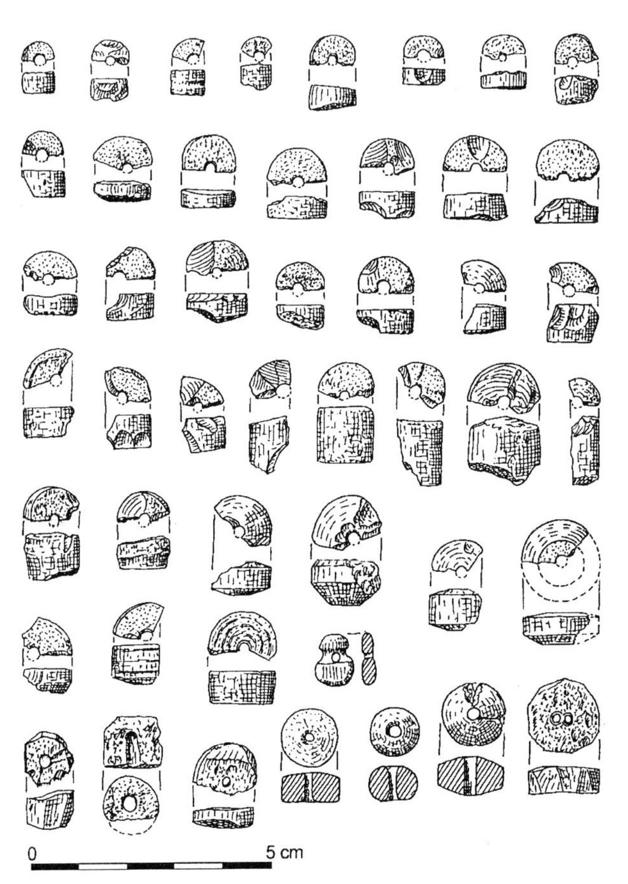

Fig. 5. Jacewo, site 4b: pieces of amber from workshop No. 1.

[8] Irena TOMYS, *Osada kultury przeworskiej w Łojewie stan. 4* [*A Settlement of the Przeworsk Culture in Łojewo, Site 4*], unpublished M.A. thesis, 1978.

[9] Cf. footnote 7.

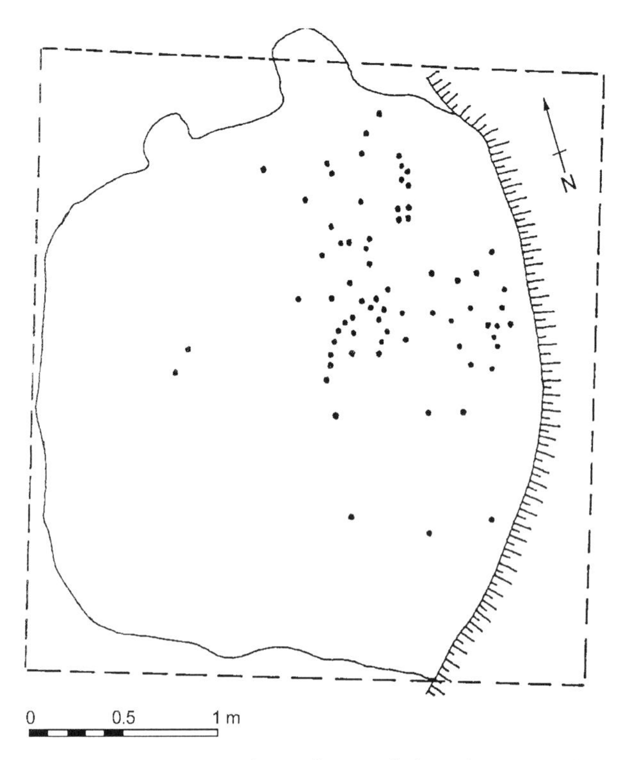

Fig. 7. Łojewo, site 4: the amber workshop (structure 2).

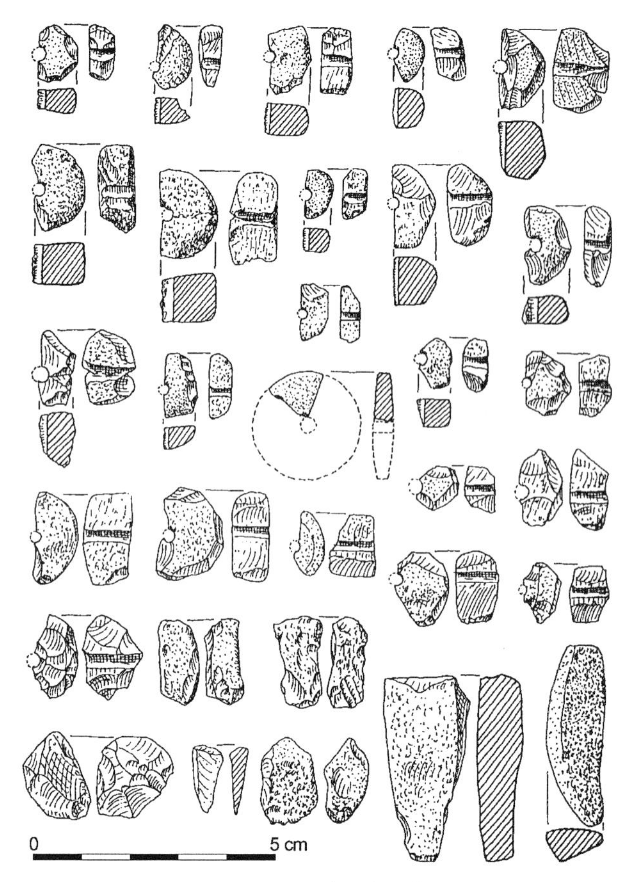

Fig. 8. Łojewo, site 4: pieces of amber from the workshop (structure 2).

assumed to have been the location of the artisan's workshop. No amber beads were identified there, although a single glass bead was discovered.

Konary, Site 28

The village of Konary is located some 15 km south-east of Inowrocław, and its prehistoric settlement was founded on the southern side of the valley of the river Bachorza, on a low sandy mound. Because of its soil, by 1975 the entire central part of the settlement had been lost to sand quarrying and only its eastern edge survived. In 1975 and 1976, rescue excavations revealed a total of 54 structures in this location, out of which only one (No. 19) is of relevance to our topic. As it was situated on the perimeter of the sand quarry, by a lucky coincidence only a small section of it had been destroyed[10]. Originally this post-built construction was apparently rectangular in plan, measuring 4 x 3.5 m (Fig. 9). Traces of five posts, 35–40 cm thick and driven deep into the ground (c. 60–70 cm), were discovered. A hollow containing a hearth of c. 60 cm in diameter was found inside the structure. Some 1,500 pieces of amber were recovered from this structure. Out of these the most numerous items were 1,346 shavings, between 1 and 8 mm in size, and 61 pieces measuring over 8 mm. The remainder consisted

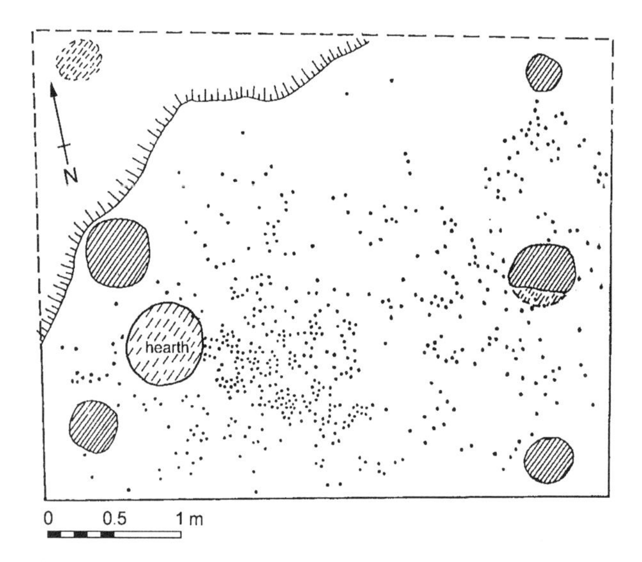

Fig. 9. Konary, site 28: the amber workshop (structure 19).

[10] Grażyna KLIJEWSKA, *Osada kultury przeworskiej w Konarach stan. 28 [A Settlement of the Przeworsk Culture in Konary, Site 28]*, unpublished M.A. thesis, 1978.

of 9 unfinished products, 77 beads broken as holes had been drilled in them, and two finished artefacts (Figs. 10 & 11). There was also a large amount of tiny grains. The total weight of this assemblage amounted to 120 g[11]. This was yet another site where pieces of amber were clustered in certain points, especially so in the lower stratum, where they were markedly concentrated within an area of 2 x 3 m. In higher strata, the pieces were more widely scattered.

In addition to amber, numerous potsherds, animal bones, pieces of pug and household objects were also discovered. Thus, this structure was most likely a dwelling which contained an amber workshop.

Gąski, Site 18

The village of Gąski is situated in the valley of the River Parchania, more than 10 km east of the outskirts of Inowrocław. This, in fact, was a locality where instead

[11] Cf. footnote 7.a

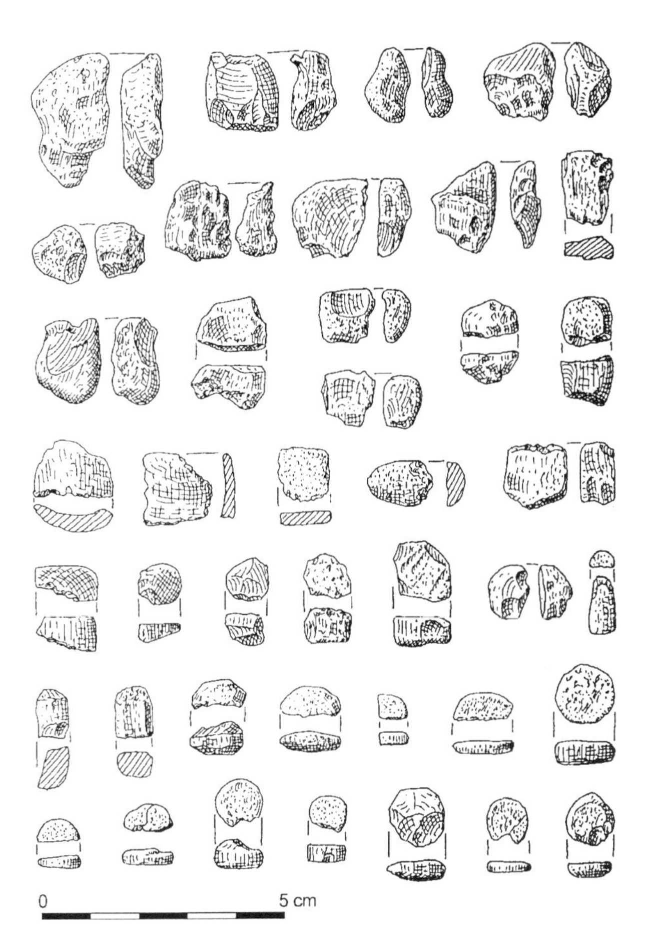

Fig. 10. Konary, site 28: pieces of amber from the workshop (structure 19).

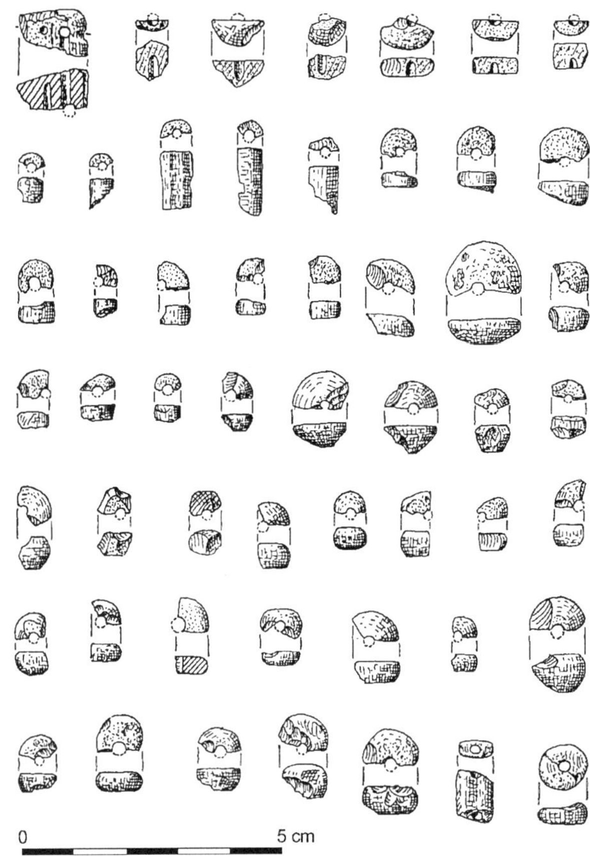

Fig. 11. Konary, site 28: pieces of amber from the workshop (structure 19).

of an individual workshop or a single community with amber workshops, a large complex of settlements was identified where this fossil resin had been worked. This is attested to by the numerous pieces of amber found not only during excavation, but also on the surface of fields covering four or five archaeological sites. Of these sites one (Gąski, site 18) was designated for scheduled excavation in the years 1984–91, another one (site 24) for reconnaissance in 1990 and 1991, and the rest (Gąski, sites 22, 25 & 19) for a detailed programme of fieldwalking. Accordingly, the greatest amount of material was recovered from site 18 in Gąski[12]. This was a relatively small settlement: only seven dwellings were revealed within the excavated area of 1,300 m^2. These included as many as three (458, 366 & 528) housing amber workshops, and another one (104) indirectly associated with this craft.

[12] Małgorzata ANZEL, *Bursztyniarstwo ludności kultury przeworskiej w rejonie Błot Gąskich* [*Amber Cutting among the Population of the Gąski Swamps in the Period of the Przeworsk Culture*], unpublished M.A. thesis, 1996.

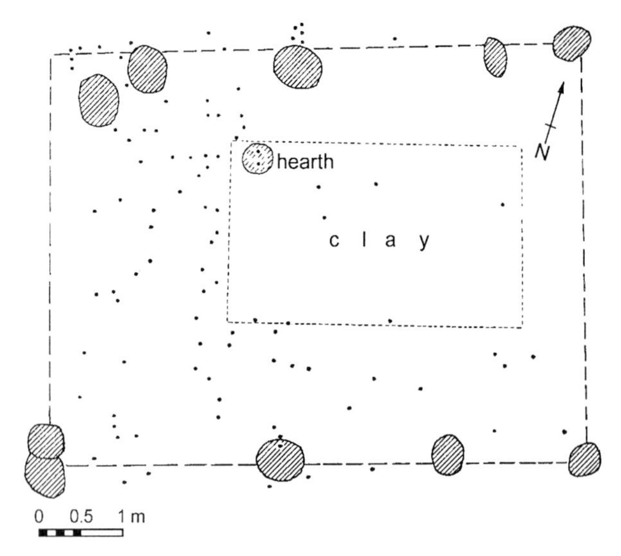

Fig. 12. Gąski, site 18: an amber workshop (structure 366).

Structure 366

This was a post-built construction: traces of four posts, 30–50 cm in diameter, outlined its rectangular floor plan. The dimensions of this structure were 4.5 x 6 m. Its central part, measuring 3.5 x 2 m, had a clay floor. In the south-western corner of this area, there was a small hearth of *c.* 30 cm diameter (Fig. 12).

Inside the structure and in its immediate vicinity, some 370 pieces of amber were recorded. These include 50 nodules, 7 specimens on which preliminary work had been carried out, 5 beads damaged during cutting, 300 shavings of various sizes and 9 clusters of small grains (Fig. 13: 4–20).

This distinctive, albeit small assemblage clearly indicates that amber was worked in this location.

Although the material was partly dispersed, it was clearly concentrated in the southern and western parts of the structure, in a strip between the walls and the clay-covered space, which may suggest that the workshop had operated in the south-western corner.

Apart from amber, items typical of dwellings were found in the structure, as well as four fragments of glass vessels and one, of a *terra sigillata* vessel (Fig. 13: 1–3).

Structure 528

Based on the extant vestiges of this structure, the layout of its interior (which was below ground level) and post-holes, we can surmise that it was rectangular in plan, measuring 4 x 5.5 m (Fig. 14). Some 90 pieces of amber, including approximately 30 nodules, fragments of two beads, one unfinished product and 50 small shavings (Fig. 15: 1, 3–6), suggest that this

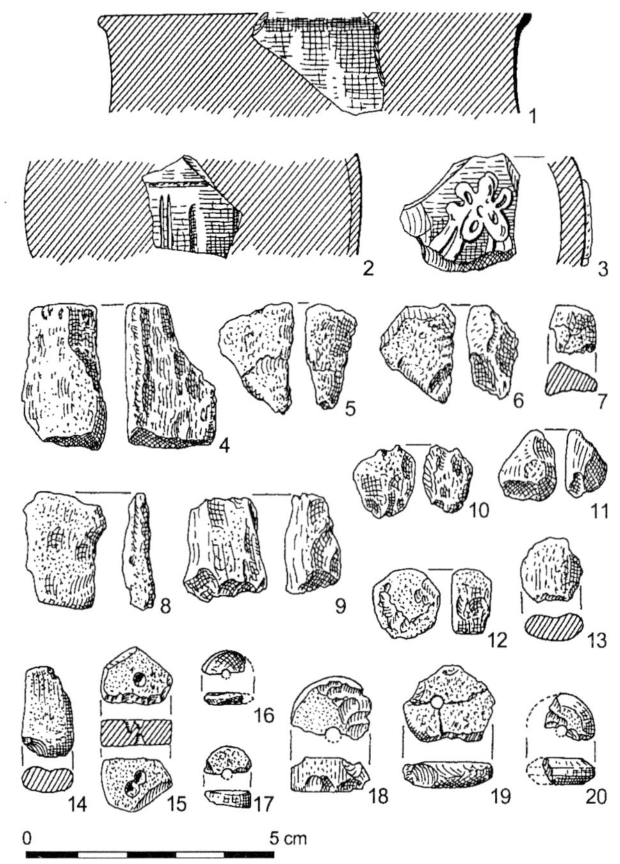

Fig. 13. Gąski, site 18: 1–3 — imported objects; 4–20 — pieces of amber (structure 366).

structure also featured an amber workshop. This was yet another instance where amber was amassed in only one part of the structure, i.e. at its north-western corner and in a strip along the western gable wall, some of it appearing immediately outside this wall.

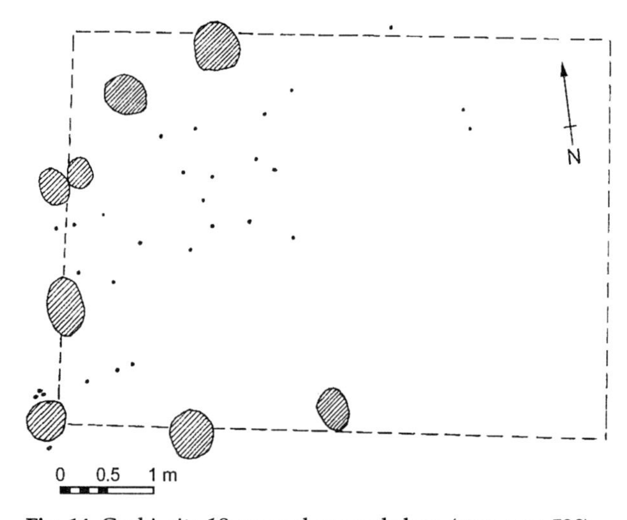

Fig. 14. Gąski, site 18: an amber workshop (structure 528).

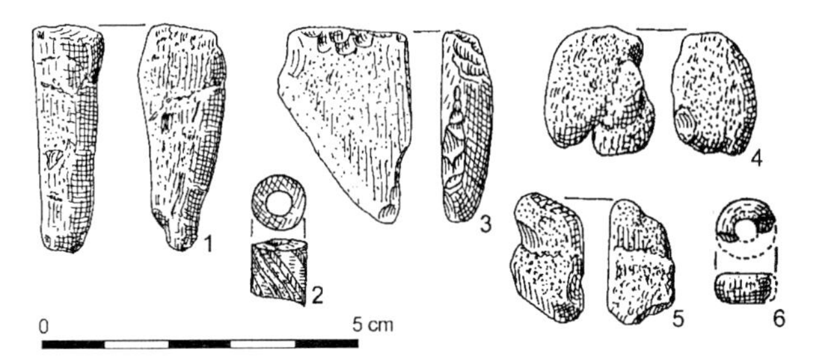

Fig. 15. Gąski, site 18: 1, 3–6 — pieces of amber; 2 — a glass bead (structure 528).

Beside typical household items, we also found imported objects: fragments of two glass beads (Fig. 15: 2) and two pieces of glass vessels, a *terra sigillata* sherd, and a Roman denarius.

Structure 458

This was severely damaged, with only traces of three posts attesting to its post-built design. Nevertheless, as with other buildings of this type, its layout was most probably rectangular (6 x 4.5 m). Both inside and immediately outside this structure, some 200 pieces of amber were identified, including 15 nodules, 2 lightly worked items, 2 unfinished beads, and approximately 150 pieces of raw material, as well as two clusters of grains (Fig. 16: 10–14).

As usual, the amber was visibly concentrated in the south-eastern part of the structure, indicating the location of the workshop of the craftsmen who had lived there.

Again, inside the structure standard household items were accompanied by foreign objects: pieces of six glass vessels, a glass bead and a *terra sigillata* sherd (Fig. 16: 1–9).

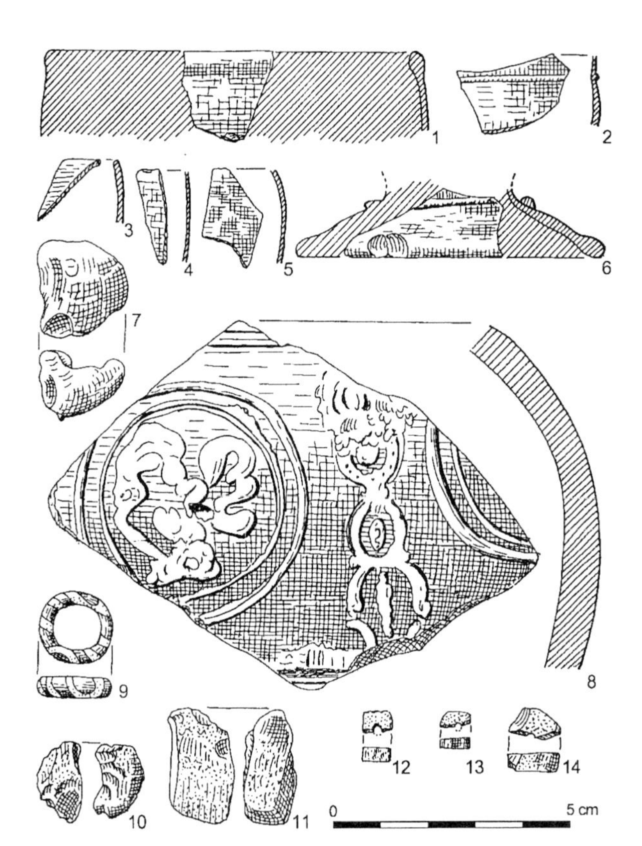

Fig. 16. Gąski, site 18: 1–9 — imported objects; 10–14 — pieces of amber (structure 458).

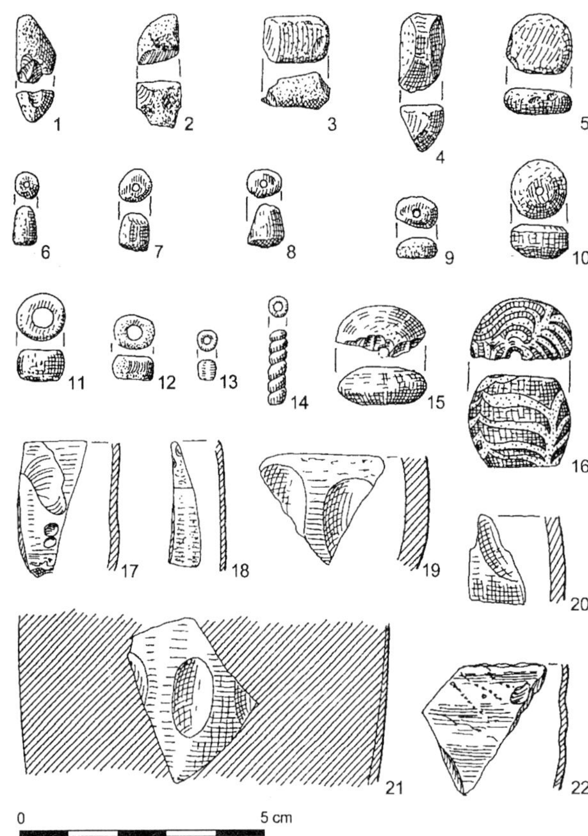

Fig. 17. Gąski, site 18: 1–10 — pieces of amber; 11–22 — imported objects (structure 104).

Structure 104

This did not house an amber workshop, yet its inhabitants — as we shall proceed to demonstrate — were closely associated with this craft. Only a very small amount of amber was recovered from this feature. Apart from a few small grains, this assemblage also included five finished beads, of a form which very rarely appears in other workshops. Further finds comprised fragments of eight glass vessels, five glass beads, two denarii, and sherds of richly decorated grey pottery (Fig. 17).

The three workshops described above yielded a total of 660 pieces of amber, exclusive of grains. In fact, this figure should be increased by a large number of pieces displaced by various types of human activity at this site in later periods of history, and now dispersed outside the excavated structures.

Gąski, Site 24

Approximately 250 m from the previous settlement of Gąski 18 was another settlement. Reconnaissance work at this site brought to light material suggesting that amber workshops had also been in operation there. During the course of two reconnaissance assignments exploring areas of 1 x 10 m and 2 x 10 m, a total of 170 pieces of amber was recorded. These comprised nodules, partly worked specimens and two beads broken as holes were drilled in them (Fig. 18: 1–14). Accompanying the amber were sherds of richly decorated greyware and a glass bead (Fig. 18: 15).

It was noted with interest that the farmer owning this land had found many Roman coins, among them a bronze piece of Antoninus Pius and two silver denarii.

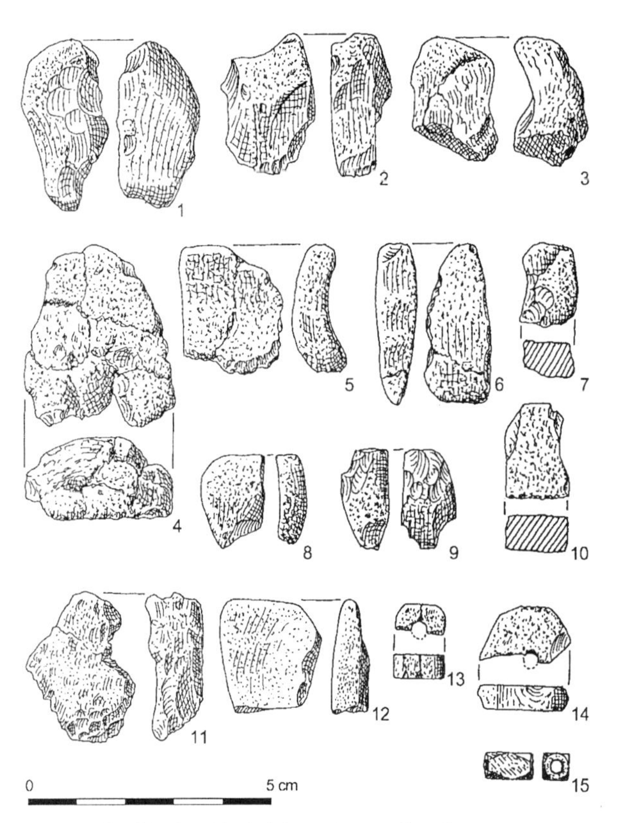

Fig. 18. Gąski, site 24: 1–14 — pieces of amber; 15 — a glass bead.

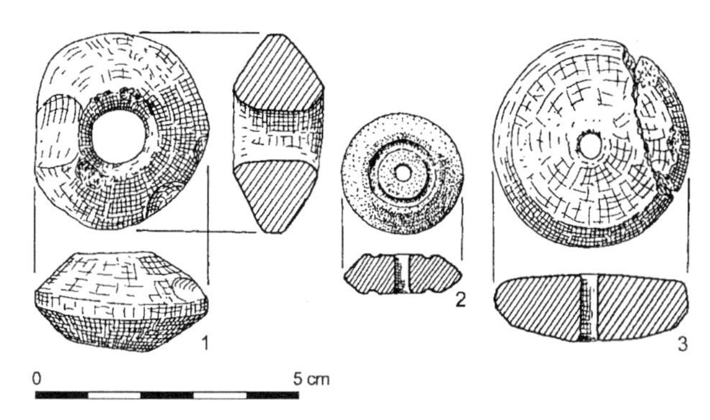

Fig. 19. Gąski, glass beads collected during surface survey: 1 — site 22; 2 & 3 — site 25.

Furthermore, numerous pieces of amber, including smaller and larger nodules and occasionally a finished bead, were identified on the surface of fields within the sites of Gąski 22 and Gąski 25 (Fig. 19). Most plausibly, these may be considered to indicate that other amber workshops had operated at these settlements. In view of all the above facts, the area of Gąski must be deemed a major centre of amber-working.

Kuczkowo, Site 1

Another settlement that provided a rich corpus of material documenting the development of local amber-working in Kuiavia, was discovered at Kuczkowo, site 1. This site is located on a mound on the northern slope of the valley of the Bachorza, some 30 km east of Inowrocław. It underwent archaeological research in 1966, when a gas pipe running

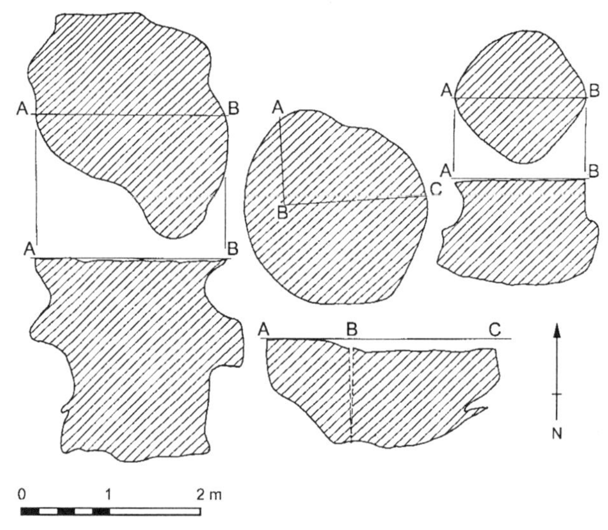

Fig. 20. Kuczkowo, site 1, pits containing pieces of amber: 1 — pit A2; 2 — pit A22; 3 — pit B13.

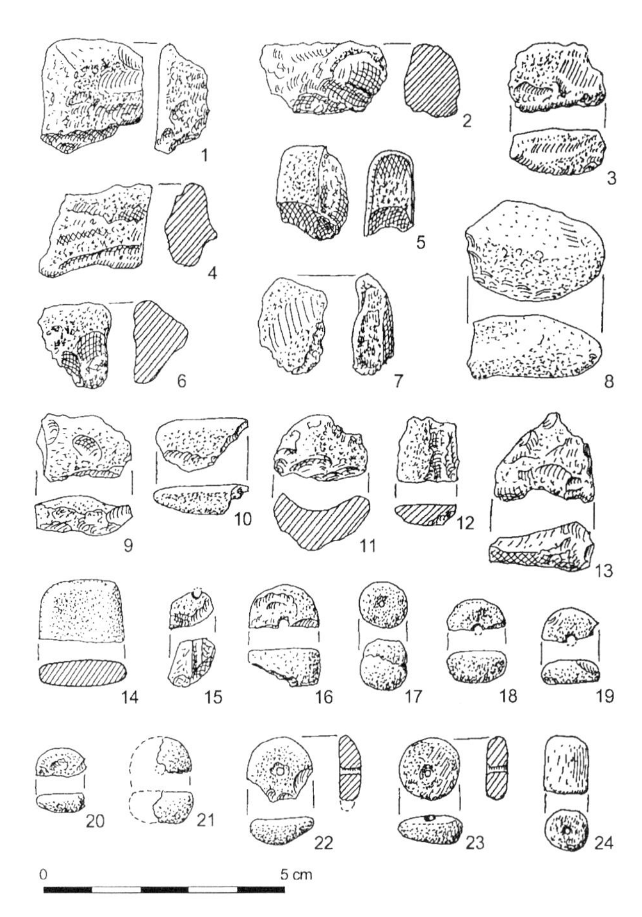

Fig. 21. Kuczkowo, site 1, pieces of amber: 1–13 — structure A2; 14 — structure A22; 15–17 — structure B52; 18–21 — structure A40; 22–24 — structure B13.

from east to west was being laid down. The latter fact accounts for the peculiar shape of the area excavated: a 13-m-wide trench was dug along the whole length of the Bachorza valley (538.4 m) in order to examine its archaeological record. Among the structures unearthed at the eastern end of the trench, five contained amber. Three of these were roughly circular pits, between 1.5 and 2 m in diameter, of a trough- or bell-shaped vertical section, and of a depth between 1 and 2 m (Fig. 20). Inside all three, potsherds, animal bones and fragments of pug were identified, as well as pieces of amber. The amber recovered consisted of: 85 pieces in pit A2, including 30 nodules and 55 fragments of various sizes (Fig. 21: 1–13); 80 pieces of various sizes in pit A22, including a fragment of a thin plate (Fig. 21: 14); and a bead with an unfinished hole in pit B13, as well as a flattened spherical bead, 20 nodules and some 330 fragments of various sizes (Fig. 21: 22–24).

The other two structures discovered at this site differed essentially from the previous three. They

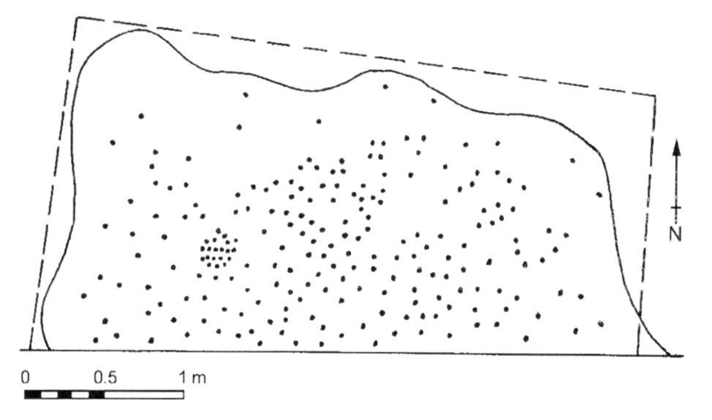

Fig. 22. Kuczkowo, site 1: an amber workshop (structure A40).

were big rectangular buildings erected on the ground. Unfortunately, because of the latter fact they had been considerably damaged.

Only part of Structure A40, covering an area of 3.5 x 2 m, was exposed, as the rest of this feature was beyond the extents of our trench (Fig. 22). This structure contained 215 pieces of amber, including four beads broken whilst holes were being drilled in them, and 210 fragments of various sizes (Fig. 21: 18–21).

Structure B52 had been almost completely destroyed. The only extant features were a small fragment of the floor (irregularly shaped and 2–3 cm thick) and traces of posts. Nevertheless, the rough outline of the structure may be reconstructed based on, among other factors, the distribution of amber finds. Thus, we may assume that the structure was rectangular in plan, measuring *c.* 6x? m (Fig. 23). An assemblage of 190 pieces of amber was found in structure B52, including an unfinished bead, two specimens broken during the process of drilling holes in them, and 185 fragments of various sizes (Fig. 21: 15–17).

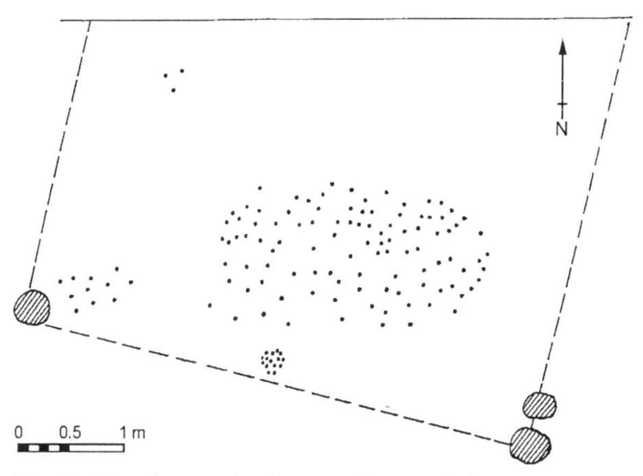

Fig. 23. Kuczkowo, site 1: an amber workshop (structure B52).

Overall, the five structures in Kuczkowo yielded a total of some 920 pieces of amber. Apart from these, only a few small fragments of this fossil resin were collected from the remainder of this extensive excavation.

It is obvious that the two structures raised on the ground (A40 & B52) were dwellings whose residents worked amber. The pits, on the other hand, had an entirely different origin and function. These were most likely refuse pits dug out near the dwellings, where craftworkers disposed of their amber goods production waste.

Inowrocław, Site 100

This site is located on the southern peripheries of the city centre. Rescue excavations have been carried out there since 1986. Of the 1,300 structures investigated, only one — a pit (structure 610) — contained pieces of amber. These consisted of: three unfinished beads, a fragment of a finished bead, and *c.* 130 small grains. The fact that so many pieces and unfinished artefacts were concentrated within a single structure may indicate that amber had been worked there, although it does not provide enough evidence that this craft had been practised as a specialist activity.

Inowrocław, Site 95

This site is located at the northern edge of the valley of the River Noteć, on the southern boundary of the city. It was inhabited by various cultures across the centuries, including several phases of settlement during the period of the Przeworsk culture. Research work was undertaken at this site in the years 1986–1991, excavating an estimated area of 15,000 m^2 and revealing 5,000 structures[13]. Again, amber was recorded in this locality, both inside and outside its structures. Never, however, did it appear in clusters. Isolated nodules of this material were usually found, except for structures 2760, where several nodules were discovered, and 300, which yielded a bead and a fragment of an amber bracelet (Fig. 24) alongside a number of nodules. Therefore, it is possible that amber was only worked on a small scale at this settlement — although, on the other hand, considerable skill was

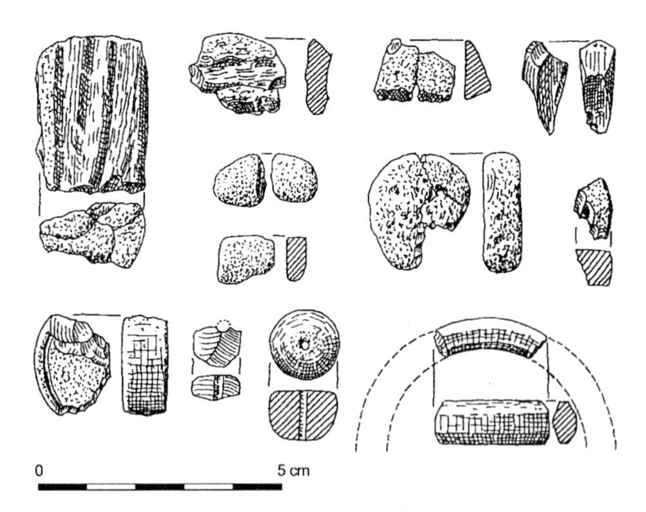

Fig. 24. Inowrocław, site 95: pieces of amber.

required to produce the bracelet with its faceted gems. Another striking fact is that amber was recorded inside so many structures: this had not previously happened at any other settlements, even those with amber workshops. This fact may imply that amber had a symbolic function in this community, e.g. that it was used during household, family or religious rites.

Krusza Zamkowa, Site 3

This is the largest settlement discovered so far in Kuiavia. It constituted the centre of a complex of several contemporary communities during the Roman period which provided an emporium for the amber trade route which passed through Kuiavia. Excavations were conducted at this site in 1973, 1974, 1976 and 1977, exploring only a small part of the settlement. Amber was found there, although only a few isolated pieces. Since this assemblage included lumps of raw material, a broken bead (Fig. 25) and a very small quantity of other specimens, some of them

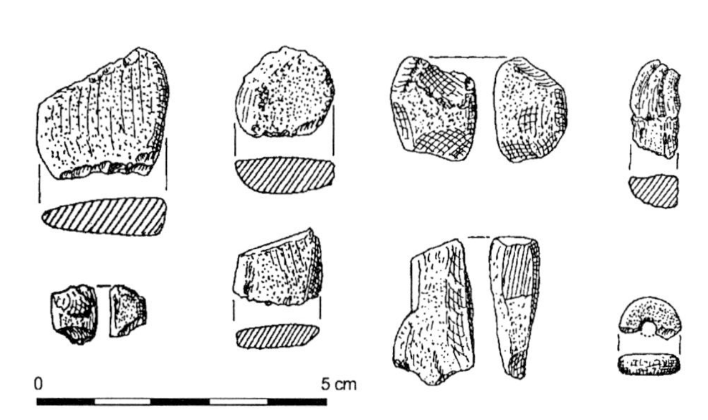

Fig. 25. Krusza Zamkowa, site 3: pieces of amber.

[13] Józef BEDNARCZYK, "Z badań sanktuarium i osady ludności przeworskiej w Inowrocławiu, woj. Bydgoszcz, stan. 95" ["Selected Issues of the Research of the Shrine and the Settlement of the Przeworsk Culture in Inowrocław, Province of Bydgoszcz, Site 95"], *Sprawozdania Archeologiczne*, **39**, 1987, 201–221.

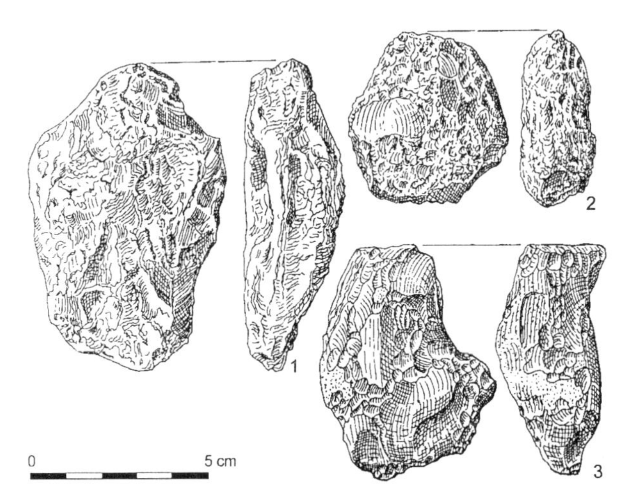

Fig. 26. Isolated pieces of amber: 1 & 3 — Inowrocław, site 55; 2 — Dziewa.

partly worked, we may assume that this fossil resin was worked in situ, but on a very small scale and only for domestic use.

All of the aforementioned finds confirm that amber-working was pursued in Kuiavia and provide data related to this activity and representing various spheres of material culture. Furthermore, several nodules of this material, dated to the period in question, have been found outside the recorded locations of amber workshops. Two nodules were discovered at the prehistoric burial ground in Inowrocław (site 55), however, these cannot be associated with the purpose of the burial structures. The pieces were fairly big, measuring 5 x 7.5 x 1–3 cm and 6 x 3.5 x 2–2.5 cm (Fig. 26: 1 & 3). Another lump of amber was found in a field in Dziewa, during the course of fieldwalking; it was circular and somehow smaller, of a diameter of 3 cm and thickness of 2 cm (Fig. 26: 2). Finally, local residents have informed us of the discovery of three more lumps, in Popowice, Opoki and Inowrocław (site 32).

Discussion of the Source Material

The above summary of the amber discovered in the Inowrocław Plain indicates that amber-working, mainly on a specialist level and on a large scale, was most probably quite a common activity in this region. Specialist workshops operated at Jacewo, Łojewo, Konary and Kuczkowo, as well as at two, or possibly four settlements in Gąski, and perhaps at Inowrocław 100. Conversely, the source material collected from Inowrocław 95 and Krusza Zamkowa may testify only to occasional practising of this craft.

In the light of various considerations, the available source material is obviously not representative of the actual volume of each workshop's production, and therefore cannot be accurately assessed in this respect. Suffice it to quote, for reference only, the total number of 8,850 identified pieces of amber, and to break this down into figures for the individual sites: Jacewo: 5,000; Konary: 1,500; Kuczkowo: 920; Gąski (18): 670; Łojewo: 400; Gąski (24): 170; Inowrocław (100): 130; and Inowrocław (95): 50.

Nevertheless, the discussed amber workshops were clearly not contemporaneous. Five of them (Jacewo, Łojewo, Inowrocław 95 & 100, and Krusza Zamkowa) operated during the early Roman period, while the rest (Konary, Kuczkowo and the area of Gąski), in the late Roman period. Their actual location is closely related to their chronology. All the earlier workshops were situated in the vicinity of the settlement complex at Krusza Zamkowa, or in the western part of the region in question, while the later ones occurred in the east, i.e. on the banks of the rivers Bachorza (in Konary and Kuczkowo) and Parchania (in Gąski). Thus, it transpires that amber-working flourished in Kuiavia over several centuries.

In spite of the long period during which the workshops were in operation, their characteristics were fairly consistent, and therefore all the workshops may be discussed as a homogeneous group.

The artisans most certainly set up their workshops in a specific section of a dwelling, usually near a wall, and on some occasions in a corner.

If there was more than one workshop in a settlement, then these were located close to one another.

Very conspicuously, the assemblages of amber found at all the workshops feature hardly any finished products. Of the largest collection of 5,000 pieces discovered at Jacewo, only 6 were finished articles. This fact makes it difficult to establish the types of bead manufactured, as the bulk of the source material constituted damaged, unfinished items, usually not yet fashioned into their ultimate shapes. We may generally assume that Kuiavian workshops produced beads in the forms of flattened spheres, double cones, cylinders, barrels, pears, figures of eight, and small flat plates. It is also striking how many tiny grains were left: this fact, juxtaposed with the evident trend to govern the shape of the finished bead by the contour and size of the lump of the raw material, indicates that the craftsmen tried to waste as little amber as possible.

The raw material used in Kuiavian workshops was very uniform. The overwhelming majority of it was translucent brandy-coloured amber, with only a small amount of dark-cherry- or honey-coloured and opaque yellow amber. A few bicoloured lumps were found, with their cores milky-white to yellow, and their outer layers, brown.

In order to offer valid conclusions on Kuiavian amber-working, it is necessary to determine the origin of the raw material used. We know that rich deposits of amber are found in the Baltic Sea and on its coast, as well as in the neighbouring regions of Pomerania, Masuria and Kurpie[14]. These do not extend as far as Kuiavia — only very limited natural resources of this resin may be expected to exist there, nodules being occasionally encountered during deep earthworks or in lakes and rivers[15]. Even so, the finds of amber reported in various publications on this subject need not imply that the local prehistoric population used native resources of this material for any purposes. On the contrary, some such finds are likely to consist of amber that may be traced back to a nearby amber-working centre. This postulate may be substantiated by the geographical location of the finds. Thus, the nodule at Dziewa was discovered close to Konary, the lumps from Inowrocław (site 32), in fields near Jacewo, and those from Popowice and Inowrocław 55, near the settlements of Inowrocław 95 & 100, on the course of the amber trade route. It is not our intention to deny that local amber might have occasionally been used for small-scale home production, although it most certainly was not the principal source of raw material for specialist workshops of the Roman period.

In view of the large scale of Kuiavian amber-working, as evidenced by the archaeological record, it must be presumed that the bulk of raw material was imported from the Baltic coast. Nevertheless, it would be far-fetched to suggest that Kuiavia was the target of long-distance amber trade, rather than a mere transit area.

Obviously, the final proof of the origin of the amber collected from Kuiavian workshops will be the results of specialist laboratory tests.

General Conclusions

The emergence of amber-working in Kuiavia during the Roman period was a unique development, not only because this region is distant from the natural deposits of this resin, but also because previous local communities were not interested in amber. To be sure, long before the epoch in question, in the Neolithic period and early Bronze Age the peoples of this region were intermediaries in the exchange of amber between the Baltic coast and the Transcarpathian regions[16], as is evident from the chemical composition of the fossil resin found in Mątwy[17], and the amber beads and pendants frequently discovered at settlements and burial grounds dating to that time. Yet, this activity did not continue in the subsequent periods of the Bronze Age or in the early Iron Age, as amber played no part in the material culture of the communities of Kuiavia throughout the period of the Lusatian culture's dominance in this area. The two small beads incorporated in the bronze necklace from the burial at Lachmirowice[18] and an inexplicit reference to amber beads appearing in the hoard of bronze artefacts from Szarlej[19], are mere exceptions to the rule.

The local communities continued to ignore amber in the late sub-period of the La Tène culture. The sum of amber finds dating from this period consists of a bead from the burial ground at Krusza Zamkowa (site 13) and three beads incorporated into a glass necklace from the cemetery at Inowrocław (site 58) (Fig. 27: 2–4). No other specimens have been found at any of the numerous excavated and recorded cemeteries of that period.

It was only in period B1b that this situation began to change. Amber beads found in the later burials of the cemetery at Gąski (sites 18 & 22) (Fig. 27: 5–7) certainly mark a shift in the population's interests. It must be emphasised that these amber finds differed from the earlier ones in their size, shape and careful

14 Barbara KOSMOWSKA-CERANOWICZ, Teresa PIETRZAK, *Znaleziska i dawne kopalnie bursztynu w Polsce* [*Findings of Amber and Ancient Amber Mining in Poland*], Warszawa, 1982, 10–41.

15 Ibid., 77, 79, 111 & 112.

16 A. KOŚKO, *Rozwój kulturowy społeczeństw Kujaw* .... Idem, "Specyfika rozwoju kulturowego..."

17 A. KOŚKO, "Badania w strefie nadnoteckiej Inowrocławia ..."

18 Krzysztof SZAMAŁEK, "Pochówki szkieletowe w Lachmirowicach nad Gopłem (woj. Bydgoszcz) a zagadnienie ich związku z grupą górnośląsko-małopolską kultury łużyckiej" ["The Burials of Unburnt Remains in Lachmirowice on the Lake Gopło (Province of Bydgoszcz) and their Relationship with the Upper-Silesia-and-Lesser-Poland Group of the Lusatian Culture"], [*in:*] *Dziedzictwo kulturowe epoki brązu i wczesnej epoki żelaza na Górnym Śląsku i w Małopolsce* [*The Cultural Heritage of the Bronze and Early Iron Ages in Upper Silesia and Lesser Poland*], *Śląskie Prace Historyczne*, **4**, Katowice, 215–222.

19 Jadwiga CHUDZIAKOWA, "Materiały kultury łużyckiej z Kujaw, cz. III" ["The Source Material on the Lusatian Culture in Kuiavia, Part III"], *Acta Universitatis Nicolai Copernici: Archeologia*, **5**, 37.

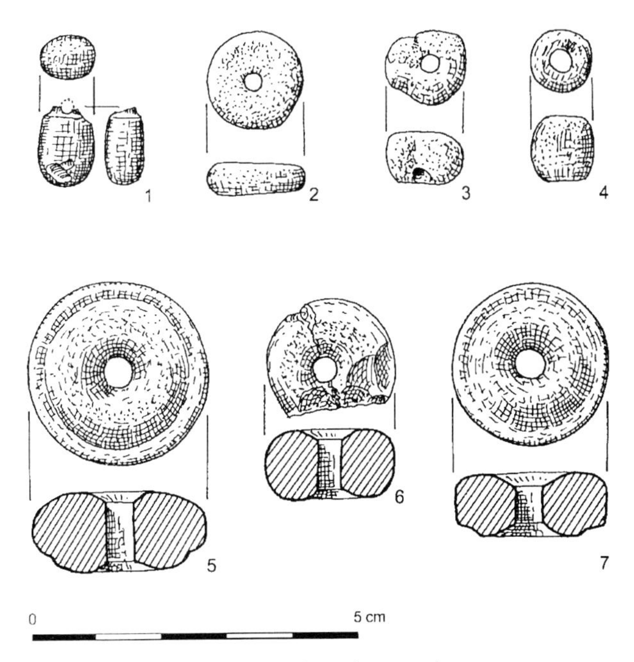

Fig. 27. Amber beads from burial grounds:
2–4 — Inowrocław, site 58; 1, 5–7 — Gąski, site 18.

working. This appreciation of amber jewellery was perhaps a legacy of the Celtic presence in this region. The Celts, specialising in transcontinental trade in various commodities, among them amber, recognised the advantageous geographical location of Kuiavia, and included this region in their new long-distance exchange route between the Baltic and Southern Europe[20]. The inhabitants of the transit area of Kuiavia, aware of the increasing importance of the route and the considerable proportion of amber among the conveyed goods, might have taken advantage of the opportunity and taken up amber-working themselves. The aforementioned beads were possibly among their earliest artefacts, although so far we have acquired no direct evidence that amber was worked locally at that time. A piece of indirect evidence may consist of the lumps of amber[21] and a bead[22] (the latter dated to the beginning of the Roman period) from a burial at the Gąski cemetery (site 23). The presence of crude amber in an amber-worker's grave, the clusters of beads at Gąski (site 18) dating to this early period, juxtaposed with the absence of similar items from other contemporaneous burial grounds, and the magnitude of the Gąski settlement, all suggest that this might well have been the first centre of amber-working in Kuiavia. As more and more goods were carried along the route in question, and the importance of the amber trade grew, so amber-working started to spread. It developed steadily until the late 4th century AD, failing to collapse even during the economic depression in Italy, when that market broke down.

Considering all of the above facts, one may ask why it was Kuiavia that became an amber-working centre, rather than any other region of Poland with a longer history of involvement with amber. One of the reasons was certainly the geographical location of Kuiavia. This region is situated at a point where the distance between the two major waterways of the Oder and the Vistula is the shortest. The favourable layout of the tributaries of these two rivers, distributed along the cartographic parallels of latitude, allow easy transportation in all directions. Another very important factor, which enhanced the significance of this region, was its natural resources, particularly its salt deposits. Furthermore, Kuiavia lies fairly close to natural deposits of amber. Finally, the local communities represented an advanced level of material culture, and boasted a long history of specialist manufacturing and concomitant long-distance trade relations. This was the second occasion, after the early Iron Age, when Kuiavia participated in transcontinental commercial exchange. In this earlier epoch, however, the primary export commodity was locally-made and richly decorated bronze jewellery. In fact, the patterns of development of bronze manufacturing and amber-working in Kuiavia were strikingly similar, as neither craft was based on indigenous raw materials. On the other hand, the difference between the two periods consisted in the directions of trade. The previous routes, running to the south-west and the south-east, were now replaced by north- and southward bound courses.

It was only at a relatively late stage that an amber trade route crossed Kuiavia. Previous routes passed through areas west of this region, e.g. during the Lusatian culture period — through Greater Poland (Wielkopolska) and sometimes through Pałuki[23],

[20] A. COFTA-BRONIEWSKA, A. KOŚKO, *Historia pierwotna społeczeństw ...*, 193–194.

[21] The collection of the Jan KASPROWICZ Museum in Inowrocław.

[22] Bonifacy ZIELONKA, "Rejon Gopła w okresie późnolateńskim i rzymskim" ["The Area of the Lake Gopło in the Late La Téne and the Roman Periods"], *Fontes Archaeologici Posnanienses*, 1969, **20**, 194.

[23] Cf. footnote 20, 160–161.

skirting Kuiavia, whose trade links at that time focused on other territories. This situation changed radically when a new route came into use, running from Silesia via Kalisz, Konin and the area of Inowrocław (Krusza Zamkowa) to the ford near Otłoczyn, and therefrom directly to the Baltic[24]. This was the main route by which crude amber was brought from the Baltic coast and the artefacts made of this fossil, to the Transcarpathian and Alpine regions, and from there to Italy. Later, during the 2nd or 3rd century AD, when the principal purchasers of amber became the Sarmatian tribes inhabiting the Transcarpathian Valley[25], the route was modified again, to proceed along the Vistula to the mountain passes in the Carpathians. The actual course of this route is evidenced by hoards from Partynice[26] (representing its earlier path) and from Basonia[27] and Świlcza[28], dated to its later period.

After the route had shifted to the Vistula, its Kuiavian section passed through the eastern, rather than the western, part of the region. Accordingly, most settlements, and therefore most centres of amber-working, were now established in the east.

Both courses of the route had a considerable impact on Kuiavia. This region combined the functions of a transit emporium of amber, a major area of amber-working and an outlet for imported articles. This is particularly evident in the local amber workshops, which contained a large number of imported items, very seldom found in other dwellings.

During the initial period of the operation of the Kuiavian section of the amber route, itinerant merchants considered this region merely a transit area for the transportation of this fossil resin from the Baltic coast to southern Europe. Later, as the demand for amber products increased in other parts of Europe, and as the volume of trade consequently rose, the local communities, who had easy access to sources of this material, developed a native amber industry.

This occupation was clearly profitable, even though the raw material had to be imported. Continuing demand enabled artisans to purchase foreign objects, which — as we have mentioned — are frequently found in their houses: principally glass vessels and beads, and *terra sigillata* vessels.

When manufacture and trade reached a certain level of intensity, it became necessary for local merchants to coordinate these two operations. Traders supplied crude amber both to the local workshops and — possibly — to their foreign counterparts visiting Kuiavia, acting as intermediaries in the sale of locally-produced amber goods, and distributing imported goods.

We assume that the dwelling excavated at the Gąski settlement (site 104) belonged to such an intermediary merchant, its outstanding feature being the accumulation of various imported objects: glass beads, pieces of glass vessels, highly decorated pottery, *terra sigillata* utensils and Roman coins — as well as amber beads. Additionally, the resident of this house buried a hoard of 138 denarii in a corner of the building[29]. In view of his profession, we can assume that this was not a reserve treasure, but a current fund necessary for ongoing trade.

To sum up the significance of local amber-working for the whole of Kuiavian material culture, we must insist that it contributed significantly to the civilizational development of the communities of this region. It was among the first branches of economy to achieve the level of specialist production. It stimulated the expansion of trade relations, and indirectly prompted the adoption of new patterns of civilization, e.g. new pottery production techniques[30]. In fact, it can hardly be considered a coincidence that the manufacturing centres of the more advanced pottery type (known as greyware) emerged so close to the identified centres of amber-working (in Parchanki near Gąski, Wróble near Konary, and Kuczkowo).

We have already stressed the fact that at virtually every excavated settlement datign from the Roman

[24] A. COFTA-BRONIEWSKA, *Grupa Kruszańska kultury przeworskiej*, 123–172.

[25] Jerzy WIELOWIEJSKI, *Kontakty Noricum i Panonii z ludami północnymi* [*The Relations of Noricum and Pannonia with Northern Peoples*], Wrocław–Warszawa–Kraków, 1970.

[26] Ibid., 222.

[27] Przemysław WIELOWIEJSKI, "Skarb bursztynu z późnego okresu rzymskiego odkryty w miejscowości Basonia, woj. lubelskie" ["The Amber Hoard from the Late Roman Period Discovered in Basonia, Province of Lublin"], *Prace Muzeum Ziemi*, **41**, 1990, 101–134.

[28] Aleksandra GRUSZCZYŃSKA, "Osada z wczesnego okresu wędrówek ludów ze Świlczy, woj. Rzeszów" ["A Settlement from the Early Period of the Migration of Peoples in Świlcza, Province of Rzeszów"], *Materiały i Sprawozdania Rzeszowskiego Ośrodka Archeologicznego*, 1976–1979 (published in 1984), 103–129.

[29] A. COFTA-BRONIEWSKA, "Badania stanowiska 18 w Gąskach ..."

[30] A. COFTA-BRONIEWSKA, *Grupa Kruszańska kultury przeworskiej*, 98–103.

period we have located evidence of local amber-working, whether professional and commercial or small-scale and intended for domestic use only. Consequently, it is obvious that this craft was of enormous importance to the communities of Kuiavia during four centuries of their history. Conversely, an entirely different opinion may be formulated if we base our suppositions on the archaeological evidence recovered from burial grounds. The total absence or occasionally very small amounts of amber found in graves would warrant the view that this fossil resin was only of marginal importance in the material culture of the local populations. Thus, two distinct categories of source material yield two radically different estimations of the civilizations of the same region, pertaining not only to the initial period of development of the craft in question, but also to the heyday of amber-working. There are a large number of burial grounds dated to the Roman period, and the majority of these have been excavated, yet amber has only been found at a few of these sites: a necklace at Inowrocław-Mątwy[31], five beads from a burial at Pruchnowo (Fig. 28)[32], one at Otłoczyn, and an unfinished piece of jewellery at Szymborze[33].

Furthermore, this general absence of amber from Kuiavian graves cannot be explained by postulating that it was burnt during cremation, as both burial rites were used in most cemeteries of that region, and in some graves only unburnt remains were buried. Neither would it be tenable to claim that the dead were not provided with personal articles, or that neck ornaments were not fashionable, as glass beads, in some instances making up impressive necklaces, are quite frequently found in graves from that period.

In point of fact, this phenomenon has its precedents. Even in Kuiavia, data pertaining to bronze jewellery from the Hallstatt period are equally divergent: these items were manufactured locally on a large scale, but never appear in the burial grounds. In turn, quite the opposite is true of amber-working in Pomerania during the Roman period: hundreds of thousands of beads, often in the form of elaborate necklaces, have been found in graves, while there is no evidence whatsoever of the local craft which inevitably must have flourished.

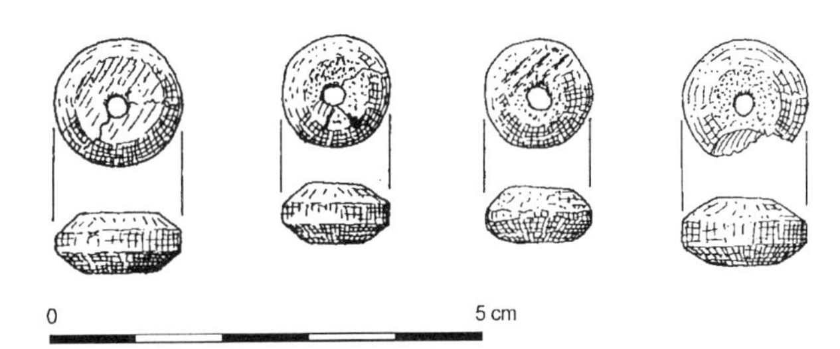

Fig. 28. Amber beads from the burial ground in Pruchnowo, site 23.

To revert to our main concern of the absence of amber from Kuiavian burial grounds, we must consider the amount of raw material available in this region and the profitability of trade. Apparently, the region's inhabitants decided to buy and wear glass beads — which were much cheaper, as they were a mass-produced item in the Roman Empire and the material from which they were made was easily procurable — in order to derive benefits from the sale of amber ornaments, which appealed to south European societies by virtue of the pleasing, exotic, rare and non-recyclable material used to produce them.

Thus, if local Stanomin bronze does not appear in Kuiavian graves from the Hallstatt period, and local amber beads, in those from the Roman period, this is doubtless a reflrection of economic necessities and a response to the need for wise management of available resources.

Recommendations

The source material examined offers abundant testimony to the significance of amber-working to the Kuiavian communities of the first five centuries AD. Nevertheless, many questions remain unanswered. Regrettably, the lack of funds required for scheduled field research has so far prevented further study of this topic. In order to gain a deeper insight into this matter, the first priority would be to conduct more extensive excavations at the complex of settlements in Gąski, which was apparently the largest centre of amber-working. Additionally, as the available evidence suggests that another amber-working centre operated in the vicinity of the village of Opoki on the river Tążyna, this area also requires

[31] J. ŻUREK, "Wykopaliska w Mątwach" ["The Excavations in Mątwy"], *Ilustrowany Kurier Polski*, 1938, **23**.

[32] Małgorzata ANDRAŁOJĆ, "Cmentarzysko ludności kultury przeworskiej w Pruchnowie stan. 23, gm. Radziejów Kujawski, woj. Włocławek" ["A Burial Ground of the Przeworsk Culture in Pruchnowo, Site 23, Commune of Radziejów Kujawski, Province of Włocławek"], *Sprawozdania Archeologiczne*, **44**, 1992, 167–180.

[33] Bonifacy ZIELONKA, "Rejon Gopła ...", 192.

investigation. Finally, laboratory analyses of all the amber assemblages discovered in Kuiavia are indispensable.

Aleksandra COFTA-BRONIEWSKA
Zakład Prahistorii Polskiej
Uniwersytet im. A. Mickiewicza
ul. Św. Marcina 68
60-528 Poznań, Poland

BURSZTYNIARSTWO U SPOŁECZNOŚCI KUJAW W OKRESIE WPŁYWÓW RZYMSKICH

Aleksandra COFTA-BRONIEWSKA

Streszczenie

Prowadzone od 30 lat przez Zespół Badań Kujaw szerokie planowe badania archeologiczne na Kujawach wykazały m.in., iż bursztyn na tym odległym od bogatych jego złóż terenie odgrywał w kulturze pradziejowych społeczeństw tego regionu istotną rolę. Zwłaszcza wyraziście ujawniają to źródła rzeczowe z okresu wpływów rzymskich, dowodzące długotrwałego rozwoju profesjonalnej obróbki tego surowca.

W studiach nad odnośną problematyką cezurującym momentem była rejestracja w 1968 roku w Jacewie, stanowisko 4b (na wschodniej granicy Inowrocławia), pierwszej kujawskiej pracowni bursztyniarskiej. W ślad za tym faktem poszły wkrótce dalsze podobne odkrycia: w 1975 roku w Łojewie (stan. 4) i Konarach (stan. 28), a w końcu lat osiemdziesiątych w Gąskach (stan. 18 i 24), w Inowrocławiu (stan. 100) i wreszcie w 1996 roku w Kuczkowie (stan. 1). Do tego należałoby dodać jeszcze jedną pracownię, nie zlokalizowaną wprawdzie punktowo, lecz znajdującą się niewątpliwie w północnozachodniej strefie Gąsek, dokumentowaną licznymi stamtąd pochodzącymi znaleziskami bursztynu.

Trzeba też zauważyć, że w poszczególnych ośrodkach istniał niekiedy więcej niż jeden warsztat bursztyniarski. Tak było na przykład w Jacewie, w Gąskach (stan. 18) i w Kuczkowie. Nadto obróbka tego surowca nie ograniczała się do wytwórczości profesjonalnej, lecz była praktykowana przez mieszkańców w różnych osiedlach na własny użytek. Ślady takich działań mamy na przykład w osadzie Inowrocław (stan. 95) i prawdopodobnie w Kruszy Zamkowej (stan. 3). Stwierdzono również, iż niekiedy obróbka bursztynu łączona była z wytwórczością wyrobów z poroży i kości (Jacewo 4b).

Dość charakterystyczne jest rozmieszczenie poszczególnych ośrodków bursztyniarskich, zwłaszcza gdy weźmie się pod uwagę ich chronologię. Wyraźnie zaznaczają się bowiem dwa obszary koncentracji pracowni o różnym czasie ich działania. Ośrodki starsze (Jacewo, Łojewo, Inowrocław), datowane na wczesny okres wpływów rzymskich, wiązały się przestrzennie i funkcjonalnie z aglomeracją osadnictwa Kruszy Zamkowej, stanowiąc jeden z najważniejszych punktów etapowych szlaku bursztynowego przechodzącego ówcześnie zachodnim skrajem Kujaw (na zachód od Inowrocławia). Pozostałe natomiast (Konary, Kuczkowo, Gąski), rozwijające się w młodszym okresie wpływów rzymskich, skupiały się na obszarach nadbachorskich i nadparchańskich. Ta odmienna lokalizacja późnych ośrodków bursztyniarskich stanowiła konsekwencję innego w tym czasie przebiegu kujawskiego odcinka wspomnianego dalekosiężnego szlaku, a mianowicie przeniesienia go około III stulecia po Chrystusie na linię Wisły. Przesunięcie do wschodniej strefy Kujaw głównego traktu szlaku znajduje odzwierciedlenie m.in. w intensyfikacji w tym czasie osadnictwa na odnośnym terenie, a także we wzroście frekwencji przedmiotów obcego pochodzenia, przy równoczesnej zmianie kierunku ich napływu.

Podstawowymi wytworami kujawskich producentów były paciorki różnej formy (krążkowe, koliste podwójnie stożkowate, cylindryczne i ósemkowate) i wielkości (od dużych paciorków o średnicy 2 cm po bardzo małe koraliki).

Przy analizie zarówno okazów gotowych, jak i półfabrykatów i odpadów produkcyjnych uderza fakt daleko posuniętej oszczędności surowca, nakazującej maksymalne wykorzystanie powstałych przy produkcji podstawowego asortymentu większych obrzynków do wykonania dostosowanych do ich wielkości i kształtów mniej efektownych okazów.

Nie ulega wątpliwości powiązanie bursztyniarstwa kujawskiego z przechodzącym przez ten teren szlakiem handlowym, który z jednej strony stanowił źródło zaopatrzenia lokalnych pracowni w niezbędny dla nich surowiec produkcyjny, z drugiej zaś gwarantował zbyt wytwarzanych w nich przedmiotów. Tutejsze paciorki stanowiły więc ważny ekwiwalent w ówczesnej wymianie, dzięki której na Kujawy napływały na przykład wyroby szklane, brązowe, a także monety. Z kolei włączenie się producentów kujawskich do dalekosiężnej wymiany stwarzało im możliwości osiągnięcia wyższego statusu majątkowego i społecznego, a pośrednio wpływało też na wzrost zamożności tutejszych społeczności.

Dalekosiężne kontakty handlowe stały się również czynnikiem inspirującym rozwój całokształtu miejscowej kultury — zarówno jej sfery materialnej, jak i symbolicznej. Mamy więc tutaj do czynienia z podobną sytuacją, jaką można obserwować na Kujawach w dobie rozwoju rodzimej metalurgii brązu w późnej epoce brązu i we wczesnym okresie żelaza.

Opowiedzenie się za wykorzystywaniem przez miejscowych bursztyniarzy surowca transportowanego znad Bałtyku na tereny południowej i środkowej Europy nie neguje posiłkowania się też okazjonalnie na miejscu znajdowanymi grudkami, tak w pobliżu zbiorników wodnych, jak i na terenach piaszczystych. Te jednak służyły raczej do wyrobu paciorków na własny użytek.

A THOUSAND YEARS OF AMBER-CRAFT IN GDAŃSK

Eleonora TABACZYŃSKA

Abstract

Amber-craft in Gdańsk a thousand years ago constitutes only a moment in the long-term history of this traditional local activity. Archaeological excavations initiated in 1948 at the castle-stronghold site yielded numerous amber finds. The analysis of these finds in their stratigraphic context has shed much light on the history of amber-working in Gdańsk. Amber ornaments (mainly pendants of various forms) and beads, were a popular commodity here and amber-working was commonly practised as a cottage industry. This is evidenced by the abundance of crude amber, production waste and variety of finished products distributed in and outside the houses, on the streets and elsewhere. A particularly interesting discovery was made within the ruler's fortified residence, where the remains of a late 10th/early 11th century specialized amber workshop were uncovered — the oldest one in "urbs Gyddanyzc".

The fascinating problem of Baltic amber, its significance in cultural and social relations between European as well as non-European countries, and its natural value, is drawing the attention of increasingly more scientists and amateur enthusiasts around the world. This is reflected in the number of publications, specialist research projects and the variety of exhibitions and events being organized at present, such as the International Amber Fair for Jewellery — AMBERIF — held every year in Gdańsk. The crafts- and tradespeople involved in this venture say that now is the time of "Rediscovering Poland's Golden Treasure", and they refer to Gdańsk as the "Amber Capital of Poland".

I am glad to be able to present this paper on amber-craft in Gdańsk a thousand years ago on the occasion of the city's Millennium celebrations. Firstly though, I would like to recall a few facts...

Amber-working in the Baltic region is as old as the first human settlements. It has been continuously practised and developed by the people who used amber as ornaments, amulets and healing "stones". Thus, the origins of amber-working in Gdańsk have their roots in the achievements of the distant past, as evidenced at a number of neighbouring archaeological sites dating from the Stone Age onwards. Amber-craft in Gdańsk a thousand years ago represents only a moment, in the long-term history of this traditional local activity.

My short presentation of this problem is based on the results of investigations and reports published by Konrad JAŻDŻEWSKI (1962; JAŻDŻEWSKI *et al.* 1966), Anna WAPIŃSKA (1967) and Andrzej ZBIERSKI (1969) — archaeologists who were involved in the excavations initiated in 1948 as part of the preparations for the state jubilee — Millennium Poloniae — celebrated in 1966 (see also: CIEŚLAK & BIERNAT 1988; KMIECIŃSKI 1959–1977). Work on amber-craft in medieval Gdańsk is still currently being carried out by Anna WAPIŃSKA, who is preparing a monograph including all the discoveries of amber finds after the year 1954 (when the largest-scale investigations were completed).

The issue of the amber-working industry in medieval Gdańsk is related to my research on the development of amber-craft technology, from its beginnings up until the Middle Ages (TABACZYŃSKA 1966; 1975). This topic has been studied under the auspices of the Polish-Italian Amber Commission, founded in Rome in 1972 (see TABACZYŃSKA, *Amber as a subject of archaeological research: the experiences of Polish-Italian collaboration*, in the volume).

We are able to reconstruct amber-craft in Gdańsk a thousand years ago by analysing amber finds in their stratigraphic context. It is well known, that Gdańsk was founded in the second half of the 10th century, at the time of the formation of the Polish state. The town was sited on the western peripheries of the delta plain of the Vistula, on the banks of the River Motława. In the first written mention of Gdańsk, known from the Life of ST. ADALBERT (*Vita prima Sancti Adalberti*), entered under the year 997, we find the name "urbs Gyddanyzc". This is the earliest confirmation we have of Gdańsk's status as a town. Archaeological research, however, indicates that at the beginning of the 10th century, and even in the 9th century, the area of Gdańsk was inhabited by people living in a small rural settlement, whose main form of subsistence was fishing.

Archaeological excavations in Gdańsk, which were among the most major investigations of those conducted in other Pomeranian towns, such as Szczecin, Wolin and Kołobrzeg, revealed the wooden relics of the castle-stronghold, its suburbs and port. The earliest phase of this complex dated from the 10th century (Fig. 1). It continued to function until the 14th century, up to the year 1308, in which Gdańsk suffered destruction by fire at the hands of the Teutonic Knights. Nevertheless, the development of this settlement centre continued in neighbouring locations, which later came to be known as the Old Town and Main Town.

The most significant site uncovered in Gdańsk (Site 1) contained 17 settlement levels dated from *c.* AD 980 to 1308, as mentioned earlier. The fortifications, houses and streets were wooden structures. The main occupations of the inhabitants were fishing and trade, with a variety of handicrafts developing under the protection of the local ruler, whose fortified residence was situated within the defensive walls of the settlement. The majority of trades and handicrafts were practised as cottage-industries. These included metallurgy, encompassing the smelting of iron and other metals such as copper, tin, lead and occasionally silver, as well as the production of bronze and brass. Carpenters, wood-workers, shoemakers, antler-workers, potters, weavers and spinners (working wool and flax) were also among the local population of craftspeople. All these occupations led to the development of both local and foreign trade, with amber-working playing an important role in this process. Amber was one of the most frequently found materials in all the Gdańsk settlement strata, distributed in and outside the houses, on the streets and elsewhere. Amber ornaments were a popular commodity and amber-working was a common

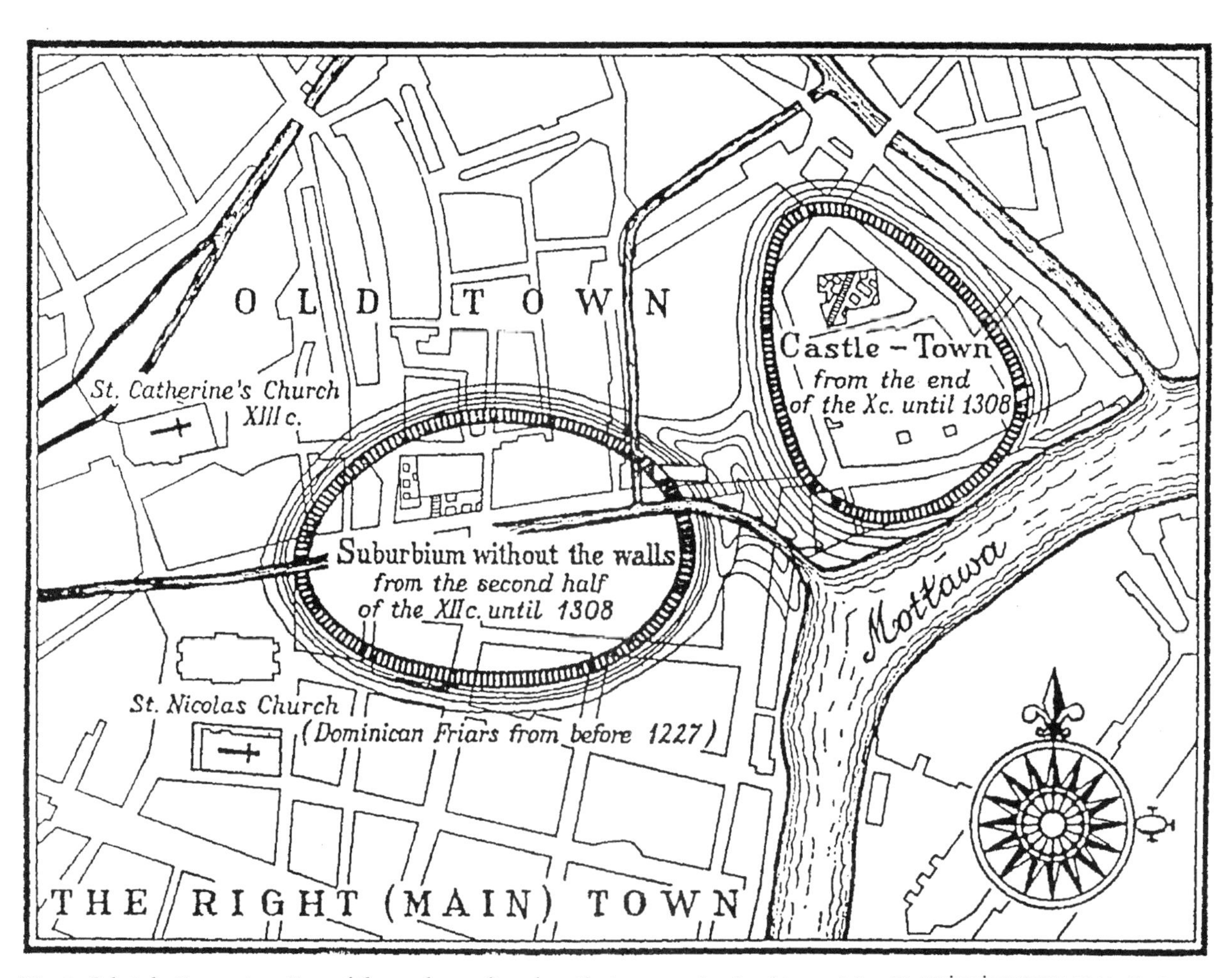

Fig. 1. Gdańsk. Reconstruction of the early medieval castle-town and suburbium. After K. JAŻDŻEWSKI 1988, [*in:*] CIEŚLAK & BIERNAT 1988.

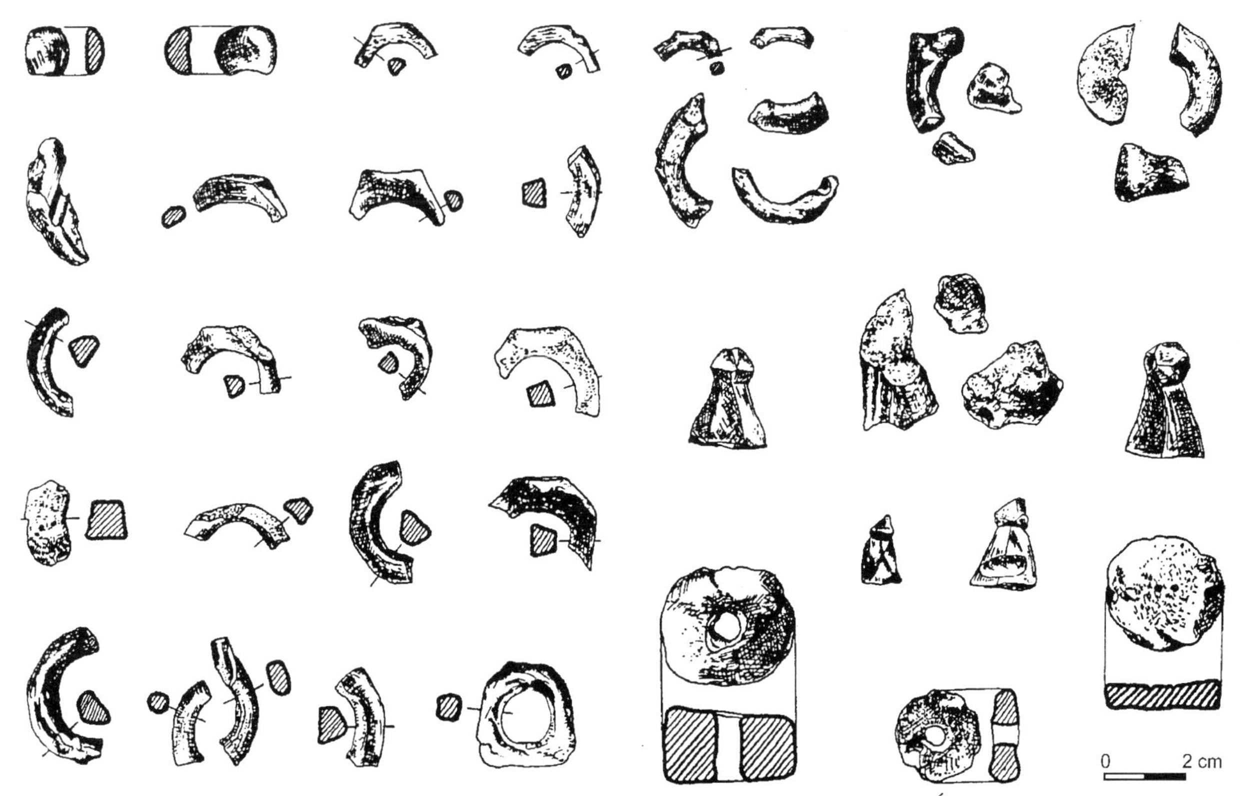

Fig. 2. Gdańsk. Site 1. Amber finds from the workshop (10th/11th century). After A. WAPIŃSKA 1967.

occupation – a cottage industry. Looking at the numerous amber finds, which include crude amber, semi-finished products and all kinds of ready-made goods — mostly a wide assortment of ornaments, such as beads, pendants, rings, necklace elements and, very occasionally, small figurines (Figs. 2 & 3) — we admire the variety of forms and colours of amber, but are left with the impression that, for the local people, the raw material itself was of greater intrinsic value than its decorative or artistic forms. As a result, the amber goods which they produced were very simply fashioned and not, in general, perfectly executed — their forms are usually irregular and their surfaces only lightly polished. Anna WAPIŃSKA has published the material from Site 1, which consisted in total of 862 objects, either complete or only half-finished, and about 21.40 kg of crude amber nodules. She has classified the ready-made products into three groups: ornaments, gaming pieces and small figurines. The ornaments comprise beads and pendants of various forms (crosses among them). Dice and chess pieces make up the second group, with the third and rarest category consisting of carved amber figures.

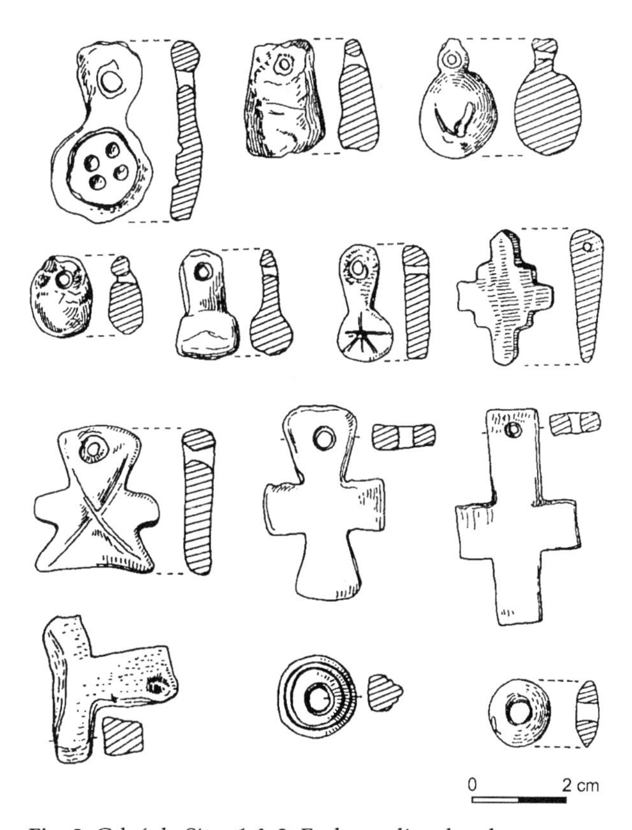

Fig. 3. Gdańsk. Sites 1 & 2. Early medieval amber pendants and beads. After A. WAPIŃSKA 1967.

Amber-working tools were very simple and similar to those used in other crafts, for instance in the working of bone and antler. Knives were the most common of these implements, followed by hatchets, planes, chisels, saws and drills with a spade-shaped end which were used to perforate the products. The majority of amber ornaments were handmade, while only about 6% exhibit traces indicating that they were turned on a lathe — these marks are visible on some of the rings and round beads. However, we do not, as yet, have any known examples of a lathe from the archaeological record. The Neolithic technique of friction using a strong thread was probably also a well-known method of cutting or dividing amber nodules. The use of some tools has been inferred from the external features and traces on the artefacts recovered and indicates that the technological level of amber-working was similar to that which is still prevalent to this today in the ethnic culture of the Polish region of Kurpie.

The remains of two specialized workshops dating from the late 10th/early 11th century were discovered in the earliest town of Gdańsk, within the aforementioned ducal residence. These workshops would have produced goods for the ruler and his court as well as for a broader market. One of these was a jeweller's workshop producing decorative wares made of silver and other metals. The second was an amber workshop which bears witness to the specialized production of amber rings with oval or round centres (similar to metal rings) and of separate central elements in amber used for metal rings, as well as the production of amber chess-pieces (and other gaming counters). This was the first amber workshop in "urbs Gyddanyzc", which paved the way for increased production and the future prosperity of the amber-working trade in Gdańsk.

Eleonora TABACZYŃSKA
ul. Daniłowskiego 6 m 98
01-883 Warszawa, Poland

OBRÓBKA BURSZTYNU W GDAŃSKU PRZED TYSIĄCEM LAT

Eleonora TABACZYŃSKA

Streszczenie

Obróbka bursztynu w „urbs Gyddanyzc", wzmiankowanym w 997 roku, wyrosła na podłożu starych tradycji sięgających epoki kamienia i należy do najstarszych (obok rybołówstwa, korabnictwa i in.) zajęć ludności zamieszkującej tereny wybrzeża bałtyckiego. Odtworzenie wczesnośredniowiecznego bursztyniarstwa w Gdańsku umożliwiają źródła archeologiczne pozyskane w trakcie zainicjowanych w 1948 roku, wieloletnich badań wykopaliskowych związanych z obchodami jubileuszu Millenium Poloniae. Rysuje się ono jako jedno z najważniejszych zajęć mieszkańców X-wiecznego grodu, uprawiane na wielką skalę, o czym świadczą duże ilości surowca, odpadków, półwytworów i różnej formy produktów znajdywane w wielu domostwach, w ich pobliżu i na ulicach odsłoniętego poziomu osadniczego. Obróbka bursztynu uprawiana była powszechnie jako rękodzieło domowe w obrębie osady rybackiej, na podgrodziu i w obrębie samego grodu. Większość produktów, na które składały się zawieszki, paciorki, pierścienie i formy figuralne, nosi ślady ręcznej obróbki przy pomocy noża i innych prostych narzędzi. Tylko 6% wytworów, głównie pierścienie i paciorki, wykazuje cechy toczenia, jak podaje A. WAPIŃSKA w swym opracowaniu bursztyniarstwa gdańskiego opublikowanym w 1967 r. Opisuje ona również interesujące odkrycie wyspecjalizowanej pracowni bursztyniarskiej o charakterze rzemieślniczym z X/XI w., zlokalizowanej w rezydencjalnym obwodzie grodu, obok reliktów warsztatu złotnika-odlewacza produkującego ozdoby srebrne. Bursztyniarz produkował tu wyłącznie pierścienie, oczka bursztynowe do metalowych perścieni oraz pionki do gry. Mamy tu do czynienia z zalążkiem rzemiosła bursztynowego w Gdańsku, którego obecny rozwój technologiczny, specjalizacja i asortyment produkowanych ozdób i przedmiotów artystycznych wzbudzają zainteresowanie w wielu krajach świata.

Bibliography

CIEŚLAK E. & BIERNAT C. (eds)
1988 *History of Gdańsk*, Wydawnictwo Morskie, Gdańsk.

JAŻDŻEWSKI K.
1962 *Gdańsk wczesnośredniowieczny w świetle wykopalisk*, Gdańsk.

JAŻDŻEWSKI K., KAMIŃSKA J. & GUPIEŃCOWA R.
1966 Gdańsk des Xe–XIIIe Siecle, *Archeologia Urbium*, Warszawa.

KMIECIŃSKI J. (ed.)
1959–1977 Gdańsk Wczesnośredniowieczny, 1–9, Gdańsk.

TABACZYŃSKA E.
1966 Problemi della produzione di oggetti d'ambra in Pomerania nell'Alto Medioevo, [*in:*] *Atti del VI*

Congresso Internazionale delle Scienze Pre- e Protostoriche, **3**, 210–212, Roma.

1975 Commento archeologico alla produzione di oggetti d'ambra in Polonia nell'Alto Medioevo, [*in*:] Atti della cooperazione interdisciplinare italo-polacca, **1**, *Studi e ricerche sulla problematica dell'ambra*, 89–96, Roma.

WAPIŃSKA A.
1967 Materiały do wczesnośredniowiecznego bursztyniarstwa gdańskiego, [*in*:] *Gdańsk Wczesnośredniowieczny*, **6**, 83–100.

ZBIERSKI A.
1969 Rzemiosło w Gdańsku wczesnośredniowiecznym, [*in*:] *Gdańsk, jego dzieje i kultura*, 363–376, Warszawa.

AMBER-WORKERS OF THE FOURTH AND FIFTH CENTURIES AD FROM ŚWILCZA NEAR RZESZÓW

Aleksandra GRUSZCZYŃSKA

Abstract

Earthworks carried out at Świlcza in 1970 led to the chance discovery of a 4th–5th-century settlement. Archaeological excavations revealed a total of nine features. Of particular interest were two houses and two workshops where amber was processed. These buildings were associated with each other. Recovered materials from the site included amber artefacts (in the form of raw material, beads and off-cuts) and a hoard of gold and silver coins and jewellery dated to the first half of the 5th century AD. These finds indicate that the inhabitants of the Świlcza settlement were actively involved in long-distance exchange both with northern and southern Europe.

A chance discovery of a 4th–5th-century settlement was made at Świlcza (Fig. 1) in 1970 by Mr Leopold SKUPIEŃ. Whilst conducting an earth moving operation prior to the widening of the road leading to his field, Mr SKUPIEŃ came across a large nodule of amber and two amber beads buried at a depth of over two metres.

The site in question lies on the high east bank of an unnamed brook (Fig. 2). Occupation levels were present in a very compact and damp soil deposit at a depth of 2 m below ground level. This situation most probably came about as a result of subsidence which brought down the upper part of the bank. The prevalent damp conditions helped the artefacts to survive in excellent condition.

A total of nine features were revealed during the course of excavation work carried out in 1970–71, 1973–77 and 1981–82 (GRUSZCZYŃSKA 1971, 82–83; 1975, 73–76; 1977, 183–188; 1982, 155; 1983, 171; 1984, 103–129). The most noteworthy of these features were two houses and two amber-processing workshops (Fig. 3).

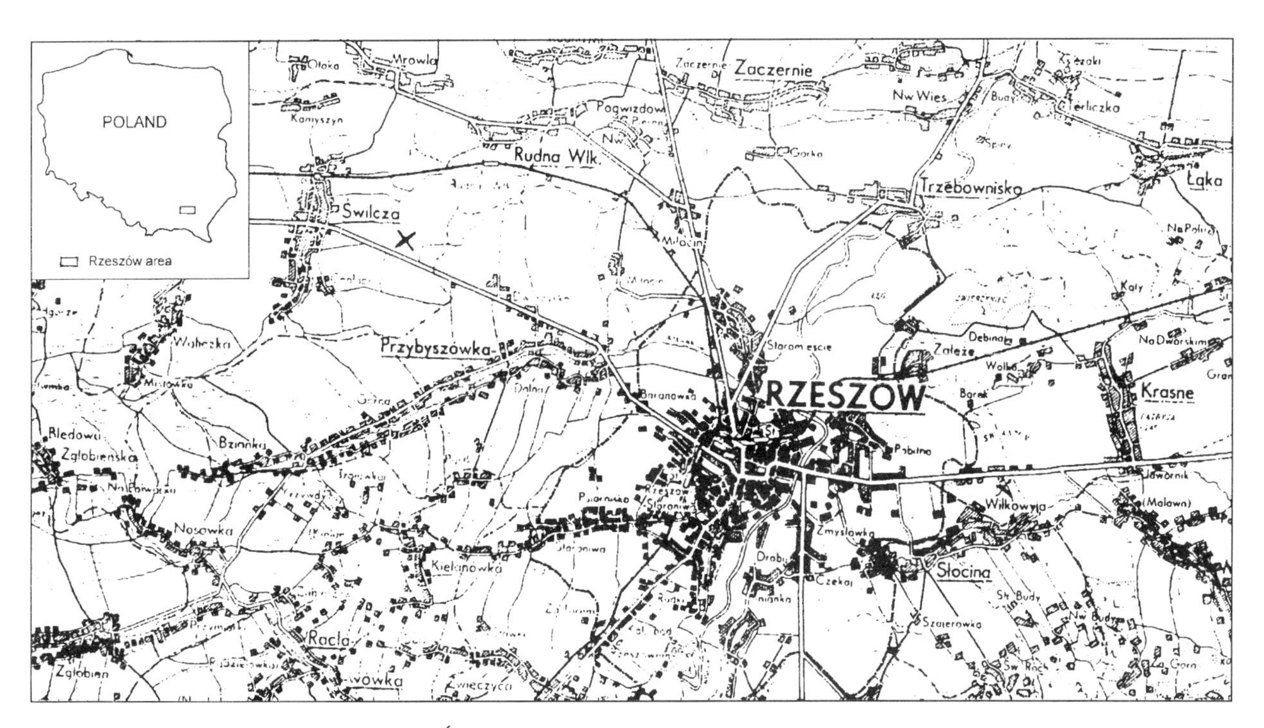

Fig. 1. Context map showing location of Świlcza and Rzeszów, SE Poland.

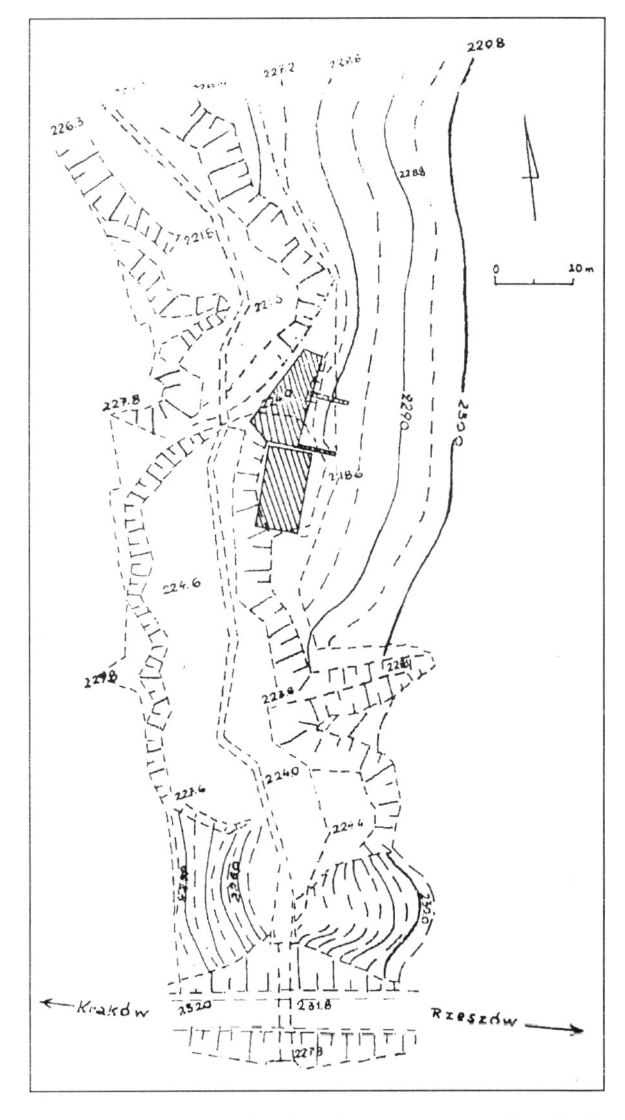

Fig. 2. Contour map showing the area excavated at site 3, Świlcza, Rzeszów voivodeship.

House 1 (initially referred to as pit 2) was a *grubenhauser*, roughly rectangular in shape, measuring 4.6 x 5.3 m. Traces of an extension to this building, measuring 2 x 3.2 m, were found on the south side of its longer wall. Post-holes were revealed in the corners of this extension and running along the length of its walls as well as within the interior of the main building. A hearth measuring 1.1 x 0.8 m was revealed at the bottom of the house. A number of well-preserved beams and planks of oak and fir were discovered in the north-east section of the building and partly beyond its boundaries at a depth of 3 m (Figs. 4 & 5). These lay in no regular order and were interspersed with sandy silt (possibly evidence of flooding). The average length of the planks ranged from 2.1 to 2.7 m, their width from 0.3 to 0.38 m and their thickness from 0.02 to 0.07 m. Axe marks were visible on the fir beams and planks. The irregular arrangement of these planks indicates that they represent the remains of a collapsed wall.

A modest quantity of pottery and a relatively large amount of animal bones and teeth were recovered from the fill of the house, together with stones, daub and charcoal. A hoard of gold and silver coins and jewellery was discovered in the annex next to the south wall of the house, at a depth of 2.8 m below ground level.

House 2 (referred to in site records as pit 8) was located at a depth of 2.6 m. This building was rectangular in plan, measuring 5.2 x 7 m, and had an annex built to the north-east of it. Evidence of four posts was discovered in the fill of this house. A very large amount of decomposed wood, arranged in regular order, was found in its north section. This probably constituted the remains of one of its walls. Traces of a hearth were revealed in the central part of the house. The fill of this building consisted mostly of a brownish-grey soil with frequent inclusions of decomposed wood and small pieces of charcoal. Finds included a limited amount of pottery sherds, stones and daub, as well as a large quantity of animal bones.

Amber workshop 1 (initially referred to as pit 1) was discovered at a depth of 2.2 m. Due to its location on the edge of an escarpment and the damage which it underwent during the earth-moving operation, it was not possible to establish the workshop's north-west extent. Its surviving section was vaguely trapezoidal in plan and measured 4.5 x 6.0 m. Its

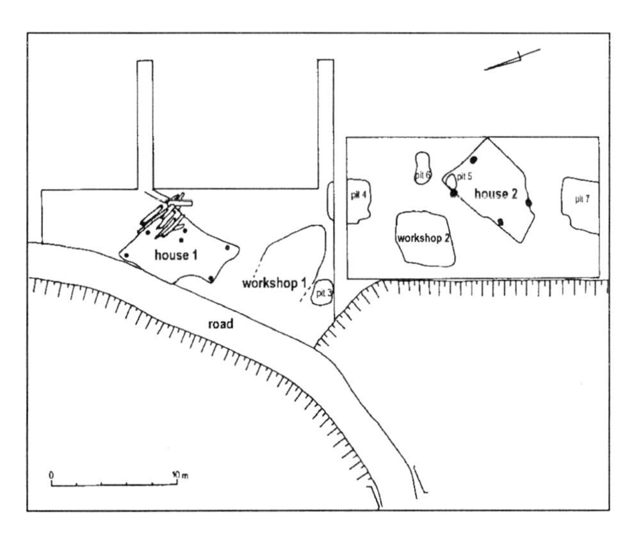

Fig. 3. Plan showing area excavated and features revealed at site 3, Świlcza, Rzeszów voivodeship. Drawing A. LUBELCZYK.

Fig. 4. Timber beams and planks from house 1 at site 3, Świlcza, Rzeszów voivodeship. Photo. A. Hadała.

Fig. 5. Timber beams and planks from house 1 at site 3, Świlcza, Rzeszów voivodeship. Photo. A. HADAŁA.

uneven floor surface lay at a depth of 2.7–3.0 m. The upper layers of fill consisted of a brownish-black soil with frequent inclusions of iron dripstones and charcoal, whilst the lower layers were predominantly grey with charcoal inclusions.

Two-hundred-and-thirty-two amber beads, 43 fragments, 13 waste pieces and a nodule of amber weighing 305 g were discovered in and around this workshop. The largest concentration of beads was found in its south-east section, with smaller concentrations in the central part and isolated items appearing across the whole workshop. Beads were present throughout the whole fill, the greatest density appearing at a depth of 2.45–2.55 m. Other finds included part of a sandstone whetstone, bone tools, small amounts of pottery sherds and a modest amount of animal bones.

Amber workshop 2 (referred to as pit 9) was revealed at a depth of 2.6 m. This building measured 4 x 4.4 m and was roughly rectangular in plan. Its fill consisted of a brownish-grey soil with numerous charcoal inclusions. One-hundred-and-fifty-six amber beads, 10 bead fragments, 12 waste pieces and part of an amber nodule were recovered. The densest concentration of beads was found in the south-west section of the workshop. Other finds included several pottery sherds and a minimal amount of animal bones.

The features described constitute an interesting example of both the architecture and function of the settlement with its two amber workshops and domestic buildings associated with them. Workshop 1 was situated 2–2.5 m from the annex of house 1, whilst workshop 2 was located 2.0–4.0 m from the north wall of house 2. The distance between the two workshops amounted to 5.0–7.0 m. The absence of any wall structures in both workshops (in contrast to the houses) points to the fact that these were seasonal open-air work areas, whilst the houses sited between them served to shield them (WIELOWIEJSKI 1991, 332, 336).

A total of 388 whole amber beads, 53 bead fragments, 25 waste pieces and an amber nodule weighing 305 g as well as part of a second nodule (with traces of preliminary working) weighing 31 g were recovered from the two workshops and their immediate vicinity.

The beads can be classified into two groups according to the way in which they were produced. **The first group** consists of hand-made beads, probably produced with the help of a knife or chisel. Most of the beads in this group show signs of having been polished. Some have holes which were drilled from one side only, which is clearly visible on those beads which were not drilled all the way through. This was most probably to prevent the bead from damage at the drill's point of exit. Drilling a hole from both sides reduces the risk of chipping the edges of the hole. Several beads had holes drilled in them in a number of different places. The group of beads in question represent a batch of material which had been prepared for further working. **The second group** comprises two items (both from workshop 1). These consist of beads which bear traces of having been turned on a lathe. Their surfaces are smooth, their tops and sides decorated with a groove motif. These beads represent the finished product, which was no doubt used as a tradable commodity. Numerous analogies of this type of product are known from around Poland. Some of the more notable examples include that of a large hoard of unworked amber nodules and beads, recovered from Basonia in the voivodeship of Lublin and dated to the first half of the 5th century (NOSEK 1951, 91); another early 5th-century hoard from Kiełpino, Gdańsk voivodeship (LA BAUME 1934, 148, Pl. 74) and a grave assemblage from a late Roman period female burial discovered in Łódź-Retkinia (KMIECIŃSKI 1952, 147, Fig. 12).

The entire assemblage from Świlcza consists of circular, cylindrical and biconoidal beads (MĄCZYŃSKA 1977, 64). The decided majority are circular, though varied in cross-section. Some have convex top and bottom surfaces, some are flat on one side and convex on the other, whilst some are entirely flat. One tubular bead was also present among these finds (Fig. 6).

The quality of amber used in the production of beads at the Świlcza settlement was good. This is evidenced by the large nodule which was found at the site as well as by the size and good state of preservation of the beads recovered. Of the five varieties of amber identified amongst the retrieved material (MAZUROWSKI 1983, 17 – see also bibliography), four were used by the amber-workers of Świlcza: transparent amber, clouded, bastard and bone amber (listed according to their degree of translucency from clearest to most opaque). These varieties were used in the following proportions (WIELOWIEJSKI 1991, 335):

workshop 1: clouded 51.06%; transparent 35.32%; bastard 11.49%; bone 2.13%;

workshop 2: clouded 50.33%; transparent 42.38%; bastard 6.63%; bone 0.66%.

The various colours of the beads from Świlcza included yellow, reddish, orange and white with a yellowish hue. The relatively large proportion of orange beads in this assemblage can be accounted for by the fact that weathering processes induce numerous alterations to the original colour of amber, one of the most common being the change from its natural yellow hue to orange or reddish (LECIEJEWICZ & MIERZEJEWSKI 1984, 35).

Infra-red analysis was carried out on the amber finds from Świlcza in order to establish which type of fossil resin they were made of[1]. The results of

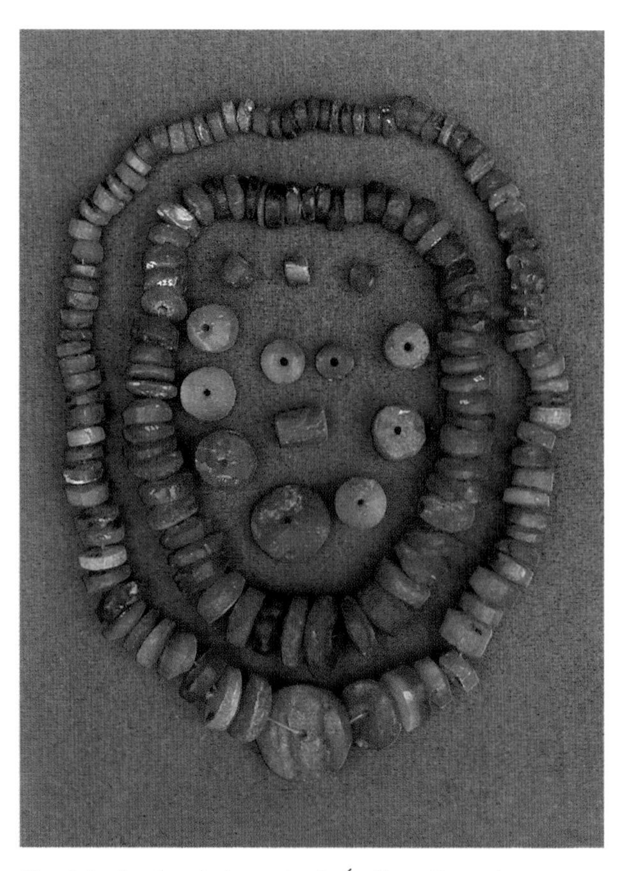

Fig. 6. Amber beads from site 3, Świlcza, Rzeszów voivodeship. Photo. S. CZOPEK.

this analysis showed that the IRS curves obtained from the material examined were characteristic of succinite.

Associated amber-working tools recovered from Świlcza included two sharpened bone fragments, part of a whetstone made of grey sandstone and one flint implement. These tools could, of course, have been used for other purposes.

The most valuable find from the Świlcza settlement was undoubtedly that of a hoard of gold and silver coins and jewellery discovered during excavations carried out in 1976. The hoard was concealed in the annex of hut 1, inside a leather pouch. The pouch contained: 2 gilded silver fibulae; 2 silver fibulae; a "Hunnish" gold pendant/ear-ring (Fig. 7); a bronze pendant; a bent silver bracelet (with one end broken) with two silver pendants threaded onto it; 5 silver-wire pendants wound onto a reel of silver; 3 pairs of silver wires (each pair consisting of 2 wire pendants joined together); 5 silver-wire pendant fragments; 10 silver denars (stacked on top of one another); a glass bead; bone and flint tools and one iron object too heavily corroded to identify. The

[1] Analysis of the amber finds from Swilcza was carried out at the Laboratories of Warsaw Polythechnic. My thanks go to Prof. Barbara KOSMOWSKA-CERANOWICZ for her interpretation of the analysis results.

fibulae were used as the primary dating material for this hoard. These were typical examples of the "Untersiebenbrunn Sösdala" type, dated to the first half of the 5th century (GRUSZCZYŃSKA 1977, 187; 1984, 120 & bibliography), or to be more precise to the mid 5th century (GODŁOWSKI 1979, 35). In this particular instance it would be misleading to take into consideration the date of issue of the coins contained in the hoard (AD 134–184). Their high silver content ensured that they remained in use in Poland during the Roman period, when the coins of the day had a reduced ore content due to the monetary reform introduced in AD 194. Further monetary devaluation followed in later years which also helped keep these earlier, better quality coins in circulation (GUMOWSKI 1958, 103; KUNISZ 1969, 90–91). The heavily worn surfaces of the coins from the Świlcza hoard bear witness to their long years of usage.

It should be stressed that only a limited quantity of pottery was found at the Świlcza settlement, this consisting of late Roman period grey wares (GRUSZCZYŃSKA 1975, 75–76; 1984, 111 & bibliography), including sherds of both handmade and wheel-thrown vessels (Fig. 8).

A large amount of animal bone was recovered in association with the pottery. The results of analysis showed that domestic species made up 97% of this bone assemblage, the most numerous being cattle (63%), horse (16%) and pig (12%). Other skeletal materials included a modest quantity of ovi-caprid and dog bones as well as those of a small species of chicken (roughly the size of a modern-day bantam). Wild animal remains (3% of the total assemblage) included bones of aurochs, deer, stag and wild boar (WOLSAN & NADACHOWSKI 1992, 193).

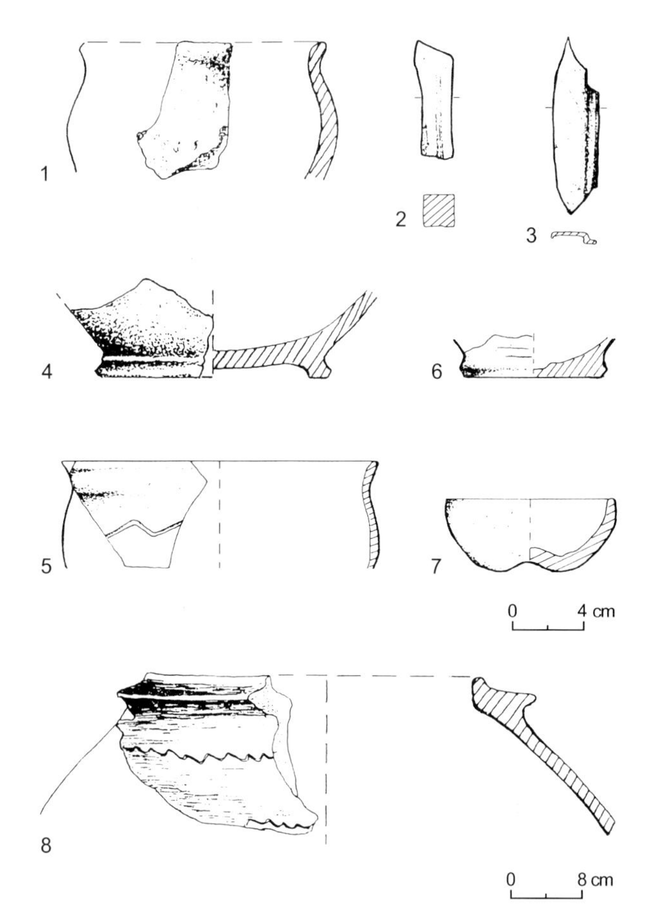

Fig. 8. Selection of finds from site 3, Świlcza, Rzeszów voivodeship: 1 — potsherd recovered from house 1; 2 — sandstone polishing stone; 3 — bone tool; 4, 8 — pottery from amber workshop 1; 5–7 — pottery from house 2.

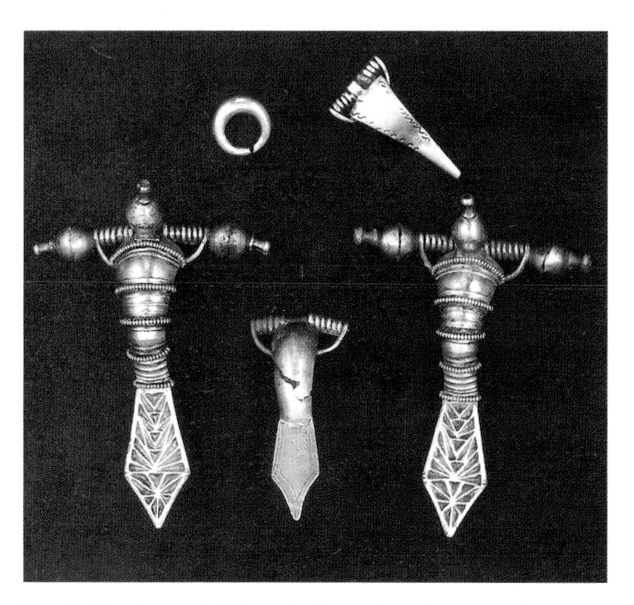

Fig. 7. Fibulae and "Hunnish" pendant from site 3, Świlcza, Rzeszów voivodeship. Photo. S. CZOPEK.

The discovery of a settlement complex with residential huts and amber workshops accompanied by a rich finds assemblage consisting of amber and gold items, a hoard of gold and silver jewellery and coins, as well as everyday items such as pottery, tools and animal bones concentrated in such a small area (the area excavated amounted to 425 m^2) attests to the wealth of the settlement's residents and to the fact that they were involved in long-distance trade, both with northern and southern Europe. It is possible that the amber-workers of Świlcza obtained their raw material from deposits in the Sambian delta which stretches along the Baltic coast from the Sambian Peninsula to the regions of Gdańsk and Karwia (KOSMOWSKA-

CERANOWICZ 1988, 173 ff). The good quality of the amber used in both workshops and the fact that Baltic amber was the most commonly used variety of this material during the Roman period further support this theory.

H. ŁOWMIAŃSKI (1963, 285) believes that during the Migration Period the amber trade gradually expanded east and south along the Vistula towards the Dukielska Pass. Evidence of this having been the case comes in the form of a large hoard of crude amber and amber beads discovered in Basonia, in the voivodeship of Lublin, which is indicative of the fact that contacts existed between the Baltic region and the Carpathian Valley, in particular with the Ostrogoths. This opinion is shared by J. WIELOWIEJSKI (1970, 149). K. GODŁOWSKI, however, believes that evidence of the Baltic region's contacts with the South during the Migration Period attests to the fact that these contacts must have been maintained, at least in part, by areas of southern Poland (GODŁOWSKI 1976, 397). The amber items recovered from Świlcza are no doubt related to the communications route in question (MĄCZYŃSKA 1977, 73), as is the amber bead found at a late Roman period settlement site in Lesko, Krosno voivodeship (BARŁOWSKA 1984, 95). To this list we can also add one further find — part of an amber ornament — discovered prior to the Second World War in Przybyszków, Rzeszów voivodeship (approximately 3 km from Świlcza)[2].

It seems more than reasonable to assume that the Świlcza settlement was directly linked to the amber trail, the finds recovered here indicating that this site served as an intermediary in the amber trade. It was here that merchants from the Baltic, who brought with them crude amber which was subsequently processed by the local amber-workers, met with those travelling from the south of Europe.

A particularly interesting problem linked to the decline of the Świlcza settlement comes in the form of the coin hoard. J. WIELOWIEJSKI considers that hoards were buried in central and eastern Europe as a result of ethnic migrations arising from periods of war (WIELOWIEJSKI 1970, 105–106). The fact that the hoard was buried around the mid 5th century leads us to conclude that this must also have been the period during which the settlement's inhabitants were forced by unforeseen circumstances to abandon their homes. The date of the hoard's deposition led K. GODŁOWSKI to propound the theory that this was an act which bore testimony to a calamity of far greater proportions, which affected a much wider area and may have ultimately been linked to the Przeworska culture's disappearance from this region. This event may have in turn been connected to the fall in *c.* AD 455 of the Hunnish Empire, whose influence most probably stretched from the Carpathian Valley up to the regions surrounding the river basin of the Upper Vistula (GODŁOWSKI 1979, 35).

Archaeological excavation results demonstrate that the Świlcza settlement played a significant role in south-east Poland from the late 4th to the mid 5th century. The territories in question were at the time under the influence of a number of different neighbouring cultural groups.

Aleksandra GRUSZCZYŃSKA
Muzeum Okręgowe
ul. 3 Maja 19
35-030 Rzeszów, Poland

BURSZTYNIARZE Z IV–V WIEKU N.E. W ŚWILCZY KOŁO RZESZOWA

Aleksandra GRUSZCZYŃSKA

Streszczenie

Osadę w Świlczy odkryto przypadkowo podczas prac ziemnych w 1970 r. Położona jest na wysokim brzegu bezimiennego potoku.

Podczas badań wykopaliskowych odsłonięto 9 obiektów, a wśród nich niezwykle ciekawe zespoły — 2 chaty oraz 2 pracownie obróbki bursztynu, usytuowane w pobliżu i jak gdyby funkcjonalnie ze sobą związane.

Materiał zabytkowy odkryty w osadzie w Świlczy to: ceramika, kości zwierzęce, kamienie, przedmioty z bursztynu, narzędzia do jego obróbki oraz skarb złotych i srebrnych ozdób i monet.

W obrębie pracowni bursztyniarskich oraz ich pobliżu znaleziono 388 całych paciorków bursztynowych, 53 fragmenty paciorków, 25 odłupków, bryłę bursztynu oraz fragment bryły ze śladami wstępnej obróbki.

[2] This find is housed at the Rzeszów District Museum (*Muzeum Okręgowe w Rzeszowie*) — Inventory No. MRP 139.

Wśród paciorków wyróżniono dwie grupy. Pierwsza z nich to paciorki wykonane ręcznie (stanowią one zapewne rodzaj półwytworów przygotowanych do dalszej obróbki). Grupa druga to paciorki obtaczane na tokarce (reprezentowana przez dwa zabytki). Pod względem kształtu wydzielono paciorki: krążkowate — obustronnie wypukłe, płasko-wypukłe, płaskie, cylindryczne, stożkowate oraz pojedynczy paciorek rurkowaty.

Bursztyniarze ze Świlczy wykorzystywali cztery odmiany bursztynu: przezroczysty, chmurzysty, bastard i kościany. Wyniki badań w świetle podczerwonym wykazały krzywe IRS typowe dla sukcynitu.

Chronologię osady ze Świlczy określa się na koniec IV i pierwszą połowę V w. n.e. Kres użytkowania osady wytycza odkryty w 1976 r. W przybudówce chaty nr 1 bogaty skarb srebrnych i złotych ozdób i monet. Skarb złożony w skórzanym woreczku zawierał: 2 fibule srebrne, pozłacane; 2 fibule srebrne; złotą zawieszkę — kolczyk; zawieszkę brązową; bransoletę srebrną zagiętą (o ułamanym jednym końcu) z nawleczonymi dwiema srebrnymi zawieszkami; fragment zakończenia wymienionej bransolety; zawieszki i druciki srebrne; 10 srebrnych denarów (ułożonych w rulonik); paciorek szklany; narzędzie z kości i krzemienia; przedmiot żelazny mocno skorodowany. Chronologię skarbu na podstawie ozdób określa się na czasy około połowy V wieku.

Teren osady stanowił zapewne miejsce pośrednictwa w handlu bursztynem. Tu najpewniej spotykali się kupcy strefy nadbałtyckiej, ci przywożący surowiec, który był obrabiany przez miejscowych bursztyniarzy, jak i podróżujący z południa Europy. Wiązało się to z przesunięciem w okresie wędrówek ludów handlu bursztynem w kierunku wschodnim wzdłuż Wisły i jej dopływów, a na południe w stronę Przełęczy Dukielskiej. Fakt wystąpienia w skarbie fibuli typu Nimburg, wskazuje również na kontakty z obszarami Europy Zachodniej, a złoty kolczyk typu „huńskiego" na powiązania z Hunami.

Ukrycie skarbu około połowy V wieku pozwala przypuszczać, iż mieszkańców osady spotkał jakiś kataklizm, zmuszając ich do pospiesznego opuszczenia swych domostw. Być może wiąże się on z większą katastrofą, która miała związek z zanikiem kultury przeworskiej w tym regionie oraz upadkiem państwa Hunów w Kotlinie Karpackiej, które to wydarzenie nastąpiło około 455 r. (GODŁOWSKI 1979, 35).

Bibliography

BARŁOWSKA A.

1984 Osada z późnego okresu wpływów rzymskich w Lesku, woj. Krosno, *Materiały i Sprawozdania Rzeszowskiego Ośrodka Archeologicznego za lata 1976–1979*, 51–101.

GODŁOWSKI K.

1976 Zagadnienie ciągłości kulturowej i kontynuacji osadniczej na ziemiach polskich w młodszym okresie przedrzymskim, okresie wpływów rzymskich i wędrówek ludów, *Archeologia Polski*, **21/2**, 378–401.

1979 Z badań nad zagadnieniem rozprzestrzenienia Słowian w V–VI w. n.e., Kraków.

GRUSZCZYŃSKA A.

1971 Bursztyny ze Świlczy, *Z Otchłani Wieków*, **37**, 82–83.

1975 Badania wykopaliskowe na stan. nr 3 w Świlczy, pow. Rzeszów w latach 1970–1971, *Materiały i Sprawozdania Rzeszowskiego Ośrodka Archeologicznego za lata 1970–1972*, 73–76.

1977 Skarb jakich mało, *Z Otchłani Wieków*, **43**, 183–188.

1982 Świlcza, woj. rzeszowskie, badania 1981, *Informator Archeologiczny*, 155.

1983 Świlcza, woj. rzeszowskie, badania 1982, *Informator Archeologiczny*, 171.

1984 Osada z wczesnego okresu wędrówek ludów w Świlczy, woj. Rzeszów, *Materiały i Sprawozdania Rzeszowskiego Ośrodka Archeologicznego za lata 1976–1979*, 103–129.

GUMOWSKI M.

1958 Moneta rzymska w Polsce, *Przegląd Archeologiczny*, **10**, 87–149.

KMIECIŃSKI J.

1952 Bogato wyposażony szkieletowy grób kobiecy z późnego okresu rzymskiego z Łodzi-Retkini, *Sprawozdania P.M.A.*, **4/3–4**, 139–148.

KOSMOWSKA-CERANOWICZ B.

1988 Niektóre złoża bursztynu i próby klasyfikacji żywic kopalnych, [*in:*] Surowce mineralne w pradziejach i we wczsnym średniowieczu Europy Środkowej, *Prace Komisji Archeologicznej*, **6**.

KUNISZ A.

1969 Chronologia napływu pieniądza rzymskiego na ziemie Małopolski, Wrocław–Warszawa–Kraków.

LA BAUME W.

1934 Urgeschichte der Ostgermanen, Gdańsk.

LECIEJEWICZ K., MIERZEJEWSKI P.

1983 Odmiany bursztynu i jego struktura, [*in:*] Bursztyn w przyrodzie, 34–38.

ŁOWMIAŃSKI H.

1963 Początki Polski, **2**, Warszawa.

MAZUROWSKI R. F.
1983 Bursztyn w epoce kamienia na ziemiach polskich, *Materiały Starożytne i Wczesnośredniowieczne*, **5**, 7–135.

MĄCZYŃSKA M.
1977 Paciorki z okresu rzymskiego i wczesnej fazy okresu wędrówek ludów na obszarze środkowoeuropejskiego Barbaricum, *Archeologia*, **28**, 61–94.

NOSEK S.
1951 Znalezisko z okresu wędrówek ludów na Lubelszczyźnie, *Sprawozdania P.M.A.*, 89–96.

TEMPELMANN-MĄCZYŃSKA M.
1985 Die Perlen der römischen Kaiserzeit und der frühen Phase der Völkerwanderungszeit im mittleeuropäischen Barbaricum, *Römisch-Germanische Forschungen*, **43**, Mainz.

WIELOWIEJSKI J.
1970 Kontakty Noricum i Pannonii z Ludami Północnymi, Wrocław–Warszawa–Kraków.

WIELOWIEJSKI P.
1991 Pracownie obróbki bursztynu z okresu wpływów rzymskich na obszarze kultury przeworskiej, *Kwartalnik Historii Kultury Materialnej*, **3**, 317–361.

WOLSAN M., NADACHOWSKI A.
1992 Szczątki zwierzęce z osady z późnego okresu wpływów rzymskich w Świlczy koło Rzeszowa, *Materiały i Sprawozdania Rzeszowskiego Ośrodka Archeologicznego za lata 1985–1990*, 193–199.

AMBER AS A SUBJECT OF ARCHAEOLOGICAL RESEARCH: THE EXPERIENCES OF POLISH-ITALIAN COLLABORATION

Eleonora TABACZYŃSKA

Abstract

The Italian-Polish Commission for Amber Route Studies, founded in Rome in 1972, was concerned with resolving various problems relating to Baltic amber, not only the question of its trade routes. The activity of this Commission and its works, presented in various publications, formed part of a common programme of large-scale interdisciplinary and comparative research on amber. This paper presents the results and experiences of the Commission's work, as realized by a team of archaeologists and representatives of other historical sciences, in collaboration with a group of specialists competent in natural, chemical and physical investigations into amber.

The *Baltic Amber and Other Fossil Resins* symposium was the first significant international meeting of amber specialists at which information was presented about the activity and results of the Polish-Italian working group on amber, whose work began in 1972. It was in that year the Italian-Polish Commission for Amber Route Studies (Commissione Italo-Polacca per lo Studio delle Vie dell'Ambra) was founded in Rome. This Commission operated within the framework of a cooperative venture between the Institute of the History of Material Culture of the Polish Academy of Sciences (now the Institute of Archaeology and Ethnology) and — in Italy — the Institute of Technologies Applied in Protection of the Cultural Heritage in the Italian National Research Council (Istituto per le Tecnologie Applicate ai Beni Culturali del Consiglio Nazionale delle Ricerche). From its inception it was concerned with resolving various problems concerning Baltic amber, not only the question of its trade routes.

It is well known that there are two main directions in current research and studies into amber. The first of these relates to the significance and role of amber in culture, in human and social life. The second examines amber in terms of its geological and natural aspects. In both these lines of enquiry scientists explore the potential of all the latest methods and techniques of analysis. All research on amber tends to be of an interdisciplinary character. We can see this in many, more or less detailed, publications about the most recent scientific investigations.

The activity of our Commission was, as it seems, a remarkable step in the development of archaeological research into amber and its role in European culture. The novelty of this enterprise — as is now acknowledged — lay in the initiation, in both participating countries, of systematic interdisciplinary investigations into a material so highly valued (on a par with gold and diamonds) by many peoples across the millennia.

Baltic amber — as is well noted — was one of the most important goods from the North to reach Italy. This fact is confirmed both by written sources and by the archaeological record. This may have been one of the reasons which led to the instigation of the extensive collaboration between Italian and Polish researchers in this broadly conceived project, which began over twenty years ago and continued until 1990. The main achievement of this collaboration was, as I see it, the creation of an international programme of large-scale comparative research into Baltic amber, considering not only strictly archaeological and historical problems but also natural and physico-chemical ones. From its very beginning, representatives of the arts and sciences from both countries participated in the realization of this programme, which seems to have provided work for more than one generation of scientists. Among those Polish institutions which took part in this project were the Museum of the Earth of the Polish Academy of Sciences, the National Ethnographic Museum, the Institute of Chemistry of Warsaw Polytechnic and others. From the Italian contingent, I shall allow myself to mention only some of the individual participants: N. NEGRONI CATACCHIO (Co-Secretary of the Commission),

G. de FOGOLARI, F. RITTATORE VONWILLER, M. FOLLIERI, M. PASQUINUCCI for archaeological and historical research, and G. GUERRESCHI, G. NICOLETTI and G. DONATO (Co-President) for the scientific and technological investigations. The Polish team consisted of W. HENSEL (Co-President), S. TABACZYŃSKI, J. WIELOWIEJSKI, T. LEWICKI, A. COFTA-BRONIEWSKA, E. TABACZYŃSKA (Co-Secretary) and other archaeologists and historians and P. SZACKI ethnographer. Further disciplines were represented by T. DZIEKOŃSKI, T. URBAŃSKI, A. SKALSKI, Z. GWIAZDA and others.

The full programme of this joint venture was published in Rome in 1975 in the book *Studi e ricerche sulla problematica dell'ambra*. Here, I shall give only a brief reminder of its main principles. The first of them concerned the complex analysis of amber finds from Poland and from Italy in all their historical, scientific and technological aspects. Arts and humanities research disciplines included archaeology, art history, cultural history, ethnography and ethnology, whilst the scientific and technological fields encompassed geology, natural sciences (biology, botany and others), organic and inorganic chemistry and physics. It was established that one of the central tasks was the development of non-destructive analytical methods for the determination of the natural and geographical provenance of amber finds.

The work of our Commission concentrated primarily on the collection and analysis of archaeological finds. Subsequently, ancient and later written sources were studied and research conducted into those cultures and regions which were of greatest interest because of the significant concentration of amber finds which occur in association with them. An essential factor was to identify diagnostic finds among the amber products of individual periods and territories, which were used as a medium of payment in the Baltic territories (eg. certain ornaments or Roman coins) and might indicate and verify the course of amber routes. The other category of joint work on amber included: research into the longtime evolution of amber-working techniques and the use of this raw material in artistic craft; ethnological studies and comparative research using ethnographic materials; studies on the heritage of legends and superstitions and on the apothropaic and therapeutic properties of amber, evidenced both in the Mediterranean world and in North European lands.

Progress reports on these projects were presented and disscussed (and sometimes altered) during annual plenary meetings organized alternately in Italy and in Poland and — when the need arose — during additional meetings of experts, or on the occasion of scholarship visits offered by both cooperating Academies. A special serial publication was established to present the current state of the Commission's work: *Atti della cooperazione interdisciplinare italo-polacca.*

The wide-ranging programme of studies and investigations into amber was realized with particular dynamism during the first three years of the Commission's work. From 1975 onwards (up until 1990 when the Commission formally ceased to exist) Italian-Polish studies on amber were very limited. The scope of this collaboration was considerably extended and the subject of amber became only one of several research projects being carried out simultaneously as part of a much enlarged programme. Therefore, the name of the Commission was changed to the Italian-Polish Interdisciplinary Working Group for Sciences Applied in Archaeology and the Protection of the Cultural Heritage. Among the new problems undertaken by this Group after the year 1975, the central ones became the theory and practice of archaeological research, conservation and protection of monuments.

The results of the collaborative work on amber have been published only in part (see Bibliography) and some enterprises are still continuing — for instance M. GULA's compilation of a bibliography and catalogue of sites where amber artefacts have been found in Poland.

This brief paper does not allow me to present all of the published works (see Bibliography). Besides the book *Studi e ricerche sulla problematica dell'ambra*, I would also like to mention those titles published in the catalogue of the *Ambra — Oro del Nord* exhibition held in Venice in 1978 by the Interdisciplinary Work Group in cooperation with the Museum of the Earth of the Polish Academy of Sciences in Warsaw and Museo Archeologico in Aquileia. Further articles containing the results of archaeological, historical, chemical and natural investigations into amber can be found in a number of different periodicals. Among those works of greatest interest, is a report on the ethnographic investigations conducted in Sicily in 1980 and 1984 by P. SZACKI from the National Ethnographic Museum in Warsaw (this publication is still in preparation). Information about traditions relating to amber in Sicilian society and culture was amassed and analysed by means of a questionnaire drawn up specifically for this purpose and distributed among the inhabitants of the island. These investigations were carried out by P. SZACKI in collaboration with Ernesto FECAROTTA (1966) from Catania, an expert on Sicilian amber, known as simetite (from the river Simeto). The questionnaire allowed these two researchers to gain some important information about the technology applied in the acquisition and working of amber, about social and mythological questions and about its use in therapy and magic — problems which have also been partially examined in Sicilian publications since the 17th century. Besides the local amber (simetite), amber finds from other lands also occur in Sicily. Some of these are most probably from Baltic region, although verification of this supposition by means of infra-red analysis has yet to be carried out.

When speaking of the results of this Polish-Italian collaboration, it has to be said that only some of the numerous joint research projects on amber, detailed in the programme aims were eventually realized. There were two main reasons for this being the case. The first, as mentioned earlier, was the substantial expansion in 1975 of the scope and aims of the cooperation programme; the second was, as usual, the cost of some planned projects, such as the very expensive infra-red spectral analysis of amber and the search for sophisticated analytical methods to identify the provenance of amber finds by non-destructive techniques.

The presentation of this selected information about the Polish-Italian cooperation on amber studies to the very competent group of specialists gathered here, seems to me a useful exercise. The experiences gained during this joint work indicate, on the one hand, the potential advantages of interdisciplinary cooperation on an international scale. On the other hand, however, they also show the risks which accompany the formulation of too extensive a programme of studies and research, even one with very reasonable intentions. The realization of a programme of this kind requires reliable research resources and stable, systematic, comprehensive financing. Neither of these conditions are easy to meet, especially in a period of political change and ever more severely limited outlays for research in the humanities.

Eleonora TABACZYŃSKA
ul. Daniłowskiego 6 m 98
01-883 Warszawa, Poland

BURSZTYN JAKO PRZEDMIOT BADAŃ ARCHEOLOGICZNYCH: DOŚWIADCZENIA POLSKO-WŁOSKIEJ WSPÓŁPRACY

Eleonora TABACZYŃSKA

Streszczenie

Utworzona w Rzymie w 1972 roku Polsko-Włoska Komisja do Badań nad Szlakami Bursztynowymi wyznacza istotny etap w rozwoju interdyscyplinarnych badań nad bursztynem. Opracowała ona globalny program wielostronnych badań uwzględniający zarówno zagadnienia ściśle archeologiczne i historyczne, jak również przyrodnicze i fizyczno-chemiczne. Podjęte zostało kompleksowe opracowanie zabytków bursztynowych występujących w Polsce i we Włoszech, z uwzględnieniem źródeł pisanych, etnograficznych i etnologicznych. Jednym z głównych zadań było wypracowanie nowych metod analitycznych, pozwalających poznać strukturę surowca zabytków bez ich zniszczenia. Prace realizowane były dynamicznie zwłaszcza w trzech pierwszych latach działalności Komisji. Od 1975 roku bowiem zakres współpracy polsko-włoskiej uległ znacznemu poszerzeniu, a problematyka bursztynu stała się odtąd jednym z kilku realizowanych równolegle nurtów badawczych w obrębie nowej struktury pod nazwą Polsko-Włoska Interdyscyplinarna Grupa Robocza Nauk Stosowanych w Archeologii i w Ochronie Patrymonium Kulturowego. Doświadczenia wynikające z tej współpracy, którą ukazują różne publikacje, dotyczą nie tylko merytorycznych wyników wspólnych badań teoretycznych i praktycznych, ale również samej organizacji międzynarodowych badań interdyscyplinarnych.

Bibliography

KOSMOWSKA-CERANOWICZ B. (ed.)
1978 *Ambra oro del Nord*, Alfieri Edizioni d'Arte, Venezia.

COFTA-BRONIEWSKA A.
1984 Amber craft in Kuiavia in the Era of Przeworsk Culture, *Archaeologia Polona*, **23**, 149–165.

FECAROTTA E.
1966 *Prefazione*, [*in*:] Dell 'ambra siciliana. Testi di antichi autori siciliani 1639–1805, a cura di C. E. Fiore, Edizioni Boemi, *Quaderni di Bibliotheca* **1**, VII–IX.

KOZIOROWSKA L.
1984 Badania nieorganicznego składu chemicznego bursztynu, *Archeologia Polski*, **29** (2), 207–232.

KRZEMIŃSKI W. & SKALSKI A.,
1983 *Pseudolimnophila Siciliana* sp. n. from Sicilian Amber (Diptera, Limoniidae), *Animalia*, **10**, 303–307, Catania.

LEWICKI T.
1984 Les sources arabes concernant l'ambre jaune de la Baltique, *Archaeologia Polona*, **23**, 121–142.

SKALSKI A.
1985 Uwagi o faunie w bursztynie sycylijskim i apenińskim, *Wiadomości Entomologiczne*, **6** (3–4), 215–218.

HENSEL W. & DONATO G. (eds)
1975 Studi e ricerche sulla problematica dell'ambra [*in:*] *Atti della Cooperazione Interdisciplinare Italo-Polacca*, **1**, CNR, Roma.

TABACZYŃSKA E.
1974 Polsko-włoskie badania nad szlakami bursztynowymi, *Archeologia Polski*, **19** (2), 537–540.

URBAŃSKI T.
1977 On ESR Signals of Amber, *Bulletin de l'Académie Polonaise des Sciences, Série des sciences chimiques*, **25** (10), 785–787.

WIELOWIEJSKI J.
1984 Il significato di "via dell'ambra", *Archaeologia Polona*, **23**, 107–119.

AMBERWEB: A PROJECT FOR AN INTERNET CENTRE ON AMBER ARTEFACTS — THE LOGICAL STRUCTURE OF THE DATA-BANK AND THE IMPLEMENTATION OF NEW DATA IN THE ARCHAEOLOGICAL SECTOR

Nuccia NEGRONI CATACCHIO, Alessandra MASSARI, Barbara RAPOSSO & Barbara SETTI

Abstract

Amberweb is an internet site offering access to a wide range of information on amber, both as a raw material and in the form of artefacts. This article offers an insight into the structure of the *Amberweb* database. Its current state is assessed and its ongoing development outlined.

Introduction

I am glad to show you the *Amberweb* project at such an important symposium on amber. *Amberweb* is a centre on amber, part of an Internet site, therefore open to everybody. I hope all the world-wide amber researchers will take part in its development.

The site (http://amberweb.cilea.it) has been on-line for about three years, and is often changed and supplemented with updated data.

Obviously, such a project is constantly developing: at present most of the forms concern archaeological amber artefacts found in Italy, but we trust that with the collaboration of all we can build a large virtual amber museum.

I am also glad to show our latest studies here because it was in Gdańsk, in the far-off May of 1973, that the scientific meeting of the "Italian-Polish Commission for Amber Route Studies" was held for the first time in Poland. Later on this Commission became the "Italian-Polish Interdisciplinary Working Group for Sciences Applied in Archaeology and the Protection of the Cultural Heritage" established in May 1972 as part of the agreement of a scientific and technical co-operation between the Polish and the Italian governments and between Consiglio Nazionale delle Ricerche (Istituto per le Tecnologie Applicate ai Beni Culturali) and the Polish Academy of Sciences (Istituto di Storia della Cultura Materiale). The deeds of the Commission and the reports of the first meetings are in the Amber Acts 1975. In this volume Eleonora TABACZYŃSKA — the scientific secretary of the Polish Commission — outlines a brief history of this collaboration. As far as I'm concerned, I would like to recall my friend Giuseppe DONATO, who died some years ago, who, together with Witold HENSEL, was the co-president of the Italian-Polish Commission. Thanks to his enthusiasm and his far-sightedness, research on amber routes — which connected Poland and Italy in prehistory, and thus the Northern world with the Mediterranean one *via* Europe — led to the development of lasting friendships among the members of the working group.

The research programme was conceived at the Istituto di Storia Antica dell'Università degli Studi di Milano with the aim of ascertaining the provenance of amber used for archaeological artefacts in Italy (NEGRONI 1967–69; GUERRESCHI 1970). The results showed that most of the amber came from the Baltic and so that a leading commercial and cultural route, connecting such distant countries, had been in use since the Bronze Age. These results were then shown to Massimo PALLOTTINO, the most important Etruscologist of our century. He gave them to Giuseppe DONATO, who was creating the "Special Programme for the Subsidiary Science of Archaeology" at the National Research Council, which later became the "Istituto per le Tecnologie applicate ai Beni Culturali". Giuseppe DONATO decided to put these results in the agreement of the scientific co-operation between Italy and Poland.

That is why I would like to dedicate this work to my colleague and to all the memories of our shared scientific activities and friendship, which have linked Italian and Polish researchers for many years.

Nuccia NEGRONI CATACCHIO

Logical Structure

Amberweb was conceived as a sort of virtual museum of amber, which can be visited by people with different interests.

It can firstly be considered as an instrument for specialist research on amber, as it offers a data-bank with documents, files on single specimens or images, as well as information about new findings and research programmes in progress. There is also a specific page where exhibitions, congresses, publications and all other related activities are announced. *Amberweb* can also be used by non-specialist browsers, who can enter a series of pleasant pictures complete with exhaustive captions.

Amberweb was designed to offer an easy way of diffusing and exchanging information. That is why we thought it useful to add to the traditional e-mail, pre-prepared forms enabling the homogeneous and systematic collection of data to update the bank. Therefore, it is possible to send information about new finds and analysis as well as contributions and images, which will then merge with the general data bank.

Because of the multidisciplinary character of amber research, data is thematically organised. At the beginning of the site, a map shows the logical structure on which the bank system is based (Fig. 1). The user can see the different research subjects and choose his path to consult data and information. The data-bank is organised into distinct sectors according to the main subjects related to amber: **archaeology, popular traditions, literary sources, chemistry, geology, palaeobotany and palaeozoology**. A bibliographic general bank and an iconographic index gather all the data put into the specific files. Each of these subjects is organised according to different standards of in-depth study and allows the user to consult written documents, files or images. All these research sectors are strictly linked, so that it is possible to pass directly from one sector to another, being conceived as hypertexts.

At present the archaeological sector is the most complex and organised structure, while the others are still simple. They will be probably be better organised in the future with the collaboration of specialists.

Each subject offers a bibliographic catalogue, which can be looked up according to the authors, titles and to a series of contributions, such as texts and written documents. In the "literary sources" subject, the user may also consult a specific passage in the original language or in the Italian translation. The geological, palaeobotanical and palaeozoological themes offer a bank of pictures and a world map of amber deposits, through which it is possible to access specific information about different varieties of amber.

In the chemical sector it is possible to consult files on chemical and physical analysis, carried out on archaeological and geological specimens, to establish the place of origin of amber.

Barbara RAPOSSO

Archaeological data management and the implementation of new data

As has already been said, *Amberweb* is based on a complex logical structure, where all works relating to amber are analysed. Each of them is strictly linked with the others and is treated as a file, on which further research may be carried out.

As we are waiting for specialists of different sectors to offer their contributions to update files, all the subjects have been simply organised and filled with general data, apart from the archaeological sector, which is still the best organised.

Entering the archaeological file you can study **general arguments** in depth (for example amber routes) through some texts and visit the virtual museum of artefacts found in Italy in the Eneolithic or in the 4th century BC. To simplify the reference to the virtual museum, some research paths have been created to supply particular interest, such as **typology, discovery location, dating (chronology)** and **storage locations**. Choosing one of them, you can access typological tables, maps, a chronological catalogue, artefact forms (**find forms**) and the way in which they were found (**forms of archaeological site**). If the user is interested in a particular typological/chronological sector or in an artefact or group of artefacts, he can independently move into the virtual museum, directly storing the file.

The file can also be organised according to the **shape** and **function** of the artefacts. This is particularly important for figured amber, which can be analysed according to its shape (human or animal aspect, male protoma etc.) or its possible usage (pendants, small plate, etc.). While the typology of unfigured ambers is based on the function (pendants, spacer rings etc.) of the artefacts as well as on their shape.

For figured ambers, *Amberweb* files contain documentation related to some southern regions, such as Marche, Campania and Puglia, although the

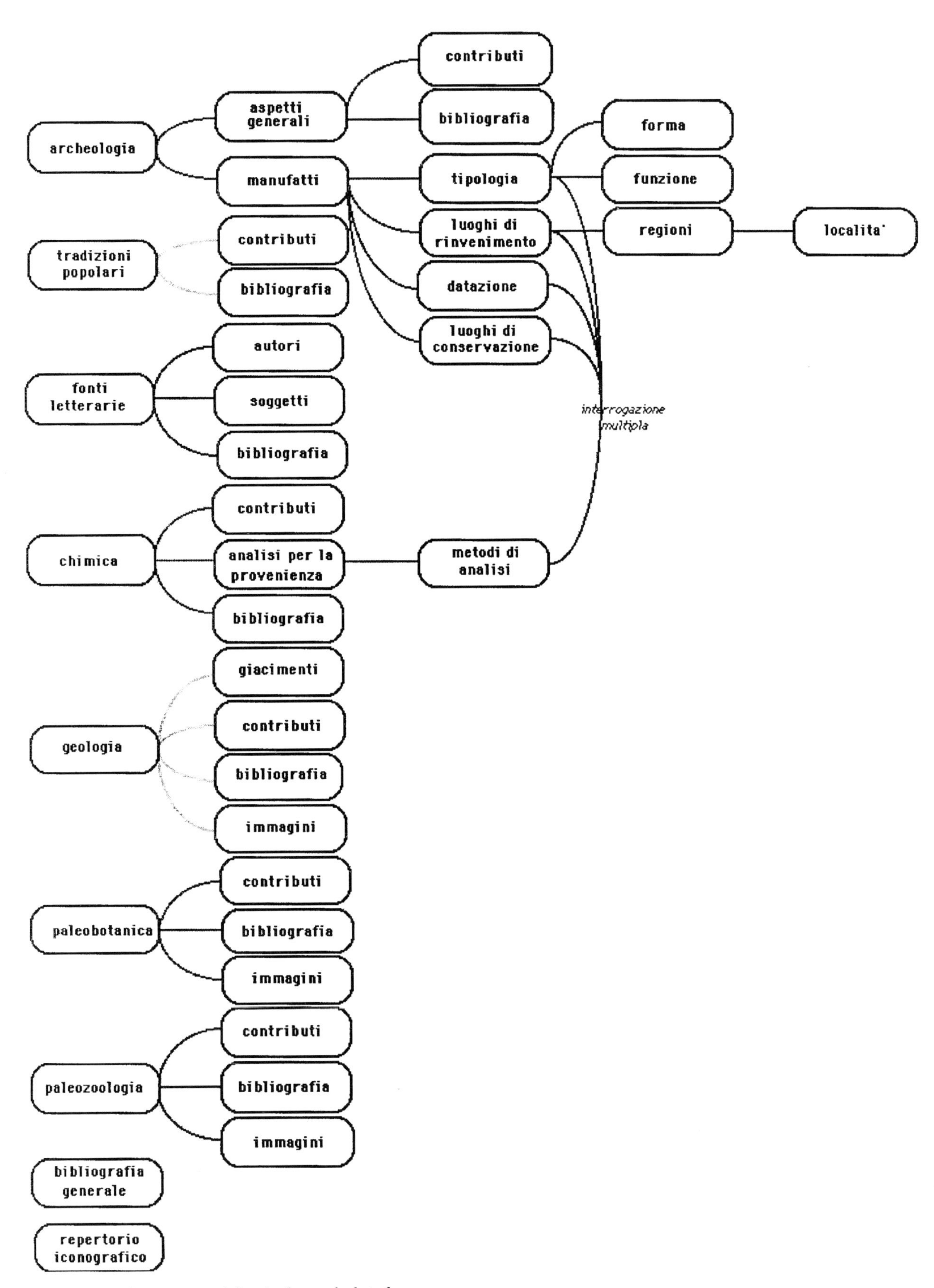

Fig. 1. Logical structure of the Amberweb data-base.

filing of these artefacts is just related to Italian territory. Many figured artefacts dating from the pre-Roman period (6th–4th centuries BC) have been found in these regions. The most representative shapes include: the male shape and protoma, human figures, sphinx-like figures and animal figures, as is evidenced by their presence in Ortona (Chieti), Rutigliano (Bari), Monte Tabor (Foggia) and the Oliveto Citra (Salerno) necropolis or nearby Pozzuoli (Napoli), Capaccio (Salerno) and Roscigno (Salerno). Each shape is shown by general tables and classified according to its iconographic characteristics, to the precision and the method of its realisation and to the position of its figure. Selecting one of them you can access a particular table containing the artefact description and list of archaeological settlements where artefacts of this type have been found. Continuing this line of research you can access **artefact forms**. Each form contains — apart from information about the discovery location — the definition of the function, the typology and the description of the artefact as well as photographs and drawings of it (views of both can be enlarged), its chronology, its bibliography and comparisons with other similar artefacts from different areas and the analysis form to establish the provenance of the raw material and where it is currently housed. Selecting a word highlighted by a different colour it is possible to move into the archaeological sector and enter settlement forms, artefact lists or pass to other *Amberweb* subjects, such as **chemistry** (if artefacts have been analysed, it is possible to view the analysis form and its graphics to establish the provenance of the raw material) or **bibliography**. Selecting "**discovery locations**", there are some maps bringing the user to the requested region, with a list of the places, where figured amber has been found. Selecting a place, you can enter the settlement and the find form. The first one contains all the information about the place (site name, town etc.), its background (necropolis, village, hoard and sporadic[1]), the settlement (tomb number), the method, the author, the discovery date, the description of the settlement, the list of finds associated with amber artefacts, the chronology, the bibliography etc. This form has words in different colours which lead to other levels of the archaeological file.

The dating path shows a chronological list, which is divided into periods — according to the pre- and protohistorical phases (Eneolithic, Ancient, Middle, Recent and Final Bronze Age) — and into centuries (8th–7th, 7th–6th, 6th–5th, 5th–4th centuries BC). Selecting one of them you can enter a list of settlements or the single forms of the settlement as well as the find.

The selection of **find site** brings one directly to maps with a list of museums, "Soprintendenze" and private collections and to their associated artefacts. Unfortunately this file is still under construction and only a few places can be visited.

The file organisation of unfigured ambers is similar to the previous one, so it is possible to move into the virtual museum through guided paths based on **typology**, **find sites**, **dating** (**chronology**) and **storage location**. We have already said that unfigured artefacts have been classified according to their function and shape, but we must add that the implementation of this file depends on those artefacts called beads which were present in large numbers in Italy as well as in the European and Mediterranean area. The items in question are Tirinto-type (cylindrically shaped with a raised rib structure in the middle) and Allumiere-type beads (cylindrically shaped with cross grooves on the body)[2]. These artefacts were present in Italian necropolii, settlements and hoards during the final phases of the Bronze Age and the first phases of the Iron Age, although we do not have any complete information about how they were found. In fact we know only where they are now housed. However, we could create a satisfying file used for our typological research[3]. Tirinto- and Allumiere-type beads differ in their height and width. These differences are referred to as: normal (A), lengthened (B), flattened (C) and highly flattened (D).

The file of those amber artefacts used as beads has recently been enriched by new data coming from the filing of finds from lake-dwellings in the Bande di Cavriana and Castellaro Lagusello (Mantova) settlements. They have been dated to between the Ancient and Middle Bronze Age and classified as disc, biconical or spheroidal-shaped beads. Some of them show traces of having been repaired in antiquity, testified by fine bronze thread or some little holes.

---

1 This definition is used for chance finds or for those artefacts, whose discovery location is unknown.

2 The filing of Tirinto- and Allumiere-type beads brought to the study of other unfigured artefacts, which are not classified, but have just a find form.

3 For more information see N. NEGRONI CATCCHIO, *Produzione e commercio dei vaghi d'ambra tipo Tirinto e tipo Allumiere alla luce delle recenti scoperte*, in *Atti del XX Convegno di Studi Etrusco-Italici*, Ottobre 1996 (forthcoming)

The census, as well as the filing work, is now concentrating on the first phases of the Bronze Age, when beads and buttons with V-shaped holes began to appear at archaeological settlements in southern and northern Italy and in Emilia Romagna. This research will be included in the *Amberweb* archaeological file.

Alessandra MASSARI

User-system interaction: updating forms, e-mail etc.

The creation of an on-line database aims to take full advantage of the interactive possibilities given by the Internet. The multidisciplinary character of amber research enables data from several fields of research, in both the humanities and sciences, to be correlated.

This is why it has been decided to offer the possibility of creating a "virtual museum" *in fieri* on amber through the reunion and the thematic and expositive location of objects world-wide as well as of new knowledge. Therefore, a number of forms have been created to allow users to send new data to the web master. The form consists of a page with a list of items already arranged to obtain all the necessary information, inserting texts or selecting some arranged items (for the computer structure see NEGRONI *et al.* 1997, pp.).

There is an icon on every *AmberWeb* screen. Selecting it, users can update the files. The following application forms can be selected:

- site form
- find form
- analysis form
- contributions
- images

Forms consist of a series of fields to be filled in and a wider text box for detailed examinations, descriptions and comments. They are already arranged in order to catch homogeneous and exhaustive information from different sources, so that the whole organic unity of the database is maintained. Graphic or pictographic images — according to the instructions given by the form — may be added to the forms for sites, analysis etc. New data can be filled in, according to a previous check made by the web master in order to guarantee the scientific nature of the information. Forms are always signed by the author, who will be able to update or modify them. There is also an e-mail (used to send messages, reports or comments etc.) on every web page.

AmberWeb is the right tool to look up in archaeological *newsgroup*, selecting the appropriate icon, or to join a *mailing list*. A brief description of the subjects and some instructions for the enrolment and the mailing of messages, have been put in each *mailing list* to clarify the reference even to unskilled users.

Barbara SETTI

Nuccia Negroni CATACCHIO
Istituto di Archeologia
Universita degli Studi di Milano
Via Festa del Perdono no 7 - C.A.P. 20122
Milano, Itay

Alessandra MASSARI, Barbara RAPOSSO
& Barbara SETTI
Centro Studi di Preistoria e Archeologia
Milano, Italy

AMBERWEB: PROJEKT INFORMACYJNEGO CENTRUM O BURSZTYNIE W INTERNECIE — LOGICZNA STRUKTURA BAZY DANYCH I WZBOGACANIE TEJ BAZY O NOWE DANE W ROZDZIALE ARCHEOLOGICZNYM

Nuccia NEGRONI CATACCHIO, Alessandra MASSARI, Barbara RAPOSSO, Barbara SETTI

Streszczenie

Amberweb jest otwartą dla wszystkich stroną internetową, z której nie tylko można korzystać, ale którą także wszyscy naukowcy badający bursztyn mogą wspólnie rozwijać, poprzez dokładanie własnych informacji. Strona ta funkcjonuje już ponad trzy lata i jest ciągle aktualizowana. W założeniu ma to być rodzaj wirtualnego muzeum bursztynu, które może być odwiedzane przez ludzi różnych zainteresowań i profesji. Umożliwia stosunkowo łatwe rozpowszechnianie i wymianę informacji.

Wszystkie dane zgromadzone w bazie mają charakter logicznej struktury, która jest prezentowana na początku strony *Amberweb*. Użytkownik może więc obejrzeć różne badania i tematy, wybierając taką ścieżkę dostępu, która zapewni mu uzyskanie odpowiednich informacji.

Główne sektory tematyczne to: archeologia, tradycja ludowa, źródła pisane, chemia bursztynu, geologia bursztynu, paleobotanika i paleozoologia. W chwili obecnej najbardziej rozbudowany jest sektor archeologiczny, podczas gdy struktura pozostałych sektorów jest stosunkowo prosta.

Bibliography

ATTI AMBRA
1975 Studi e ricerche sulla problematica dell'ambra, *Acts of the Italian-Polish interdisciplinary cooperation*, vol. I, CNR, Roma.

BECK C. W. & BOUZEK J. (eds)
1993 *Amber in Archaeology*, Praha.

GUERRESCHI G.
1970 La problematica dell'ambra nella protostoria italiana: metodo sperimentale per la determinazione della provenienza, *Studi Etruschi* **38** (serie II).

LOSI M., RAPOSSO B. & RUGGIERO G.
1993 *The production of amber female heads in pre-roman Italy*, [*in*:] Beck C.W. & Bouzek J. (eds.) 1993, 203–211.

MASSARI A., RAPOSSO B. & SETTI B.
1997 La diffusione dell'ambra nel Bronzo Antico in Italia, *L'antica etŕ del bronzo in Italia*, Atti del Convegno, Viareggio 1995, Firenze, 620–621.

NEGRONI CATACCHIO N.
1967–69 Il problema dell'ambra nella protostoria italiana: metodo sperimentale per la determinazione della provenienza, *Sibrium* **9**, 377–387.

1970a La problematica dell'ambra nella protostoria italiana: diffusione dell'ambra in Italia e suoi rapporti con il mondo culturale preistorico, *Sibrium* **10**, 275–288.

1970b La problematica dell'ambra ecc.: le ambre intagliate delle culture protostoriche dell'area lombardo-veneto-tridentina, *Memorie del Museo Civico di Storia Naturale di Verona* **18**, 319–336.

1970c La problematica dell'ambra nella protostoria italiana: cenni introduttivi, *Studi Etruschi* **38** (serie II), 165–168.

1972 La problematica dell'ambra ecc.: le ambre intagliate di Fratta Polesine e le rotte mercantili nell'Alto Adriatico, *Padusa* **8**, n. 1–2, 1–18.

1973 La problematica dell'ambra ecc.: ancora sulle ambre di Frattesina di Fratta Polesine, *Padusa* **9**, nn. 2–3–4, 1–13.

1975 La problematica dell'ambra ecc.: manufatti in ambra dell'ambiente modenese, *Emilia preromana* **7**, 375–388.

1976 Le vie dell'ambra, i passi alpini orientali e l'Alto Adriatico, *Aquileia e l'arco alpino orientale*, *Atti della VI settimana di Studi Aquileiesi*, Centro Antichitŕ Alto Adriatico, Aquileia, 21–59.

1985 L'ambra, il Baltico, i Veneti e gli scrittori antichi, *I tesori dell'antica Polonia*, Catalogo della Mostra, Padova, 41–46.

1986 Il vago tipo Tirinto, [in:] D. Cocchi Genick, *Il riparo dell'ambra*, Viareggio, 199–202.

1989 L'ambra: produzione e commerci nell'Italia preromana, *Italia, omnium terrarum parens*, Milano, 659–696.

1993 The production of amber figures in Italy from the 8th to the 4th centuries BC, [in:] Beck C.W. & Bouzek J. 1993, 191–202.

In press Produzione e commercio dei vaghi d'ambra tipo Tirinto e tipo Allumiere alla luce delle recenti scoperte, *Protostoria e storia del «Venetorum angulus», Atti della XX Conferenze di Studi etrusco-italici, 1996*, Firenze.

NEGRONI CATACCHIO N., PADULA M., MASSARI A., RAPOSSO B., ROSAFALCO M., SETTI B. & TOSI M.L.
1997 AmberWeb: progetto di un polo Internet sui manufatti in ambra, *Atti del III Convegno Internazionale di Archeologia e Calcolatori, Roma 1995*, Firenze, 1011–1026.

INVESTIGATIONS INTO AMBER FROM DEPOSITS IN POLAND USING ELECTRON PARAMAGNETIC RESONANCE (EPR) AND POSITRON ANNIHILATION SPECTROSCOPY (PASCA)

Adam JEZIERSKI, Michał SACHANBIŃSKI, Franciszek CZECHOWSKI & Jan CHOJCAN

Abstract

Characterization of 14 samples of amber from different localities in Poland (Baltic Coast near Gdańsk, Bełchatów Tertiary brown coal deposit, and Jaroszów clay mine) was performed using electron paramagnetic resonance spectroscopy (EPR) and positron annihilation spectroscopy for chemical analysis (PASCA). The free radical EPR signals of similar concentrations and g parameter values were found for all succinite samples. Pronounced differences in these parameters were, however, observed for interior and exterior parts of the distinct, weathered (cherry-like) samples. The single line at $g = 2.004$ is characteristic for the exterior part while the anisotropic line (effective $g_z = 2.02$ and $g_{x,y} = 2.01$) was detected for the interior part. The differences revealed are probably due to different chemical environments stabilizing the free radicals. The structural differences in the polymeric matrix of exterior and interior parts of this amber sample were confirmed by the PASCA and Fourier Transform Infrared (FT-IR) investigations.

Introduction

Chemical reactions or radiations can generate molecular fragments of organic molecules with an excess or a deficiency of electrons in various natural materials. These fragments containing unpaired electrons (called free radicals, paramagnetic centres, radical centres, electron or hole centres, spin centres, etc.) can be measured by the technique of electron paramagnetic resonance (EPR, ESR). Such anlyses are used in the dating of minerals, rocks, sediments, and natural organic materials (IKEYA 1993).

Various plant resins emit intensive EPR signals (ROBINS *et al.* 1985). Particle tracks (resulting from radiation leading to the formation of paramagnetic centres) in amber samples were recognized by UZGRIS & FLEISCHER (1971). URBAŃSKI (1977) has also detected EPR signals in Baltic amber.

The presence and stabilization of free radicals (electron/hole centres) in a polymeric matrix depends on both the radical type and the matrix properties. Furthermore, the polymeric skeleton of amber is relatively resistant and alters very slowly during diagenetic processes. Over relatively long periods of time the exterior of an amber nodule changes under the influence of weathering, indicating changes in the chemical constitution of its polymeric matrix. Therefore, it was intriguing to investigate the paramagnetic centres detected by EPR in spin probing of different parts of the polymeric matrix of amber.

One of the characteristic features of amber is its relatively low porosity. It can, however, be investigated using several techniques, among which a family of positron annihilation methods is known as a useful tool (CZECHOWSKI *et al.* 1996). We applied positron annihilation spectroscopy for chemical analysis (PASCA) to study various parts of the amber polymeric matrix as a supplementary tool to the EPR technique.

Experimental

Fourteen samples of amber from different localities in Poland were investigated: 2 samples from the Baltic Coast near Gdańsk, 10 from Tertiary Bełchatów brown coal (occluded in detrital material) and 2 from Jaroszów (occurring in clay minerals associated with detritus of highly altered plant remnants). The locations of the samples and basic physicochemical investigations have in earlier work been described (CZECHOWSKI *et al.* 1996).

The quantitative EPR technique (QEPR) was applied (JANCZAK *et al.* 1995). EPR measurements were made on an ESP 300E BRUKER spectrometer using a nuclear magnetometer and frequency counter. Standards for g parameter measurements (nitroxyl radicals, Li/LiF standard) as well as for spin concentration (4-Hydroxy-TEMPO and ultramarine, see BARANOWSKI *et al.* 1995) were used. Microwave power of 2 mW, modulation amplitude of 0.1 mT and samples of 30.0 mg in standard quartz tubes were applied.

Measurements of lifetime spectra and one-dimensional angular correlations of photons coming from the two-photon annihilation of an electron-positron pair (PASCA) were carried out. The details of the measurement methodology are described elsewhere (JERIE *et al.* 1983).

Results and Discussion

1. Spin concentration in amber samples

Small pieces of the investigated amber (1–5 mg grains) show a spin concentration of about 10^{16} spins/gram. Practically the same concentrations were detected for the amber samples from the Baltic coast, Bełchatów and Jaroszów. This observation is consistent with the conclusion that the investigated ambers from different localities represent a common class of fossil resin: succinite Ia class (ANDERSON *et al.* 1992; CZECHOWSKI *et al.* 1996). The observed data are consistent with earlier data for amber occurring in Poland (KOSMOWSKA-CERANOWICZ 1986).

Pulverization of the samples (or even grinding in an agate mortar) results in a strong increase of the spin concentration up to $n \cdot 10^{17}$ spins/gram. A similar phenomenon has previously been observed (URBAŃSKI 1977). Moreover, some proportion of the newly observed radicals (after the grinding procedure) has a "transient" character and we found their gradual decay over time within days and months.

2. Influence of weathering on the EPR spectra

Well-defined differences in the EPR spectra were observed for interior and exterior parts of the cherry-like amber samples. We investigated 6 samples from Bełchatów (spherical pieces of 3–4 cm diameter) with homogenous, colourless or pale-yellow interiors (cores) and weathered, brown or cherry-like exteriors (rinds). It was shown that the polymeric matrix of the exterior exhibits lower infra-red absorbances due to ester group vibrations and higher absorbances

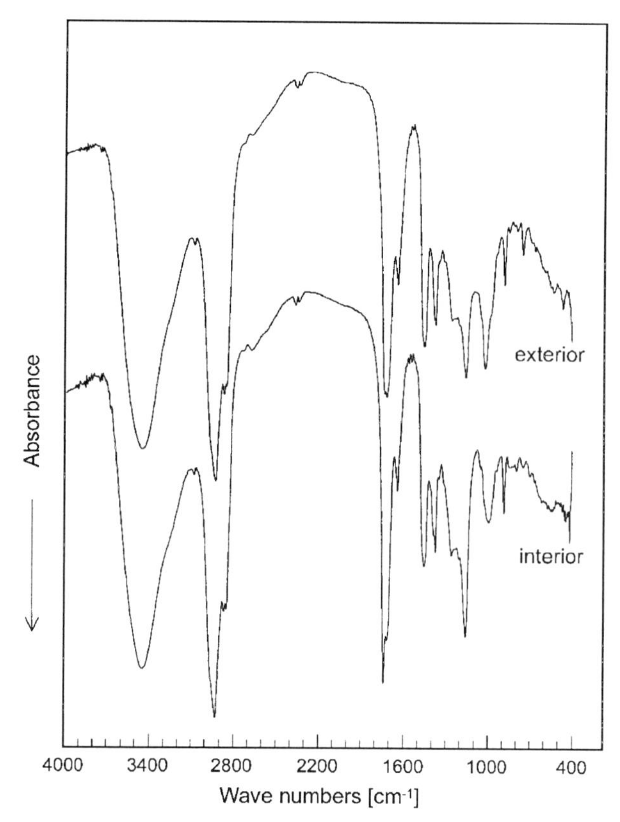

Fig. 1. FT-IR spectra for the interior and exterior parts of cherry-like, weathered amber from Bełchatów brown coal deposit.

attributed to hydroxyl, methyl and methylene groups in comparison to the interior region (Fig. 1; Czechowski *et al.* 1996). The weathering process probably causes hydrolysis of esters and removal of the released water-soluble acids from the resinous polymeric matrix.

The characteristic EPR spectra of the exterior and interior parts are shown in Fig. 2. The single line at g = 2.0044–2.0047 is characteristic for the weathered, exterior part while the anisotropic line at $g_z = 2.0188$ and $g_{x,y} = 2.009$ is observed for the interior part. The EPR line of the interior, indicated by a distinct arrow, is characteristic for the admixture of the weathered part. This line is very weak immediately after grinding and it gradually increases during storage of the sample in air. The intensity increase of the line indicated by the arrow after 2 months of storage in air appears comparable to the effect obtained over a shorter period of time, i.e. under laboratory oxidation by gaseous oxygen at elevated pressure (2.5 MPa) and temperature (330 K) over a period of 40 h (Fig. 3).

Both of the observed EPR radical signals are characteristic for radicals containing oxygen. The signal

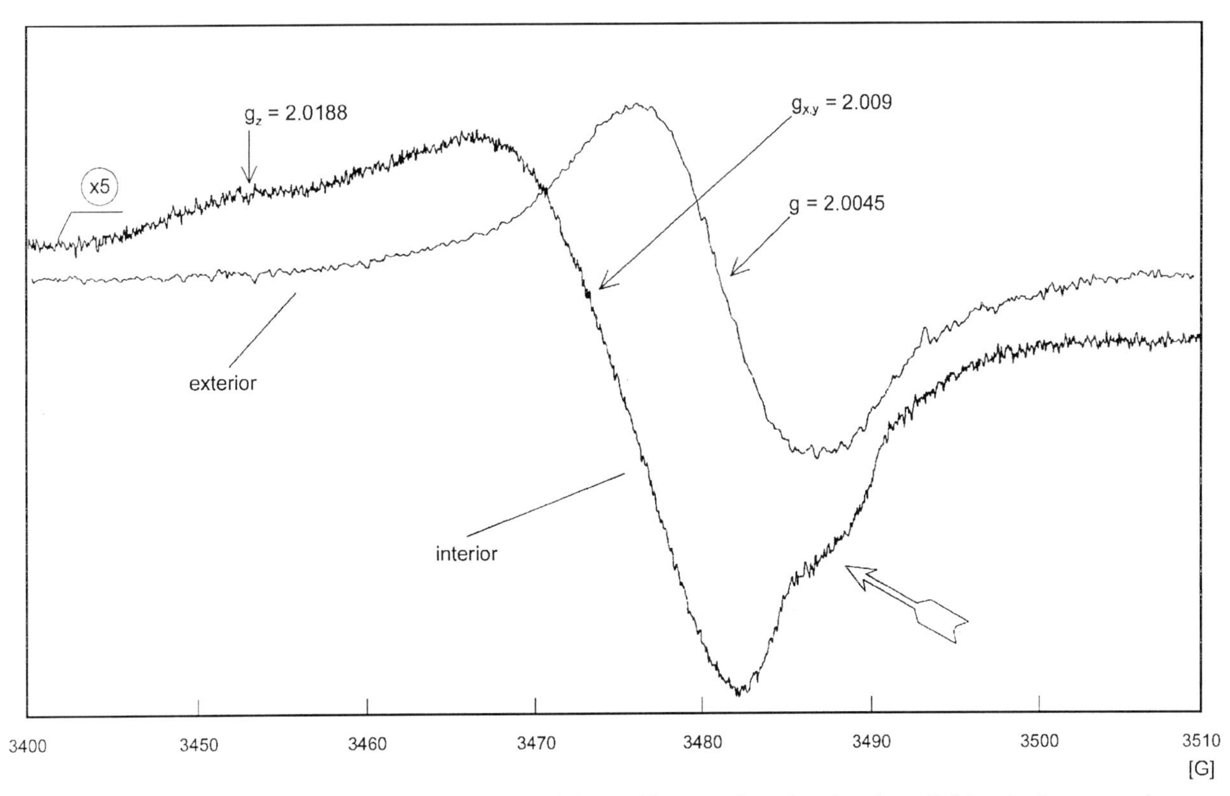

Fig. 2. EPR spectra of the interior and exterior parts of cherry-like, weathered amber from Bełchatów brown coal deposit. T = 297 K. Magnetic field is on X axis. 1G = 0.1 mT.

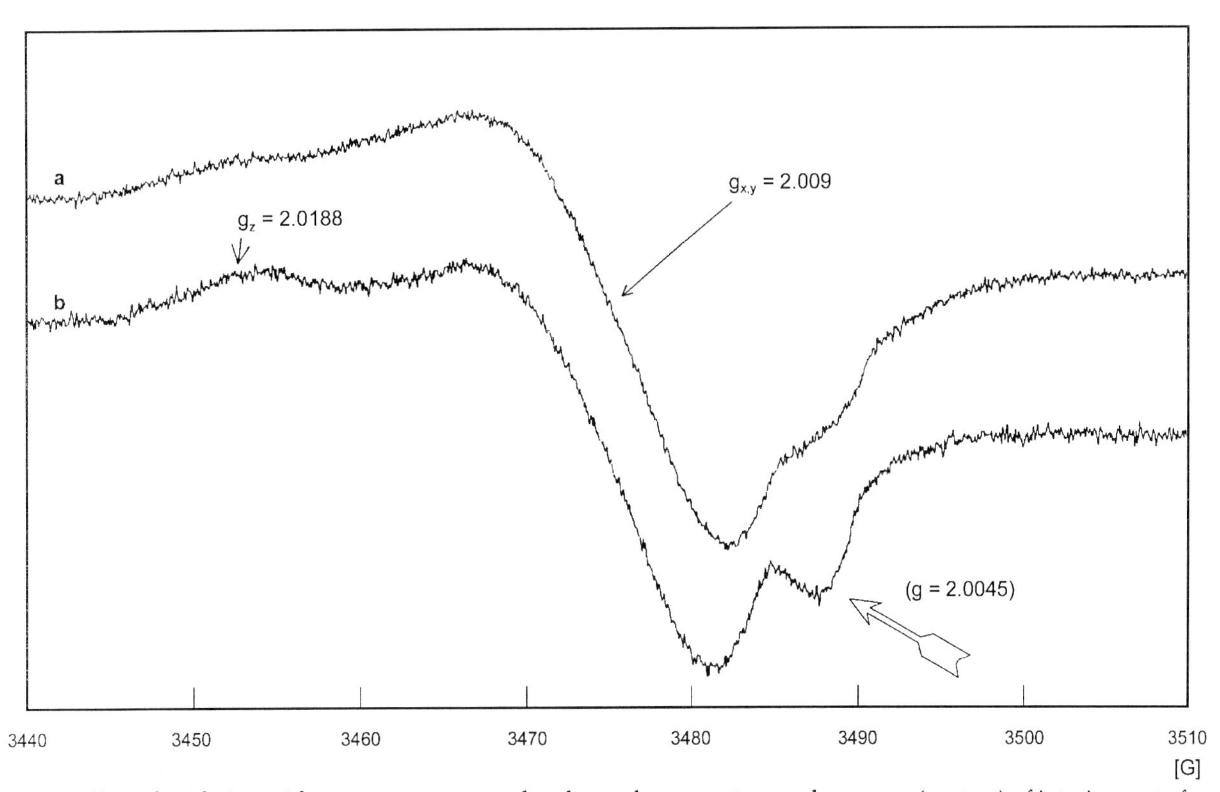

Fig. 3. Effect of oxidation with gaseous oxygen under elevated temperature and pressure (see text) of interior part of amber on EPR signals: a — before oxidation, b — after oxidation.

at g = 2.004 is typical for semiquinone-like radicals observed in coals, humic substances, etc. (CZECHOWSKI & JEZIERSKI 1997), while the second signal at g_z = 2.02 and $g_{x,y}$ = 2.01 is more commonly observed for various paramagnetic forms of oxygen (e.g. O^-) in solid matrices (IKEYA 1993). Weathering modifies the amber matrix, which is reflected by the formation of more stable new radicals.

The PASCA measurements also showed noticeable differences between the interior and exterior parts of the amber samples (Fig. 4; CZECHOWSKI *et al.* 1996). The estimated sizes of the pores in the samples examined were similar within the range of 0.8 to 0.9 nm, however, their concentration in the exterior constituted only half of the value found for the interior part. Thus, the ester components in the resin matrix may be responsible for spheres where positronium atoms can be formed. The distinct characteristics in the PASCA data for the exterior and the interior are illustrated in Fig. 4.

3. Stability of radical centres in amber

Radical centres in amber show distinct activity in the solid matrix which is demonstrated by: (i) the essential changes of radical type upon weathering, (ii) the appearance of radicals as a result of grinding (and subsequent decay of the „transient" radicals), and (iii) solvent action. We observed a sudden decay of the radical signals of the powdered sample of amber (interior or exterior parts) after the addition of toluene, saturating porosity of the sample. The change of the characteristic EPR signal for the exterior is illustrated before (Fig. 5a) and 20 minutes after toluene addition (Fig. 5b). Twenty-four hours after toluene addition, only 3% of the initial radical intensity was observed.

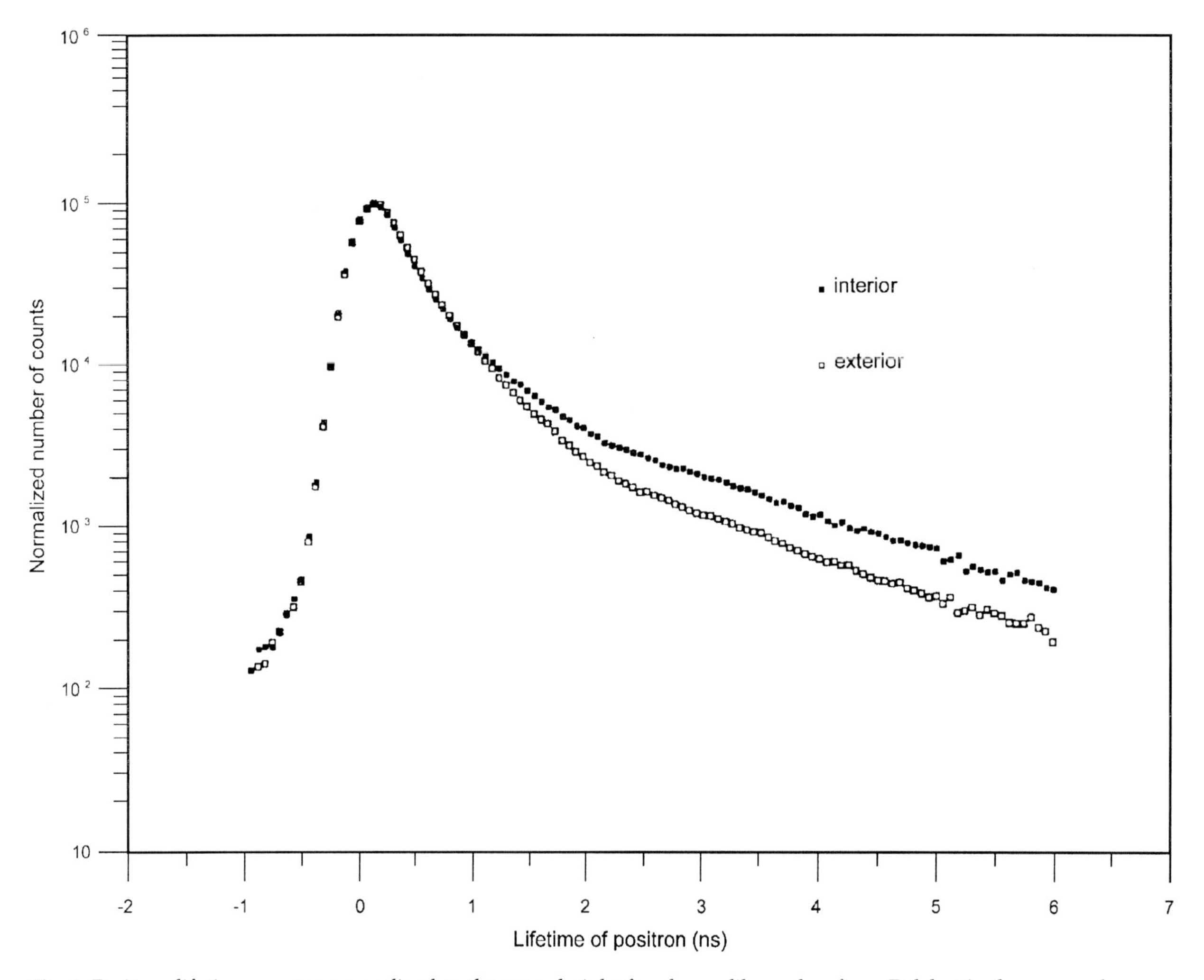

Fig. 4. Positron lifetime spectra normalized to the same height for cherry-like amber from Bełchatów brown coal deposit.

This phenomenon probably relates to an increase in the mobility of the amber structure in the presence of toluene, allowing progressive recombination of the radicals. This observation remains in contrast to the behaviour of different radicals-containing natural materials. Namely, we observed that in the case of bitumens, treatment with organic solvents caused a pronounced increase in the concentration of radicals.

Adam JEZIERSKI
Faculty of Chemistry, Wrocław University
F. Joliot-Curie St. 14
50-383 Wrocław, Poland

Michał SACHANBIŃSKI
Institute of Geological Sciences, Wrocław University
Cybulskiego St. 30
50-202 Wrocław, Poland

Franciszek CZECHOWSKI
Institute of Organic Chemistry,
Biochemistry and Biotechnology
Wrocław University of Technology
Wybrzeże Wyspiańskiego St. 27
50-370 Wrocław, Poland

Jan CHOJCAN
Institute of Experimental Physics, Wrocław University
Borna Sq. 9
50-204 Wrocław, Poland

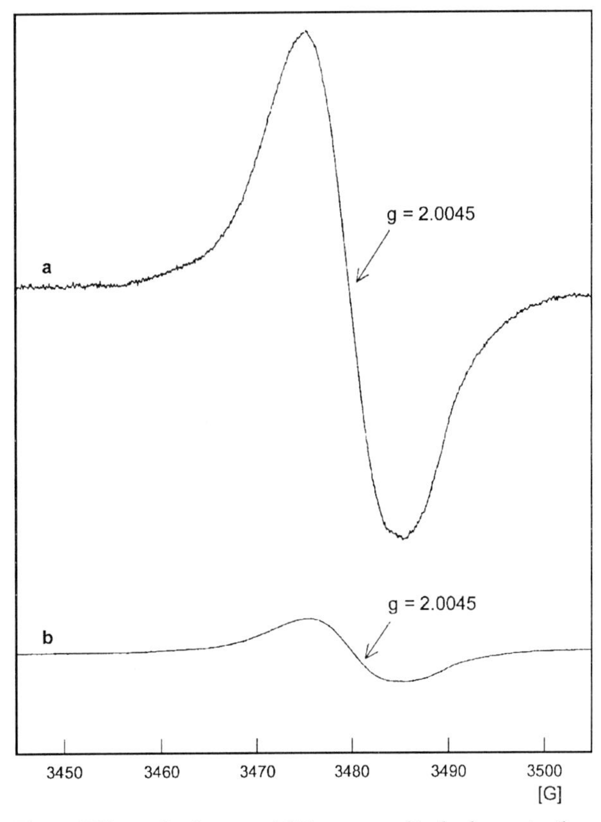

Fig. 5. Effect of toluene addition on radicals decay in the weathered exterior part of cherry-like amber: a — before toluene addition, b — 20 minutes after toluene addition.

BADANIA ŻYWIC KOPALNYCH ZE ZŁÓŻ POLSKICH METODĄ ELEKTRONOWEGO REZONANSU PARAMAGNETYCZNEGO (EPR) I ANIHILACJI POZYTONÓW

Adam JEZIERSKI, Michał SACHANBIŃSKI, Franciszek CZECHOWSKI, Jan CHOJCAN

Streszczenie

Metodami elektronowego rezonansu paramagnetycznego (EPR) i anihilacji pozytonów zbadano właściwości 14 próbek bursztynu z trzech regionów Polski: z wybrzeża Bałtyku z rejonu Gdańska, z kopalni węgla brunatnego w Bełchatowie oraz z rejonu występowania iłów ogniotrwałych w Jaroszowie. Charakter chemiczny badanego bursztynu jest podobny. Istotne różnice w stężeniach estrów i grup hydroksylowych stwierdzono dla zewnętrznych i wewnętrznych zwietrzałych części okazów bursztynu; zawartość grup estrowych jest wyraźnie mniejsza w warstwie zewnętrznej.

W oparciu o badania anihilacji pozytonów wykazano wyraźne różnice w tworzeniu się pozytu w tych częściach okazów. Grupy estrowe w polimerycznej sieci żywicy wydają się tworzyć (w wewnętrznej części okazów) dobre warunki dla powstawania pozytu (mikropory).

Ilościowe badania EPR (QEPR) wykazały, że stężenie centrów paramagnetycznych o charakterze wolnorodnikowym nie zależy od miejsca pochodzenia bursztynu i wynosi 10^{16} spinów/gram próbki dla bursztynu o masie ziaren około 1–5 mg. Stężenie to wzrasta o rząd w trakcie rozcierania bursztynu, przy czym jest to wzrost przemijający (następuje stopniowa powolna rekombinacja rodników powstałych wskutek rozdrabniania).

Zaobserwowano wyraźne różnice w widmach EPR części wewnętrznych i zewnętrznych (zwietrzałych) okazów: widmo EPR części zwietrzałych stanowi jedną linię z parameterem g od 2,0044 do 2,0047; część wewnętrzna wykazuje widmo anizotropowe, $g_z = 2{,}0188$, $g_{x,y} = 2{,}009$.

W stosunkowo krótkim czasie (kilku miesięcy) dla przechowywanych na powietrzu zmielonych próbek z wewnętrznych części okazów pojawia się wyraźne widmo EPR odpowiadające części zewnętrznej, zwietrzałej. Proces ten można przyspieszyć w wyniku działania tlenu w podwyższonej temperaturze (330 K) i ciśnieniu (2,5 MPa).

Rodniki w sieci bursztynu wykazują znaczną dynamikę — ulegają zmianom w czasie przemian chemicznych (wietrzenia), czynników mechanicznych (mielenie) oraz działania rozpuszczalników organicznych. Zwilżenie zmielonej próbki bursztynu toluenem prowadzi do szybkiego (rzędu minut) zmniejszenia stężenia wolnych rodników o rząd wielkości. Jest to związane ze wzrostem ruchliwości struktur zawierających rodniki i postępującymi procesami ich rekombinacji.

Bibliography

ANDERSON K. B., WINANS R. E. & BOTTO R. E.
1992 The nature and fate of natural resins in the geosphere, II, Identification, classification and nomenclature of resinites, *Organic Geochemistry*, **18**, 829–841.

BARANOWSKI A., DĘBOWSKA M., JERIE K., JEZIERSKI A. & SACHANBIŃSKI M.
1995 Ultramarine, lazurite and sodalite studied by positron annihilation and EPR methods, *Acta Physica Polonica A*, **88** (1), 29–41.

CZECHOWSKI F., SIMONEIT B. R. T., SACHANBIŃSKI M., CHOJCAN J. & WOŁOWIEC S.
1996 Physicochemical structural characterization of ambers from deposits in Poland, *Applied Geochemistry*, **11**, 811–834.

CZECHOWSKI F. & JEZIERSKI A.
1997 EPR studies on petrographic constituents of bituminous coals, chars of brown coals group components, and humic acids 600°C char upon oxygen and solvent action, *Energy & Fuels*, **11**, 951–964.

IKEYA M.
1993 *New Applications of Electron Spin Resonance; Dating, Dosimetry and Microscopy*, 67–102 & 379–392, World Scientific Publishing Co. Pte. Ltd.

JANCZAK J., KUBIAK R. & JEZIERSKI A.
1995 Synthesis, crystal structure, and magnetic properties of indium (III) diphthalocyanine, *Inorganic Chemistry*, **34**, 3505– 3508.

JERIE K., BARANOWSKI A., ROZENFELD B., ERNST S. & GLIŃSKI J.
1983 Positron annihilation in and compressibility of water-organic mixtures. I. The system water-tetrahydrofuran, *Acta Physica Polonica A*, **64** (1), 77–92.

KOSMOWSKA-CERANOWICZ B.
1986 Bernsteinfunde und Bernsteinlagerstätten in Polen, *Zeitschrift der Deutschen Gemmologischen Gesellschaft*, **35**, 21–26.

ROBINS G. V., SALES K. D. & ODUWOLE A. D.
1984 Electron spin resonance of plant resins: an assessment of ESR dating possibilities, *ESR Dating and Dosimetry*, *Ionics*, Tokyo, 435–438.

URBAŃSKI T.
1977 On ESR signals of amber, *Bulletin l'Academie Polonaise des Sciences*, *Serie des sciences chimiques*, **25**, 785–787.

UZGRIS E. E. & FLEISCHER R. L.
1971 Charged particle registration in amber, *Nature*, **234**, 28–29.

APPLICATION OF ANALYTICAL PYROLYSIS TO THE EXAMINATION OF AMBER OBJECTS FROM THE ETHNOGRAPHIC COLLECTIONS OF THE ISRAEL MUSEUM

Alexander M. SHEDRINSKY, Ester MUCHAWSKY-SCHNAPPER, Zeev AIZENSHTAT & Norbert S. BAER

Abstract

Pyrolysis gas chromatography (PyGC) and pyrolysis gas chromatography mass spectrometry (PyGC-MS) were applied to the analysis of amber and amber-look-alike objects from the collections of the Ethnographic Department of the Israel Museum (Jerusalem). In the course of this investigation, forty-one analyses of eighteen objects were performed. We found four different types of amber substitutes: bakelite and modern phenolic resins, polystyrene, polymethyl methacrylate, and copal. In addition, numerous examples of pressed amber (ambroid) were found. We demonstrated that while PyGC may be the method of choice to identify synthetic materials used to fake amber objects, in the case of copals, the use of PyGC-MS is preferable. A brief summary of the development of modern synthetic materials used as substitutes for amber is provided. It is demonstrated that such material history is a useful tool in the dating of ethnographic objects.

Introduction

An aura of mystery has always been associated with amber, a material known even in prehistoric times. Ancient peoples believed in its healing power. Further, since amber could gain an electrostatic charge and so attract small objects, it was thought to endow the wearer with magic powers, attracting such good things as love and health. Based on the principle that an evil could be counteracted by something that was in some way similar to it, it was also believed that the wearing of amber prevented jaundice by virtue of its yellow color. As a result of these beliefs, in the Middle East amber was an obligatory part of the Muslim bride's wedding ensemble. Even among the little "bride-dolls" in the Israel Museum's RATHJENS Collection, the Muslim brides wear imitation amber beads.

In addition to its magical aspects, amber jewellery fulfilled a social function by reflecting personal and social status. It may also have served as a symbolic token in the shape of a gift given on certain ceremonial occasions, such as an engagement or wedding.

The cultural setting of a jewellery piece is of great value for the ethnographer. The more complete and authentic the information about its origin and use, the more useful it becomes.

Material, technique, and style are of course, also subject to analysis. Here is where the knowledge of the scientist, craftsman, and art historian becomes essential. These concrete data can tell a great deal about the state of technological development of the culture, as well as its trade ties and contacts with other cultures.

Dates of production may also be ascertained by technical and material analysis, while ethnographic knowledge depending on oral transmission of information may be limited to a few generations (though ethnographic data from written records is occasionally extant). In the case of amber-like materials, i.e. for certain plastics, dates earlier than their invention, the so-called *ante quam non*, are excluded.

Amber imitations

When one addresses the process of "faking amber", the terminology used should be defined precisely. In the 20th century this mainly means the substitution of amber with amber-look-alike plastics (e.g. phenol-formaldehyde resins, polyesters, polystyrene or epoxy resins) for creating a convincing imitation of amber jewellery or fake amber inclusions.

Thousands of years ago, it was mainly copals or other, relatively speaking, similar "fresh" resins which were used to imitate very rare and thus highly prized pieces of amber containing small animals, such as lizards and frogs, or insects, such as bees, beetles, flies and mosquitoes. There is a murky area between these two extremes of genuine versus fake, which was derived from the first German patent in 1879 describing a preparation of ambroid, or pressed amber-material made from small pieces of real Baltic amber "glued" together under high pressure and temperature (FRAQUET 1987). This technology, which initially created an easily recognized poor imitation of real amber, today permits the production of an excellent material which can be distinguished from the real amber only by the highly trained eye of an expert or by careful measurements of hardness and light scattering. Needless to say, *any* chemical analysis is useless in the ambroid case because it will unquestionably reveal the presence of the "authentic" Baltic amber.

Plastic imitations and substitutes

The story of the plastic imitation of amber begins in 1868–1870, when John Wesley HYATT received several patents related to the production of celluloid (solution of nitrocellulose in camphor) and the word "celluloid" was registered as a trademark in the United States Patent Office. On January 28, 1871, the Celluloid Manufacturing Company was organized in Albany. It moved to New Jersey in late 1872. The full story of this plastic is beyond the scope of this chapter; readers can refer to a very detailed treatment of the subject by FRIEDEL (1983). It is curious to note that in spite of their flammability, the celluloid amber imitations have been widely used for pipe stems. At the famous 1885 exhibition *Novelties*, sponsored by the FRANKLIN Institute in Philadelphia, the Celluloid Manufacturing Company proudly displayed, among other articles, imitation amber mouthpieces. The high quality of the imitations could be judged by the fact that the Committee on Science and Art of the FRANKLIN Institute recommended awarding a gold medal to the Company, specifically in praise of the celluloid as „a very desirable substitute" for amber. But according to the same account, imitation ivory, more than amber, was the most important celluloid product, and remained so through the 1920s, followed by imitation tortoiseshell, amber, and pearls. Perhaps this is the reason why celluloid amber imitation is not as common today as are bakelite, phenol-formaldehyde, polystyrene, and unsaturated polyester imitations.

Phenol-formaldehyde resins

One of the most successful chapters in amber imitation history began in 1907 when Leo BAEKELAND patented his new material "bakelite" and the first mass-production plant was built in Germany. The very first samples of the newly invented phenol-formaldehyde resin were quite dark in color (usually reddish or chocolate brown), leading to the legend of "very rare red Baltic amber". On a recent trip to Russia, one of the authors of this chapter spotted a familiar type of reddish necklace in a jewellery store in the center of St. Petersburg. Surprised by the high price (approximately 300 USD), he asked the sales clerk why the plastic cost so much. In reply, she provided a whole lecture on the rarity of this kind of "red Russian amber".

Unfortunately, greed and conscious desire to deceive, as well as ignorance, may play a role in many cases of amber forgeries. In Asia, there is a huge market for medium-sized sculptures (from 10–40 cm) made of burgundy-red bakelite and intentionally sold as "rare Burmese amber".

It should be stressed that the term "bakelite" is used indiscriminately in reference to all phenolic resins. These could be very different structurally depending on the starting materials and curing process.

It took many years of intensive research to overcome two drawbacks of the earliest phenolic resins: they were dark brown or dark red (sometimes even black), and therefore could not be used where bright colors or transparency were required, and they gave off an odor of phenol when heated. Even in 1937, a managing director of Bakelite Ltd. wrote a pessimistic account: "The production of natural amber-like color has been the subject of so much work and so many investigations, and so far it has not been entirely overcome" (POTTER 1937).

It was only at the end of the 1930s that the first examples of phenol-formaldehyde resin with a high formaldehyde ratio were produced, and since then, the majority of amber imitations have been produced from these resins. These could be cast in a mold and cured by oven heating. This type of resin could be made in a variety of colors, transparent or opaque. Opaque appearance is usually achieved with additives.

An amber collector can easily find extremely convincing imitations of milky amber made of phenol-formaldehyde resins throughout the entire continent of Africa, the bazaars of the Middle East, and now even in the United States. It is obvious that this material was produced by European and American companies, yet this fake amber is now returning under such names as "African amber" or "Nepalese amber", and as such is being sold for very handsome prices in some of the finest art galleries of the Old and New Worlds.

In this investigation of a portion of the ethnographic collections at the Israel Museum, Jerusalem, we found that one of the most expensive, and largest, „amber" necklaces (No. 548.71-3 — Fig. 1) was actually made of phenolic resins. Fortunately, in an ethnographic collection it often does not matter what an object is made of, what is important is the history behind who wore or who used it, at what time, and for what purpose.

Casein plastics

Casein is the main product of natural cow's milk, in which it occurs as a calcium compound forming approximately 3% of the whole. Biochemically, casein is a phosphoroprotein, i.e. it contains a significant amount (0.71% by weight) of organically bound phosphorus as the phosphate ester.

Fig. 1. Necklace made of phenolic resin (No. 548.71-3). Photo. A. Shedrinsky (courtesy of the Israel Museum).

The most successful methods of precipitation of casein from skimmed milk are: precipitation by acid (HCl in the United States), and coagulation by the rennet enzyme, rennin.

Plastic casein grains prepared under the influence of heat and pressure in the presence of moisture should be treated with formaldehyde. Following this, the newly formed material will acquire hard, horn-like properties. Casein, cross-linked with formaldehyde, has been widely used for the manufacturing of plastics since before World War I. It was mainly used for tortoiseshell imitations, but some amber-look-alike articles made from casein-formaldehyde plastics are also known.

Relatively speaking, these imitations are very uncommon today, keeping in mind that most casein production in the second half of the 20th century has shifted to food processing.

Acrylics

This group of plastics includes polymers and copolymers of acrylic acid, methacrylic acid, and their esters or acrylonitrites. The final products can result in a range of rather soft and flexible to very hard and stiff thermoplastics and thermosets. They can be produced in a wide range of forms including sheets, rods, tubes, films, pellets, beads, solutions, and latexes. They can be easily dyed in any color of the full spectrum, and to different degrees of transparency, from outstandingly crystal-clear to absolutely opaque.

It took a quarter of a century from the time RÖHM noted such remarkable properties of acrylic polymers, while preparing his dissertation in 1901 until the first limited production of acrylates begun by the RÖHM and HAAS Company in Darmstadt, Germany in 1927 (RIDDLE 1954). But polymethyl methacrylate (PMMA) was produced even later (1933–1936), and very soon in 1938 was recognized as one of the most successful polymers of this class. Currently the production of PMMA far exceeds the combined volume of all the other methacrylates.

Acrylics can be very good materials for amber imitation, ranging from honey-colored transparents (for fake inclusions) to milky yellow translucents. Polymethyl methacrylates, because of their marked hardness, tend to be used as shaped objects. Polyacrylates are much softer, and therefore tend to be used in applications that require flexibility or extensibility.

The Slocum Laboratory, in Royal Oak, Michigan (USA), produces a convincing amber substitute under the trade name *Slocum Amber*. They have even managed to introduce induced stress rings into the final product. However, Py-GC and Py-GC-MS analyses reveal that this material is a pure polymethyl methacrylate (see Fig. 2). Under pyrolysis conditions (650°C, 10 seconds), this material decomposes by an unzipping mechanism into pure monomer methyl methacrylate (peak 1), with small amounts of trimers of this compound (peak 2), and traces of the thermal decomposition of stearic acid, used as a lubricant for the suspension polymerization of PMMA (*Encyclopedia* ... 1985).

Polystyrene

Around 1937, a new material, polystyrene, entered the polymer market. Over time its application grew, from rubber for car tires to foam for insulation and packing. Amber forgers learned to add colorants to obtain very convincing amber-look-alike materials of various colors (lemon-yellow to quite pronounced brown). However, this substitute did not provide a transparency resembling that of real amber. This created a serious limitation on attempts to prepare forged amber inclusions. It is interesting to mention that a current examination of one of the necklaces in the Ethnographic Department of the Israel Museum revealed that the necklace's amber and coral beads were actually made of polystyrene. Pyrolysis analysis of the coral beads yielded a powder of coral-colored inorganic pigment (Fig. 3).

Polyesters and epoxy resins

In the period from 1942–1947, two new materials entered the polymer market: unsaturated polyesters and epoxy resins. These synthetic polymers created a small-scale revolution in amber forgeries, particularly in the area of forged inclusions.

Unlike any of the polymers described earlier, these could be used in any "kitchen laboratory" because their starting materials are readily available in any hardware store. Using a few organic colorants, one can prepare convincing imitations of large, transparent amber pieces with a wide variety of inclusions (e.g. ants, bees, mosquitoes, and even lizards and frogs). It should be noted that the price of amber pieces with inclusions, especially large ones, has skyrocketed in recent years. As a result, a recent flood of forged amber with inclusions has appeared for sale in gem and mineral shops and at fossil shows, and many have been purchased by private collectors for as prices as high as thousands of dollars.

Another widespread application of unsaturated polyesters is their use as amber look-alike beads for necklaces. A necklace from the P. RICE's collection made of this material was purchased from an Indian tribe in the United States, but examples of similar fakes have been spotted in New York under the name of "African Amber". These beads could sometimes be very large in diameter (3–5 cm) and look very convincing.

Analysis of unsaturated polyesters mainly reveals the presence of phthalic acid along with significant amounts of styrene and its dimers and trimers. The most prevalent components for unsaturated polyesters are propylene glycol, phthalic anhydride and maleic anhydride. For casting resins, a molar composition of propylene glycol (1.0), phthalic anhydride

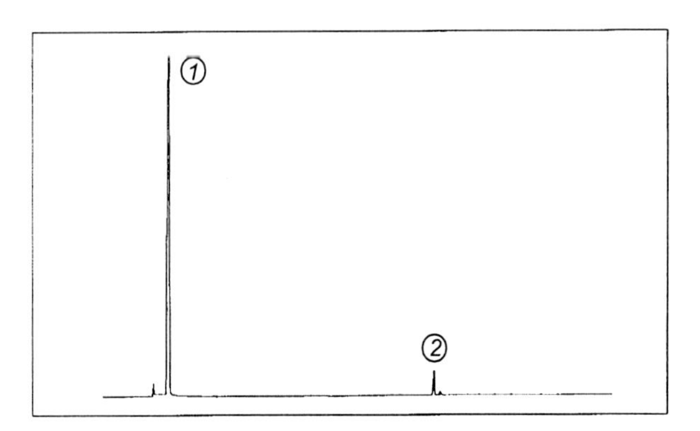

Fig. 2. Pyrogram of polymethyl-methacrylate ("Slocum amber).

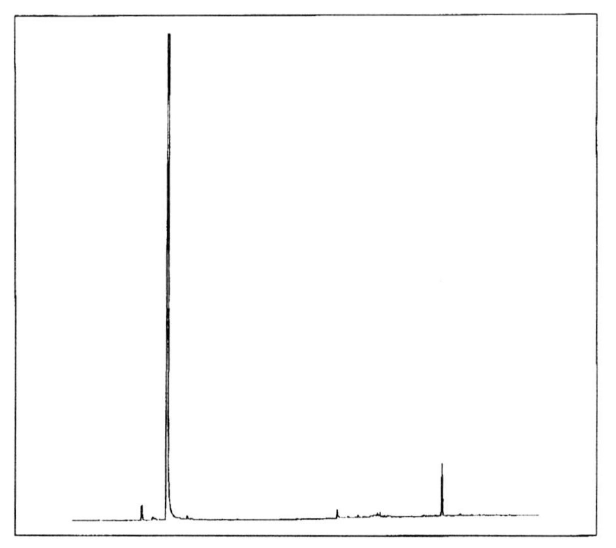

Fig. 3. Pyrogram of polystyrene (necklace #B-95.0764, Israel Museum).

(0.66), maleic anhydride (0.33) and styrene (0.7) in the ratio 3:2:1:2 is usually employed (SHEDRINSKY *et al.* 1993). The unsaturated polyester resins sold in stores could be easily transformed from a stabilized liquid into a rigid plastic state in the presence of free-radical catalysts (methyl ethyl ketone peroxide MEKP is most frequently used) which initiate a controlled crosslinking reaction between the fumarate polymer and styrene monomer. The great convenience of this process is that MEKP cures the unsaturated polyester resin blend at room temperature.

Fig. 4. Necklace with a copal bead in the middle (No. M 2931-11-63). Photo. A. Shedrinsky (courtesy of the Israel Museum).

Experimental methodology

In all Py-GC experiments, the Pyroprobe 120 (CDS Analytical Inc.) and Perkin-Elmer Model 8500 GC were used. The GC instrument was equipped with a flame ionization detector (FID) and 50 m x 0.25 mm (0.5 μm film thickness) SE-54 (Quadrex Corp.) fused silica capillary column. The samples were injected in the split mode (split ratio 50:1) with the column temperature held at 50°C for 1 min, then programmed at 8°C/min to 325°C where it was held for 10 min. The carrier gas (He) flow rate was 4 ml/min. The usual pyrolysis temperature was 650°C unless otherwise specified. In all our experiments the temperature of the interface was held at 250°C. In the case of the object M-2931-11-63 (No. 1 Bead) (Fig. 4) it was necessary to apply PyGC-MS which was performed with the following conditions.

Pyrolysis Pyroprobe 2500 autosampler
Interface: 300°
Pyrolysis: 750° for 15 seconds

Chromatography HP 6890/MSD
Injector: EPC, 300°
Column: HP-5, 30 m, 0.25 mm
Carrier: He, 6 psi, split 75:1
Program: 40° for 2 minutes, then 6°/minute to 295°
Director: MSD, scanning from 35 to 550

Results and discussion

Since analytical pyrolysis (PyGC, PyGC-MS) was introduced as a method of choice for differentiation among different ambers and amber-look-alike materials (POINAR & HAVERKAMP 1985; SHEDRINSKY *et al.* 1989/90/91), we have analysed numerous amber substitutes and identified all of the major plastics used for this purpose (SHEDRINSKY *et al.* 1993). The only new one which has come to our attention as a result of this most recent investigation was polymethyl methacrylate in the object No. M-2931-11-63. Since these analyses were completed, we received a request from Dr. P. RICE, the author of *Amber. The Golden Gem of the Ages* (1980) to analyze the so-called Slocum Amber which also turned out to be polymethyl methacrylate. Similar results were obtained in 1996 in the analysis of an "amber" cross bought in Venice (CARLSEN *et al.* 1996). It appears that an acrylic substitute of amber that quickly gained popularity and soon was widely imitated. It should be noted that for many years the most frequently applied scientific method for amber analysis was IR-spectrometry (KOSMOWSKA-CERANOWICZ & KRUMBIEGEL 1989; 1990; LANGENHEIM & BECK 1965; SAVKEVICH 1970). This method could be used to distinguish between Baltic amber and non-Baltic fossil resins and also identify some synthetic amber forgeries. But the discrimination between non-Baltic ambers and different copals was quite poor (for example, Canadian and Baltic amber have very similar IR spectra). Also, the presence of highly oxidized crust, which is often the case in archaeological samples, could make the IR spectra practically "unreadable".

This paper is an inappropriate place to discuss all the "pros" and "cons" of IR-spectrometry in comparison to PyGC or PyGC-MS. But one example related to the material under investigation should be given. When we analysed object No. L.80.100 from the Ethnographic Collection of the Israel Museum, the fingerprints of one of the beads turned out to be very uncharacteristic. It was clear that the material was a man-made polymer, but to determine the precise nature of the material was difficult. We took FTIR of the sample and received the spectra suggesting that the material is Acryloid B-48N. But the analysis of a standard sample of this acryloid done by PyGC immediately produced fingerprints completely different from the material under investigation. And only a comparison of pyrograms of a standard sample of regular PMMA (ROHM & HAAS) with the bead from the Ethnographic Collection proved that they were made from the same materials.

In the majority of analyses there is good reason to simply use PyGC analysis instead of much more expensive and complicated equipment such as PyGC-MS. PyGC provides easily readible and distinguishable "fingerprints" for the majority of natural resins and their artificial substitutes (SHEDRINSKY *et al.* 1989/90/91; 1993). The major drawback of this method is that it is always "comparitive", i.e. the researcher must have a large library of reliable samples for comparison. If a new material never before analysed would emerge, PyGC would be of little help. In this case PyGC-MS can solve the problem.

Altogether we analysed forty-one samples from eighteen objects of the Ethnographic Collection of the Israel Museum (see Table 1). All analyses of the museum objects revealed the presence of natural (or pressed) Baltic amber together with some beads made of synthetic materials. As usual, the dominant plastics were early bakelite and modern phenolic resins (Fig. 5) followed by polystyrene and polymethylmethacrylate (see Fig. 2, 3). It is worth noting that we also analysed one of the beads imitating coral (B-95.0764, 4) and it turned out to be made of polystyrene. One of the beads (M-2931-11-63 (Morocco), bead No.1) was made of natural copal, probably of African origin. The PyGC-MS analysis confirmed that this material is a natural resin, but did not precisely pinpoint the origin of its copal.

The greatest surprise of this investigation was the high number of authentic amber beads in analysed objects (Fig. 6). Comparing the notes taken, for example during the sampling of No.2864.11.63, No.2582-10-63 and No.3089.11.63, with the results of the analyses it is hard to accept the chemical data just at their face value. All above-named samples were much harder than natural amber should be, and look quite artificial. The only reasonable explanation for this contradiction relies on the fact that these beads were made of pressed amber or "ambroid".

What should be made clear about the general possibility to differentiate among different amber and amber-look-alike materials using PyGC or PyGC-MS is that it is impossible to distinguish between natural Baltic amber and pressed Baltic amber. Chemically they are identical and as a result they produce the same fingerprints in the Py-GC. The only way to differentiate between these two materials is to measure their hardness. In the Mohs' Scale, the hardness of Baltic amber is around 2.3 and the hardness of pressed amber is about 3. The first production of pressed amber started in the second part of the 19th century which means that the presence of beads made of this material will date them between the 19th and 20th centuries.

Also, it will take a trained eye or instrumental hardness measurements to distinguish between pressed and natural amber. Nevertheless, the systematic analysis of ethnographic objects could play a very important and positive role in our understanding of the materials themselves and also their applications. One should interject a word of caution about conclusions made on the basis of chemical analysis. It should remembered that the separate

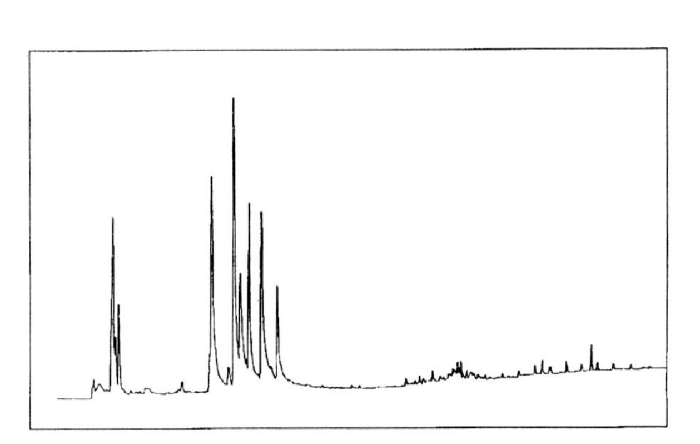

Fig. 5. Pyrogram of phenolic resin.

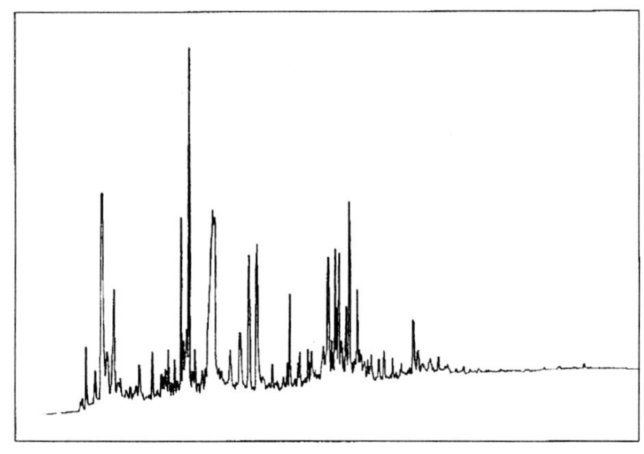

Fig. 6. Pyrogram of pressed Baltic amber.

ACCESSION NO.	BEAD NO.	MATERIAL
2582-10.63	#1	amber
2582-10.63	#2	amber
2582-10.63	#3	amber
2582-10.63	#4	amber
2582-10.63	#5	amber
2582-10.63	#6	amber
2582-10.63	#7	amber
3082.11.63	left	amber
3082.11.63	right	amber
3089.11.63	#1	amber
3089.11.63	#3	amber
3089.11.63	#6 (brownish)	amber
M-2931-11-63	#1	copal
M-2931-11-63	#2	amber
M-2931-11-63	#3	amber
M-2931-11-63	#4	amber
B-95.0764	#1	amber
B-95.0764	#2	phenol-formaldehyde
B-95.0764	#3 (slightly reddish)	polystyrene
B-95.0764	#4 (coral-look-alike)	polystyrene
A-569-71-3		amber
548-71-3		phenolic resin
L.80.100	#1	polymethylmethacrylate
L.80.100	#2	amber
2-4909	#1	amber
2-4909	#2	amber
3092.11.63	#2	amber
3092.11.63	#3	amber
2515.10.63	right	amber
2515.10.63	left	amber
2847.11.63		amber
2847.11.63	extreme left, lower level	amber
2507.10.63	#1	amber
2507.10.63	#2	amber
2864.11.63	#1	amber
2864.11.63	#2	amber
2864.11.63	#3	amber
B-569-71-3		amber
510.71	center bead	amber
B-93.04.72	#1	phenol-formaldehyde
B-93.04.72	#2	phenol-formaldehyde
B-93.04.77	#1	phenol-formaldehyde

Table 1. Sample designations and descriptions of objects from the Ethnographic Collection of the Israel Museum, Jerusalem.

parts of all these decorations were generally interchangeable. They were quite often restrung and, in the case of damage, reassembled from two or more different strands ("different" in terms of age and source of production). That is why it is feasible when a label attached to a metal veil in one of a museum's display windows says "dated to the second part of the 19th century", despite the presence of some obvious plastic beads on some of these strings attached to the metal frame. These strings were easily breakable and could be repaired with modern materials though the core of this decoration could be very old.

Acknowledgements

The authors are deeply grateful to the Senior Curator Rivka GONEN and the staff of the Department of Ethnography of the Israel Museum for their enthusiastic support of this work. One of the authors (AMS) also gratefully acknowledges an L.I.U. Faculty Released Time Grant and the support of the FORCHHEIMER Foundation in making his three-month stay in Israel possible. We are also all thankful to Prof. Barbara KOSMOWSKA-CERANOWICZ for her kind invitation to present this work at the Symposium in Gdańsk and for editing the manuscript. The generous support of the Samuel H. KRESS Foundation in making this trip to Poland possible is gratefully acknowledged.

Alexander M. SHEDRINSKY
Conservation Center, IFA
New York University
14 East 78th Street
New York, NY 10021-1745, USA

Ester MUCHAWSKY-SCHNAPPER
Ethnography Department
The Israel Museum, P.O.B. 1299
91012 Jerusalem, Israel

Zeev AIZENSHTAT
The Hebrew University of Jerusalem
Casali Institute of Applied Chemistry
Givat Ram Campus
Jerusalem 91904, Israel

Norbert S. BAER
Conservation Center, IFA
New York University
14 East 78th Street
New York, NY 10021-1745, USA

ZASTOSOWANIE PYROLIZY DO BADANIA OBIEKTÓW BURSZTYNOWYCH Z KOLEKCJI ETNOGRAFICZNEJ MUZEUM IZRAELA

Alexander M. SHEDRINSKY, Ester MUCHAWSKY-SCHNAPPER, Zeev AIZENSHTAT, Norbert S. BAER

Streszczenie

Metody pirolitycznej chromatografii gazowej (PyGC) i pirolitycznej chromatografii gazowej- spektrometrii masowej (PyGC-MS) zostały zastosowane do identyfikacji bursztynowych i bursztynopodobnych okazów z kolekcji Działu Etnograficznego Muzeum Izraela w Jerozolimie. Wykonano 41 analiz 18 obiektów, na podstawie których stwierdzono cztery rodzaje imitacji bursztynu: bakelit i nowoczesną żywicę fenolową, polistyren, polimetakrylan metylu i kopal. Ponadto zostały zidentyfikowane liczne okazy prasowanego bursztynu (ambroidu). Starano się wykazać, że do identyfikacji syntetycznych materiałów używanych do imitacji (fałszerstwa) bursztynu może być użyta metoda PyGC, natomiast w przypadku kopalu najlepiej stosować PyGC-MS. Opisano rozwój nowoczesnych syntetycznych materiałów używanych jako imitacje bursztynu.

Kolekcje bursztynowej biżuterii w Muzeum Izraela zawierają głównie okazy z Jemenu i Maroka przywiezione przez imigrantów albo zakupione przez muzeum w ostatnich latach. Biżuteria ta nie tylko była ozdobą upiększającą właściciela i określającą jego stan posiadania, lecz także powinna mieć symboliczne i magiczne znaczenie. Muzułmańskie kobiety noszą paciorki z bursztynu importowanego, prasowanego albo z imitacji bursztynu.

Bursztyn był także obowiązującą częścią biżuterii noszonej przez muzułmańskie narzeczone. Wśród narzeczonych-lalek, w kolekcji RATHJENA Muzeum Izraela, muzułmańskie lalki zdobią paciorki z imitacji bursztynu.

Wartość etnograficzna obiektów nie jest uzależniona od właściwości samego materiału. Dokładne określenie materiału jest jednak ważne dla datowania względnego danego obiektu. Na przykład imitacje bursztynu wykonane z bakelitu nie mogły być produkowane przed 1909 rokiem, a „bursztynowe" paciorki z poliestru mogły powstać dopiero po roku 1942–1947.

Bibliography

CARLSEN L., FELDTHUS A., KLARSKOV T. & SHEDRINSKY A. M.
1996 (January) Det Er Rav! – Eller Er Det?, *(Dansk Kemi) Danish Chemistry*, **1**, 8–13.

ENCYCLOPEDIA ...
1985 *Encyclopedia of Polymer Science and Engineering*, **1**, 272.

FRAQUET H.
1987 *Amber*, Butterworth's Gem Books, Butterworth & Co., London.

FRIEDEL R.
1983 *Pioneer Plastic*, The University of Wisconsin Press, Wisconsin.

KOSMOWSKA-CERANOWICZ B. & KRUMBIEGEL G.
1989 Geologie und Geschichte des Bitterfelder Bernsteins und anderer fossiler Harze, *Hallesches Jahrbuch für Geowissenschaften*, **14**, 1–25.
1990 Bursztyn bitterfeldzki (saksoński) i inne żywice kopalne z okolic Halle (NRD), *Przegląd Geologiczny*, **38** (9), 394–400.

LANGENHEIM J. & BECK C. W.
1965 Infrared spectra as a means of determining botanical sources of amber, *Science*, **149**, 52–55.

POINAR G. O. & HAVERKAMP J.
1985 Pyrolysis mass spectrometry in the identification of amber samples, *J. Baltic Stud.*, **16**, 210–222.

POTTER H. V.
1937 Artifical Resins, *J. R. Soc. Arts*, **LXXXV**, 249.

RICE P.
1980 *Amber. The Golden Gem of the Ages*, Van Nostrand Reinhold, New York.

SAVKEVICH S. S.
1970 *Amber* [in Russian], 13, Nedra, Leningrad.

RIDDLE E. H.
1954 *Monomeric Acrylic Esters*, Reinhold Publishing Corp., New York.

SHEDRINSKY A. M., GRIMALDI D. A., BOON J. J. & BAER N. S.
1993 Application of PyGC and PyGC-MS to the unmasking of amber forgeries, *Journal of Applied Analytical Pyrolysis*, **25**, 77–95.

SHEDRINSKY A. M., GRIMALDI D., WAMPLER T. P. & BAER N. S.
1989/90/91 Amber and Copal: Pyrolysis Gas Chromatographic (PyGC) Studies of Provenance, *Wiener Berichte der Naturwissenschaftlichen Kunst*, **6/7/8**, 37–63.

GAS INCLUSIONS IN BALTIC AMBER

Gennadij S. KHARIN

Abstract

Results of new analyses of gas inclusions in Upper Eocene Baltic amber are given and the literary data on this topic are summarized in this paper.

Introduction

Gas inclusions in amber can be found as small bubbles, single or in clusters, often accumulating in the vicinities of vegetable and animal inclusions. Some researchers (KAZANSKIY *et al.* 1969) expressed an opinion that gases enclosed in bubbles, may be the relics of ancient atmosphere, preserved in amber. Gas-liquid and liquid inclusions are less common. The composition of gas inclusions in minerals is very difficult to determine. Therefore, researchers study gases in special thermobarogeochemical laboratories. Two research laboratories, namely the Geological Institute (VSEGEI, St. Petersburg) and the Institute of Crystallography and Mineralogy (Novosibirsk), have carried out studies of gas inclusions in Baltic amber. The results of this research are described below. The possibility of using gas inclusions in amber to estimate the Late Eocene atmosphere composition is also considered. For comparative purposes, the findings on gas composition in amber inclusions discovered in some other areas, are also quoted.

Material and analytical techniques

Samples of amber were taken by S. M. ISATCHENKO from an underwater field in the vicinity of Cape Taran, on the Sambian Peninsula (Kaliningrad region). The same authors also prepared these samples for analysis. From a great many specimens 17 pieces of transparent amber (succinite), containing inclusions of large (0.2–2 mm) gas bubbles were chosen. Studies of gas inclusions in amber were carried out according to procedures developed in the Institute of Geology and Geophysics, Siberian Department, Russian Academy of Science (KAZANSKIY *et al.* 1969). The amber-gas problem was researched also in Canada and Japan (CREIG & HORIBE 1994). Hermetically sealed gas cavities, isolated from each other and with no signs of fissures were chosen for analysis. In all of the inclusions examined the gas pressure turned out to be close to the present-day atmospheric pressure.

Analyses results

For the majority of amber specimens collected at the underwater slope of the Sambian Peninsula it was found that gas composition is characterized by high concentrations of CO_2 and N_2. The latter (N_2) contains impurities in the form of inert gases, amounting to less than 1.3% of the total volume (NESMELOVA & KHABAKOV 1967). The results of chromatographic and volumetric analyses of gases, liberated from the total mass of ground, milk-white osseous amber, are different from those quoted above (Table 1). This type of amber is characterized by a large amount of gas — up to 631 ml per 1 kg of amber or, on average, about 543 ml/kg. In contrast to transparent amber, where gas inclusions are in the form of large (up to 2–3 mm) and infrequent bubbles, gas inclusions in osseous amber are numerous, but of microscopic size (order of microns), and scattered over the entire body of the specimen. The concentration of bubbles determines the intensity of the amber's white colour. The composition of gas liberated from ground amber is characterised by a high content of nitrogen (up to 93.3%), with much lower concentrations of other gases, including CO_2. Maximum concentrations of the latter were found in Baltic amber (up to 40.6%). It should be noted that gases from Cretaceous amber (Southern Siberia, Kiya River) contain a large amount of hydrocarbons (up to 66.8%), while Cenomanian amber from Northern Siberia (Pyasina River) is noted for its high content of hydrogen (22.1%) and carbonic oxide (10.5%) (ZAKHAROV *et al.* 1992).

Correlation analysis shows that there are no firm connections between the size of the bubbles and gas composition (Table 2).

Quoted laboratory data regarding gas composition in amber inclusions testify that this composition is notably different from that of the present-day near-

Group	Type of amber	Occurrence of specimen	Ø of bubble, mm	Content of gases, %							N_2/O_2
				H_2S, SO_2 NH_3	CO_2	O_2	CO	H_2	HC	N_2^+ (Ar,Ne,He)	
I	Transparent succinite	Underwater of slope of Sambian Peninsula	0.02	0	18	16	0	0	0	66	4.21
		"	0.24	0	15	13	0	0	0	72	5.54
		"	0.033	0	20	16	0	0	0	64	4
		"	0.034	0	21	17	0	0	0	62	3.64
		"	0.04	0	22	18	0	0	0	60	3.33
		"	0.03	0	19	16	0	0	0	65	4.06
	Average		0.03	0	19.17	16	0	0	0	64.8	4.12
II	Honey-yellow transparent succinite	"	0.032	0	17	15	0	0	0	68	4.5
		"	0.052	0	12.5	12.5	0	0	0	75	6.25
		"	0.082	0	15	15	0	0	0	70	4.67
		"	0.09	0	16	16	0	0	0	68	4.25
		"	0.07	0	15	15	0	0	0	70	4.67
	Average		0.07	0	15.1	14.7	0	0	0	70.2	4.87
III	Succinite with plant debris	"		0	38	0	0	0	0	58.2	
		"		0	32.6	0	0	0	0	67.4	
		"		0	36.5	0	0	0	0	64.5	
		"		0	39.2	0	0	0	0	63.2	
		"		0	37.5	0	0	0	0	52.4	
		"		0	59.4	0	0	0	0	59.4	
	Average			0	40.39	0	0	0	0	60.85	
IV	Milk-white amber	"			36	0.5	0	0	0	87.6	175
		"			10.2	0.6			6.5	93.3	185

Table 1. Composition of gas inclusions in Baltic amber specimens. **Note:** Data on milk-white amber was taken from NESMELOVA & KHABAKOV (1967), the remainder comes from research carried out by N. SHUGUROVA (Novosibirsk).

surface atmosphere. A nitrogen–oxygen ratio, close to that in the present-day atmosphere, was found only in 2 specimens. Change in the concentrations and ratio of gas components in ancient inclusions is the natural course of events. The most intensive variations are inherent to sediments and rocks containing a large amount of organic material. Decomposition of the latter leads to the liberation of gases (CO_2, CO, H_2 etc.) and to the absorption of oxygen.

Most commonly the amber bubbles, as was mentioned earlier, are concentrated near the plant and animal inclusions. This juxtaposition undoubtedly exerted an influence on the composition of the palaeoatmospheric gases which were caught in the pine resin together with the organic inclusions. Multiple diurnal heating and cooling of resin resulted in the migration of gas bubbles and the mixing of primary atmospheric gases with secondary ones, which were liberated from both the resin and organic inclusions. When the burial of resin in the floor of an „amber forest" had been completed, the transformation process of the trapped gases continued, but under different conditions. Most probably, these processes occurred in swampy and stagnant surroundings, which provided an oxygen free and carbon dioxide rich environment. It appears that these conditions influenced the composition of gas inclusions in Baltic amber. The next stage of amber formation — the ripening of resin in marine sediments — is characterized by an increased amount of alkaline components, in particular glauconite. This process often took place in a slightly stagnant environment,

Group I

	D	CO_2	O_2	N_2
D	1.00	-0.78	-0.85	0.81
CO_2		1.00	0.96	-0.99
O_2			1.00	-0.99
N_2				1.00

Group II

	D	CO_2	O_2	N_2
D	1.00	-0.45	0.12	0.25
CO_2		1.00	0.82	-0.98
O_2			1.00	-0.93
N_2				1.00

Table 2. Correlation matrices of Baltic amber groups I and II.

e.g. a lagoon delta (KHARIN 1995), and has evidently left its mark on gas composition in inclusions.

As a result, a secondary nitrogen–carbon dioxide atmosphere formed within inclusions, which does not reflect the real composition of the Late Eocene atmosphere. Gas in milk-white amber inclusions has been transformed to the smallest extent. However, such specimens, due to their small size and failure to open each individual inclusion, have been insufficiently studied. Both primary and secondary gases appear to be present in the total mass of gases liberated from the ground down material. Publications announce from time to time sensational information about the highly oxygenated atmosphere of the Cretaceous period (LANDIS 1993). This conclusion is drawn on the basis of information about the composition of gas inclusions in amber. To judge by the results obtained, at the end of the Cretaceous era (67 mln years ago) the oxygen concentration in the atmosphere had fallen from 35% to 28% (the present-day atmosphere has an oxygen concentration of 21%). This supposedly resulted in the extinction of the dinosaurs, which, in contrast to mammals, were unable to adapt to the dramatically reduced levels of oxygen in the air. This hypothesis undoubtedly has the right to exist, but it should be studied more closely in order to obtain comprehensive evidence. Taking gas inclusions in Baltic amber as an example, one can even suggest that a carbon dioxide–nitrogen atmosphere really existed in the Late Eocene (43–37.5 mln years ago). But this would be an absolutely erroneous conclusion. As we have already shown, the composition of gases in amber depends on many factors. The composition of ancient air inside the trapped bubbles changed considerably during the transformation processes which turned resin into amber. In a reduction medium the growth of oxygen percentage may take place. A specific and, for the time being, undetermined set of conditions is needed to preserve the remnants of ancient atmosphere without any changes. In conclusion, it should be noted that not enough research has been carried out thus far on gas and gas-liquid inclusions in amber. Further studies in this field should be undertaken, making use of a wider range of material, different types of amber and the technique of electron microscopy. It is also necessary to carry out analyses of trapped gases in modern resins, in order to establish whether or not chromatographic gas separation has taken place.

Acknowledgements

The author wishes to thank S. ISATCHENKO for permittng access to specimens of amber with gas bubbles, Dr. N. SHUGUROVA for the chemical analysis of gas inclusions, Prof. E. E. EMELYANOV for reading this paper, Dr. A. KRYLOV — president of the Sea Venture Bureau (SVB) — and V. SIVKOV (director of SVB) for their financical support. This project was also supported by the Russian Academy of Sciences and RFFR (projects 96-05-65639 and 96-15-98336).

Gennadij S. KHARIN
Atlantic Branch of P.P. Shirshov Institute of Oceanology
of the Russian Academy of Sciences
Prospekt Mira 1
236000 Kaliningrad, Russia
e-mail: kharin@geology.ioran.kern.ru

INKLUZJE GAZOWE W BURSZTYNIE BAŁTYCKIM

Gennadij S. KHARIN

Streszczenie

Przedstawione zostały wyniki własnych badań inkluzji gazowych w bursztynie bałtyckim i dane z literatury.

Badaniom poddano trzy odmiany bursztynu bałtyckiego: przezroczysty (I grupa), przezroczysty miodowożółty (II grupa), sukcynit z detrytusem roślinnym (III grupa), które zestawiono z mleczno-białymi (IV grupa). Dane dotyczące IV grupy zacytowano z literatury (NESMELOVA & KABAKOV 1967). Okazy do badań zostały pobrane z Bałtyku, ze skłonu Półwyspu Sambijskiego.

Niektórzy badacze wyrażają opinię, jakoby gaz zamknięty w pęcherzykach bursztynu stanowił pozostałości dawnej atmosfery. Rezultaty przeprowadzonych analiz (tab. 1 i 2) przedstawiono wraz z dyskusją. Dla porównań zacytowane zostały również dane z literatury dotyczące składu gazu z inkluzji w okazach żywic kopalnych znalezionych w kilku innych rejonach („bursztynu" kredowego z południowej Syberii znad rzeki Kiya oraz „bursztynu" cenomańskiego z północnej Syberii znad rzeki Pjasina).

W wyniku badań pęcherzyków w sukcynicie uzyskano złożony skład gazu (typowy przez znaczną koncentrację CO_2 i N_2), który jest **niepodobny** ani do składu współczesnej, ani dawnej atmosfery.

Bibliography

CREIG H. & HORIBE Y.
1994 Atmospheric gases in Amber, *Abstracts 16th General Meeting IMA*, 85–86, Pisa.

KAZANSKIY Y. P., KATAYEVA V. N., SHUGUROVA N. A.
1969 Experience of study of gas-liquid inclusions as relics of ancient atmosphere and hydrosphere [in Russian], *Geology and Geophysics*, **11**, 39–43, Nowosybirsk.

KHARIN G. S.
1995 Geological conditions of the amber-bearing deposits originating in the Baltic Region, *Amber & Fossils*. **1**, 47–54, Kaliningrad.

LANDIS G.
1993 *New Scientist*, **148**, No. 998, 11.

NESMELOVA Z. N. & KHABAKOV A. B.
1967 Gas inclusions in the Baltic amber, *Proceedings of VSEGEI*, **11**, issue 110, 225–230.

ZAKHAROV V. A., KAZANSKIY Y. P. & SHUGUROVA N. A.
1992 Composition of gas inclusions in Cenomanian and Eocene amber [only summary in English], *Geology and Geophysics*, **12**, 89–93, Nowosybirsk.

FOSSIL RESINS FROM AUSTRIA: BIOMARKERS DETECTED IN ROSTHORNITE (EOCENE, CARINTHIA), KÖFLACHITE (MIOCENE, STYRIA) AND A RESIN FROM THE LOWER CRETACEOUS OF SALZBURG

Norbert VÁVRA

Abstract

Some fossil resins from Austria (köflachite — Miocene, Styria; rosthornite — Eocene, Carinthia, and an unnamed resin from the Lower Cretaceous of Salzburg) have been studied by means of computer-aided gas liquid chromatography/mass spectroscopy. According to these studies köflachite can no longer be regarded as a resin; it represents a mixture of hydrocarbons, a number of which could be identified on the basis of their mass spectra: phyllocladane, eudesmane, sandaracopimarane, norpimarane, dehydroabietane, simonellite and retene. In the soluble fraction of rosthornite two different amyrins could be identified. In the resin from the Lower Cretaceous of Salzburg degradation products of ß-sitosterol have been tentatively identifed. On the basis of these results the botanical origin of köflachite and rosthornite is discussed.

Introduction

In contrast to countries like Poland, Russia, Lithuania, the Dominican Republic etc. Austria does not have any occurrences of fossil resins — "amber" — which have ever been mined or could be mined in future, or perhaps used in some small local industry for any purpose like the production of jewellery etc. As is the case with many other minerals, one can also apply the old statement that Austria is "rather rich in poor finds" to fossil resins — this means that there are a lot of places (22 or even more) — where fossil resins have been reported but the amounts are usually tiny. Only on one occasion (Lower Cretaceous of Salzburg, near Golling) a few hundred kilograms have been found. There are, however, some convincing reasons for the study of such finds.

(1) Identification and revision of single mineral specimens:

Quite a number of organic minerals have been described from Austrian localities, places which now rank as "type localities". The first to be described was hartite (a hydrocarbon, phyllocladane) from Hart near Gloggnitz (Lower Austria). Others included: ixolyte (a possible fossil resin) from a coal mine — now disused — in the same area; jaulingite from Jauling near St. Veit (within easy reach of Vienna) and rosthornite from Carinthia — to give but a few examples.

(2) Identification of biological markers and determination of their possible botanical origin:

The search for chemical fossils which could yield some information as regards their botanical source can sometimes be promising; however, in many cases results have to be confirmed by finds of fossil leaves or pollen.

(3) Studies of diagenesis of organic substances:

Data from organic minerals can generally be useful for the benefit of organic geochemistry — e.g. for the study of the aromatization of terpenes being the most probable source for aromatic compounds in oil.

Rosthornite — a fossil elemi resin from Carinthia

Author: HÖFER (1871)

Type locality: Knappitsch coal mine at the Sonnberg, near Guttaring, Carinthia

Geological age: Early Eocene (Palaeogene — Zone 13)

Name: in honour of F. v. ROSTHORN, who dedicated his life to the mineralogical and geological exploration of Carinthia and also published one of the early descriptions of the Palaeogene of the area of

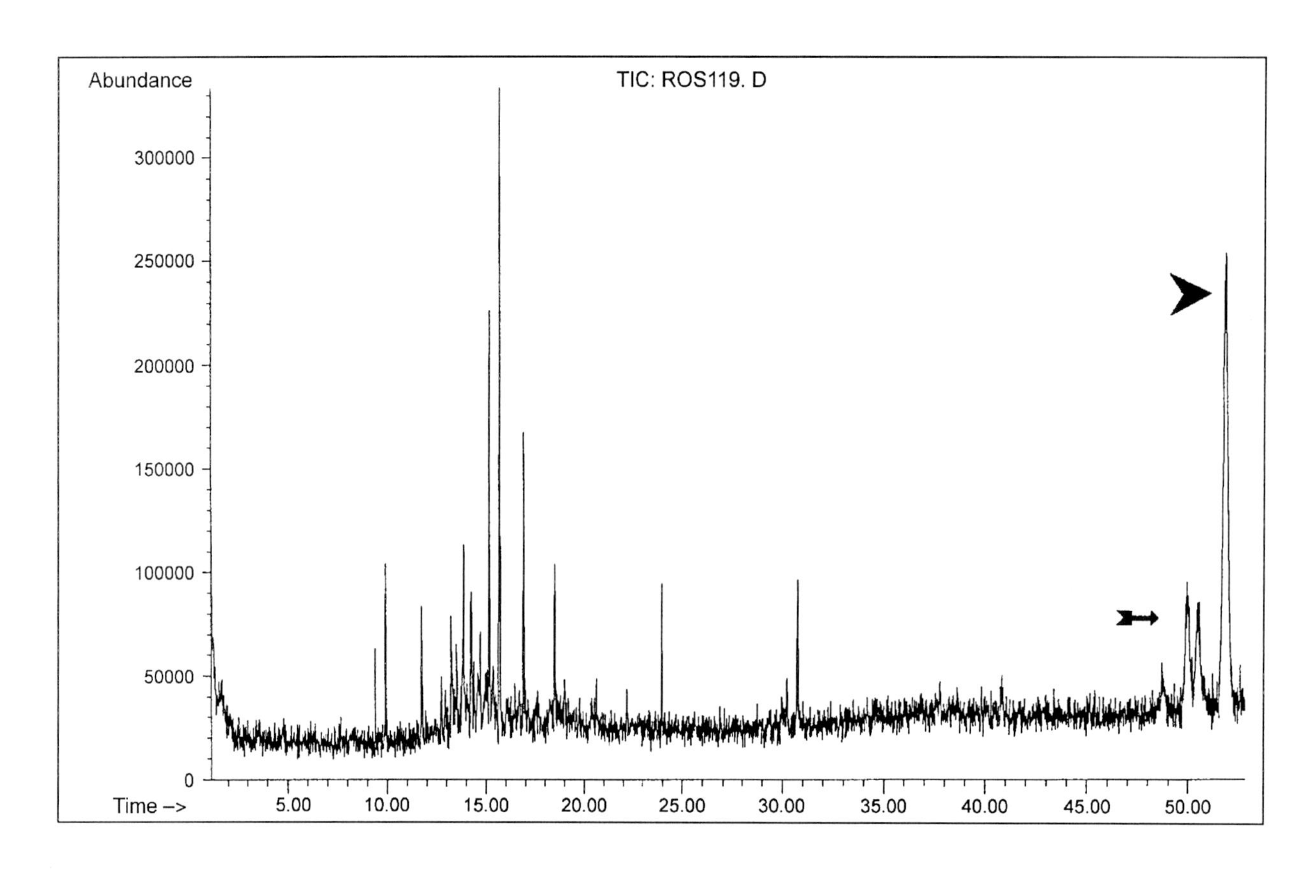

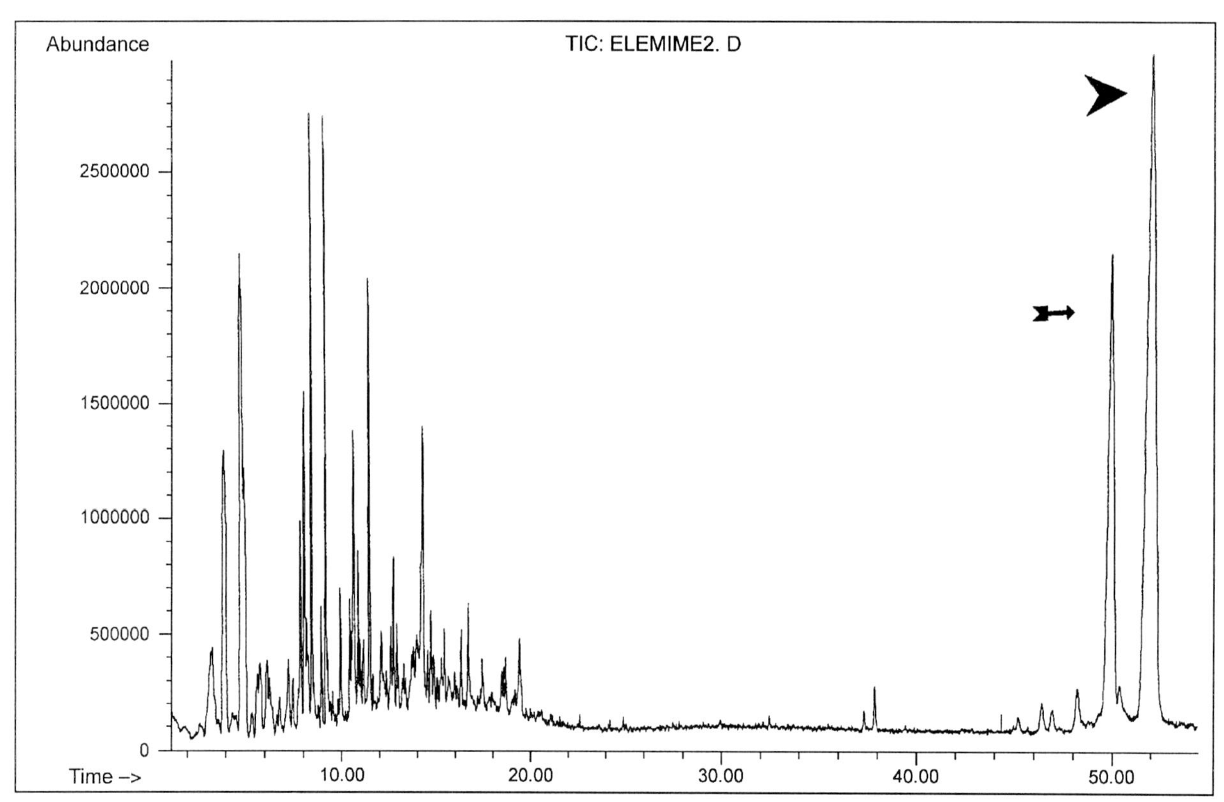

Fig. 1. Gas liquid chromatograms (better: total ion chromatograms "TIC") of a soluble fraction (solvent: methylene chloride) of rosthornite (Eocene, Carinthia; above) and of a Recent elemi resin ("Elemi electum", solvent: methylene chloride; below). Arrows: ß-amyrin (retention time: approx. 50 min.) and α-amyrin (ret. time approx. 52 min.)

Guttaring (ROSTHORN & CANAVAL 1853)

Occurrence: probably at the type locality only

Botanical origin: probably Burseraceae

Specific gravity: 1.076

Chemistry: C — 84.42%, H — 11.01%, O — 4.57% (average values from HÖFER [1871])

Rosthornite contains α- and ß-Amyrin

In the Eocene basin of Krappfeld (NNE Klagenfurt, Carinthia) there are three localities where coal seams are known to occur: Althofen, Guttaring and Klein-St. Paul. These are the oldest coal deposits known from the central Alpine area in Carinthia. They were deposited at the southern edge of the Gosau-formation during the marine advance of the Lower Eocene transgression, being part of a worldwide Early Eocene transgressive cycle (STEININGER *et al.* 1988/89). On the basis of palynological studies they have been attributed to the Palaeogene — Zone 13 (Early Eocene).

At one of the aforementioned localities (Guttaring) a small coal mine (e.g. in 1902 — 26 miners only) existed between the years 1773 and 1933, situated at the Sonnberg — a low hill close to this village. HÖFER mentions the name "Knappitsch" for this mine, probably derived from the owner. From this coal mine HÖFER (1871) reported the find of a resin ("rosthornite") occurring in lens-shaped nodules being one inch thick and having a diameter up to six inches. A rather detailed description of the geology is given by REDLICH (1905), who mentions two different coal seams (whereas older literature mentioned three or even four). REDLICH reports that this fossil resin was rather common and gives an exact location of the finds in a geological section. HÖFER tried to do his best to make careful comparisons with other different fossil resins (jaulingite, euosmite, pyroretine, bucaramangite, succinite and retinite) on the basis of methods which were available at the time — studies resulting finally in the opinion that rosthornite was distinctly different from all other material studied and that it was well justified to establish a new mineral variety. Later authors show a tendency, however, to emphasize the similarities of rosthornite with one or other fossil resin: HINTZE (1933) regarded rosthornite as being close to euosmite and similar to jaulingite, stating, moreover, a relationship to bucaramangite; PACLT (1953) mentions rosthornite among "Other Caenozoic Resins derived probably from Coniferous Woods" without any more specific remark in respect to its relationship to other fossil resins. HEY (1962), however, he regards rosthornite — together with many other fossil resins — as a variety of resinite only.

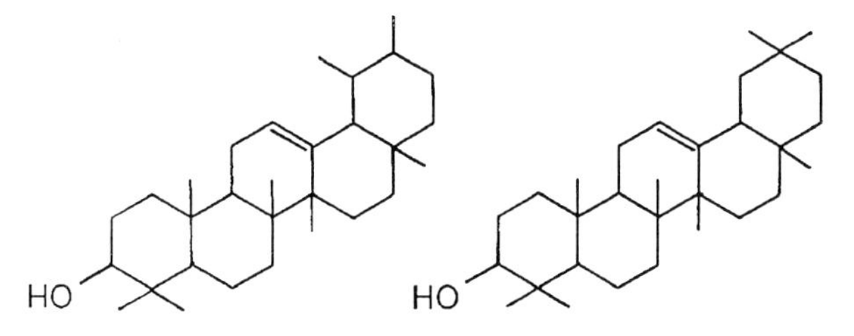

Fig. 2. Formulae of α-amyrin (left) and ß-amyrin (right).

Analysis of soluble fractions of samples of rosthornite (e.g. methylene chloride being the solvent) by means of computer-aided gas liquid chromatography/mass spectroscopy (Fig. 1) has not only confirmed the presence of traces of possible lower terpenes and their products of diagenetic degradation (like isopropyldimethyl tetralin and isopropyldimethylnaphthalene) but also the occurrence of amyrins. Thus, the presence of ß- and α-amyrin (Fig. 2) could be confirmed on the basis of mass spectra and retention times and by careful comparisons with chromatograms of Recent elemi resin and authentic samples of amyrins.

The chemotaxonomical significance of amyrin

Resins containing triterpenes are generally regarded as being derived from angiosperms (BRACKMAN *et al.* 1984 and other references given there); amyrins, however, are unfortunately rather widespread in the plant kingdom. Thus, the α-amyrin (= viminalol, α-amyrenol or urs-12-en-3ß-ol), for instance, has been reported to occur in *Ficus* as well as in members of the Balanophoraceae and Erythroxylaceae families. ß-amyrin (= olean-12-en-3ß-ol, ß-amyrenol) occurs as its palmitate ("balanophorin") in the same plants together with the α-compound (WINDHOLZ *et al.* 1976). Among resin-producing plants, the family Burseraceae ("frankincense family") can be regarded as the most probable source for amyrin. However, *Amyris* (Rutaceae), another resin-producing genus (from which the name "amyrin" is in fact derived!), cannot be excluded with certainty (BOITEAU *et al.* 1964 and other references given there). The family Burseraceae is represented by 16–20 genera and about 600 species of trees or, less often, shrubs with prominent schizogenous resin ducts. Members of this family occur in tropical America and northeastern Africa. Well-known products obtained are frankin-

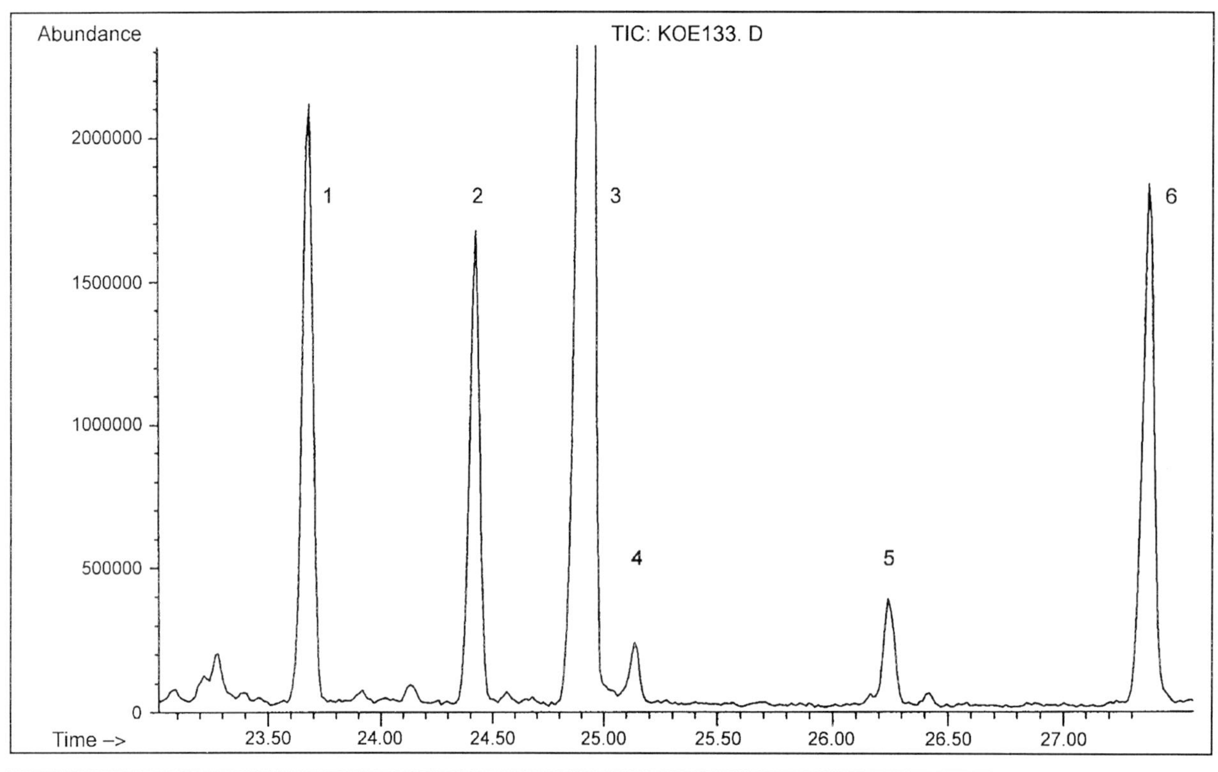

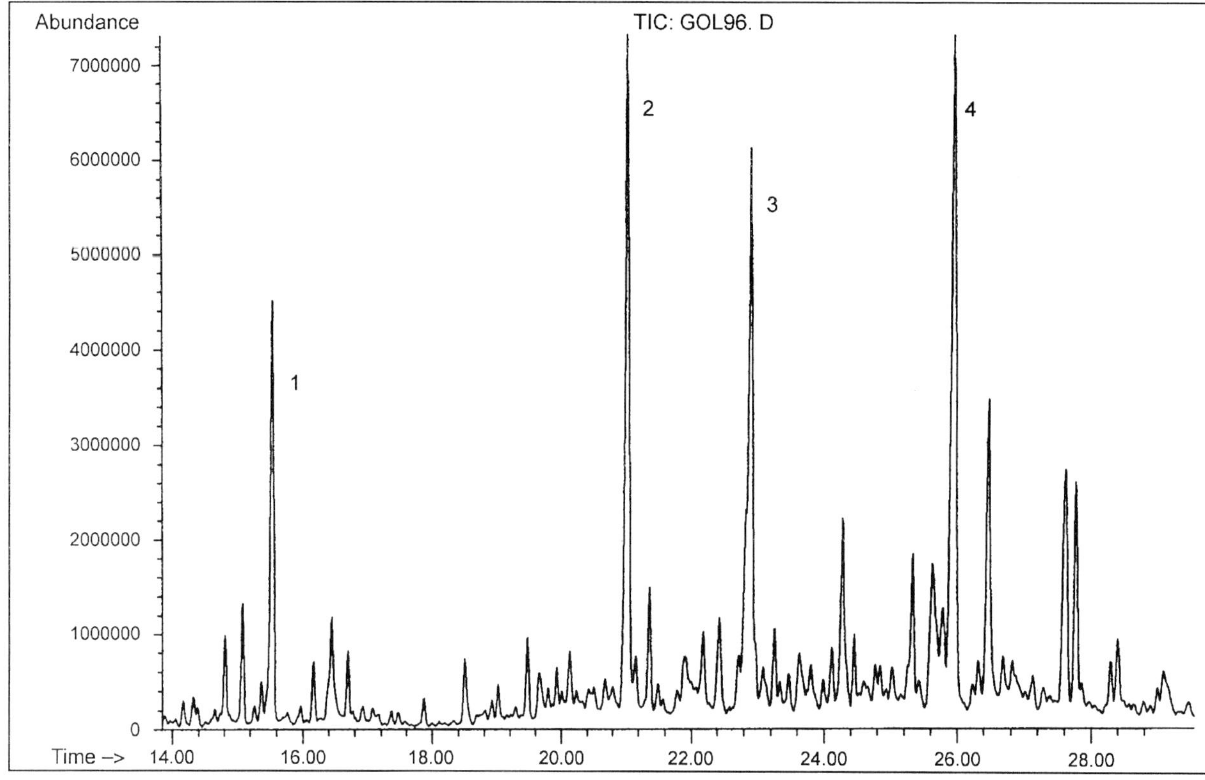

Fig. 3. Part of gas liquid chromatograms (better: total ion chromatograms "TIC") of a solution of köflachite (above) and of a soluble fraction of the resin from the Lower Cretaceous of Golling (Salzburg; below). PEAKS IDENTIFIED: köflachite — (1) norpimarane; (2) sandaracopimarane; (3) phyllocladane (formula: Fig 4. 3); (4) dehydroabietane (formula: Fig. 4. 4); (5) simonellite (formula: Fig. 4. 5); (6) retene (formula: Fig. 4.6). Soluble fraction of the resin from Golling — (1) see Fig. 7, formula 1; (2) see Fig. 7, formula 2; (3) see Fig. 7, formula 3; (4) see Fig. 7, formula 4 — tentative identifications by mass spectra only.

cense (from *Boswellia carteri* and related species) and myrrh (from *Commiphora*). Other members of the Burseraceae produce different sorts of gum, resin and balsam (CONQUIST 1981). The family itself originated not later than the Eocene; fruits from the English Palaeogene (London Clay) have been attributed to them (*Protocommiphora, Palaeobursera, Bursericarpum* — COLLINSON 1983) as well as some endocarps of a fossil *Canarium* species from Karlovy Vary (Czech Republic; GREGOR & GOTH 1979) and some macrofossils from the western USA (CONQUIST 1981).

The very first identifications of fossil resins as elemi resins based on of the detection of amyrin were made by FRONDEL on the basis of x-ray diffraction of crystalline amyrin and by means of comparisons of thin-layer chromatograms (FRONDEL 1967, 1969). Thin-layer chromatograms of fossil resins from Whetstone (Hertfordshire), Richmond (Surrey) and of Highgate copalite from the London Clay were found to be identical with the chromatogram of glessite from the Baltic area. On the basis of such comparisons, a genetic relationship was suggested between Highgate copalite, glessite and modern Burseraceae resins, including even the fossil guayaquilite (FRONDEL 1969). For glessite from Bitterfeld (Saxonia, Germany) the occurrence of ß- and α-amyrin has been established on the basis of studies by gas liquid chromatography/mass spectroscopy (KOSMOWSKA-CERANOWICZ *et al.* 1993); For this fossil resin it was confirmed that the relationship between ß- and α-amyrin (1:2.01) was rather similar to the relationship found in recent elemi resins (1:1.90).

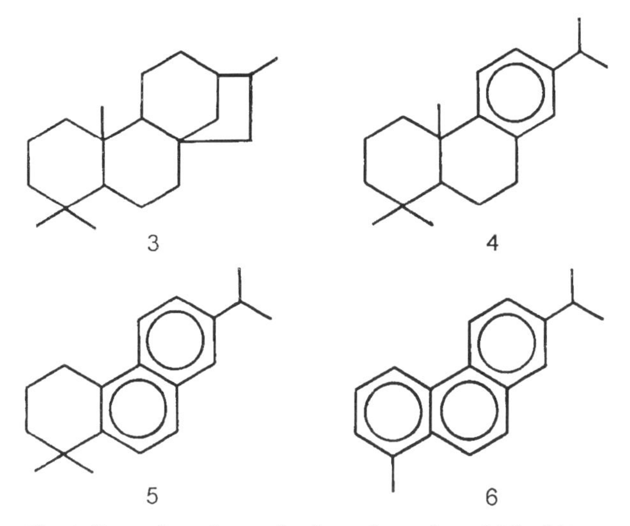

Fig. 4. Formulae of some hydrocarbons from köflachite (Miocene, Styria): (3) phyllocladane; (4) dehydroabietane; (5) simonellite; (6) retene.

One has to bear in mind that far more detailed studies of all these finds of fossil resins are still urgently needed; nevertheless, it seems to be justified to regard the aforementioned resins — including rosthornite from Carinthia — as a group of angiosperm resins with a remarkably close genetic relationship. As a final remark, it seems worth mentioning that HELM (1881) had already noticed a similarity between glessite and recent "Gummiharzen" like myrrh (*Commiphora*, Burseraceae) — this shows that he was already then very close to the present opinion.

Köflachite

Author: DOELTER (1878)

Type locality: Lankowitz, Hangendstollen, Gottesgabenschacht

Geological age: Miocene (Karpatian/Badenian)

Name: derived from Köflach, a small city close to the type locality

Occurrence: in the coal-mining area near Köflach only

Botanical origin: probably Taxodiaceae, possibly *Sciadopitys* and/or *Cryptomeria*

Chemistry: mixture of hydrocarbons (phyllocladane, eudesmane, sandaracopimarane, norpimarane, dehydroabietane, simonellite, retene etc.)

From a coal mine near Köflach (Styria) DOELTER (1878) reported the find of a fossil resin which he believed to belong to the retinites, although it also exhibited similarities to jaulingite. Later authors accepted the name suggested ("köflachite") and attributed it either to the retinites (e.g. HEY 1962) or emphasized that it shows similarities to jaulingite (HATLE 1885; MEIXNER 1950). PACLT (1953) mentioned köflachite together with many other fossil resins under the title "Other Caenozoic resins derived probably from coniferous woods" and suggested a species of *Taxodioxylon* as a possible botanical source for köflachite. Five years ago different hydrocarbons could be identified in köflachite (Figs. 3 & 4) by the use of GLC/MS (VÁVRA 1992). However, with regard to the material used for these studies there exists one problem: I have never had in my hands the authentic material described by DOELTER — I do not even know where it is kept. My material was from Köflach and had been kept in the "Joanneum", the local museum of Styria. Meanwhile (VÁVRA 1993) a total of seven different hydrocarbons have been identified by their mass spectra, making up 60–71% (area percentage!) of all the substances being volatile enough for gas liquid chromatography. If any compounds remain which are not volatile — and therefore not detectable by gas liquid chromatography — they will be studied in future by HPLC. One of the most remarkable things about this material, however, is the fact that it is completely soluble in common solvents — unusual for a fossil resin.

Another remarkable observation was the fact that the melting point determined does not match the mp. published by DOELTER (DOELTER: 98°C; mine: no melting until 230°C!!) — perhaps changes of some sort occurred (polymerisation??) during the storage of this mineral at the museum.

Among all the hydrocarbons studied, the most remarkable find was phyllocladane — this means that "hartite" (which has also been found in the same coal mine) is the main component of köflachite. Other substances identified were: eudesmane, sandaracopimarane, norpimarane, dehydroabietane, simonellite and retene (cf. Fig. 4).

Hartite

From a coal mine at Hart (also: "Oberhart") near Gloggnitz (S Vienna) HAIDINGER (1841) described an organic mineral by the name of "hartite" which later turned out to be phyllocladane — a hydrocarbon with a rather complex history of scientific study. Details concerning this (now abandoned) coal mine have been reported by SIGMUND (1937) and others. The age of the coal seams has been determined as Karpatian (STEININGER *et al.* 1988/89). ROLLE (1856) had already realized that hartite was a crystallized hydrocarbon, three sets of analytical data being published between 1856 and 1885 (HATLE). A detailed study of finds from Styria was published by SOLTYS (1929); other different organic minerals were finally recognized as junior synonyms: iosen, josen, bombiccite, hofmannite and branchite (e.g. HINTZE 1933). Not until 1937 (BRIGGS) was the exact chemical structure determined: α-dihydrophyllocladene (= phyllocladane). STREIBL & HEROUT (1969) emphasized that this was a hydrocarbon mineral which is rather common in numerous coal mines. The mass spectrum as published by PHILP (1985) was also obtained by our own studies, using material from Styria. This mineral has been known since 1855 in this federal country.

Chemofossils from köflachite

The interpretation of the hydrocarbons identified in köflachite as chemofossils ("fossil biomarkers") may give a few hints in respect to the botanical origin of this material: eudesmane can be regarded as a product of diagenesis of eudesmole (= Cryptomeradole), an alcohol occuring in *Cryptomeria* (Taxodiaceae) as well as in genera of Cupressaceae; dehydration followed by hydrogenation of the double bond yields eudesmane.

The next hydrocarbon, sandaracopimarane, can be formed either through hydrogenation of the corresponding unsaturated hydrocarbon or may also be formed from the corresponding resin acid. Its value as a chemical biomarker is, however, very low.

The value of the next hydrocarbon, dehydroabietane, as a chemofossil must also generally be regarded as very low. This is likewise true of retene and simonellite. Retene, especially, is a hydrocarbon which is very widespread in peats, lignites and brown coals. Realizing however that the main compound in this mixture is α-phyllocladane, the most probable source for the formation of dehydroabietane, simonellite, and retene is phyllocladane itself. A mechanism explaining the formation of these three hydrocarbons has been suggested by ALEXANDER *et al.* (1987).

There remains the question concerning the botanical origin of phyllocladane. A substance occurring in different recent plants is phyllocladene, the unsaturated version of this hydrocarbon skeleton. It occurs in *Phyllocladus* (name!) a genus of the Podocarpaceae (Coniferopsida) yielding an essential oil in which phyllocladene is reported as the main constituent. Phyllocladene also occurs in the genus *Dacrydium* (Podocarpaceae). An isomer hydrocarbon (kaurenem = podocarprene) occurs in some species of the genus *Podocarpus* itself. Among the Taxodiaceae, *Sciadopitys* yields an essential oil with 5% phyllocladene. Corresponding isomeric hydrocarbons (e.g. isophyllocladene) or a corresponding alcohol (phyllocladanole) — forming phyllocladene easily by dehydration — occur in some other genera of Coniferopsida (e.g. *Cryptomeria*).

Following all these informations to be found in literature (e.g. HEGNAUER 1962; OTT 1996 and references given there) Taxodiaceae as well as Podocarpaceae and even Araucariaceae may have produced phyllocladene, having then by diagenetic processes been transformed to phyllocladane. By reasons of biogeography of plants and on the basis of paleobotanical data, Podocarpaceae and Araucariaceae have to be excluded. The genus *Sciadopitys* (recent occurrence: southern Japan) has been reported as a common constituent of brown coal floras of the European Miocene (e.g. Rhine area,

Lausitz, Vistula area): its grass-like needles are sometimes so common that a special variety of coal has been called "grass-coal". On the other hand, *Cryptomeria* having also been confirmed, the present state of knowledge allows us only to suggest that some species of the genus *Sciadopitys* and/or *Cryptomeria* can be regarded as the botanical source of phyllocladane. Bearing in mind, however, that phyllocladanes have even been detected in coals from the Carboniferous period and in Permian sediments (NOBLE *et al.* 1985; SCHULZE & MICHAELIS 1990) — facts suggesting that ancestral conifers of the Palaeozoic were able to produce diterpene compounds of the phyllocladane/kauran type — makes the whole situation very complex indeed. For a discussion of this problem see also OTT (1996).

Hydrocarbons derived from terpenoids — some remarks

The temptation exists to make some generalization in this respect: terpene compounds obviously have two different possibilities with respect to their diagenetic pathways. Either the aerobic way — under the influence of oxygen — yielding after polymerisation and polycondensation fossil resins; or — the anaerobic way — transformation of terpenes into (mixtures of) hydrocarbons, migrating within coal seams and becoming enriched in some locations, forming minerals like hartite or mixtures of hydrocarbons like köflachite.

A similar material — duxite — from Bohemian brown coals has recently been re-described (VÁVRA *et al.* 1997). In this case, the presence of drimane, labdane, simonellite and retene, as well as a C_{16}-bicyclic sesquiterpane and a C_{18}-tricyclic diterpane were confirmed. A different botanical origin has to be suggested here, nevertheless, the pattern seems to be the same: anaerobic conditions during the diagenesis of a complex mixture of terpenes, yielding saturated, partly aromatized hydrocarbons. In any case, neither köflachite nor duxite can be regarded as fossil resins any longer; further studies (using HPLC) will be necessary to establish whether they contain any type of resin-like material at all.

Resin from the Lower Cretaceous of Salzburg

Remarks concerning finds of fossil resins from Salzburg can be found in relatively early publications (e.g. FUGGER 1878); however, the first (short) report on a fossil resin from the Lower Cretaceous of this region was not published until 1968 (STRASSER). These remarkable finds were made in an area to the south of the city of Salzburg, situated a few kilometres east of Golling, in the so-called "Weitenau". The first specimens, found in 1962, were exceptional for at least three reasons from the very beginning:

(1) their geological age: Lower Cretaceous (Neocomian, "Roßfeld-strata") — in the publication by STRASSER (1968) the finds had been erroneously described as Jurassic!; (2) the considerable size of some of the resin nodules: up to 18 cm in diameter; and (3) the remarkable total amount of resin collected in this area — approximately 500–800 kg were collected from the most important outcrop in the years 1979–1982 (SCHLEE 1985). The colour of the resin was described as dark brown, reddish-brown, greenish, violet and even (rarely!) amber-yellow/completely transparent. The darker varieties exhibit transparency only in the case of thin fragments. The hardness was given as approximately 2; the specific weight as 1.12–1.16 (STRASSER 1968). Further detailed descriptions have also been given by SCHLEE (1984; 1985).

Despite this resin being of the greatest scientific interest, it has until now undergone only casual research. With regard to its botanical origin ROTTLÄNDER & MISCHER (1970) suggested, on the basis of comparative chemical studies, that Araucariaceae produced this resin — a suggestion later confirmed by BANDEL & VÁVRA (1981). In his summarizing description SCHLEE (1985) also reported finds of a very few inclusions (Ceratopogonidae, Cecidomyiidae, and fragments of a representative of Hymenoptera). It is a pity that most of the material disappeared into many different private collections without ever having been studied for fossils at all: mindful of the geological age of this material, this would have been of general palaeontological interest.

As regards the material from the area of Golling, selected results of chemical research will be presented here. Gas liquid chromatography (see Fig. 3) of the soluble part of this Cretaceous resin has confirmed the presence of four main constituents. On the basis of the mass spectra, structural formulae (Fig. 7) can be tentatively suggested for at least three of these compounds. These identifications need further confirmation, however. At the moment these formulae have been established based on the following:

(1) Comparing one compound (Figs 5 & 7; MW: 188) with the authentic spectrum of a related compound (pentamethyl-2,3-dihydroindene, Wiley library of mass spectra, HEWLETT PACKARD) confirms the presence of the dihydroindene-ring system;

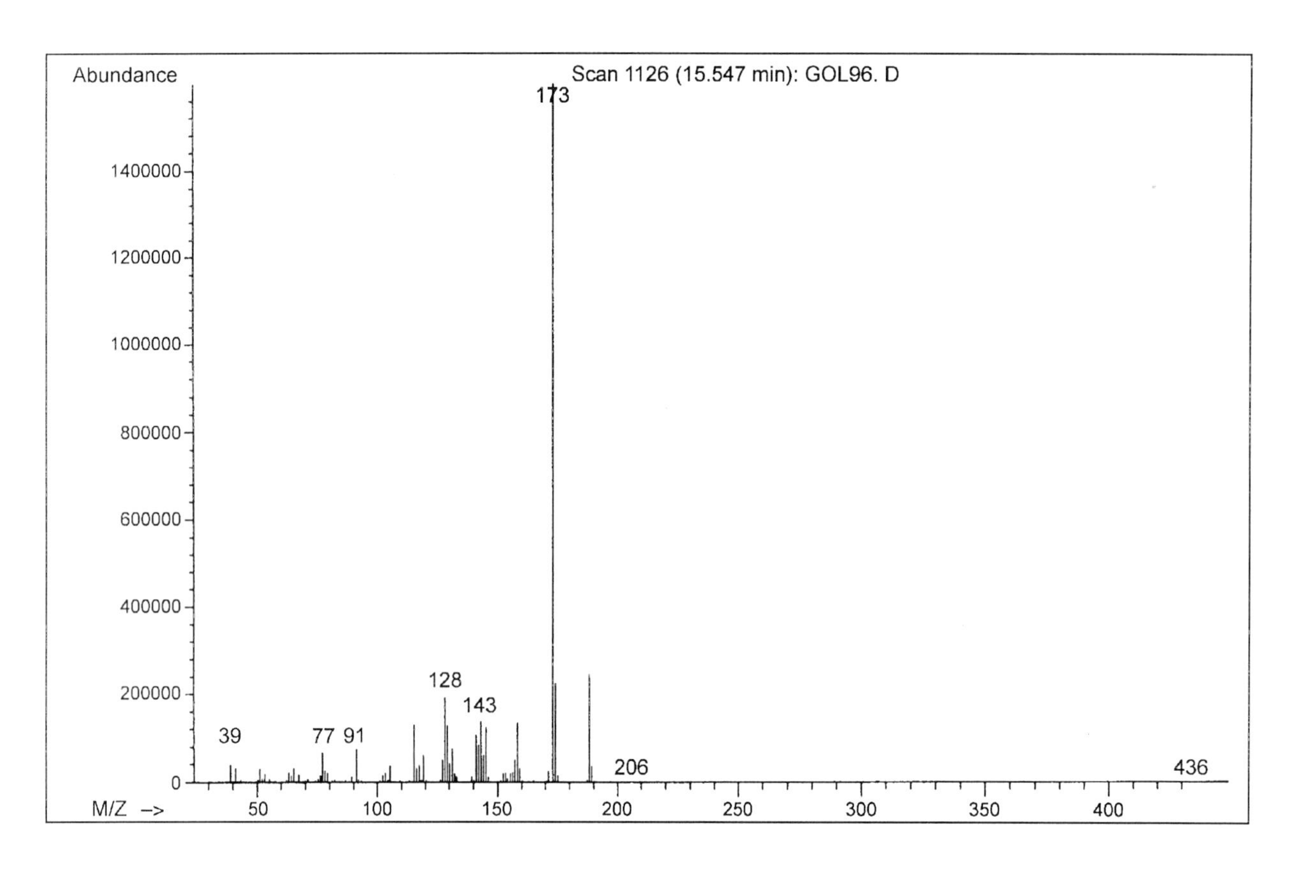

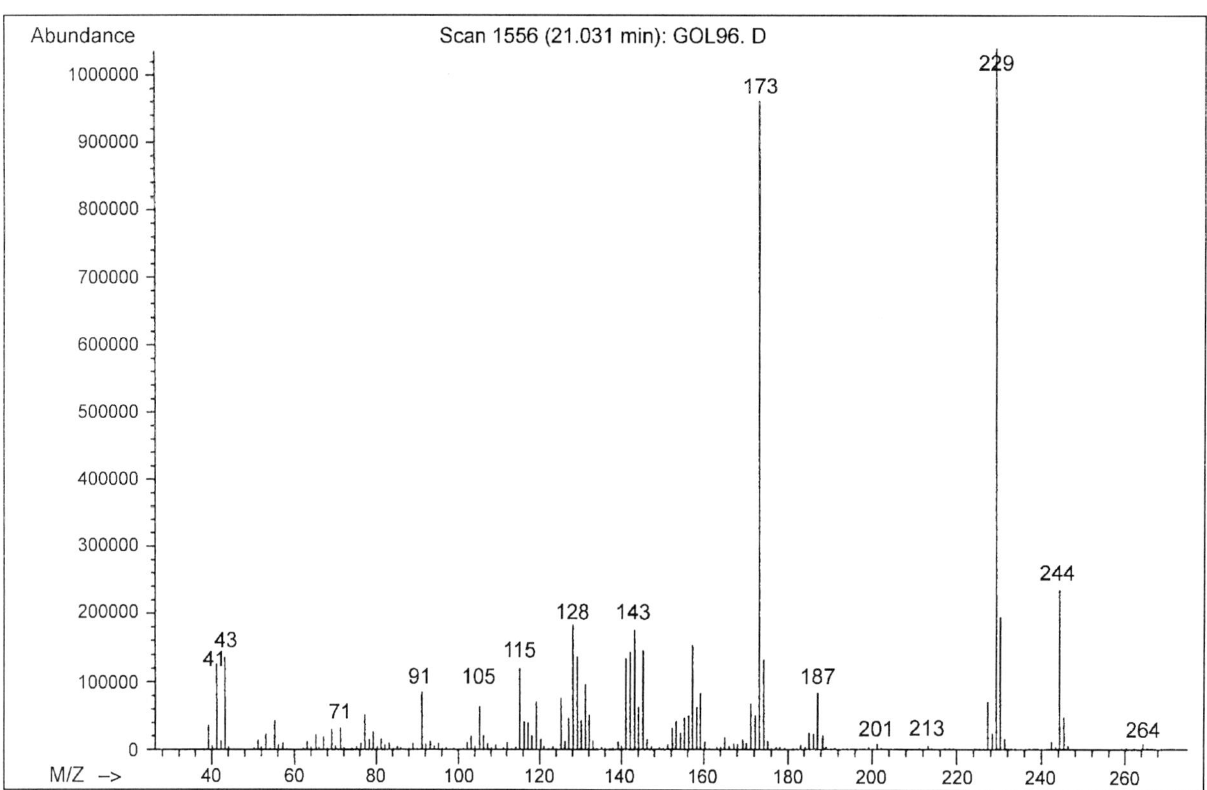

Fig. 5. Mass spectra of two main components of the resin from the Lower Cretaceous of Golling (Salzburg): two substituted dihydroindenes with a molecular mass of 188 (above, tentative identification: formula — Fig. 7.1) resp. mol. mass 244 (below, tentative identification: see Fig. 7.2).

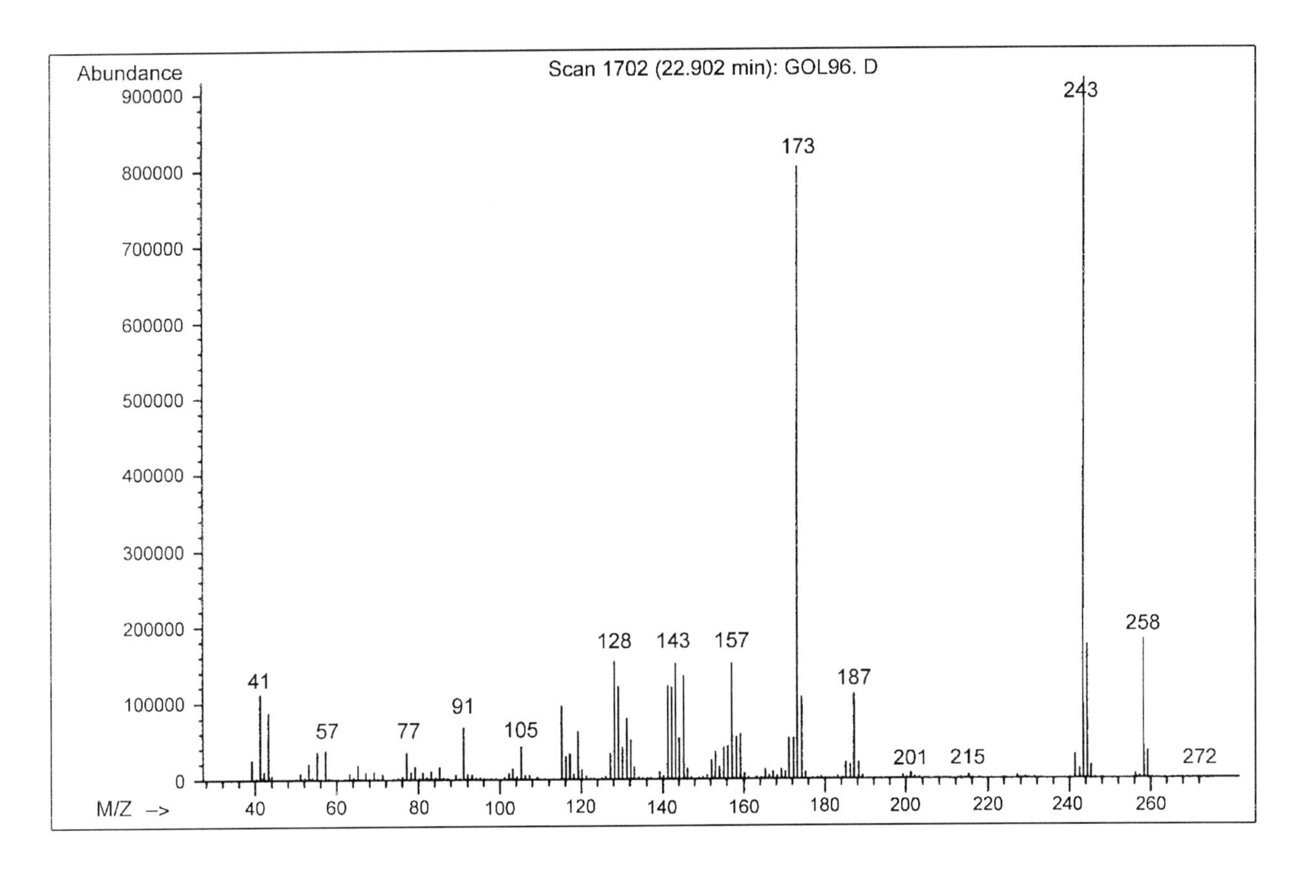

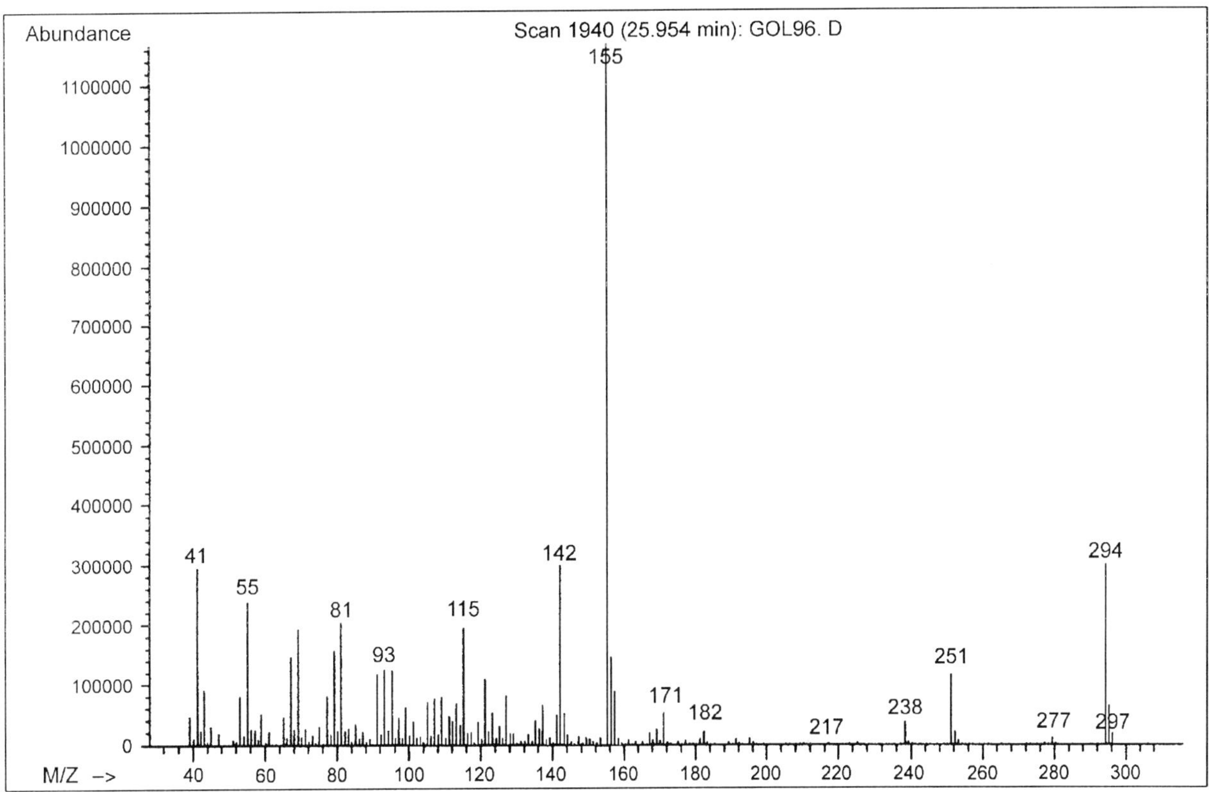

Fig. 6. Mass spectra of two main components of the resin from the Lower Cretaceous of Golling (Salzburg): a substituted dihydroindene (above; tentative identification: formula — Fig. 7.3) and one more hydrocarbon (mol. mass: 294), for which the formula given in Fig.7.4 seems possible (below).

(2) the relationship between retention times and molecular weight shows that the compounds must be closely related (in terms of their chemistry);

(3) the base peak in the mass spectrum of a substance with a molecular weight of 188 is 173; the same peak is also very prominent in two of the compounds under consideration (MW 244 resp. 258; Figs 5 & 6), suggesting a dihydroindene structure for them (Fig. 7: 2, 3). The last substance (MW 294) shows a strikingly different mass spectrum (Fig. 6). However, to suggest a detailed formula for this compound would be mere guesswork at the moment — it could possibly be represented by formula 4 (Fig. 7).

All four of the substances identified in a soluble fraction of the resin from the Lower Cretaceous of Salzburg can be regarded as degradation products of ß-sitosterol — a plant steroid detected in Baltic amber by SZYKUŁA *et al.* (1990). A lot of products of aromatization and other diagenetic changes of steroids have been described (e.g. MACKENZIE *et al.* 1982; LEEUW & BAAS 1986). However, examples for the loss of the A-ring or A- and B-ring of steroids during the diagenesis seem to have been virtually unknown until now, though such a possibility has been referred to (KILLOPS & KILLOPS 1993).

Experimental details

Samples were refluxed in solvent (usually methylene chloride) for one hour and used for gas liquid chromatography without any further separation.

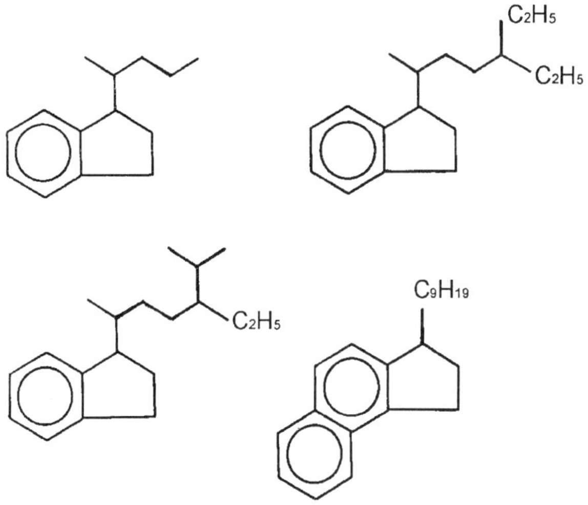

Fig. 7. Formulae for hydrocarbons tentatively identified in the soluble fraction (methylene chloride) of the resin from the Lower Cretaceous of Golling (Salzburg).

Gas liquid chromatography:

Column: HP-1 (methylsilicone rubber, "cross-linked"), 12 m x 0.2 mm x 0.33 μm film.

Carrier gas: helium (30 psi)

Injector: 250°C

Oven: Initial temperature: 60°C (resp. 100°C)

Initial time: 2 minutes

Increase of temperature: 6°C/min.

Final temperature: 250°C

Mass spectrometer:

Transfer Line: 280°C

Mass range: (mostly) 35–520 daltons.

Sampling: 2

Threshold: 400

Configuration of equipment (HEWLETT-PACKARD):

GC 5890, series II.

Automatic sampler: 7673

MSD: 5971 A

Computer: Vectra QS/20

Acknowledgements

The financial support of the "Hochschuljubiläumsstiftung der Stadt Wien" for research on chemofossils, as part of the project "Organische Geochemie fossiler Harze und anderer organischer Minerale unter Berücksichtigung entsprechender Inhaltsstoffe einheimischer Kohlen" (= Organic geochemistry of fossil resins and other organic minerals under consideration of corresponding substances occurring in Austrian coal-mines), is gratefully appreciated. The author also owes sincerest thanks to: Prof. Dr. F. EBNER (Leoben), Dr. W. POSTL (Steirisches Landesmuseum Joanneum, Graz), Dr. F. H. UCIK (Landesmuseum Kärnten, Klagenfurt), Mr. M. WANK (Wolfsberg, Kärnten), and Mr. L. ZIMA (Salzburg) for samples of different organic minerals and — last but not least — to Mrs. B. KOSMOWSKA-CERANOWICZ for translating the summary into Polish.

Norbert VÁVRA
Institut für Paläontologie
Geozentrum
Althanstraße 14
A-1090 Wien/Vienna, Austria

ŻYWICE KOPALNE Z AUSTRII: BIOMARKERY WYKRYTE W ROSTORNICIE (EOCEN, KARYNTIA), KEFLACHICIE (MIOCEN, STYRIA) I ŻYWICY Z DOLNEJ KREDY SALZBURGA

Norbert VÁVRA

Streszczenie

Niektóre żywice z obszaru Austrii (keflachit — miocen, Styria; rostornit — eocen, Karyntia i nienazwana jeszcze żywica z dolnej kredy Salzburga) zostały przebadane metodą chromatografii gazowo-cieczowej / spektroskopii masowej ze wsparciem komputerowym. Wyniki tych badań wskazują, że keflachit nie może być dłużej uważany za żywicę; reprezentuje on mieszaninę węglowodorów, których liczba może być identyfikowana na podstawie krzywych spektroskopii masowej: phyllocladanu, eudesmanu, sandaracopimarenu, norpimaranu, dehydroabietanu, simonelitu oraz retenu.

We frakcji rozpuszczalnej rostornitu zidentyfikowano dwie różne amyriny. W żywicy z dolnej kredy Salzburga próbnie oznaczono niższy jakościowo produkt b-sitosterolu. Na podstawie uzyskanych wyników przedyskutowano botaniczne pochodzenie keflachitu i rostornitu.

Bibliography

ALEXANDER G., HAZAI I., GRIMALT J. & ALBAIGES J.
1987 Occurrence and transformation of phyllocladanes in brown coals from Nograd Basin, Hungary, *Geochimica et Cosmochimica Acta*, **51**, 2065–2073.

BANDEL K. & VÁVRA N.
1981 Ein fossiles Harz aus der Unterkreide Jordaniens, *Neues Jahrbuch für Geologie und Paläontologie, Monatshefte*, 1981, 19–33.

BOITEAU F., PASICH B. & RATSIMAMANGA A. R.
1964 *Les Triterpénoides en physiologie végétale et animale*, 1–1370, Gauthier-Villars, Paris.

BRACKMAN W., SPAARGAREN K., VAN DONGEN J. P. C. M., COUPERUS F. A. & BAKKER F.
1984 Origin and structure of the fossil resin from an Indonesian Miocene coal, *Geochimica et Cosmochimica Acta*, **48**, 2483–2487.

BRIGGS L. H.
1937 The identity of dihydrophyllocladene with iosene, *Journal of the Chemical Socitey*, 1937, 1035–1036.

COLLINSON M. E.
1983 Fossil Plants of the London Clay, [*in*:] *Palaeontological Association Field Guides to Fossils*, **1**, 1–121, Palaeontological Association, London.

CONQUIST A.
1981 *An Integrated System of Classification of Flowering Plants*, Columbia University Press, New York.

DOELTER C.
1878 Ueber ein neues Harzvorkommen bei Köflach, *Mitteilungen des naturwissenschaftlichen Vereines für Steiermark*, **1877**, 93– 96.

FRONDEL J. W.
1967 X-Ray Diffraction Study of fossil Elemis, *Nature*, **215**, 1360–1361.
1969 Fossil Elemi Species Identified by Thin-Layer Chromatography, *Naturwissenschaften*, **5**, 280.

FUGGER E.
1878 *Die Mineralien des Herzogthumes Salzburg*, (= XI. *Jahres-Bericht der k.k. Ober-Realschule in Salzburg*, 1–124, published by the author, Salzburg.

GREGOR H. -J. & GOTH K.
1979 Erster Nachweis der Gattung *Canarium* STICKMAN 1759 (Burseraceae) im europäischen Alttertiär, *Stuttgarter Beiträge zur Naturkunde, Ser.B*, **47**, 1–15.

HAIDINGER W.
1841 Ueber den Hartit, eine neue Species aus der Ordnung der Erdharze, *Annalen der Physik und Chemie*, **54** (= 130), 261–265.

HATLE E.
1885 *Die Minerale des Herzogthums Steiermark*, 1–212, Leuschner & Lubensky, Graz.

HEGNAUER R.
1962 Chemotaxonomie der Pflanzen. Eine Übersicht über die Verbreitung und die systematische Bedeutung der Pflanzenstoffe, Band 1, 1–517, [*in*:] *Lehrbücher und Monographien aus dem Gebiete der exakten Wissenschaften, Chemische Reihe*, **14**, Basel and Stuttgart, Birkhäuser.

HELM O.
1881 Glessit, ein neues in Gemeinschaft von Bernstein vorkommendes Harz, *Schriften der naturforschenden Gesellschaft in Danzig*, **5** (1/2), 291–293.

HEY M. H.
1962 *An Index of Mineral Species & Varieties arranged chemically*, I–XXIV & 1–728, 2nd ed., British Museum, London.

HINTZE C.
1933 *Handbuch der Mineralogie*, 1. Band, 4. Abtlg. (2), Phosphate, Arseniate, Antimoniate, Vanadate, Niobate und Tantalate, 2. Teil Arsenite und Antimonite, Organische Verbindungen, 1–1446, 204 fig., W. de Gruyter, Berlin, Leipzig.

HÖFER H.
1871 Studien aus Kärnten, I. Rosthornit, ein neues fossiles Harz, *Neues Jahrbuch für Mineralogie, Geologie und Palaeontologie*, **1871**, 561–566.

KILLOPS S. D. & KILLOPS V. J.
1993 *An Introduction to Organic Geochemistry*, 1–265, Longman & J. Wiley, New York.

KOSMOWSKA-CERANOWICZ B., KRUMBIEGEL G. & VÁVRA N.
1993 Glessit, ein tertiäres Harz von Angiospermen der Familie Burseraceae, *Neues Jahrbuch für Geologie und Paläontologie*, **187** (3), 299–324.

LEEUW J. W. DE & BAAS M.
1986 Early-stage Diagenesis of Steroids, 101–123, [*in*:] Johns R. B., *Biological Markers in the Sedimentary Record (= Methods in Geochemistry and Geophysics*, **24**), 1–364, Elsevier, Amsterdam etc.

MACKENZIE A. S., BRASSELL S. C., EGLINTON G. & MAXWELL J. R.
1982 Chemical Fossils: The Geological Fate of Steroids, *Science*, **217**, 491–504.

MEIXNER H.
1950 Über "steirische" Mineralnamen, *Der Karinthin*, Folge **11**, 242 ff.

NOBLE R. A., ALEXANDER R., KAGI R. I. & KNOX J.
1985 Tetracyclic diterpenoid hydrocarbons in some Australian coals, sediments and crude oils, *Geochimica et Cosmochimica Acta*, **49**, 2141–2147.

OTT A.
1996 Paläobotanische und organisch-geochemische Untersuchungen an zwei oligozänen Tonprofilen aus dem Weißelster-Becken, Sachsen, [*in*:] *Berichte aus der Geowissenschaft*, 1–179, Shaker Verlag, Aachen.

PACLT J.
1953 A System of Caustolites, *Tschermaks Mineralogische und Petrographische Mitteilungen*, **3** (4), 332–247.

PHILP R. P.
1985 Fossil Fuel Biomarkers. Application and Spectra, [in:] *Methods in Geochemistry and Geophysics*, **23**, 1–294, Elsevier, Amsterdam, Oxford, New York, Tokyo.

REDLICH K. A.
1905 Die Geologie des Gurk- und Görtschitztales, *Jahrbuch der kaiserlich-königlichen Geologischen Reichsanstalt*, **55**, 327–348.

ROLLE F.
1856 Die tertiären und diluvialen Ablagerungen in der Gegend zwischen Graz, Köflach, Schwanberg und Ehrenhausen in Steiermark, *Jahrbuch der kaiserlich-königlichen Reichsanstalt*, **7**, 35– 602.

ROSTHORN F. V. & CANAVAL J. L.
1853 Beiträge zur Mineralogie und Geognosie von Kärnten, *Jahrbuch des naturhistorischen Landesmuseums von Kärnten*, **2**, 113–176.

ROTTLÄNDER R. & MISCHER G.
1970 Fossil-Lagerstätten, Nr.11: Chemische Untersuchungen an libanesischem Unterkreide-Bernstein, *Neues Jahrbuch für Geologie und Paläontologie, Monatshefte*, **1970**, 668–673.

SCHLEE D.
1984 Notizen über einige Bernsteine und Kopale aus aller Welt, 29–37, [*in*:] Bernstein-Neuigkeiten, *Stuttgarter Beiträge zur Naturkunde, Serie C*, **18**, 1–100.

1985 Der Österreichische Bernstein von Golling, *Goldschmiede Zeitung*, **8/85**, 70–73.

SCHULZE T. & MICHAELIS W.
1990 Structure and origin of terpenoid hydrocarbons in some German coals, *Organic Geochemistry*, **16**, 1051–1058.

SIGMUND A.
1937 *Die Minerale Niederösterreichs*, 1–247, Deuticke, Wien, Leipzig.

SOLTYS A.
1929 Über das Josen, einen neuen Kohlenwasserstoff aus steirischen Braunkohlen, *Monatshefte für Chemie*, **53–54**, 175–184.

STEININGER F. F., RÖGL F., HOCHULI P. & MÜLLER C.
1988/89 Lignite deposition and marine cycles. The Austrian Tertiary lignite deposits — A case history, *Sitzungsberichte der Österreichischen Akademie der Wissenschaften, Mathematisch-naturwissenschaftliche Klasse, Abt.I*, **197** (5–10), 309–332.

STRASSER A.
1968 Über den Neufund eines fossilen Harzes in der Weitenau bei Golling/Salzburg, *Aufschluß* **19** (1), 17.

STREIBL M. & HEROUT V.
1969 Terpenoids — Especially Oxygenated Mono-, Sesqui-, Di-, and Triterpenes, 401–424, [*in*:] Eglinton G. & Murphy M. T. J. [eds.] *Organic Geochemistry. Methods and Results*, 1–828, Springer, Berlin, Heidelberg, New York.

SZYKUŁA J., HEBDA C., ORPISZEWSKI J., AICHHOLZ R. & SZYNKIEWICZ A.
1990 Studies on Neutral Fraction of Baltic Amber, *Prace Muzeum Ziemi*, **41**, 15–20.

VÁVRA N.
1992 Analyse einiger organischer Mineralien der Steiermark mittels kombinierter Kapillargaschromatographie/Massenspektrometrie, *Mitteilungen der Österreichischen Mineralogischen Gesellschaft*, **137** (Jgg.1991), 216–218.

1993 Organische Mineralien aus der Steiermark, I. (Hartit, Köflachit, Retinit, Trinkerit), *Matrixx Mineralogische Nachrichten aus Österreich*, **2**, 24–38.

VÁVRA N., BOUŠKA V. & DVOŘAK Z.
1997 Duxite and its geochemical biomarkers ("chemofossils") from Bilina open-cast mine in the North Bohemian Basin (Miocene, Czech Republic), *Neues Jahrbuch für Geologie und Paläontologie, Monatshefte*, **1997** (4), 223–243.

WINDHOLZ M., BUDAVARIS S., STROUMTSOS L. Y. & FERIG M. N.
1976 *The Merck Index. An encyclopedia of chemicals and drugs*, 1–1313, Merck & Co., Rathway, N.J., USA.

BECKERIT AUS DEM TAGEBAU GOITSCHE BEI BITTERFELD (SACHSEN-ANHALT, DEUTSCHLAND)

Günter KRUMBIEGEL

Kurzfassung

Beckerit, auch als „Braunharz" bezeichnet, ist ein braunes an ein hartes Stück inkohltes Holz erinnerndes fossiles Harz.

In der Bernsteinlagerstätte bei Bitterfeld (Bitterfelder oder Sächsischer Bernstein) findet er sich seit 1990 in den „Bitterfelder Schichten" (Bitterfelder Glimmersand) im Liegenden des Bitterfelder Hauptflözes (Braunkohle) im Tagebau Goitsche. Die kleine Beckerit-Sammlung besteht aus 7 Stücken von ca. 3–7 cm Durchmesser.

Sämtliche Beckerit-Proben aus dem Tagebau Goitsche wurden lithologisch beschrieben und nach der Methode der Absorptions-Infrarotspektroskopie (IRS) untersucht. Die Diagramme wurden mit eindeutig determinierten Beckerit-Proben aus dem Samland verglichen und als nahezu identisch erkannt. Der baltische Beckerit dient als Grundlage für die Bestimmung des sächsischen Beckerits.

Historisches

Beckerit, auch als „Braunharz" bezeichnet, ist ein braunes, an ein hartes Stück inkohltes Holz erinnerndes fossiles Harz. Es wurde bereits 1867 von KÜNOW, dem Konservator des Zoologischen Museums der Albertus-Universität Königsberg in der Blauen Erde bei Großkuhren (Primorje) und Kleinkuhren (Filino) entdeckt. Möglicherweise gibt es aber aus „Gräbereien" und unter dem von der See ausgeworfenen Bernstein noch ältere Funde von Braunharz (vgl. CASPARY 1881).

Diese noch älteren Schwarz- und Braunharzfunde sind nach CASPARY (1881) unter der Bezeichnung „Schwarzer Braunstein" zu erwarten. In den Privatsammlungen in Königsberg sind aus der Zeit von 1551–1835 eine Reihe derartiger Funde bekannt gewesen (CASPARY 1881). 1871 wurde Beckerit auch in Warnicken (Lesnoje) und 1872 in Palmnicken (Jantarnyj) gefunden.

Im Braunharz beobachtete KÜNOW Pflanzenreste und -abdrücke, die von CASPARY (1881) und CASPARY & KLEBS (1906) beschrieben wurden, aber zu keiner rezentbotanischen Zuordnung des Beckerits führten. Es fanden sich Reste von:

- *Sequoia sternbergii* GOEPPERT, 1836 (Fam. Sequoiaceae),
- *Carpolithus paradoxus* CASPARY, 1881,
- *Zamites sambiensis* CASPARY, 1881 (Fam. Cycadaceae),
- *Proteacites pinnatipartitus* CASPARY, 1881 (Fam. Proteaceae),
- *Alethopteris* CASPARY, 1881 (Filices),
- *Phyllites lancilobus* CASPARY (*inc. sedis*) Cycadaceae?, Proteaceae?

Die Beschreibung der physikalischen Eigenschaften von Beckerit und Stantienit (Schwarzharz) erfolgte dann 1880 durch PIESZCZEK, ebenso die Namensgebung des Braunharzes als Beckerit, zu Ehren von Moritz BECKER, dem zweiten Mitinhaber der Bernsteinfirma STANTIEN und BECKER in Königsberg. CASPARY (1881, 31) zweifelte jedoch diese letzteren Benennungen aus Gründen der Priorität der Entdeckung an und schlug vor, wenigstens das Schwarzharz als „Künowit" zu bezeichnen. Es kam aber nicht zu einer Umbenennung des Beckerits bzw. Stantienits, denn in der späteren Literatur finden sich keine diesbezüglichen Hinweise.

Erst SCHUBERT (1961, 109–110) kommt im Zusammenhang mit Untersuchungen über den Bau und das Leben der Bernsteinkiefern zu dem Ergebnis, daß es sich bei Beckerit um einen kontaminierten Succinit handelt. „Es ist der durch den Harzfluß aus einem Fraßgang von Insekten oder ihrer Larven ausgespülte und dann im Harz feinst verteilte Kot."

Die physikalisch-chemischen Unterschiede beruhen auf mehr oder weniger kontaminiertem „Fraas" (vom deutschen „fressen"), einer pulverigen Mischung von zersetztem Holz und Insektenexkrementen (Material von Bohrhöhlenresten). SCHUBERT (1961) faßt daher seine Untersuchungsergebnisse über den

Beckerit wie folgt zusammen: „Damit ist der Beckerit als ein durch Schwemmgut der Bohrgänge verunreinigtes, sonst aber normales Bernsteinharz erkannt. Keinesfalls stammt es von einem spezifischen Harzbaum, und es entfällt die Berechtigung, den Beckerit in Hinsicht auf seine Herkunft weiterhin als ein besonderes Harz anzusprechen" (109–110).

Zum gleichen Ergebnis kommen BECK *et al.* (1986), die im Beckerit kein fossiles Harz sehen, sondern lediglich einen durch zerriebenes Holz und Insektenfraßreste verunreinigten Succinit. Sie beweisen dies mit Hilfe der Infrarotspektrographie und der Carbon-13-Magnet-Resonanz-Spektroskopie (Tab.1).

Eigenschaften	Beckerit	Succinit
Härte	1,126[a]	1,050–1,096[b]
Löslichkeit in Ether	1,2%[a]	16–23%[b]
	4,8%[d]	5,04%[d]
Löslichkeit in Chloroform	1,0%[a]	20,6%[b]
Löslichkeit in Azeton	2,32%[d]	8,42%[d]
Löslichkeit in Essigsäure	7,65%[d]	18,72%[d]
Löslichk. in ethanolhaltiger KOH	unlösl.[a]	
	3,45%[d]	35,05%[d]
Schwefel	vorhanden[a]	0,4%[c]
Asche	5,5%[a]	keine[c]
Kohlenstoff	67,86%[a]	77,33–78,96[b]
Wasserstoff	8,56%[a]	9,01–10,51%[b]
Bernsteinsäure	keine[a]	3–8%[b]
	0,0005%[d]	
Säurezahl	19,6[d]	42,06[d]
		15–35[c]
Verseifungszahl	81,2[d]	137,2[d]
		87–125[c]

Werte sind entnommen aus: a — PIESZCZEK (1880); b — SCHMID (1931); c — KAUFMANN (1948); d — KLEBS (1889; 1897).

Tabelle 1. Eigenschaften von Beckerit und Succinit (nach BECK *et al.* 1986, 411, Tab. 1).

Auf den pflanzlichen Ursprung des Beckerits gibt es bisher in der Literatur wenige, unsichere Hinweise.

ANDRÉE (1951) stellt fest, daß die Herkunftspflanze von Beckerit eine andere ist, als die von Succinit. PACLT (1953) nennt *Pinites succinifera*, d.h. den gleichen Ursprung wie Succinit. Eine andere Meinung vertritt HEY (1962); er stellt Beckerit zu den Retiniten.

RICE (1987) nimmt dagegen an, daß die Herkunftspflanzen des Beckerit-Harzes Hülsenfrüchtler (Fam. Leguminosae) sind. Dies ist aber wenig wahrscheinlich.

Auch unter den fossilen Harzen aus dem Tagebau Goitsche im Bitterfelder Braunkohlenrevier (Sachsen-Anhalt, Deutschland) wiesen FUHRMANN & BORSDORF (1986, 310, Tab. 1 & 2, 313–314, Abb. 7) das Vorkommen von vermutlichem Beckerit nach. Sie interpretieren aber fälschlicherweise den Siegburgit als Beckerit. Er findet sich dort selten und nur in kleinen Stücken. Von KOSMOWSKA-CERANOWICZ & KRUMBIEGEL (1989, 21), die damals weder Beckerit noch Stantienit gefunden hatten, wurde jedoch die Bezeichnung Beckerit und das Vorkommen in Bitterfeld angezweifelt.

KOSMOWSKA-CERANOWICZ (1996) teilte jedoch mit, daß „letztens" in der Bitterfelder Bernsteinlagerstätte „auch Beckerit bestimmt wurde". Dieses Ergebnis wurde auf Grund lithologischer Vergleiche durch KRUMBIEGEL (1994/1995) und durch eine IRS-Analyse an vermutetem Beckerit-Material aus der Sammlung KRUMBIEGEL, Halle/S. in Warschau durch KOSMOWSKA-CERANOWICZ (1996) festgestellt. WEITSCHAT (1996, 72) weist dagegen nochmals darauf hin, „daß Beckerit und Stantienit ... bisher nicht in Bitterfeld nachgewiesen" wurden.

Auf diese neuen Beckerit-Funde soll im Folgenden eingegangen werden.

Die Schichtenfolge des Tertiärs im Tagebau Goitsche

Das bernsteinführende Tertiär (Oberoligozän bis Untermiozän) von Bitterfeld und Bad Schmiedeberg (Sachsen-Anhalt) liegt im Bereich der Halle-Wittenberger Scholle und der Flechtingen-Roßlauer Scholle zwischen Hallescher Störung und dem Wittenberger Abbruch (KRUMBIEGEL & KOSMOWSKA-CERANOWICZ 1989; KRUMBIEGEL 1995; 1996).

Das geologische Alter der Bernstein-Lagerstätte und den Bernstein betreffend stehen sich derzeit zwei Thesen gegenüber.

These 1

Der Bitterfelder Bernstein wurde auf einer parautochthonen Lagerstätte gebildet. Hinsichtlich auf Genese und Alter des Vorkommens sowie auf das Inventar der begleitenden Harze ist er eine eigenständige Ablagerung. Der Bitterfelder Bernstein wurde im Untermiozän abgelagert, und die bernsteinführenden Schichten haben ein absolutes Alter von 22 Mio Jahren. So nimmt RÖSCHMANN (1996; 1997) zwar an, daß Baltischer und Bitterfelder Bernstein mehr oder weniger gleichzeitig entstanden sind. Der ökofaunistische Vergleich der Sciaridae und Ceratopogonidae (vgl.

auch SZADZIEWSKI 1993) weist aber auf zwei geographisch getrennte bernsteinproduzierende Waldgebiete hin: eines im Norden und eines im Süden der epikontinentalen europäischen Nord-Ostsee im Tertiär (BARTHEL & HETZER 1982; KOSMOWSKA-CERANOWICZ & KRUMBIEGEL 1989; SCHUMANN & WENDT 1989; KRUMBIEGEL 1996; KRUMBIEGEL & KRUMBIEGEL 1996; RÖSCHMANN 1996; 1997).

These 2

Der Bitterfelder Bernstein ist ein mehrfach umgelagerter Baltischer Bernstein, „zumindest bereits auf dritter Lagerstätte" (WEITSCHAT 1996) und aufgrund der paläoentomologischen Identität der fossilen Fauna des Bitterfelder Bernsteins mit der des Baltischen Bernsteins (WUNDERLICH 1983; WEITSCHAT 1987; KRZEMIŃSKI 1994 [*in*: SZADZIEWSKI *et al.* 1994]; WICHARD & WEITSCHAT 1996) und aufgrund neuer paläoklimatischer und paläogeographischer Fakten wird eine Eigenständigkeit abgelehnt (WEITSCHAT 1996). Die fossilen Harze in den bernsteinführenden Schichtenfolgen der Lagerstätten werden als mehrfach allochthon angesprochen.

Umfangreiche und detaillierte stratigraphische und sedimentpetrographische Untersuchungen durch das Geologische Landesamt Sachsen-Anhalt im größeren regionalgeologischen Rahmen werden jedoch erst in der Perspektive zur Klärung dieses genetischen Problems beitragen. Zahlreiche Beobachtungen sprechen aber schon heute für die Gültigkeit der These 1 (vgl. Abb. 1)

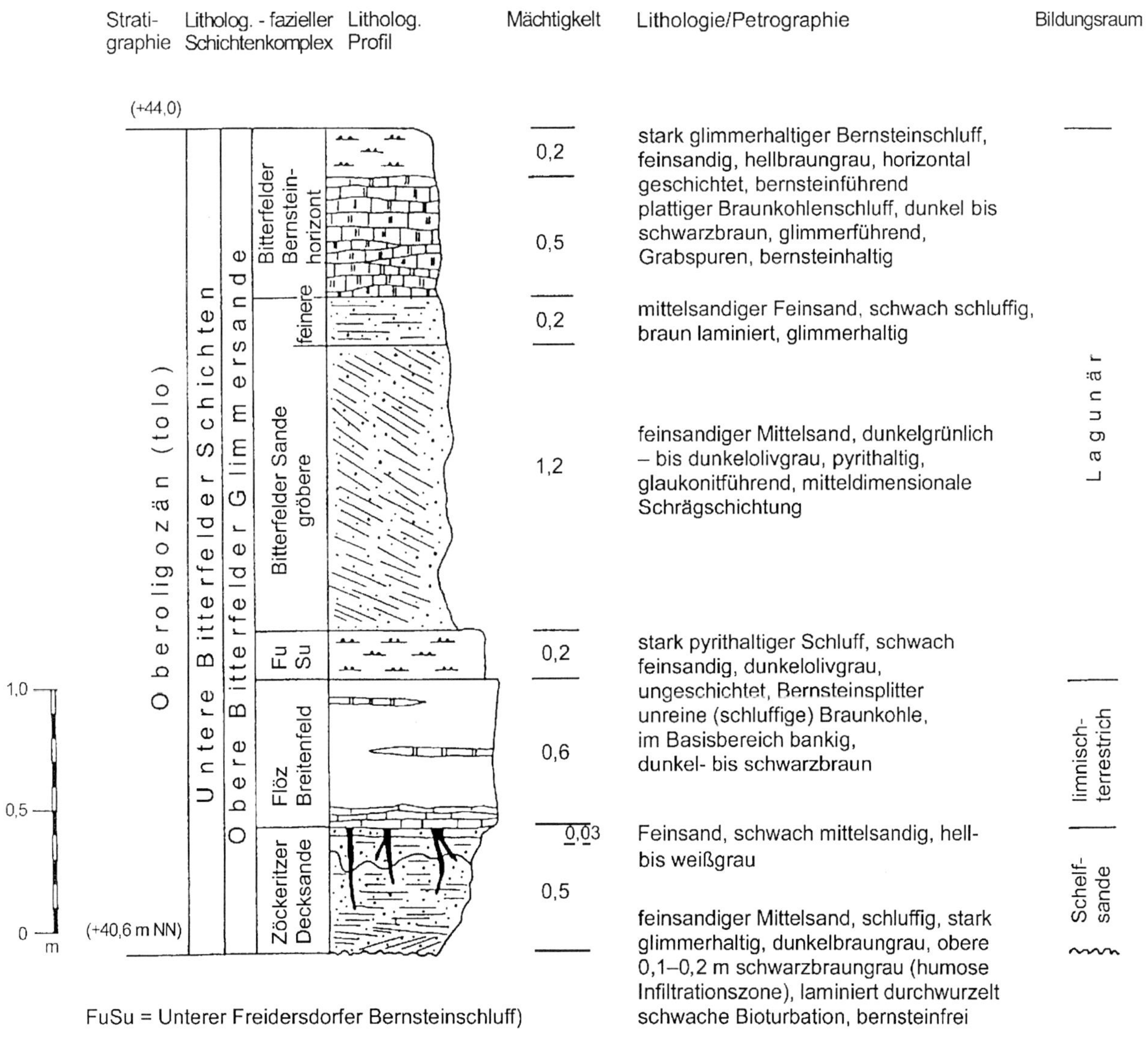

Abb. 1. Aufschlußprofil 01/95 Tagebau Goitsche [Baufeld Niemegk]; nach R. WIMMER, unveröffentliches.

Das Fundmaterial und die makroskopischen Eigenschaften des fossilen Harzes (Beckerit) (Neuaufsammlungen)

Das vorliegende Fundmaterial stammt mit Sicherheit aus den sog. „Bitterfelder Schichten" (Bitterfelder Glimmersande) im Liegenden des Bitterfelder Hauptflözes und im Hangenden des Flözes Breitenfeld. Neuere feinstratigraphische Untersuchungen an Bohrungsmaterial deuten aber daraufhin, daß die Braunkohle unter dem Bitterfelder Hauptflöz (BiO I u. II) im Tagebau Goitsche-Niemegk (Profil am Bärenholzrücken) zur Bitterfelder Unterbank (BiU) gehört und damit eine gesicherte Zeiteinordnung (s.o.) angenommen werden kann. Die bernsteinführenden Horizonte sind etwa 4–6 m mächtige, schwarzgraue, sandig-schluffige Lagen und Linsen mit Beimengungen von reichlich Muskovit. Eine genaue feinstratigraphische Entnahme der Beckerit-Proben erfolgte nicht. Die fossilen Harze wurden vielmehr aus den Harzrückständen der Naßaufbereitung nach der Trocknung bei der Handsortierung verlesen (Abb. 2–5).

Das neuaufgesammelte Fundmaterial wurde in die Sammlung KRUMBIEGEL, Halle/S. eingeordnet. Es erhielt die Inventarnummern KRU 16, 40, 41, 42, 44 und 45. Größe und Gewicht der einzelnen Stücke sowie die Nummern der IRS-Spektrogramme sind Tabelle 2 zu entnehmen.

Abb. 2. Beckerit-Funde aus den Glimmersanden der Bitterfelder Schichten im Tagebau Goitsche bei Bitterfeld. Durchmesser der Stücke etwa 3–7 cm. Numerierung der Stücke: Inventarnummern Sammlung KRUMBIEGEL, Halle/Saale, Deutschland. 2 Stücke rechts: Beckerit aus der „Blauen Erde" des Samlandes (Sammlung Otto HELM, ehem. Westpreußisches Provinzialmuseum, Danzig). (Inventar-Nr. 46 ist ein rot- bis dunkelbrauner Glessit als Vergleichsprobe). Fot. G. KRUMBIEGEL.

Inventar-Nr.	Größe in cm Breite, Höhe, Dicke	Gewicht in g	Nr. des IRS (vgl. Abb. 6)
KRU 16	6,5 x 6,0 x 1,5	29	IRS/MZ/245
KRU 40	4,2 x 3,0 x 1,5	8	IRS/MZ/460
KRU 41	4,0 x 2,5 x 1,0	6	IRS/MZ/445
KRU 42	4,4 x 3,9 x 1,0	9	IRS/MZ/461
KRU 44	3,0 x 2,8 x 0,8	3	IRS/MZ/440
KRU 45	3,5 x 2,5 x 0,5	3	IRS/MZ/444

Abkürzungen: KRU — Sammlung KRUMBIEGEL, Halle/Saale, Deutschland; IRS — Infrarotspektrum; MZ — Muzeum Ziemi, Polska Akademia Nauk, Warszawa, Polen.

Tabelle 2. Neuaufsammlungen von Beckerit aus dem Tagebau Goitsche bei Bitterfeld (Sachsen-Anhalt, Deutschland) seit 1990.

Die Harzstücke haben folgende makroskopische Eigenschaften:

Das Harz hat eine dunkel- bis mittelbraune Oberflächenfarbe, einige Stücke (KRU 16) weisen weiße Flecken auf. Die Bruchflächen sind hellbraun, erdigbraun und matt. Die Form ist knollenförmig (vgl. JENTZSCH 1892) bis tropfenförmig, z.T. aber auch plattig. Die Oberfläche ist unregelmäßig und gerunzelt und weist vermutliche Trockenrisse bzw. Schrumpfungsrisse auf. Die kleineren Stücke (KRU 42, 44, 45) haben Längsvertiefungen und sind löchrig; vermutlich handelt es sich um Rindenabdrücke. Das Material ist hart und spröde (Abb. 4, 5).

Abb. 3. Wie Abb. 2 — Rückseite der Beckerit-Proben. Fot. G. KRUMBIEGEL.

Abb. 4. Die größten Beckerit-Funde aus den Glimmersanden der Bitterfelder Schichten im Tagebau Goitsche bei Bitterfeld Durchmesser der Stücke etwa 4–7 cm. 2 Stücke rechts: Beckerit-Proben aus der „Blauen Erde" des Samlandes (Sammlung Otto HELM, ehem. Westpreußisches Provinzialmuseum, Danzig). Fot. G. KRUMBIEGEL.

Abb. 5. Beckerit mit Schrumpfungsrissen auf der Oberfläche (größter Neufund; Durchmesser 7 cm). Fundort Tagebau Goitsche. Fot. G. KRUMBIEGEL.

Die äußeren Merkmale sind denen bei PIESZCZEK (1880, 434) beschriebenen Eigenschaften des baltischen Beckerits sehr ähnlich (Tab. 3). Irgendwelche Pflanzenabdrücke, wie sie CASPARY (1881) am Braunharz beobachtete, waren nicht feststellbar.

Elementaranalytische und andere chemische Untersuchungen konnten in neuerer Zeit am Bitterfelder Material noch nicht durchgeführt werden. Eine Elemantaranalyse nach PLONAIT (1935 [*in*:] SANDERMANN 1960) zeigt folgende Zusammensetzung: C — 63%, Asche — 5%, H — 8%, Schwefel vorhanden.

Eigenschaften	Erstbeschreibung PIESZCZEK (1880)	Neufunde aus dem Tagebau Goitsche bei Bitterfeld KRUMBIEGEL (1990)
Größe	erbsen- bis hühnereigroß	3–7 cm
Farbe dicke Scheibe (10 mm) Pulver	hell, graubraun, matt schwer pulverisierbar; graubraun	dunkel- bis mittelbraun; erdigbraun; z.T. weißfleckig mittelbraun
Durchsichtigkeit		undurchsichtig
Bruch	erdig; selten muschelig, nicht glänzend, graubraun	trüb; hellbraun, matt
Sprödigkeit	zäh	hart; spröde
Struktur	knollen- und tropfenförmig; lamellenartige Stücke; Längserhabenheiten und -vertiefungen	unregelmäßige Oberfläche, gerunzelt; Trocken- und Schrumpfungsrisse; Löcher und Längsvertiefungen
elektrische Aufladung		schwach
Verwitterungsrinde		sehr dünn, meist keine
Polierbarkeit		schlecht
Elementaranalyse	C: 67,86%; H: 8,56%; O: 23,58%; S: vorhanden; Asche 5,7%	
Löslichkeit	in Ether: 1,2%; in Chloroform: 1,0%; jeweils klarer, firnisartiger, aromatisch	
organische Reste	pflanzliche Abdrücke und Reste; vereinzelt Frucht von *Carpolithus paradoxus*	keine pflanzlichen Reste

Tabelle 3. Eigenschaften des Beckerits nach PIESZCZEK (1880) (Erstbeschreibung) mit den makroskopischen Befunden an den Beckerit-Neufunden (seit 1990) aus dem Tagebau Goitsche bei Bitterfeld.

Untersuchungen des Beckerits mit Hilfe der Absorptions-Infrarot-Spektroskopie (IRS) (Neuaufsammlungen)

Die Methode der IRS ist heute ein häufig angewandtes Verfahren bei der Untersuchung fossiler Harze. Das Problem der schlechten Löslichkeit fossiler Harze in allen bekannten Lösungsmitteln wird umgangen, indem bei der Vorbereitung der Proben für die spektroskopischen Untersuchungen *ca* 2 mg Harzsubstanz mit Kaliumbromat zu Preßlingen geformt und untersucht werden kann (HECK 1997).

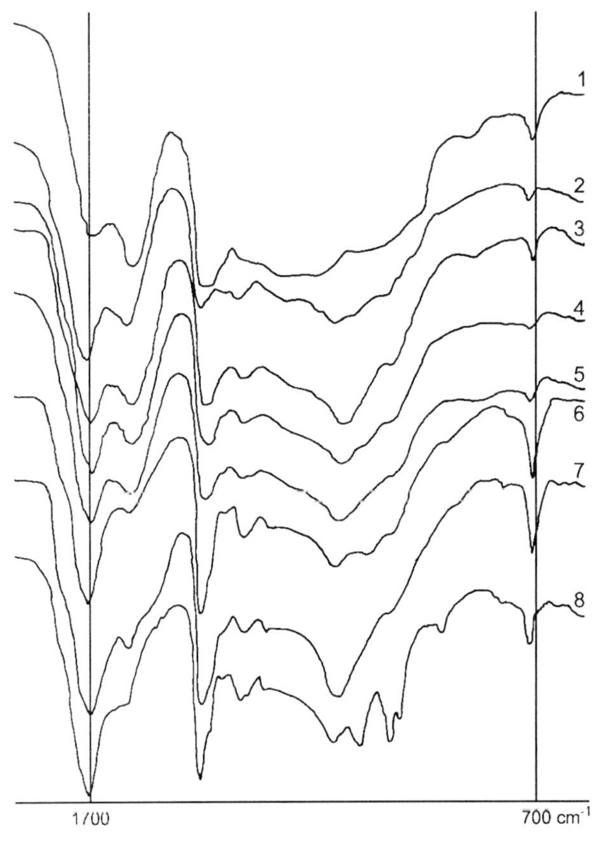

Abb. 6. Infrarotspektrogramme von Stücken des Beckerits aus dem Tagebau Goitsche bei Bitterfeld (KRU-Proben) im Vergleich mit dem Beckerit aus der „Blauen Erde" des Samlandes). 1 — IRS 436; Inventar-Nr. 5 WIENHAUS/HELM-Nachlaß (Archäologisches Museum Danzig), 2 — IRS 461; Inventar-Nr. KRU 42, 3 — IRS 190; Beckerit aus der Königsberg-Sammlung des Geol.-Paläont. Inst. u. Museum der Univ. Göttingen, 4 — IRS 245; Inventar-Nr. KRU 16 (untypische Kurve), 5 — IRS 460; Inventar-Nr. KRU 40, 6 — IRS 445; Inventar-Nr. KRU 41, 7 — IRS 444; Inventar-Nr. KRU 45, 8 — IRS 440; Inventar-Nr. KRU 44.

Die Interpretation der Absorptionsbanden der IR-Kurven ermöglicht eine Identifikation der fossilen Harze aber auch eine Unterscheidung bzw. Korrelation der Harze aus den unterschiedlichen Regionen. Die Anwendung anderer Methoden zur Harzuntersuchung ist jedoch notwendig und empfehlenswert für eine sichere Harzdetermination.

Von den Neufunden wurden sechs Proben mittels IRS bestimmt (Abb. 6).

Die Proben KRU 16 (IRS/MZ 245), KRU 40 (IRS/MZ 460), KRU 41 (IRS/MZ 445), KRU 42 (IRS/MZ 461), KRU 44 (IRS/MZ 440) und KRU 45 (IRS/MZ 444) stammen aus den „Bitterfelder Schichten" (Bitterfelder Glimmersande) des Tagebaues Goitsche/Bitterfeld. Zum Vergleich wurden eine Beckerit-Probe (coll. HELM/WIENHAUS; (IRS/MZ 436) und andere Beckerit-Proben aus KOSMOWSKA-CERANOWICZ (IRS-Katalog) dem Bereich der Ostseeküste des Samlandes untersucht (Abb. 7). Die Vergleichsprobe (IRS 436) ist eine eindeutig determinierte Harzart und stammt aus der Sammlung Otto HELM (1904) des Westpreußischen Provinzialmuseums in Gdańsk. Sie befanden sich im Nachlaß von H. WIENHAUS (1923), Miltitz b. Leipzig (Firma SCHIMMEL & Co.) (vgl. KRUMBIEGEL *et al.* 1997). Auch der Vergleich der IR-Banden zwischen 1330 bis 1100 cm^{-1} der IR-Spektren bei BECK *et al.* (1986) weist keine Ähnlichkeiten mit den vorliegenden, 1995 angefertigten IR-Spektren von Beckerit auf.

Ergebnisse

Während einer Sonderausstellung im Naturkundemuseum Leipzig im Jahre 1995 wurden im Bernstein-Nachlaß des Chemieprofessors Dr. H. WIENHAUS, Leipzig-Miltitz (vgl. KRUMBIEGEL *et al.* 1997a, b) Beckerit-Proben aus der Bernsteinsammlung des Bernsteinforschers Otto HELM (1826–1902), einem Mitarbeiter des ehemaligen Westpreußischen Provinzialmuseums Danzig, entdeckt.

Die Hinweise auf den Etiketten des Sammlungsmaterials von H. WIENHAUS: „aus der Sammlung O. HELM, Westpreußisches Provincialmuseum in Danzig" und der Vermerk „von Dr. LA BAUME" sprechen dafür, daß die Beckerit-Proben im Nachlaß WIENHAUS aus Danzig bzw. aus dem Samland stammen.

Diese Tatsache sowie der lithologische Materialvergleich und der Vergleich mit den Daten der Erstbeschreibung des Beckerits nach PIESZCZEK (1880) ermöglichten zuerst eine makroskopische lithologische Determination des durch die Neuaufsammlungen seit

1990 im Tagebau Goitsche bei Bitterfeld gewonnenen zunächst fraglichen Beckerit-Materials (Tab. 3).

Auf dieser Grundlage wurden alle Beckerit-Stücke der Neuaufsammlungen sowie der Beckerit aus der Sammlung O. HELM, nach der IRS-Methode untersucht (Abb. 6).

Die Untersuchungen an den sieben Beckerit-Proben erbrachten folgende Ergebnisse:

Die Beckerite weisen eine schwache Absorption im Wellenbereich bei 1710 cm^{-1}, der Carbonylgruppe, auf. Kräftiger ist dagegen der Bereich bei 1615–1620 cm^{-1} (H_2O). Sehr deutlich, jedoch in unterschiedlicher Intensität ist die Absorption bei 720 cm^{-1}, dem Bereich der Terpene, CH_2-aromatische Ringsysteme.

Innerhalb der Beckerit-Proben ist an den Spektrogrammen im Bereich von 720 cm^{-1} noch eine weitere Differenzierung erkennbar. So weisen die IRS 460, 461, 445 der Proben KRU 40, KRU 42 und KRU 41 aus dem Tagebau Goitsche bei Bitterfeld eine schwächere Absorption bei 720 cm^{-1} auf. Die IRS 125 und 436 der Beckerite aus dem Samland sowie die IRS 245, 444 und 440 der Proben aus dem Tagebau Goitsche bei Bitterfeld zeigen dagegen bei 720 cm^{-1} eine etwas stärkere Absorption.

Die für Succinit typische Absorption bei der Bande 890 cm^{-1} fehlt in den Beckerit-Spektren oder ist nur sehr schwach angedeutet.

Zusammenfassend läßt sich aufgrund der absorptionsinfrarotspektroskopischen Untersuchungen feststellen, daß Beckerit und Succinit nicht, wie BECK *et al.* (1986) konstatierten, einunddasselbe Harz, sondern unterschiedliche Harze sind.

Es wäre aber empfehlenswert, die Beckerit-Proben aus dem Tagebau Goitsche bei Bitterfeld zusätzlich noch mit den Methoden der Gaschromatographie und der Massenspektroskopie zu überprüfen, um einmal die definitiven Herkunftspflanzen, vor allem von Beckerit zu ermitteln und ferner den Nachweis des Vorhandenseins zweier verschiedener Harze erneut zu bestätigen.

Danksagung

Für die Unterstützung und Beratung bei den wissenschaftlichen Untersuchungen des Beckerit-Materials aus der Bernsteinlagerstätte Bitterfeld gilt mein Dank nachfolgenden Institutionen und Personen: Mitteldeutsche Braunkohlengesellschaft mbH Bitterfeld-Theißen, Prof. Dr. B. KOSMOWSKA-CERANOWICZ (Muzeum Ziemi, PAN, Warszawa/Polen), Dipl.-Geol. R. BAUDENBACHER (Naturkundemuseum Leipzig), Prof. Dr. O. WIENHAUS (Technische Universität Dresden, Institut für Pflanzenchemie und Holzforschung

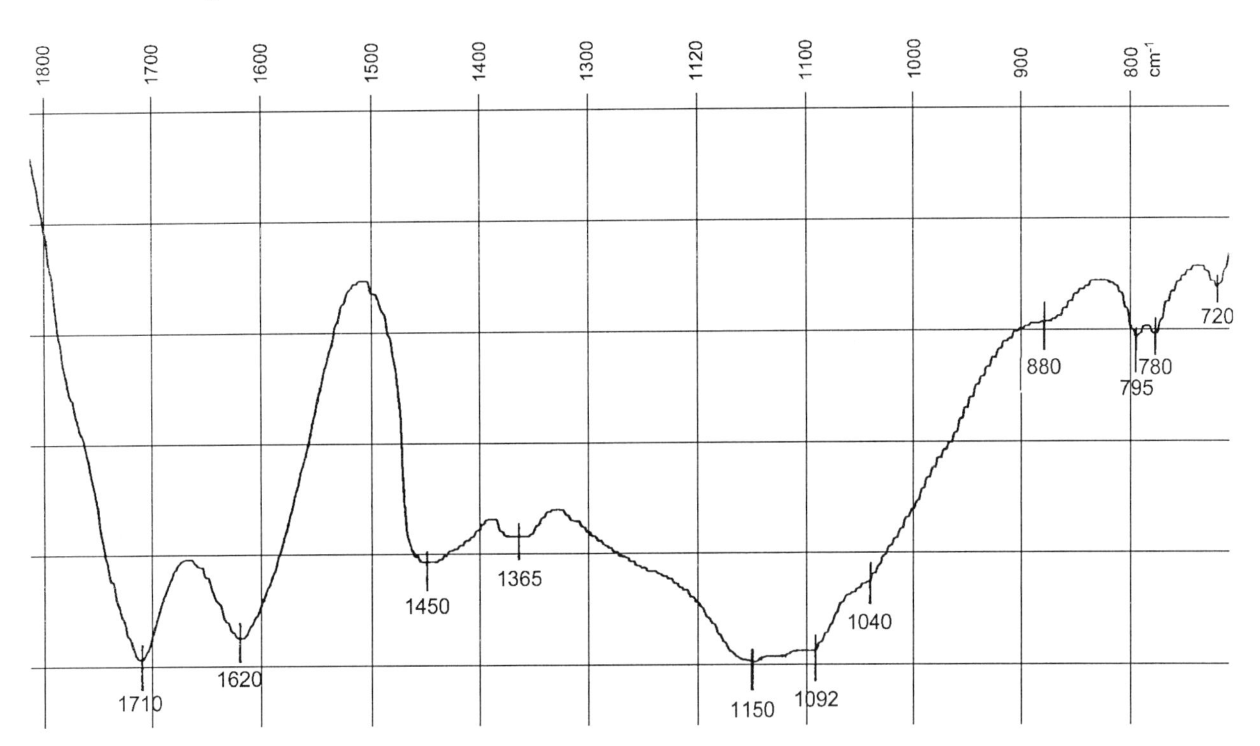

Abb. 7. Typisches IRS von einem Beckerit aus dem Samland. Probe aus der Kollektion W. SIMON des Museums für Naturkunde der Humboldt-Universität Berlin in der Sammlung des Muzeum Ziemi PAN in Warszawa (vgl. KOSMOWSKA-CERANOWICZ & KRUMBIEGEL 1989, Farbtafel, Foto 10). (Invent.-Nr. 20691/MZ/PAN Warszawa).

Tharandt/Sachsen), Dipl.-Ing. R. WIMMER (Geologisches Landesamt Sachsen-Anhalt, Halle/S.) und Dr. A. KRUMBIEGEL (Halle/S.).

Günter KRUMBIEGEL
Klara-Zetkinstr. 16
D-06114 Halle/Saale, Germany

BEKERYT Z KOPALNI GOITSCHE KOŁO BITTERFELDU (SAKSONIA, NIEMCY)

Günter Krumbiegel

Streszczenie

Bekeryt, zwany też brunatną żywicą, jest podobny do kawałków zwęglonego drewna. Przez KÜNOWA — konserwatora Instytutu Zoologicznego Uniwersytetu Alberta w Königsbergu — został odkryty w 1867 r. w niebieskiej ziemi w Großkuren (Primore) i Kleinkuren (Filino), w 1871 r. znaleziono go w Warnicken (Lesnoye) i w 1872 r. w Palmnicken (Jantarnoje).

Pierwszy opis właściwości fizycznych bekerytu pochodzi od PIESZCZKA z 1880 r.. Nazwał on tę kopalną żywicę **bekerytem**, na cześć Moritza BECKERA, drugiego właściciela firmy obróbki bursztynu „STANTIEN & BECKER" w Königsbergu. CASPARY w 1881r. na zasadzie pierwszeństwa wywiódł dla brunatnej żywicy nazwę „Künowit". Nie doszło jednak do przemianowania.

Również w niemieckim złożu bursztynu koło Bitterfeldu (bursztynu bitterfeldzkiego albo saksońskiego — sukcynitu) spotyka się obok dominującego sukcynitu i akcesorycznych kopalnych żywic, takich jak gedanit, glessyt, goitszyt, zygburgit i innych, również bekeryt (KRUMBIEGEL 1996). Występuje on w warstwach bitterfeldzkich (bitterfeldzkie piaski łyszczykowe, górny oligocen do najniższego miocenu) w spągu głównego pokładu dolnomioceńskich węgli brunatnych. Tu występuje stosunkowo często w kawałkach o średnicy około 3 do 7 cm. Barwy powierzchni są od jasno- do ciemnobrązowych. Na przełamie są jasnobrunatne, ziemistobrunatne i matowe. Bryłki mają kształt bulwiasty lub kroploforemny. Powierzchnia zewnętrzna jest nieregularna, z nabrzmieniami i zadziorami, czasem ze śladami wysychania. Materiał jest twardy i kruchy.

Zewnętrzne cechy są bardzo podobne do opisanych przez PIESZCZKA (1880, 134) cech bekerytu bałtyckiego. Potwierdza to również badanie bekerytu z kolekcji HELMA.

Wszystkie okazy bekerytu z kopalni Goitsche zostały zbadane w podczerwieni (metodą IRS). Krzywe zostały porównane z materiałem z Sambii, co uznano za podstawę oznaczenia bekerytu saksońskiego.

Przedstawionego przez BECKA, LAMBERTA i FRYEGO (1986) poglądu, że bekeryt nie jest czystą kopalną żywicą, ale sukcynitem zanieczyszczonym rozdrobnionym drewnem i resztkami nadjedzonych owadów, nie potwierdzają wyniki badań autora.

Otwarte pozostaje pytanie o macierzyste drzewo żywicy bekerytowej.

Literatur

ANDRÉE K.
1951 Der Bernstein. Das Bernsteinland und sein Leben, 1–96, *Kosmos*, Franck'sche Verlagshandlung, Stuttgart.

BARTHEL M. & HETZER H.
1982 Bernstein-Inklusen aus dem Miozän des Bitterfelder Raum, *Zeitschrift für Angewandte Geologie*, **28** (7), 314–336, Berlin.

BECK C. W., LAMBERT J. B. & FRYE J. S.
1986 Beckerite, *Physics and Chemistry of Minerals*, **13**, 411–414, New York.

CASPARY R.
1881 Neue fossile Pflanzen der blauen Erde, d.h. des Bernsteins, des Schwarzharzes und des Braunharzes, *Schriften der physikalisch-ökonomischen Gesellschaft zu Königsberg*, **22**, 22–31, Königsberg.

CASPARY R. & KLEBS R.
1906 Die Flora des Bernsteins und anderer fossiler Harze des ostpreußischen *Tertiärs, Abh. Königl. Preuß. Geol. Land.-Anst. Berlin, N.F.*, **4**, 1–182, Berlin.

FUHRMANN R. & BORSDORF R.
1986 Die Bernsteinarten des Untermiozäns von Bitterfeld, *Zeitschrift für Angewandte Geologie*, **32** (12), 309–316, Berlin.

HECK G.
1997 Analyse und Herkunftsbestimmung, *Archäol. i. Dtschl.*, 3, 28–30, Stuttgart.

HEY M. H.
1962 *An index of mineral species and varieties arranged chemically*, British Museum, London.

JENTZSCH A.
1892 *Führer durch die Geologischen Sammlungen des Provinzialmuseums der Physikalisch-Oekonomischen Gesellschaft zu Königsberg*, 1–106, Kommiss. W. Koch, Königsberg/Pr.

KAUFMANN H. P.
1948 *Analyse der Fette und Fettprodukte einschließlich der Wachse, Harze und verwandter Stoffe*, Springer Verlag, Berlin.

KLEBS R.
1889 *Der Bernstein. Seine Gewinnung, Geschichte und geologische Bedeutung. Erläuterung und Catalog der Bernstein-Sammlung der Firma Stantien & Becker*, 1–32, Druck v. G. Laudin, Königsberg/Pr.

1897 Cedarit, ein neues bernsteinähnliches fossiles Harz Canadas und sein Vergleich mit anderen fossilen Harzen, *Jahrbuch der Königlich Preussischen geologischen Landesanstalt und Bergakademie zu Berlin*, **17**, 199–230, Berlin.

KOSMOWSKA-CERANOWICZ B.
1996 *Bernstein-Lagerstätten*, 4–8, [*in*:] Kosmowska-Ceranowicz B. *et al.*, *Bernstein — Schatz urzeitlicher Meere*, Sadyba, Warszawa,

KOSMOWSKA-CERANOWICZ B.
& KRUMBIEGEL G.
1989 Geologie und Geschichte des Bitterfelder Bernsteins und anderer fossiler Harze, *Hallesches Jahrbuch für Geowissenschaften*, **14**, 1–25, Gotha.

KRUMBIEGEL G.
1995 *Der Bitterfelder Bernstein (Succinit)*, [*in*:] Weidert W. K., *Klassische Fundstellen der Paläontologie*, **3**, 11–12, 191–204, 268–269, Goldschneck Verlag, Weinstadt.

1996 *Bernstein (Succinit) — Die Bitterfelder Lagerstätte*, [*in*:] Ganzelewski M. & Slotta R. (Hrsg.), Bernstein — Tränen der Götter, *Veröffentlichungen aus dem Deutschen Bergbau-Museum Bochum* **64**, 89–100, Bochum.

KRUMBIEGEL G. & KOSMOWSKA-CERANOWICZ B.
1989 Der Bitterfelder Bernstein — Geschichte, Geologie und Genese, *Fundgrube*, **25** (2), 34–39, 4. Umschlags., Berlin.

KRUMBIEGEL G. & KRUMBIEGEL B.
1996 Bernstein — Fossile Harze aus aller Welt, *Fossilien, Sonderband*, 2. Auf., 1–112, Weinstadt.

KRUMBIEGEL G., KRUMBIEGEL B.
& KOSMOWSKA-CERANOWICZ B.
1997a *The "amber" legacy of Wienhaus*, [*in*:] International Interdisciplinary Symposium "Baltic amber and other fossil resins. 997 Urbs Gyddanyzc — 1997 Gdańsk", Gdańsk, 2.–6. Sept. 1997. *Museum of the Earth, Scientific conferences*, ***Abstracts***, **9**, 40–41, Warsaw.

1997b *Bursztynowa spuścizna Wienhausa*, [*in*:] Międzynarodowe interdyscyplinarne Sympozjum „Bursztyn bałtycki i inne żywice kopalne. 997 Urbs Gyddanyzc — 1997 Gdańsk", 2–6 września 1997, *Muzeum Ziemi, Konferencje Naukowe, Streszczenia referatów*, **8**, 41–42, Warszawa.

PACLT J.
1953 A system of caustolites, *Tschermaks Mineral. u. Petrograph. Mitt.*, **3** (4), 332–347, Wien.

PIESZCZEK E.
1880 Ueber einige neue harzähnliche Fossilien des ostpreußischen Samlandes, *Archiv der Pharmacie*, **17** (3); **59** (1880), 433–436, Halle.

PLONAIT C.
1935 Entstehung, Bau und chemische Verarbeitung des Bernsteins, *Zeitschrift für Angewandte Chemie*, **48** (38), 605–607, Berlin.

RICE P.
1987 *Amber — The golden gem of ages*, 1–289, 2. Auf., „The Kościuszko Foundation, Inc.", New York.

RÖSCHMANN F.
1996 Friedrichsmoor. Fossil insects in amber and sedimentary rocks, *Inclusion-Wrostek*, **26**, 8, Cracow.

1997 Ökofaunistischer Vergleich von Nematoceren-Faunen (Insecta; Diptera; Sciaridae und Ceratopogonidae) des Baltischen und Sächsischen Bernsteins (Tertiär, Oligozän-Miozän), *Paläontologisches Zeitschrift*, **71** (1/2), 79–87, Stuttgart.

SANDERMANN W.
1960 *Naturharze, Terpentinöl, Tallöl. Chemie und Technologie*, 96–99, Springer Verlag, Berlin, Göttingen, Heidelberg.

SCHMID L.
1931 *Bernstein*, [*in*:] Dölter C. & Leitmeyer H. (Hrsg.), *Handbuch der Mineralchemie*, **4** (3), 842–943, Dresden u. Leipzig.

SCHUBERT K.
1961 Neue Untersuchungen über Bau und Leben der Bernsteinkiefern (*Pinus succinifera* [CONW.] emend.). Ein Beitrag zur Paläohistologie der Pflanzen, *Geologisches. Jahrbuch., Beih.*, **45**, 1–149, Hannover.

SCHUMANN H. & WENDT H.
1989 Zur Kenntnis der tierischen Inklusen des sächsischen Bernsteins, *Deutsche Entomologische Zeitschrift, N. F.*, **36** (1–3), 33–34, Berlin.

SZADZIEWSKI R.
1993 Biting midges (Diptera, Ceratopogonidae) from miocene Saxonian amber, *Acta zoologica cracoviensis*, **35**, 603–656, Cracow.

SZADZIEWSKI R., KRZEMINSKI W. & KUTSCHER M.
1994 A new species of *Corthrella* (Diptera, Corethrellidae) from Miocene Saxonian amber, *Acta zoologica cracoviensis*, **37**, 87–90, Cracow.

WEITSCHAT W.
1987 *Bernstein der Insel Sylt*, [*in*:] Hacht U. (Hrsg.), *Fossilien von Sylt.*, 109–121, Hamburg.

1996 Bitterfelder Bernstein — ein eozäner Bernstein auf miozäner Lagerstätte, *METALLA, Veröffentlichungen aus dem Deutschen Bergbau-Museum Bochum*, **66**, 71–84, Bochum.

WICHARD W. & WEITSCHAT W.
1996 Wasserinsekten im Bernstein. Eine paläobiologische Studie, *Entomol. Mitt. Löbbecke-Museum u. AQUAZOO, Beih.*, **4**, 1–22, Düsseldorf.

WUNDERLICH J.
1983 Zur Konservierung von Bernstein-Einschlüssen und über den „Bitterfelder Bernstein", *Neue Entomologische Nachrichten*, **4**, 11–13, Keltern.

CEDARITE AND OTHER FOSSIL RESINS IN CANADA

Alicja M. ZOBEL

Abstract

Analyses of secondary metabolite of plant debris in fossil resins from two very different locations in Canada — Axel Heiberg Island and southern Alberta — have been begun. In the plant debris, only coumarin has been identified, with several unidentified peaks also appearing on HPLC chromatograms.

I am aware of two sites of fossil resin deposits in Canada, one of which I visited personally in the summer of 1997. The second deposit known to me was described by Drs BASINGER and LEPAGE (BASINGER *et al.* 1988) from the Geological Sciences department of the University of Saskatchewan.

The Arctic fossil resins

Several years ago there was an item in several newspapers on the discovery, by observation from the air, of well-preserved trunks of trees belonging to the Taxoidaceae-dominated swamp forest (ANDERSON & LEPAGE 1995) on Somerset and Axel Heiberg Islands, near Ellesmere Island in the Arctic Ocean, north of the Canadian mainland (Fig. 1, X). Two scientists from the University of Saskatchewan investigated this area and found that, as a result of global warming and the retreat of glaciers, there was an exposed area with pieces of *Pinus, Pseudolarix* and *Metasequoia* trunks *c.* 45 million years old (BASINGER 1986; BASINGER *et al.* 1988; LEPAGE & BASINGER

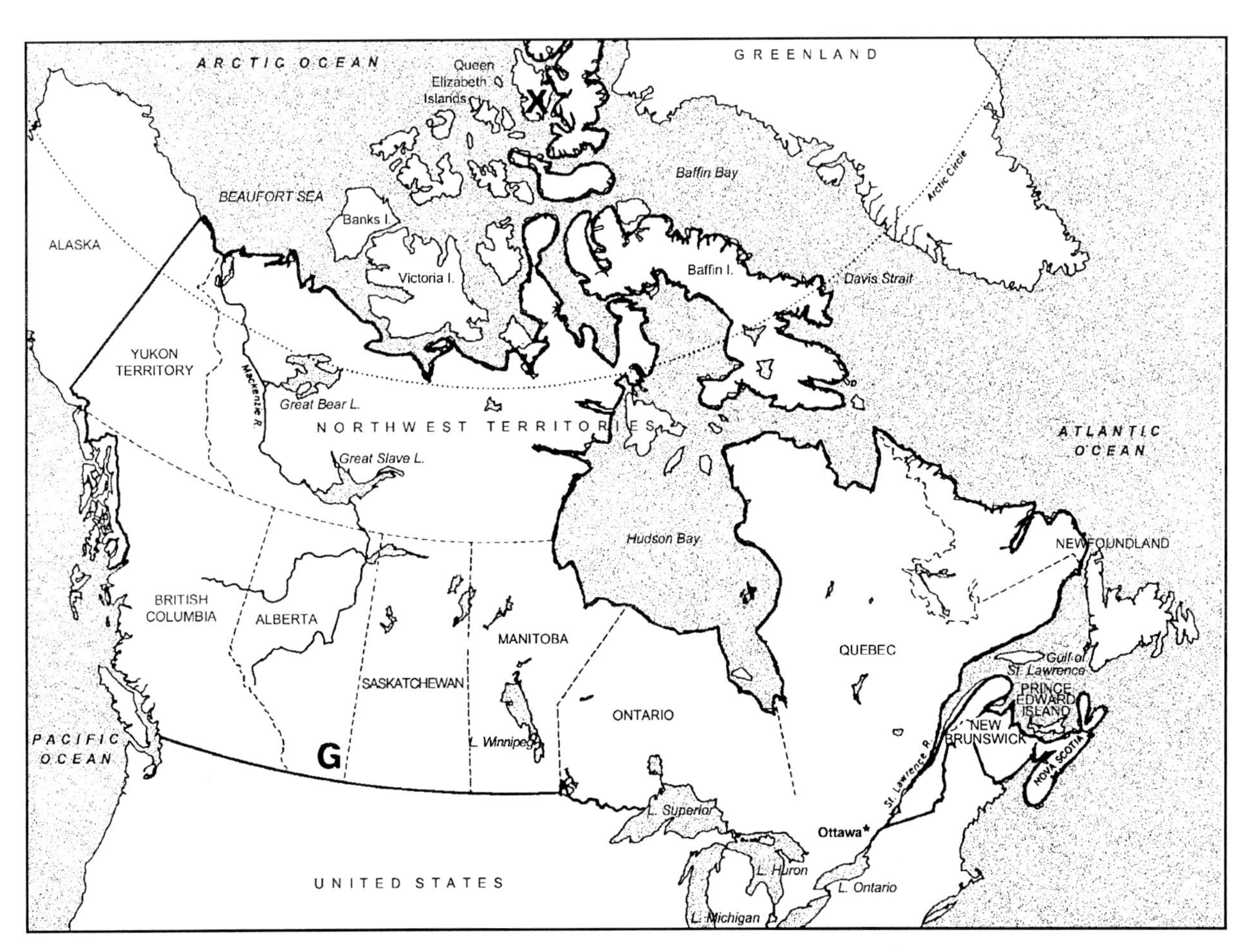

Fig. 1. Location of two sites of fossil resins in Canada: Arctic (X) and next to Grassy Lake (G).

Fig. 2. Fossil resins from Axel Heiberg Island. Photo. A. ZOBEL.

1991). In plant debris and in the trunks themselves there were deposits of fossil resin, some of which (Fig. 2) were sent to me for chemical analysis. The structure of these pieces was rod-like, with a white cortex surface. The interior showed spiral, almost concentric, yellow and darker brown bands. Other shapes were also visible, but only tiny pieces of these were obtained from the larger chunks, possibly because these fossil resins are very brittle, and were easily fragmented. Such sand-like debris is visible in the centre of the picture shown in Fig. 2 (arrow).

Investigation of plant material (*c.* 1 kg) from this site and from present-day sequoias growing in California showed both quantitative and qualitative chemical differences in coumarins. Dichloromethane extracts of the samples were subjected to reverse phase high performance liquid chromatography by a method developed in our laboratory for analysis of coumarins (THOMPSON & BROWN 1984; ZOBEL & BROWN 1988). The fact that only traces of coumarin itself, as well as several unidentified peaks, were observed in the fossil debris may have been due either to originally low concentrations of these compounds or to their slow subsequent removal over long periods by leaching with water, even though they are only water-soluble with difficulty. We have found that in modern seeds and fruits of rutaceous and leguminous plants these compounds play a defensive role against microbial attack and form a shield against ultraviolet radiation, possibly the same role played in the palaeobotanical species. We now intend to work on the terpenoid content of the plant debris, after suitable methods of extraction, purification and isolation have been developed. We still have several unidentified peaks on chromatograms which may possibly be terpenoids. ANDERSON & LEPAGE (1995) studied the resin from Axel Heiberg Island using a pyrolysis-gas chromatography-mass spectrometry procedure, and found several phenolic compounds, predominantly terpenoids, which, together with our findings, suggest a very complicated phytochemistry of plants embedded in fossil resins.

Fig. 3. Shore of Grassy Lake in southern Alberta. Photo. A. ZOBEL.

Figs 4 & 5. Resin pieces on the surface of small soil protrusions next to brown coal pieces by Grassy Lake. Photo. A. ZOBEL.

In previous preliminary studies using thin-layer chromatography and histochemical techniques, coumarin had been identified in the ethanol extract of modern pine and *Metasequoia* (ZOBEL & NIGHSWANDER 1990).

Pieces of xylem tissue from fossil debris and modern *Metasequoia* were embedded in Spur resin (Epon 812 resin) and prepared for analysis under the transmission electron microscope (TEM), but sectioning of fossil tissue with a diamond knife was unsuccessful because the resin and the enclosed tissue were too brittle. Investigations under the scanning electron microscope, when the broken pieces needed only to be sputter-coated with platinum, were more successful, and we observed morphological similarities between the xylem vessels of the fossil and modern plants. We are still working on embedding the fossil material in resins harder than Spur resin, which would allow us to cut the very thin sections necessary for observation in the TEM.

The Grassy Lake fossil resin

In July, 1997 I personally investigated the second site of fossil resins, which lie on the shore of Grassy Lake (Fig. 3) near Bassano, in southern Alberta (Fig. 1, G). The first description of the chemistry of this resin was by KLEBS (1897), and it was reported to exist in Manitoba, Saskatchewan and Alberta. MCALPINE (LANGENHEIM 1969) has stated that the resins are 80 million years old. LANGENHEIM (1969) has stated that cedarite from Cedar Lake came from the Araucariaceae. According to the IRS method, resin from Cedar Lake may be compared with cedarite from Grassy Lake (KOSMOWSKA-CERANOWICZ, personal communication). The resins occur in brown coal of about that age (Figs 4 & 5) on a small protrusion of the ground next to growing shrubs (Fig. 5, arrow). The pieces are small (Fig. 6), the largest being *c.* 5 cm in diameter, and they easily disintegrated into still smaller fragments after exposure on the surface, owing to prolonged weathering. We visited the area after a heavy rainfall, and found the resin pieces protruding from the surface after erosion of the surrounding plant debris and, in some cases, soil (Fig. 5). The pieces can be in the shape of very long, thin rods,

Fig. 6. Several pieces of differently shaped fossil resins from shore of Grassy Lake. Photo. A. ZOBEL.

which can form longitudinal clusters (Fig. 6, arrow). Sometimes individual rods or shorter, drop-like forms were observed (Fig. 6, double arrow). The pieces are dark brown in colour, much darker and clearer than the Arctic island resins, and lack a heavy cortex surface. The other forms found were flat, round or oval discs with clean surfaces, very often showing imprints of plant debris (Fig. 6, triple arrow). Such pieces were observed wedged between strata of plant debris that resembled the pages of a book.

On a visit to the nearby Royal Tyrrell Museum at Drumheller we saw several insect inclusions, the most intriguing of which was a "fly with hairy wings", as the label described it. Although the exhibits date only from 1995, we were informed that over 100 species of insects have already been identified, most of their descriptions being included in the PhD thesis of Dr Ted PIKE from the University of Calgary.

Little is known about the Canadian amber discovered in the Yoho Valley at the end of the 19th century (HARRINGTON 1891; TYRRELL 1890 *vide* POINAR & POINAR 1994), which has been used for furniture varnish by the Hudson's Bay Company, and is no longer available for investigation (POINAR & POINAR 1994). POWELL, in his book on the Burgess shale in 1997, makes no mention of amber deposits in the Rockies.

Prospective studies

As very little has been done on flora in amber (LANGENHEIM 1969) we are currently establishing contacts with both Saskatchewan University and the Archaeology Department of the TYRRELL Museum, seeking access to fossil plant material. My hope is to compare plants and animals embedded together in the same pieces. As a phytochemist, a person working on chemicals in plants, my interests are in the recognition of compounds both in the fossil resins themselves and in plant debris. Comparative chemosystematics of modern and fossil species has great potential for identification of the latter. In modern plants some 100,000 different natural products have been recognized (HARBORNE 1988; ZOBEL 1997), playing the role of plant defence against environmental changes, microbial and herbivore (insect and mammal) attacks. As some of these compounds have been used by insects as attractants, and as oviposition stimulators, some plants and insects can coexist at a higher probability than others. Chemoecology (HARBORNE 1988) will in future allow us to reconstruct, with a high degree of probability, the ecosystems in which resins were formed, trapping coexisting plants and animals. We must remember that secondary metabolites were produced and extruded to the plant surface for protection of the plants and were involved in their response to the environment, which included insects (ZOBEL & BROWN 1995). I foresee a great possibility, using modern methods of analysis of very small amounts of material, to pin-point the precise location of different inclusions in the same sample, e.g. coexisting microbes, plants and animals. We have used HPLC, gas chromatography/mass spectrometry, nuclear magnetic resonance and ion traps (TKACZYK *et al.* 1993; ZOBEL *et al.* 1991) in the reconstruction of such material, but work is at a very preliminary stage. It will be a huge task, but well worth undertaking.

Alicja ZOBEL
Department of Chemistry, Trent University
Peterborough, Ontario, Canada K9J 7B8

CEDARYT I INNE ŻYWICE KOPALNE W KANADZIE

Alicja M. ZOBEL

Streszczenie

Rozpoczęto badania wtórnych metabolitów w szczątkach flory zachowanej w dwóch rodzajach żywic kopalnych z Kanady: (1) z wysp Somerset i Axel Heiberg w pobliżu wyspy Ellesmere na Oceanie Arktycznym i (2) z okolic jeziora Grassy, niedaleko Bassano w południowej Albercie (fig. 1). Metodą ciekłej chromatografii (HPLC) została zidentyfikowana kumaryna.

Na wyspach żywice występują wśród paleogeńskich skamieniałych pni metasekwoi, odsłaniających się spod topniejącego lodu. Próbki pobrali Dr BASINGER i Dr LEPAGE z uniwersytetu w Saskatchewan. Badaniom poddano ksylem z drzew skamieniałych i współcześnie rosnących w Kalifornii.

W południowej Albercie żywice kopalne występują na obszarze stepowym, na terenie dawnej kopalni odkrywkowej węgla kredowego (wg MCALPINE, vide LANGENHEIM 1969, wiek osadów szacuje się na 80 mln lat). Według badań w podczerwieni żywice te można porównać z cedarytem znanym z rejonu Jeziora Cedar (KOSMOWSKA-CERANOWICZ wiadomość ustna).

Fragmenty żywic kopalnych w postaci małych dysków lub sopli i innych niewielkich form naciekowych po ulewnych deszczach pokazują się na powierzchni wciśnięte pomiędzy resztki skamieniałych roślin.

Dział Archeologii Muzeum Tyrrell'a i Uniwersytet w Saskatchewan zajmują się badaniem inkluzji organicznych w żywicach kanadyjskich. Celem autorki są badania składników chemicznych zarówno inkluzji roślinnych, jak i samego bursztynu, w którym dana inkluzja została zamknięta, i zgromadzenie danych do systematyki chemicznej żywic kopalnych.

Bibliography

ANDERSON K. B. & LEPAGE B. A.
1995 Analysis of fossil resins from Axel Heiberg Island, Canadian Arctic, *American Chemical So*ciety, **260**, 170–192.

BASINGER J. F.
1986 Our "tropical" arctic, *Canadian Geographic*, **106**, 28–37.

BASINGER J. F., MCIVER E. E. & LEPAGE B. A.
1988 The fossil forests of Axel Heiberg Island, *Musk-Ox*, **36**, 50–55.

HARBORNE J. B.
1988 *Introduction to ecological biochemistry*, Academic Press, London.

HARRINGTON B. J.
1891 On the so-called amber of Cedar Lake, N. Saskatchewan, Canada, *American Journal of Science*, **42** (3), 332–338.

KLEBS R.
1897 Cedarit, ein neues bernsteinähnliches Harz Canadas und sein Vergleich mit anderen fossilen Harzen, *Jahrbuch der Königlich Preussischen Geologischen Landesanstalt und Bergakademie zu Berlin*, **17**, 199–230 (1996).

LANGENHEIM J. H.
1969 Amber: the botanical inquiry, *Science*, **163**, 1157–1169.

LEPAGE B. A. & BASINGER J. F.
1991 *Early tertiary Larix from the Buchanan Lake Formation, Canadian Arctic Archipelago, and a consideration of the phytogeography of the genus*, [*in:*] Christie R. L. & McMillan N. J. (eds), Tertiary Fossil Forests of the Geodetic Hills, Axel Heiberg Island, Arctic Archipelago, *Geological Survey of Canada, Bulletin* **403**, 67–82.

POINAR G. & POINAR R.
1994 *The quest for life in amber*,1–219, Addison Wesley Publishing Company.

POWELL W.
1997 *The Burgess shale*, Sunding Printers Ltd., Calgary.

THOMPSON H. J. & BROWN S. A.
1984 Separation of some coumarins in higher plant by liquid chromatography, *Journal of Chromatography*, **314**, 323–336.

TKACZYK M., ZOBEL A. M., PLOMLEY J. B. & MARCH R. E.
1993 Mass spectrometric study of selected coumarins and psoralens. I. Negative ions, *Organic Mass Spectrometry*, **28**, 1148–1154.

TYRRELL J. B.
1890 *Ann. Rept. Geol. Survey Canada, N. S.*, **5**, 30–39.

ZOBEL A. M.
1997 *Phenolic compounds in defense against air pollution*, [*in:*] Junus M. & Iqbal M. (eds), *Plant Response to Air Pollution*, 100–130, Wiley, Chichester.

ZOBEL A. M. & BROWN S. A.
1988 Determination of furano coumarins on the leaf surface of *Ruta graveolens* with an improved extraction technique, *Journal of Natural Products*, **51**, 941–946.

1995 Coumarins in the interaction between the plant and its environment, *Allelopathy Journal*, **2**, 9–20.

ZOBEL A. M. & NIGHSWANDER J. E.
1990 Accumulation of phenolic compounds in the necrotic areas of austrian and red pine needles due to salt spray, *Annals of Botany*, **66**, 629–640.

ZOBEL A. M., WANG Y., MARCH R. E. & BROWN S. A.
1991 Identification of eight coumarins occurring with psoralen, xanthotoxin and bergapten on leaf surfaces, *Journal of Chemical Ecology*, **17**, 1859–1870.

RESTE DER BERNSTEINSAMMLUNG OTTO HELM IM NACHLASS VON HEINRICH WIENHAUS

Günter KRUMBIEGEL, Brigitte KRUMBIEGEL & Barbara KOSMOWSKA-CERANOWICZ

Kurzfassung

Der Bernsteinnachlaß enthält: 10 Proben Bernstein und andere fossile Harze aus der Sammlung Otto HELM des ehemaligen Westpreußischen Provinzialmuseums in Danzig; Schriftverkehr (4 Briefe) aus dem Jahre 1926; zahlreiche Photos über die Bernsteingewinnung (Palmnicken).

Die Proben aus der HELM-Sammlung, 1902 umfaßte sie etwa 5000 Exponate, sind heute die einzigen Reste dieser früher zweitgrößten Bernsteinsammlung der Welt. HELM-Proben wurden lithologisch neu beschrieben und mit Hilfe der Infrarot-Absorptions-Spektroskopie neu determiniert.

Einleitung

Im Naturkundemuseum Leipzig fand vom Mai bis Oktober 1995 eine Sonderausstellung zum Thema „Bernsteinsplitter" statt. Diese Ausstellung zeigte u.a. die Vielseitigkeit der Aspekte unter denen man das Thema Bernstein betrachten kann. Vor allem wurde immer wieder deutlich, welch enge Verbindungen zwischen Naturwissenschaften und Kulturgeschichte in der Geschichte des Bernsteins bestehen und ihn daher ständig in neuem faszinierenden Licht erscheinen lassen sowie zu weiteren wissenschaftlichen, insbesondere wissenschaftshistorischen Forschungen von oft grundlegenden bisher ungeklärten Problemen anregen (BAUDENBACHER 1995; 1996; KRUMBIEGEL & KRUMBIEGEL 1995).

Eine Vitrine dieser Ausstellung enthielt eine Spezialsammlung fossiler Harze aus dem Baltikum (als Reste der Danziger Bernsteinsammlung von Otto HELM), 4 Briefe, die sich mit interessanten Problemen der Verwendung von Bernstein in der Chemischen Industrie beschäftigen sowie historische Photos der Bernsteinwerke Königsberg i. Pr. Diese Objekte stammen aus dem Nachlass von Prof. Dr. H. WIENHAUS, Professor der Organischen Chemie und Leiter des wissenschaftlichen chemischen Labors der Firma SCHIMMEL & Co in Miltitz bei Leipzig (vgl. LA BAUME 1935, 11). Die Exponate dieses Nachlaßmaterials waren jetzt Gegenstand wissenschaftshistorischer und naturwissenschaftlicher Untersuchungen, deren Untersuchungsergebnisse im folgenden aufgezeigt werden.

Nachlas Material von Heinrich WIENHAUS

Heinrich WIENHAUS (1882–1959) war Professor für Organische Chemie in Göttingen (1922–1925) und in

Abb. 1. Prof. Dr. Heinrich WIENHAUS * 26. Oktober 1882 in Beckinghausen / Westf. + 5. September 1959 in Tharandt / Sachsen (Bildnis aus der Zeit um 1925).

Leipzig (1928–1935). Ab 1.4.1935 bis 1954 hatte er den Lehrstuhl für forstliche Pflanzenchemie an der Technischen Hochschule Dresden inne und war gleichzeitig Direktor des Institutes für Pflanzenchemie und Holzforschung an der damaligen Forstakademie Tharandt bei Dresden (*Miltitzer Berichte*... 1959; *Wienhaus Festschrift*... 1992; FISCHER 1982; SANDERMANN 1982). Hier wirkte er maßgeblich als Hochschullehrer mit hoher Disziplin und als Gelehrter, aber ebenso in der Naturwissenschaftlichen Gesellschaft ISIS zu Dresden (Abb. 1).

Das vorliegende Nachlassmaterial, das heißt Bernstein, sowie der Schriftverkehr WIENHAUS — Bernsteinwerke Königsberg und die Photos stammen aus WIENHAUS' Tätigkeit in der Zeit von 1925 bis 1935. Im Jahre 1925 übernahm er die Leitung des wissenschaftlichen Laboratoriums von SCHIMMEL & Co, Fabrik für aetherische Öle in Miltitz bei Leipzig, die er bis zu seiner Berufung am 28.3.1928 als a.o. Professor an die Univerität Leipzig inne hatte. Während dieser Zeit führte er auch die im folgenden Schriftverkehr festgehaltene Gutachtertätigkeit über eine Nutzung ätherischer Destillationsprodukte von Baltischem Bernstein aus.

A. Proben fossiler Harze

Wissenschaftlich und vor allem wissenschaftshistorisch interessant aus der Sammlung WIENHAUS sind die Proben fossiler Harze. Das Inventar der Sammlung WIENHAUS (das auf einer Mineralienbörse in Dresden von Herrn A. GROBER, Leipzig erworben wurde und der es freundlicher Weise für die wissenschafthistorischen Unteruchungen zur Verfügung stellte) umfaßt folgendes Material (Abb. 2):

– Fossile Harze aus der Sammlung O. HELM, Danzig, gesammelt September 1923, 10 Gläser (Abb. 3). Die Etiketten der zehn untersuchten Proben des Sammlungsmaterials sind mit folgenden Original-Beschriftungen versehen:

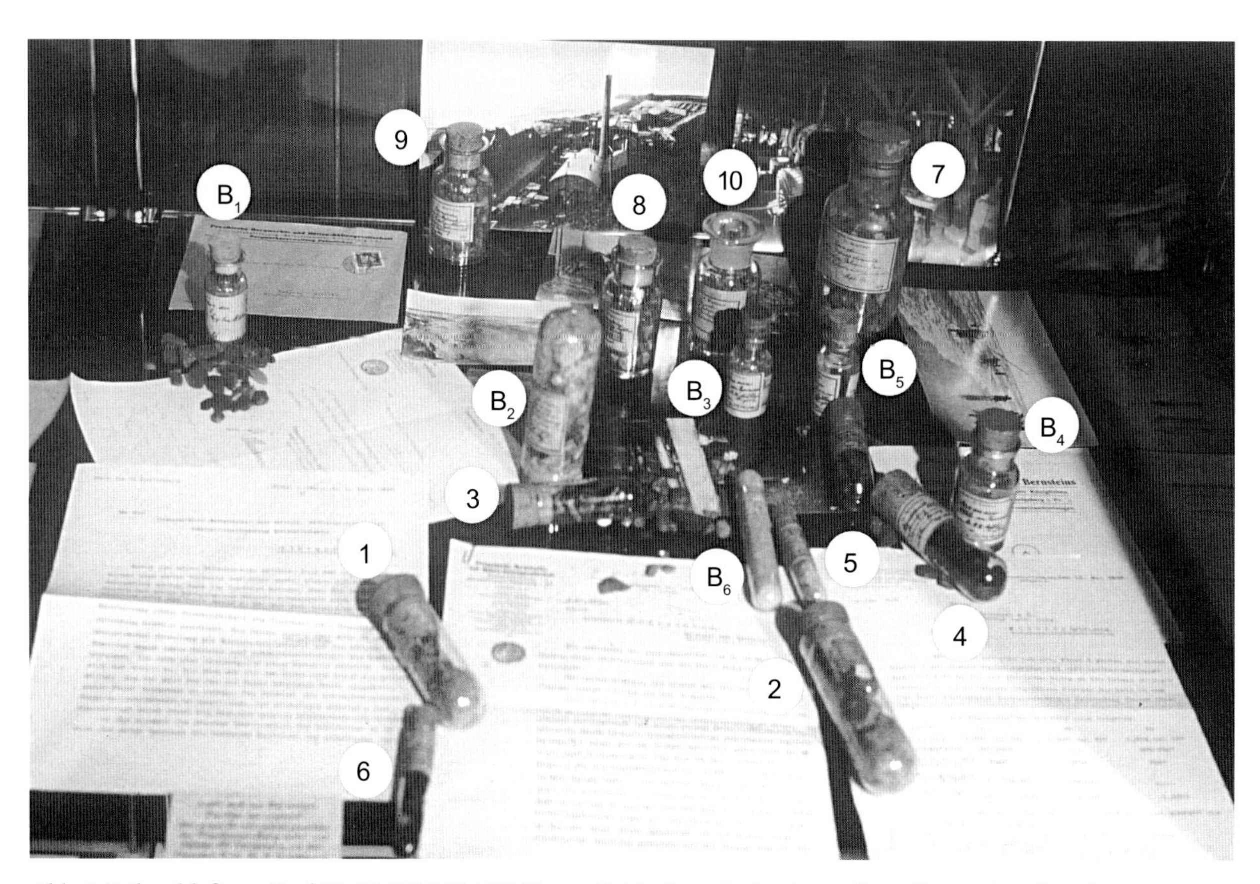

Abb. 2. Teilnachlaß von Prof. Dr. H. WIENHAUS Tharandt / Sachsen in der Ausstellung Bernsteinsplitter im Naturkundemuseum Leipzig 1995 mit fossilen Harzproben aus der Sammlung des Apothekers Otto HELM aus dem ehemaligen Westpreußischen Provinzialmuseum für Naturkunde und Vorgeschichte in Danzig (Nrn. 1–10). Proben B1, B3, B4, B5 sind miozäne fossile Harze aus der Braunkohle von Borneo. Die Proben B2 und B6 sind Pyropissit sog. Wachskohle bzw. Schwelkohle und stammen aus dem 1. und 2. Flöz der obereozänen Braunkohle des ehemaligen Tagebaues Stedten (Grube Walters Hoffnung) im Mansfelder Seekreis (Sachsen-Anhalt).

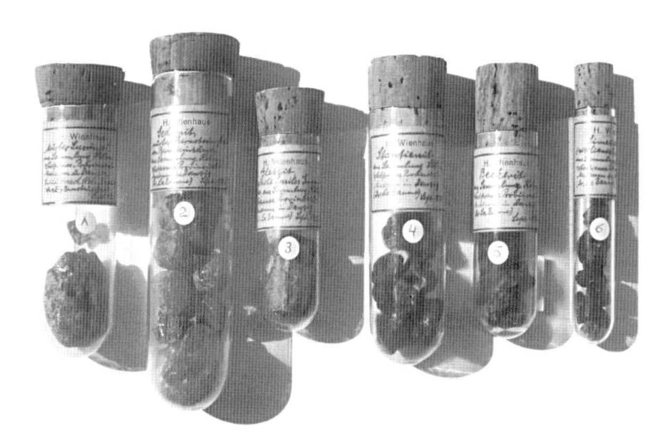

Abb. 3. Fossile Harze aus der Sammlung O. HELM, coll. 1923, Danzig (Proben 1–6).

Proben 1–6:

1. „H. Wienhaus / „Mürber Succinit" / aus Sammlung Helm, Westpreuss. Provincial-Museum in Danzig. Enthält nach Helm wenig (etwa 2%) Bernsteinsäure. Sept. 1923";
2. „H. Wienhaus / Gedanit, mürber Bernstein frei von Bernsteinsäure / aus Sammlung Helm, Westpreuss. Provincial-Museum in Danzig. (Dr. La Baume) Sept. 1923";
3. „H. Wienhaus / Glessit vielleicht fossiles Gummiharz, aus Sammlung Helm, Westpreuss. Provincial-Museum in Danzig. (Dr. La Baume) Sept. 1923";
4. „H. Wienhaus / Stantienit / aus Sammlung Helm, Westpreuss. Provincial-Museum in Danzig. (Dr. La Baume) Sept. 1923";
5. „H. Wienhaus / Beckerit / aus Sammlung Helm, Westpreuss. Provincial-Museum in Danzig. (Dr. La Baume) Sept. 1923";
6. „H. Wienhaus / Simetit / (Sicilianischer Bernstein) aus Sammlung Helm, Westpreuss. Provincial-Museum in Danzig. (Dr. La Baume) Sept. 1923";

Baltischer Bernstein von der Ostseeküste, Frische Nehrung und Samlandküste, 4 Gläser (Abb. 4, Flaschen mit Proben 7–10):

7. „H. Wienhaus / Bernstein vom Ostseestrande zwischen Palmnicken und Brüsterort (Samland), Sept. 1923 / 165 g";
8. „H. Wienhaus / Bernstein ausgesuchte hellere und klare Stücke vom Ostseestrande zwischen Palmnicken und Brüsterort, Sept. 1923, 38 g";
9. „H. Wienhaus / Bernstein von der Ostseeküste der Frischen Nehrung zwischen Bodenwinckel und Kahlberg, Aug. 1923 / 48 g";
10. „H. Wienhaus / Bernstein v. d. Ostseeküste (Samland) für die ersten Versuche, / 38 g".

Die Beschriftung erfolgte durch Prof. WIENHAUS handschriftlich bei Einordnung der Proben in seine Sammlung (vgl. aufgedruckt auf den Etiketten „H. Wienhaus").

Daß diese hier zitierten Proben „echt" sind, d.h. das fossile Harz-Material der HELM'schen Sammlung entstammen, geht aus den Behältnissen (Glasflaschen, Glasröhrchen) und deren Originaletiketten und der Beschriftung durch WIENHAUS hervor. Hier findet sich überall der handschriftliche Hinweis „aus Sammlung Helm, Westpreuss. Provincial-Museum in Danzig". Der Vermerk „von Dr. La Baume" deutet daraufhin, daß das Material vom ehemaligen Kustos (1911–1923) und seit 1923 Direktor des Westpreußischen Provinzialmuseums, Prof. Dr. LA BAUME, an Prof. H. WIENHAUS zur Untersuchung und Begutachtung übergeben wurde. Das Übergabedatum ist ebenfalls vermerkt (Jahreszahl 1923 und Monate August und September) (LA BAUME 1935; 1957; KLEEMANN & KÜHN 1956; BOHNSACK 1965; BAHR 1975).

Es sind im Nachlaß WIENHAUS auch noch folgende Proben vorhanden:

– Baltischer Bernstein von der Ostseeküste zwischen Danzig Umgebung (2 x), von Hartmannsche Ziegelei in Danzig (1 x), Preußisch Stargard (1 x) und der Kurischen Nehrung (Samland, Tagebau Palmnicken — Kraxtepellen 1 x, gesammelt September 1923); 4 Gläser (Flaschen) und lose Proben B8/2, B9, B10, B11, B12.

– Pyropissit (sog. Wachskohle bzw. Schwelkohle) aus dem Tagebau Stedten, Oberröblinger Revier (Sachsen-Anhalt), gesammelt 1923 (September), 4 Gläser: B2, B6, B7, B8/1.

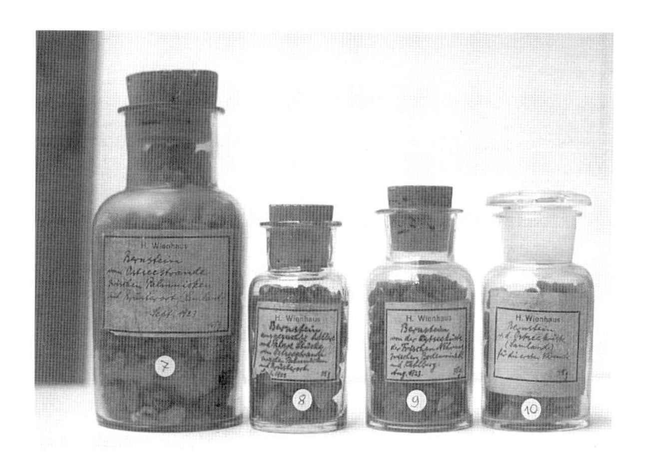

Abb. 4. Fossile Harze von der Ostseeküste, der Frischen Nehrung und der Samlandküste im Nachlaß H. WIENHAUS (Proben 7–10).

Diese Proben stehen vermutlich im Zusammenhang mit den Untersuchungen von Wienhaus mit Arbeiten an Lignin und Montanwachs (FISCHER 1982, 12) und der Isomerie der Coniferen-Harzsäuren in der Zeit von 1919–1924 (ABRAHAM *et al.* 1996);

– Harze aus Borneo (vgl. HILLMER, WEITSCHAT & VÁVRA 1992, 337, Abb. 5; HILLMER, WEITSCHAT & VÁVRA 1992), gesammelt 1933, 4 Gläser: B1, B3, B4, B5, B13;

– Retinit, Braunkohle von Löderburg b. Staßfurt (Sachsen-Anhalt);

– Egeln – Staßfurter-Mulde, 2 Stücke: B15, B16;

– Erdwachs/Galizien, Handstück B14.

In die wissenschaftliche Bearbeitung einbezogen wurden im folgenden nur die Harzproben 1–10 aus der Sammlung O. HELM, Danzig. Die anderen Proben Baltischen Bernsteins wurden zum Vergleich herangezogen.

Das übrige Material bleibt unberücksichtigt, da es keine Besonderheiten in seiner lithologischen Beschaffenheit aufweist gegenüber den aus Mitteldeutschland von KRUMBIEGEL & KOSMOWSKA-CERANOWICZ (1992) beschriebenen fossilen Harzen aus den subherzynen Braunkohlenlagerstätten. Es ist aber wertvolles Sammlungs-Belegmaterial für fossile Harze aus heute nicht mehr zugängigen geologischen Aufschlüssen und Tagebauen in Mitteldeutschland.

Die Proben 1–10 sowie B8/2, B9, B10, B11, B12 sind hinterlegt in der Bernstein-Sammlung des Archäologischen Museums in Gdańsk (Mariackastr. 25/26, PL 80-958 Gdańsk). Alle übrigen Proben befinden sich im Besitz von Herrn A. GLOGER, Leipzig bzw. sind über ihn oder über das Naturkundemuseum Leipzig zugängig.

B. Schriftverkehr

Vier Briefe aus dem Jahre 1926 (12. Mai–1. Juli) deuten auf das Interesse und die Bestrebungen hin, den Bernstein in der Parfümfabrikation zu nutzen (Vewendung von ätherischen Destillationsprodukten aus Baltischem Bernstein). Es ist das erste Dokumentationsmaterial, das wir in der Literatur zeigen können.

1. Technisch Chemisches Institut der Technischen Hochschule Berlin –Charlottenburg, Laboratorium für Öle und Fette — Prof. Dr. HOLDE; gerichtet an die Firma SCHIMMEL & Co, Fabrik für aetherische Öle, Miltitz bei Leipzig, v. 12. Mai 1926, Signatur: H/Ba. 905.

[WIENHAUS] – HOLDE (Schreib. v. 12. Mai 1926)

Prof. HOLDE schickte die Forschungergebnisse der drei Proben von einem aus einem Edelharz gewonnenen rektifizierten und raffinierten Ölen zur Überprüfung und bat um ein Urteil „ob und zu welchen (...) Preisen die Oele in den von Ihnen betriebenen Industriezweigen Verwendung finden können". Randbemerkung „Dem Geruch nach Bernsteinölfraktionen. Für uns ohne Interesse"

„Die Eigenschaften der Öle sind folgende:

Öl Nr	1a	2a	3a
Spez. Gew.	0,907 bei 18°	0,930 bei 20°	0,974 bei 18°
Siedegrenze	170–260°	–	295–350°
Mittlerer Siedepunkt*	240°	271°	318°
Optische Drehung**	+12,3°	+34,96°	–
Brechungskoeffiz.***	1,5053	1,5230	1,537

* Diejenige Temperatur, bei der 50% des Öles überdestilliert sind. ** [α] D^{20}. *** n^{20} D."

Das Ergebnis lautete, daß es sich bei den Ölen dem Geruch nach zwar um Bernsteinöl-Fraktionen handelt, die aber für den Industriezweig „Ätherische Öle" ohne Interesse sind und daher für Parfümfabrikation nicht in Frage kommen, insbesondere wegen ihres speziellen Geruchs nach Schwefelverbindungen. D.h. Destillationsprodukte von Bernstein spielten Ende der 20er Jahre des 20. Jahrhunderts als Grundstoff in der Parfümindustrie keine Rolle.

2. Preussische Bergwerks- und Hütten-Aktiengesellschaft; Zweigniederlassung: Bernsteinwerke Königsberg i. Pr. gerichtet an Prof. WIENHAUS, Miltitz (in Firma SCHIMMEL & Co) Königsberg (Pr)

Brief v. 26. Mai 1926, Sign.: J. – Nr. 1507.

Das Schreiben enthält einen erneuten Hinweis, dass Prof. HOLDE im Auftrage der Preussag „einige Untersuchungen am Bernsteinöl durchgeführt und dabei einige Destillationsprodukte hergestellt hat". Und weiter, dass „Prof. Holde es für möglich hielt, dass dieses als Fixativ für die Parfümfabrikation in Betracht kommen könnte". Die Preussag wollte mit WIENHAUS in Verbindung treten und ihn um seine Meinung darüber bitten „Inzwischen hören wir von Herrn Prof. Holde, daß er das Oel Ihrer Firma übersandt und von dieser einen ablehnenden Bescheid erhalten hat. Wir nehmen deshalb für richtig an, dass das erwähnte Oel von Ihnen selbst geprüft worden ist, und somit erübrigt sich eine nochmalige Anfrage an Sie. Wir wären Ihnen aber sehr dankbar, wenn Sie uns Ihre Ansicht über das erwähnte Oel mitteilen wollten (...). Ihr Anerbieten, sich mit gewissen Fragen unseres Preisausschreibens zu beschäftigen, nahmen wir dankend zur Kenntnis. Wir haben unsere Bergwerksverwaltung Palmnicken beauftragt, Ihnen zunächst von unseren 7 Kolophoniumsorten je 1/2 kg zugehen zu lassen, (...) mit denen Sie weitere Versuche vornehmen wollen (...) und von welcher Kolophoniumsorte wir Ihnen grössere Mengen zur Verfügung stellen sollen".

Für einen Vortrag von WIENHAUS zum Thema „Bernsteingewinnung und -verarbeitung" übersenden die Bernsteinwerke Königsberg i. Pr. 15 Lichtbilder (Fotos) und eine Druckschrift: „Der Bernstein und seine Wirtschaft" (N. N. 1936, 6. Aufl.).

3. Staatliche Berkwerksverwaltung Palmnicken, gerichtet an Prof. WIENHAUS, Miltitz (in Firma SCHIMMEL & Co)

Brief v. 28. Mai 1926, Sign. J. – Nr 4111/kl.

Das Schreiben enthält die Mitteilung der Absendung von Bernsteinproben (50 kg!) und Kolophoniumsorten (0,5 kg). Diese Proben sind im Nachlass WIENHAUS nicht mehr vorhanden. Vermutlich wurden sie bei den chemischen Untersuchungen in der Parfümfabrik SCHIMMEL & Co in Miltitz aufgearbeitet. Über diese Untersuchungen sind keine Ergebnisse bekannt.

4. Schreiben von Prof. WIENHAUS gerichtet an Preussag als die Beantwortung des Schreibens vom 26.5.1926. Brief v. 1. Juli 1926.

In dem Brief lesen wir: „die Proben wurden [...] in unserer Analytischen Abteilung als Bernsteinoel-Fraktionen erkannt und von unseren Parfümeuren wegen ihres Geruchs nach Schwefelverbindungen als ungeeignet befunden".

C. Photos der Bernsteinwerke Königsberg i. Pr.

Zur Vorbereitung eines Vortrages durch H. WIENHAUS über das Thema „Gewinnung und Aufbereitung des Bernsteins an der Samlandküste (Westküste Ostpreußens)" im ehemaligen Tagebau Palmnicken [Jantarnyj, Russische Föderation, Oblast Kaliningrad (Königsberg)] wurden ihm durch die Preussag, Bernsteinwerke Königsberg i. Pr. mit Schreiben von 26. Mai 1926 14 Bilder (Photos) zur Verfügung gestellt, von denen im Nachlass WIENHAUS noch sieben Photos (b, c, d, e, f, g) vorhanden sind. Die restlichen acht Photos a, i, k, l, m, n, o fehlen. Die Photos werden hier nicht veröffentlicht, sondern es wird Bezug genommen auf Literatur, wo diese Bilder bereits publiziert sind (BÖLSCHE 1927; ANDRÉE 1951). Im September 1996 wurde umfangreiches Photomaterial auf einer Bernsteinausstellung des Deutschen Bergbaumuseums in Bochum zum Thema *Bernstein — Tränen der Götter* gezeigt (GANZELEWSKI & SLOTTA 1996). Die Photos besitzen aber großen Dokumentationwert für die geologischen Aufschlußverhältnisse im Gebiet des ehemaligen Bernsteinbergbaues des Raumes Palmnicken – Kraxtepellen und der dortigen Verarbeitungsstätten.

Auf den historischen Photos war folgendes im einzelnen sichtbar (nach der Beschreibung im Brief vom 26. Mai 1926):

a — „Geologisches Profil an der samländischen Steilküste mit der bernsteinführenden „Blauen Erde" (...)". [Hier handelt es sich vermutlich um das schematische Profil der Samlandküste bei Palmnicken].

b — „Bernsteingewinnung am samländischen Strande durch Schöpfen".

c — „Die Tagesanlagen der im Jahre 1923 stillgelegten Tiefbaugrube „Anna" bei Kraxtepellen. Rechts im Hintergrund ist die Blaueerdewäsche zu sehen".

d — „Gewinnung der Blauen Erde im bergmännischen Tiefbau in der früheren Grube „Anna".

e — „Gesamtübersicht über den Tagebau, in welchem jetzt die Gewinnung der Blauen Erde stattfindet (= Palmnicken) (...)". [Gewinnung im Hoch- und Tiefbaggerbetrieb etwa Ende der 1920er Jahre].

f — „Blaue Erde-Wäsche (Rohwäsche) (...). Aufbereitung des Rohmaterials aus dem Tagebau mit Wasserkanonen über einer Siebanlage".

g — „Ein Blick in die Reinwäsche, in welcher der Bernstein in rotierenden Trommeln mit Wasser und Sand von einem Teil der ihn umgebenden Verwitterungsrinde befreit und in einige verschiedene Korngrößen zerlegt wird".

h — „Einige Bernsteinsorten verschiedener Größe".

Fehlende Photos i–l — „Einige wertvolle Bernsteingegenstände aus unserer kunstgewerblichen Sammlung und zwar":

i — „Geschnitzter Pokal aus dem Besitz der Herzöge von Kurland";

k — „Kleine geschnitzte Bernsteinfigur, wahrscheinlich chinesische Arbeit, etwa 18. Jahrhdt";

l — „Geschnitzter Bacchus aus einem Stück, etwa 15 cm hoch, wahrscheinlich 17. Jahrhdt";

m — „Schmuckkästchen aus Bernstein; moderne Arbeit aus dem Betriebe der Staatlichen Bernsteinmanufaktur G. m. b. H. [Königsberg i. Pr]";

n — „Geschnitztes Schachspiel, ebenfalls von der Staatlichen Bernsteinmanufaktur G. m. b. H.

hergestellt. Die Figuren stellen die Römer und Germanen dar";

o — „Unterscheidung von Bernstein und Kunstharzen nach dem spezifischen Gewicht".

Otto HELM — seine Bernsteinforschungen und -sammlungen in Danzig

Weiten Kreisen bekannt wurde Otto HELM durch seine chemischen Untersuchungen von Bernstein und von prähistorischen Metallfunden (s. Schriften von O. HELM 1881; 1882; 1896; OTTO 1981 mit Porträt; LAKOWITZ 1928).

HELM (Abb. 5) wurde am 21.02.1826 in Stolp (Słupsk) geboren und starb am 24.03.1902 in Danzig (Gdańsk). Nach einer Apothekerausbildung erwarb er im Jahre 1855 für 91.500 Mark die „Polnische Apotheke" (Hendewerk'sche Apotheke) in Danzig, die er jedoch 1874 wieder verkaufte. Bis zu seinem Tode lebte er in hohem Ansehen in Danzig. Er war lange Zeit Stadtrat. Er erwarb sich als Privatsammler ferner große Verdienste in der Naturforschenden Gesellschaft Danzig, der er seit 1865 angehörte und im Westpreußischen Botanisch-Zoologischen Verein. Auch war er ehrenamtlicher Mitarbeiter des Westpreußischen Provinzialmuseums.

Abb. 5. Apotheker und Chemiker Dr. Otto HELM * 21. - Februar 1826 in Stolp, + 24. März 1902 in Danzig. 1855–1874 Besitzer der Polnischen Apotheke in Danzig, danach chemischer Sachverständiger und Gutachter.

Seine Persönlichkeit als Mensch faßt ein Nachruf in der *Apotheker-Zeitung* (1902, 17, 215) mit folgenden Worten zusammen: „Dr. Helm war ein liebenswürdiger Mensch, der viele Freunde, aber wohl kaum einen Feind besessen hat".

Das Bernsteinthema beschäftigte HELM Zeit seines Lebens. In den Jahren 1877–1902 veröffentlichte HELM 22 Arbeiten (BECK *et al.* 1966; 1967). Regelmäßig und ausführlich veröffentlichte er *Mitteilungen über Bernstein* (1878–1891) in den *Schriften der Naturforschenden Gesellschaft in Danzig*. Er galt damals als bester Kenner des Bernsteins und genoß internationales Ansehen auf Grund seiner Untersuchungen nicht nur einheimischer fossiler Harze, sondern auch von weltweiten fossilen Harzfunden (Rumänien, Sizilien, Appeninen, Spanien, Japan, Birma, Bosnien, Libanon, Grönland). HELM ist auch als Erstautor bekannt geworden; er hat folgende neue Harzarten beschrieben: Gedanit (= spröder Bernstein; 1878), Glessit (1881), Rumänit (1881; 1891), Burmit (1894). Auch Simetit wurde nach der Erwähnung im 17. Jahrhundert (CARRERA 1639) durch HELM (1881; 1882), HELM & CONWENTZ (1886) ausführlich bearbeitet.

Seine guten entomologischen und botanischen Fachkenntnisse spiegeln sich in den Untersuchungen von Bernsteininklusen wieder. Eine Pflanze und zwei Käfer tragen seinen Namen: *Stephanostemon helmi* CONW., *Palaeomastigus helmi* SCHAUF, *Arthropterus helmi* SCHAUF.

Für seine wissenschaftlichen Forschungsarbeiten sowie für seine Verdienste um die wissenschaftliche Erforschung Westpreußens verlieh ihm die Philosophische Fakultät der Albertus-Universität in Königsberg 1899 die Würde eines Ehrendoktors.

Die ehemalige Bernstein-Sammlung von Otto HELM, neben denen von Anton MENGE (1808–1880) (OHLERT 1881), Hugo CONWENTZ (früher — CONVENT; 1855–1922) und Paul DAHMS (1866–1922), war eine der größten Sammlungen in Danzig, wurde nach seine Tode durch Schenkung in den Besitz des Westpreußischen Provinzialmuseums für Naturkunde und Vorgeschichte in Danzig übergeführt (HENSCHE

1865; JENTZSCH 1892; KUHSE 1924; LA BAUME 1935; LIPPKY 1980). Die ehemalige Bernstein-Sammlung von Otto HELM, bestand aus zwei Einzelkollektionen. Eine regionale Kollektion enthielt das Originalmaterial seiner chemischen und physikalischen Forschungen sowie die Formen und Varianten des Succinits. Die zweite Kollektion enthielt die organischen Inklusen — 5000 Stücke (N. N. 1902).

Alle vier Bernsteinsammlungen bildeten in den 30ger Jahren dieses Jahrhunderts, neben der Königsberger Albertus-Universitäts-Sammlung im Geologisch-Paläontologischen Institut, die zweitgrößte Bernstein-Sammlung einer wissenschaftlichen Institution der Welt. Die Ergebnisse der Forschungen mit diesem Material wurden von oben genannten Wissenschaftlern zumeist in den *Schriften der Naturforschenden Gesellschaft zu Danzig* veröffentlicht (LA BAUME 1957) und deren Originale gingen später in die genannten Museen und Institute über.

Die Sammlung HELM ging aber bei der Zerstörung der Stadt Danzig während des zweiten Weltkrieges in den Jahren 1942 und 1944 und während der Auslagerung in den Weichselwerder 1941–1945 verloren (LA BAUME 1957, 8; LIPPKY 1980). Über das Schicksal der Danziger Sammlungen des Provinzialmuseums am Kriegsende berichtet im einzelnen LIPPKY (1980, 113–115; Abb. 6 & 7).

Das Westpreußische Provinzialmuseum ab 1. April 1923 umbenannt in Staatliches Museum für Naturkunde und Vorgeschichte, wurde am 18. September 1880 gegründet (CONWENTZ 1905; LA BAUME 1930; LIPPKY 1980) und hatte seine Heimstätte im Grünen

Abb. 6. Rekonstruiertes „Grünes Tor" am Langen Markt (Długi Targ) in Danzig, Sitz des ehemaligen Westpreußischen Provinzialmuseums.

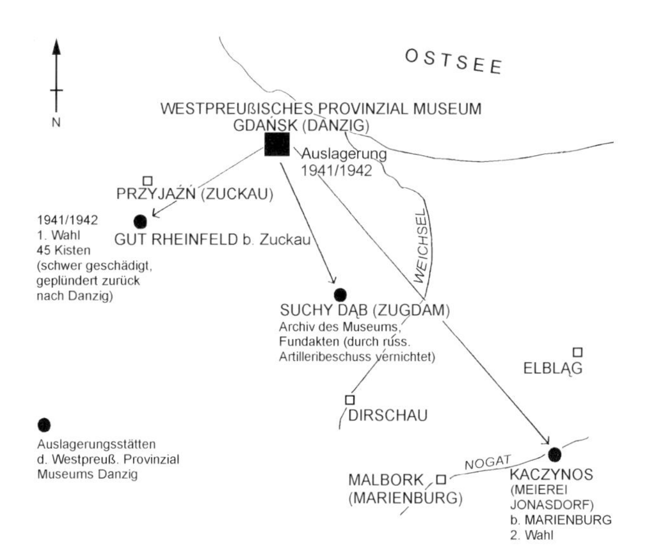

Abb. 7. Auslagerungsstätten des Archiv- und Sammlungsmaterials, darunter auch die Sammlung O. HELM des Westpreußischen Provinzialmuseums Danzig, während des 2. Weltkrieges (1941, 1942) (nach Unterlagen von LIPPKY 1980, Entwurf KRUMBIEGEL 1997).

Tor (Zielona Brama) an der Mottlau (Stara Motława) am Ende des Langen Marktes (Długi Targ) in Danzig. Hier im Obergeschoß des Tores, war bereits von 1743–1869 der Sitz der *Naturforschenden Gesellschaft zu Danzig* untergebracht (SCHUMANN 1893; MAMUSZKA 1993). Nach dem Wiederaufbau von Danzig nach 1945 befinden sich aber hier die Werkstätten der weltberühmten polnischen Restaurateure (Pracownia Konserwacji Zabytków) sowie der Verein der Baltischen Städte (Związek Miast Bałtyckich).

Materialbeschreibung der fossilen Harze aus der Sammlung WIENHAUS-HELM

Ergebnisse der Forschungen

Der Teilnachlaß von H. WIENHAUS (aus Sammlung HELM) enthält nachfolgend aufgeführte 10 Proben von fossilen Harzen, von denen 10 lithologisch und 6 Proben mit Hilfe der Infrarotspektroskopie überarbeitet wurden.

A. Materialbeschreibung

Probe 1. „Mürber Succinit" (HELM 1896; det. STOUT *et al.* 1994; 1995) nach O. HELM / H. WIENHAUS, Fundort nicht bekannt.

Menge 12 g, 2 St.: 3,5 x 2,5 x 2,1 cm, ∅ 1cm großes Bruchstück; Verlust 2 g; IRS 438, 447 — Gedanit (?), IRS 446 (1995) — Gedanit.

Rotbraunorange mit Verwitterungsrinde; durchsichtig bis durchscheinend; spröder glänzender Bruch; stellenweise weißgelblich angewittert.

Probe 2. „Gedanit, mürber [?], frei von Bernsteinsäure" (HELM 1878, det. STOUT *et al.* 1995) nach O. HELM / H. WIENHAUS, Fundort nicht bekannt.

Menge 25 g, 17 St., ∅ 2–3 cm; Verlust 3 g; IRS 434 (1995) — Succinit.

6 Stücke — honigbraun mit schwacher Verwitterungsrinde, durchscheinend; muschliger glänzender Bruch. 7 Stücke — durchsichtig klar, splittriger Bruch honigbraun. 4 Stücke — honigbraun, splittrig, mürbe, stark angewittert.

Probe 3. „Glessit, vielleicht fossiles Gummiharz" (HELM 1881), nach O. HELM / H. WIENHAUS, Fundort nicht bekannt.

Menge 6 g, 2 St.: (a) 2,4 x 1,6 x 0,8 cm; (b) 3,2 x 2 x 1,6 cm; Verlust 3 g; Das Stück IRS 437 (1995) — Glessit (?).

(a) dunkelbraun undurchsichtig, matter Bruch; (b) rotbraun, gebändert undurchsichtig, stark mürbe, gelblichorange Verwitterungsrinde, fast zuckerkörnig, glänzend muschliger Bruch.

Probe 4. „Stantienit" (PIESZCZEK 1880, sog. Schwarzharz; HELM 1880; DAHMS 1922) nach O. HELM / H. WIENHAUS, Fundort nicht bekannt.

Menge 12 g, 5 St.: 3 x 1,2 x 1,4 cm; 2,1 x 1,8 x 1 cm; 2,2 x 1,8 x 0,9 cm; 2,7 x 1,6 x 1,4 cm; 1,7 x 1,5 x 0,7 cm; Verlust 2 g; IRS 435 (1995) — Stantienit.

Schwarzbraun, undurchsichtig, schwarzer muschliger Bruch, z. T. rissige bräunliche Verwitterungsoberfläche, kantengerundet.

Probe 5. „Beckerit" (PIESZCZEK 1880: sog. Braunharz; HELM 1880; BECK *et al.* 1986) nach O. HELM / H. WIENHAUS, Fundort, nicht bekannt.

Menge 4 g, 2 St.: 5 x 1,7 x 1,2 cm; 2,6 x 1,6 x 0,9 cm; Verlust 1 g; IRS 436 (1995) — Beckerit.

Dunkelbraun, undurchsichtig, matte, wulstige z. T. gerippte Oberfläche, spröde, z. T. hellbraun verwittert.

Probe 6. „Simetit (Sizilianischer Bernstein)" (HELM 1882; KOHRING & SCHLÜTER 1992) nach O. HELM / H. WIENHAUS, Fundort nicht bekannt.

Menge 4 g; 16 St. — 14: 0,7–1,5 cm ∅; Verlust 1 g; IRS 439 (1995) — Simetit (?) verwittert.

Schwarz, undurchsichtig, rot durchscheinend; glatte, mattglänzende Oberfläche 2 Stke. rötlichbraun, durchsichtig.

Zusätzliche Proben, die Prof. WIENHAUS vermutlich von Prof. LA BAUME für chemisch-analytische Untersuchungen zur Verfügung gestellt wurden, die aber nicht in die eigenen Untersuchungen einbezogen wurden.

Probe 7. Bernstein nach H. WIENHAUS; Okt. 1995 vom Ostseestrand zwischen Palmnicken und Brüsterort (Majak), Samland, Russia.

Menge 165 g.

Hellgelbe bis hellbraune, mit Verwitterungsrinde, abgeschliffen, löchrig angewittert.

Probe 8. Bernstein nach H. WIENHAUS; Okt. 1995 vom Ostseestrand zwischen Palmnicken und Brüsterort (Majak), Samland, Russia.

Menge 37 g.

Hönigbraun bis hellbraun, z.T. durchsichtig bis durchscheinend, splittriger Bruch, z.T. mit schwacher Verwitterungsrinde, gelbliche Schlieren.

Probe 9. Bernstein nach H. WIENHAUS; Oktober 1995, von der Ostseeküste der Frischen Nehrung (Zalew Wiślany) zwischen Bodenwinkel (Kąty Rybackie) und Kahlberg (Krynica Morska).

Menge 49 g.

Honigbraun, gelb dunkelbraun mit Verwitterungsrinde, abgerundete Bröckchen, durchscheinend bis gelblich trübe und undurchsichtig.

Probe 10. Bernstein nach H. WIENHAUS; Oktober 1995, von der Ostseeküste (Samland).

Menge 39 g.

Eigenschaften wie Probe 9.

B. Infrarotanalysen

Alle Infrarot-Kurven, von den Proben der HELMSCHEN Sammlung sind untypisch. Im allgemeinen weisen alle Spektren auf sehr verwittertes Material hin: sie sind platt (Abb. 8).

Probe 1. „Mürber Succinit".

Mürber Bernstein (HELM 1896), durchsichtige gelbe Probe, zeigt zwei mehr dem Gedanit ähnliche Kurven (IRS 438, 447), die dunkelgelblichen Stückchen und eine — (IRS 446) des hellgelben Materials haben ein typisches Gedanit-Spektrum ergeben.

Probe 2. „Gedanit, mürber [?]* Bernstein frei von Bernsteinsäure".

Gedanit (= spröder Bernstein, HELM 1878). Diese ganz durchsichtige und honiggelbe Probe, hat leider eine typische Succinit Kurve (IRS 434), mit typischer Baltischer Schulter zwischen 1150 und 1250 cm^{-1} gezeigt.

Probe 3. „Glessit, vielleicht fossiles Gummiharz".

Glessit (HELM 1881). Obwohl die Stücke dunkel rot und etwas dem Glessit ähnlich sind, kann man die

* Diese Nachschrift wird schon in: A. Proben fossile Harze hinein gestellt. Nach HELM wäre es: „spröder Bernstein" diesen Fehler hat vielleicht LA BAUME gemacht (?).

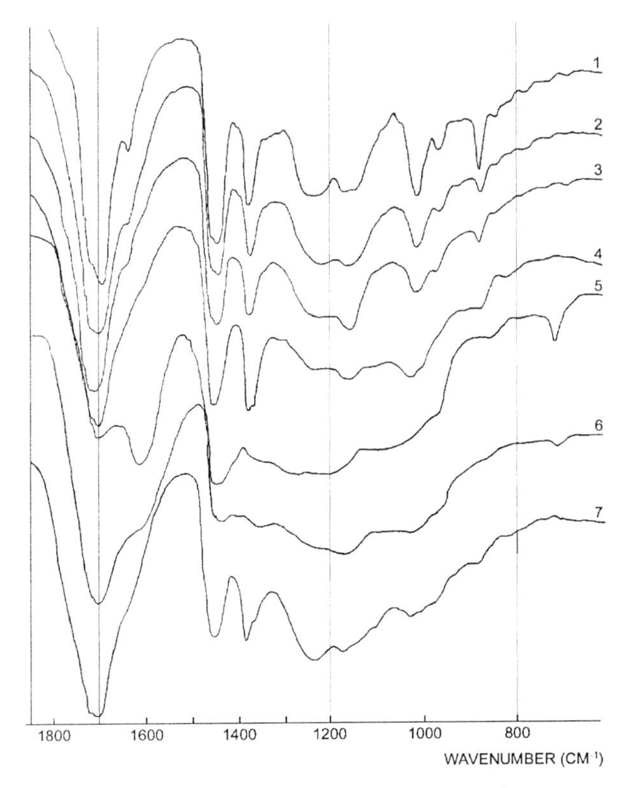

Abb. 8. Infrarotspektren von fossilen Harzen des Samlandes aus der Sammlung des Apothekers O. HELM im Teilnachlaß von H. WIENHAUS, Tharandt. 1, 2 — Mürber Succinit, IRS 438, 446; 3 — Gedanit, IRS 434; 4 — Glessit, IRS 437; 5 — Stantienit, IRS 435; 6 — Beckerit, IRS 436; 7 — Simetit, IRS 439 (die Nummern der IRS sind im Museum der Erde Warschau im IRS-Katalog eingeordnet).

erhaltene Kurve (IRS 437) auch als Succinit (?)-Kurve interpretieren. Eine ähnliche untypische Kurve hat die rote Variante des Glessits aus dem Tagebau Lohsa (IRS 209)**. Sie zeigt eine sehr starke Absorption bei 1700 cm^{-1} (Carbonyl-Gruppe) und zwischen 1100–1250 cm^{-1} und fast keine Absorption bei 888 cm^{-1} (exocyclic methylene-Gruppe).

Probe 4. „Stantienit".

Stantienit (PIESZCZEK 1880). Die Kurve IRS 435 weist auf Stantienit hin.

Probe 5. „Beckerit".

Beckerit (PIESZCZEK 1880). Die Kurve IRS 436 weist auf Beckerit hin. Beckerit wurde mit schwacher Absorption der Welle bei 1710 cm^{-1} (Carbonylgruppe), kräftiger im Bereich 1615–1620 cm^{-1} (H_2O) und sehr deutlicher (obgleich mit verschiedener Intensität) bei 720 cm^{-1} (Terpene, CH_2) an 7 Stücken (aus dem Samland und aus dem Tagebau Goitsche) erkennbar. Inzwischen wurde bei den Beckerit-Proben noch eine Differenzierung erkannt — mit kleiner oder grosser Absorption der Welle bei 720 cm^{-1}. Eine typische (inzwischen für Succinit) Bande bei 890 cm^{-1} ist in den Beckeritkurven sehr wenig oder gar nicht zu beobachten.

Probe 6. „Simetit (Sicilianischer Bernstein)".

Simetit (CARRERA 1639; HELM 1881; 1882; HELM & CONWENTZ 1886). Der Simetit ist so stark verwittert, daß die Kurve 439 schwer zu interpretieren ist. Als Vergleichsmaterial wurden zwei Kurven von Simetit aus der Berliner Bernsteinsammlung von coll. SIMON (IRS 79) und aus der Bernsteinsammlung in Warschau (Inv. Nr 3862 — IRS 399) gezeigt (Abb. 8).

Kurve Nr.	IRS Nr.	nach HELM	nach IRS
1.	446	mürber Bernstein	Gedanit
2.	438	mürber Bernstein	Gedanit
-	447	mürber Bernstein	Gedanit
3.	434	Gedanit	Succinit
4.	437	Glessit	?
5.	435	Stantienit	Stantienit
6.	436	Beckerit	Beckerit
7.	439	Simetit	?Simetit

Tabelle 1. Infrarotspektren von HELMSCHEN Proben aus dem Nachlass WIENHAUS (siehe Abb. 8).

Schlussfolgerungen

Die im Sammlungsnachlaß WIENHAUS aufgefundenen Harzproben sind heute von besonderem Wert für die Bernsteinforschung, vor allem als Vergleichsmaterial. Es sind wahrscheinlich die einzigsten erhaltengebliebenen Proben der Sammlung HELM, aus dem Westpreußischen Provinzialmuseum, die aber im Jahre 1902 wenigstens 5000 Stücke enthielt.

Das meiste vorliegende Harz ist „Mürber Bernstein". Beschreibungen des mürben Bernsteins aus der Baltischen Region erfolgten hauptsächlich in Fach- und Lehrbüchern der Mineralogie und stammen aus der ersten Hälfte des 20. Jahrhunderts (DAHMS 1901; KAUNHOWEN 1914). In der neueren Literatur findet man fast keine Hinweise über mürben Bernstein (PACLT 1953; STRUNZ 1977). Diese Art von Harzen, wie sie im Nachlass WIENHAUS gefunden und zur Verfügung gestellt wurden, gab es in Sammlungen bisher nicht (das heisst: die Autoren fanden bei der Recherche nach Vergleichmaterialien in vielen Museen

** 13. GLESSIT; IRS 209, Braunkohlen Tgb. Lohsa, aus Quartär Sedimenten, Tiefe ca. 50 m, coll. SAUER, Litschen (Lausitz). Geiseltalmuseum Halle (Saale): GM/O/18, KRU 14 (4,5 x 7,5 cm); C: 79,84–80,28 (80,06)%; H: 11,64–11,92 (11,78)%; S: < 0,3%, Vergl. mit IRS 194 (Göttingen) und IRS 238 (Tgb. Golpa).

den mürben Bernstein nicht). Neue Forschungsergebnisse, wie auch der Name Gedano-Succinit für den mürben Bernstein, wurden erst durch S. SAVKEVITCH (1970, 1983) bekannt (ober Vergleichsmaterial herangezogen hat, ist unbekannt). In der Bernsteinsammlung des Museums der Erde in Warschau befindet sich Gedano-Succinit [Invent. Nr. 2222], der von SAVKEVITCH beschriftet ist und als **Neotypus** verwendet wird. Die Bestimmung des Gedano-Succinits zog weitere Forschungen nach sich. Die Verschiedenheit zwischen dem mürben Bernstein, dem Gedanit und dem Succinit, die HELM in botanischer Herkunft sah, haben neue Interpretationen in den neuesten Arbeiten (SAVKEVITCH 1983; STOUT *et al.* 1995) gefunden.

Bei den Forschungen an den HELMschen Stücken, haben die Autoren leider nicht die erhoffte Bestätigung gefunden, daß Mürber Bernstein und Gedano-Succinit fossile Harze ein und derselben Art sind. Infolge der divergenten Ergebnisse der Infrarotspektroskopie muß die Frage offen bleiben.

Die übrige Proben haben leider wegen des hohen Verwitterungsgrades des Materials auch nicht die erwarteten Ergebnisse gezeigt: z.B. sind die IR-Kurven sehr plattig, oder das untersuchte Stück Gedanit ergab ein Succinit-Spektrum.

Der historische Wert der Briefe von 1926 aus dem WIENHAUS-Nachlaß ist auch von großer Bedeutung: es ist das erste Archivmaterial über die Verwendung von Bernstein in der chemischen Industrie, die wir in der Bernstein-Literatur finden konnten. Diese Forschungen führten gleichzeitig Prof. WIENHAUS in Miltitz bei Leipzig und Prof. HOLDE in Berlin durch. Beide Wissenschaftler nahmen diese Untersuchungen im Auftrag und im Interesse der Preussag vor.

Ihre Ergebnisse zeigten, daß es sich bei den Ölen dem Duft nach zwar um Bernsteinöl-Fraktionen handelt, die aber für den Industriezweig „Ätherische Öle" ohne Interesse sind und daher für Parfümfabrikation nicht in Frage kamen, insbesondere wegen ihres speziellen Geruchs nach Schwefelverbindungen. Trotzdem sollte man auf dieses Problem zurückkommen, obgleich sich zeigte, daß die Destillationsprodukte von Bernstein Ende der 20er Jahre des 20. Jahrhunderts als Grundstoff in der Parfümindustrie keine Rolle spielten. Heute jedoch, wo man nach Naturrohstoffen fahndet und im Samland mehr als 700 Tonnen Rohbernstein pro Jahr abgebaut werden, wovon aber nur etwa 13% für Schmuck, Kunsthandwerk und Industrie genutzt werden, ist die Bedeutung des Bernsteins in der Chemieindustrie ebenso wichtig, wie viele andere Wirtschaftsfragen. Dafür spricht nach SLOTTA (1997) der Hinweis, daß die „AG Russischer Jantar" aus Jantarnyj in der Lage ist, jährlich 200 Tonnen Kolophonium, 300 Kilogramm Bernsteinsäure* und 20 Tonnen Bernsteinöl für industrielle Zwecke zu liefern, eine bedeutende Rohstoffreserve für die Perspektive des 21. Jahrhunderts!

Danksagung

Für die Möglichkeit das vorliegende wissenschaftshistorische fossile Harzmaterial von Otto HELM auszuleihen, untersuchen und bearbeiten zu können, gilt unser Dank nachfolgenden Institutionen und Personen: Prof. Dr. O. WIENHAUS (Technische Universität Dresden, Inst. f. Pflanzenchemie und Holzforschung), Dipl.-Geol. R. BAUDENBACHER (Naturkunde-Museum, Leipzig), H. AHLHEIM (Dresden), A. GLOGER (Leipzig), O. PRIESE (Genthin), Kustos Dr. J. BARFORD (Ostpreußisches Landesmuseum Lüneburg), Stiftung Martin-Opitz-Bibliothek (Herne/Westfalen), Museumsdirektor H. J. SCHUCH (Westpreußisches Landesmuseum Münster).

Günter KRUMBIEGEL & Brigitte KRUMBIEGEL
Clara-Zetkin-Str. 16
D-06114 Halle/Saale, Germany

Barbara KOSMOWSKA-CERANOWICZ
Muzeum Ziemi PAN
Aleja Na Skarpie 27
00-488 Warszawa, Poland

POZOSTAŁOŚCI ZBIORÓW BURSZTYNU OTTO HELMA W SPUŚCIŹNIE WIENHAUSA

Günter KRUMBIEGEL,
Brigitte KRUMBIEGEL,
Barbara KOSMOWSKA-CERANOWICZ

Streszczenie

Spuścizna Heinricha WIENHAUSA eksponowana w 1995 roku na wystawie pt. *Bernsteinsplitter* w Lipsku obejmuje: (1) 10 próbek bursztynu i innych żywic

* Bernsteinsäure diente schon im 19. Jahrhundert (siehe: *Archiv der Pharmazie*) zur Arzneifabrikation.

kopalnych z nieistniejących już dawnych gdańskich zbiorów Westpreussisches Provinzial-Museum in Danzig (powstałego w 1880 r.), przemianowanego w 1923 r. na Staatliches Museum für Naturkunde und Vorgeschichte, z kolekcji Otto HELMA, które zostały przekazane WIENHAUSOWI przez LA BAUMA w 1923 r.; (2) 4 listy z 1926 r. dotyczące problemu wykorzystania bursztynu w przymyśle chemicznym (do wyrobu perfum); (3) 7 fotografii sprzed 1926 r. z rejonu kopalni Palmnicken ze zbiorów PREUSSAG (Preussische Bergwerks- und Hütten- Aktiengesellschaft Bernsteinwerke Königsberg i. Pr.)

Próbki z kolekcji HELMA, która liczyła w 1902 r. co najmniej 5000 okazów, stanowią dziś jedyny ślad drugich co do wielkości po Königsbergu zbiorów bursztynu w przyrodniczym muzeum, które istniało w Gdańsku do 1945 r. Otto HELM, aptekarz gdański, w latach 1877–1902 opublikował 22 prace dotyczące chemicznych i fizycznych właściwości bursztynu i innych żywic kopalnych. Po raz pierwszy opisał: **gedanit** (1878), **glessyt** (1881), **rumenit** (1881), **birmit** (1894), bursztyn **kruchy** (mürber) (1896).

Autorzy szczegółowo opisali wszystkie próbki HELMA ze spuścizny WIENHAUSA, a badaniom w podczerwieni poddali glessyt (IRS 437), gedanit — na etykiecie zapisano: **łamliwy** *(spröder)* bursztyn (434), ***kruchy*** *(mürber) bursztyn* (438, 446, 447), stantienit (435), bekeryt (436) oraz symetyt (439). Najcenniejszą żywicą kopalną jest ***kruchy*** *(mürber) bursztyn.* Ten rodzaj żywicy, aż do chwili udostępnienia spuścizny WIENHAUSA, nie zachował się w żadnych zbiorach (to znaczy w poszukiwaniach materiałów porównawczych autorzy dotychczas go nie znaleźli). Uzyskane wyniki, częściowo ze względu na silny stopień zwietrzenia materiału, były niestety różne od spodziewanych: krzywe są silnie spłaszczone, nie uzyskano krzywej gedanitu (próbka opisana jako gedanit okazała się sukcynitem). Krzywa *kruchego (mürber) bursztynu* dała krzywą gedanitu, co stawia pod znakiem zapytania porównywanie go z gedanosukcynitem.

Jak wynika z zachowanych listów, w 1926 r. badania oleju bursztynowego pod kątem jego zastosowania jako płynu utrwalającego przy wyrobie perfum prowadzili równocześnie prof. H. WIENHAUS z Miltiz-Lipsk i na zlecenie firmy PREUSSAG prof. HOLDE w Berlinie. Frakcje oleju bursztynowego według opinii WIENHAUSA ze względu na zapach związków siarki nie spełniały zaplanowanych wymogów. Jak wynika z korespondencji, WIENHAUS nie zaniechał wykonywania analiz, a zakłady w Königsbergu zainteresowane uzyskaniem pozytywnego wyniku, dostarczały surowego bursztynu i różnych gatunków kalafonii do dalszych badań.

Literatur

ABRAHAM J., BERGER D., FRELLSTEDT H., SCHRÖDER W. & SLOTTA R.

1996 Braunkohle mit Montanwachs — zum Verbrennen zu schade, [*in*:] Ganzelewski M. & Slotta R., Bernstein — Tränen der Götter, *Veröffentlichungen aus dem Deutschen Bergbau-Museum Bochum*, **64**, 493–504.

ANDRÉE K.

1951 Der Bernstein — Das Bernsteínland und seín Leben, *Kosmos*, 1–96, Franck'sche Verlagshandlg, Stuttgart.

BAHR E.

1975 *La Baume, Wolfgang. Altpreußische Bíographie, Bd. III u. Ergänzgg. zu Bd. I u. II*, 1–988, Forstreiter K. & Gause F. (Hrsg.), , N. G. Elwert Verlag Marburg/ Lahn.

BAUDENBACHER R.

1995 Bernsteinsplitter, *Faltblatt — Sonderausstellung i. Naturkundemuseum Leipzig 1995*, 1–8, Leipzig.

1996 „Bernsteinsplítter" — ein Rückblick, *Veröff. Naturkundemuseum Leipzig*, **14** (1996), 141—148, Leipzig.

BECK C. W., GERVING & WILBUR E.

1966, 1967 The provenience of archaeological Amber-Artifacts. An Annotated Bibliography. Teil I: 8th Century- B.C. to 1889; *Art and Archaeology techn. Abstracts*, (1966), **2**, 215–302. Teil II: 1900 to 1966; *AATA*, **6** (1967) (3), 201–280.

BECK C. W., LAMBERT J. & FRYE J. S.

1986 Beckerite, *Physics and Chemistry of Minerals*, **13** (1986), 411–414, New York.

BÖLSCHE W.

1927 Im Bernsteinwald, *Kosmos*, 1–78, Franck'sche Verlagshandlg, Stuttgart.

BOHNSACK D.

1965 Professor Dr. Wolfgang La Baume zum 80. Geburtstag, *Das Ostpreußenblatt*, **16** (6), 1–10, v. 6. Febr. 1965.

CARRERA P.

1639 *Memorie historici della cita di Catania*, Catania.

CONWENTZ H.

1905 *Das Westpreußische Provinzial-Museum 1880 bis 1905*, Danzig.

DAHMS P.

1901 Mineralogische Untersuchungen über Bernstein VII, : Ein Beitrag zur Constitutionsfrage des Bernsteins, *Schriften der Naturforschenden Gesellschaft zu Danzig*, **10** (2–3), 243–257, Danzig.

1922 Schwarzharz und Ostseebernsteine, *Schriften der Naturforschenden Gesellschaft zu Danzig*, **15** (3/4), 57–68, Danzig.

FISCHER F.
1982 Heinrich Wienhaus — Werk und Wirken, *Wiss. Z. Techn. Univ. Dresden*, **31** (1982), 187–193; Festkolloquium 100. Geburtstag von Heinrich Wienhaus, 7–20, Dresden.

GANZELEWSKI M., R., SLOTTA R., (HRSG.) U.A.
1996 Bernstein — Tränen der Götter, *Veröffentlichungen aus dem Deutschen Bergbau-Museum Bochum*, **64**, 1–585, Bochum, edition Glückauf, Verlag Glückauf, Essen 1997.

HELM O.
1878 Gedanit, ein neues fossiles Harz, *Archiv der Pharmacie*, N.F., **10** (1878) (6), 503–507, Halle.

1880 Ueber einige neue harzähnliche Fossilien des ostpreussischen Samlandes, *Archiv der Pharmacie*, N.F., **17**, (3); **59** (1880), 433–436, Halle.

1881 Glessit, ein neues in Gemeinschaft von Bernstein vorkommendes fossiles Harz, *Schriften der Naturforschenden Gesellschaft zu Danzig* N.F., **5**(1881) (1/2), 291–296, Danzig.

1882 Ueber sicilianischen Bernstein, *Schriften der Naturforschenden Gesellschaft zu Danzig*, N.F., **5**(1882), (3), 8–14, Danzig.

1891 Über Rumänit, ein in Rumänien vorkommendes fossiles Harz, *Schriften der Naturforschenden Gesellschaft zu Danzig, N. F.*, **7** (4), 186–187.

1894 Ueber Birmit, ein in Oberbirma vorkommendes fossiles Harz, *Schriften der Naturforschenden Gesellschaft zu Danzig, N. F.*, **8** (3/4), 63–66.

1896 Ueber den Gedanit, Succinit und eine Abart des letzteren, den sog. mürben Bernstein, *Schriften der Naturforschenden Gesellschaft zu Danzig* , N.F., **9** (1896) (1), 52–57, Danzig.

HELM O. & CONWENTZ H.
1886 Sull 'ambra di Sicilia, *Malphighia*, **1** (2), 49–56, Messina.

HENSCHE A.
1865 Beilage. C. Bericht über die Sammlung der Physikalisch-ekonomischen Geselschaft 1865, *Schriften der physikalisch-ökonomischen Gesellschaft zu Königsberg*, **6**, 21–23, Königsberg.

HILLMER G., WEITSCHAT W. & VÁVRA N.
1992 Bernstein aus dem Miozän von Borneo, *Naturwiss. Rundsch.*, **45** (2), 72–74, Stuttgart.

HILLMER G., VOIGT P. C. & WEITSCHAT W.
1992 Bernstein im Regenwald von Borneo, *Fossilien*, **9** (6), 336–340, Weinstadt.

JENTZSCH A.
1892 *Führer durch die Geologischen Sammlungen des Provinzialmuseums*, Königsberg.

KAUNHOWEN F.
1914 Der Bernstein in Ostpreußen, *Jahrbuch der Preußischen Geologischen Landesanstalt*, **34** (2), 1–80, Berlin.

KLEEMANN O. & KÜHN H.
1956 Wolfgang La Baume. Curriculum vitae und Bibliographie. Rhein. 'Forschgn. zur Vorgesch., **5**, 137–143, *Documenta Archaeologica*. Ludw. Röhrscheid Verl., Bonn.

KOHRING R. & SCHLÜTER T.
1992 Der Simetit — das fossile Harz Siziliens, *Fossilien*, **9** (4), 221–226, Weinstadt.

KRUMBIEGEL G. & KOSMOWSKA-CERANOWICZ B.
1992 Fossile Harze der Umgebung von Halle (Saale) in der Sammlung des Geiseltalmuseums der Martin-Luther-Universität Halle-Wittenberg. *Wiss. Z. Univ. Halle*, XXXXI '92 M, **6**, 5–35, Halle.

KRUMBIEGEL G. & KRUMBIEGEL B.
1995 „Bernsteinsplitter", czasowa wystawa bursztynu w Lipsku. Muzeum Ziemi PAN. Sekcja Owadów Kopalnych PTE, XII Spotkanie „Inkluzje organiczne...", 27.10.1995. (Streszczenia referatów) 18–19, Warszawa.

KUHSE F.
1924 Die Geologie in der Naturforschenden Gesellschaft in Danzig, *Schriften der Naturforschenden Gesellschaft zu Danzig*, N.F. **XVI** (2), 62–80, Danzig.

LA BAUME W.
1930 Allgemeines. S. V – IX, [*in*:] *50 Jahre Museum für Naturkunde und Vorgeschichte (Westpreußisches Provinzial-Museum in Danzig (1880–1930))*, I-L, Druck v. Julius Sauer, Danzig.

1935 Zur Naturkunde unnd Kulturgeschichte des Bernsteins, *Schriften der Naturforschenden Gesellschaft zu Danzig, N.F.*, **20** (1), 5–48, Danzig.

1957 Die Bernsteinsammlungen in Königsberg und Danzig, *Ostdeutsche Monatshefte*, 1957, **23** (2), 85–91, Stollhamm (Oldb.).

LAKOWITZ K.
1928 50 Jahre Westpeußischer Botanisch-Zoologischer Verein, Unsere Toten. – Helm (S. 22–23), Menge (S. 16), Conwentz (S. 35-38), Dahms (S. 44–45). *50. Jubiläumsbericht des Westpreußischen Botanisch-Zoologischen Vereins Danzig 1878–1928*. 1928, 1–50, Danzig.

LIPPKY G.
1980 Das Westpreußische Provinzial-Museum in Danzig 1880–1945 und seine vier Direktoren, *Westpreußen-Jahrbuch*, **30**, 104–115, Münster.

MAMUSZKA F.
1993 *Danzig und Umgebung. Laumanns Reiseführer*, 1–222, Laumann Verl. Dühnen, Dühnen.

MILTITZER BERICHTE FIRMA SCHIMMEL & CO
1959 *Nachruf und Veröffentlichungen Heinrich Wienhaus (mit Porträt)*, Leipzig.

N. N.
1902 *Amtlicher Bericht über die Verwaltung der naturhistorischen, archäologischen und ethnologischen Sammlungen des Westpreußischen Provinzislmuseums*, Danzig.

N. N.
1936 *Der Bernstein und seine Wirtschaft*, 1–38,6. Aufl, Preußische Bergw. u. Hütten – AG, Königsberg.

OHLERT B.
1881 Nekrolog des Herrn Professor ANTON MENGE, *Schriften der Naturforschenden Gesellschaft zu Danzig*, N. F. **5** (1), XXXX–XXXXVIII, Danzig.

OTTO H.
1981 Otto Helm... [*in*:] Das Chemische Laboratorium der Königlichen Museen in Berlin, *Berliner Beitr. z. Archäometrie*, **4** (1979), 130–147, Berlin.

PACTL J.
1953 A system of caustolites, *Tschermaks Mineral. u. Petrograph. Mitt.*, **3** (4), 332 –347, Wien.

PIESZCZEK E.
1880 Ueber einige harzähnliche Fossilien des ostpreussischen Samlandes, *Archiv der Pharmacie*, **17** (3); **59** (1880), 433–436, Halle.

SANDERMANN W.
1982 *Erinnerungen an meinen Lehrer Heinrich Wienhaus. TU Dresden u. Chem. Ges. DDR; Festkolloquium 100. Geburtstag von H. Wienhaus*; 21–33, Dresden.

SAVKEVITCH S. S.
1970 *Jantar*, 1–192, „Nedra" Verlag, Leningrad.

1983 Change processes of amber and other amber like fossil resins, depending on condition of origin and occurence in nature, *Izw. Acad. Nauk SSR*, Ser. Geol., **12**, 96–106, Moskau.

SCHUMANN E.
1893 Geschichte der Naturforschenden Gesellschaft in Danzig 1743–1892, *Schriften der Naturforschenden Gesellschaft zu Danzig*, N. F., **VIII** (2), 1–150, Taf. I–IX, Danzig.

SLOTTA R.
1997 Bernstein und seine bergmänische Gewinnung, [*in*:] *Saarbrücker Bergmannskalender* 1997, 51–66, Saarbrücken.

STRUNZ H.
1977 *Mineralogische Tabellen*, 6. Aufl., Akadem. Verlagsges., Leipzig.

STOUT E. C., BECK C. W.
& KOSMOWSKA-CERANOWICZ B.
1995 Gedanite and Gedano-Succinite, [*in*:] Anderson K. B. & Crelling J. C. (Hrsg) Amber, Resinite, and Fossil Resins, *Americ. Chem. Soc. Washington, ACS Symposium Series*, **617**, 130–148, Washington.

WIENHAUS-FESTSCHRIFT (1882–1982)
1992 *Festkolloquium anläßlich 100. Geburtstages von Heinrich Wienhaus,* Techn. Univ. Dresden, Sekt. Forstwirtschaft Tharandt u. Chem. Ges. d. DDR, Fachverband Naturstoffchemie. Dresden.

PEASANT GOLD — AMBER JEWELLERY IN FOLK CULTURE

Joanna DANKOWSKA

Abstract

This paper deals with the problem of the multifunctional role of amber jewellery in the traditional culture of villages in the regions of Pomerania, Kashubia, Kurpie and Kujawy at the turn of the 19th century. One can distinguish many different functions of amber jewellery (prestige, economic, social, aesthetic, magical and protective, and votive). They are all closely connected with one another and it is impossible to determine which is the most important. This paper points out the apotropaic significance of amber. The analysed material clearly suggests that amber jewellery in folk beliefs was much more than just an adornment — a fact further exemplified by folk legends and tales.

Amber jewellery has, for centuries, been an important and valued addition to human and in particular women's attire. It has been used in various forms by rich and poor alike, both in towns and villages, on everyday as well as on festive occasions; by people who used only very simple tools as well as by those acquainted with more advanced technologies.

As archaeologists note, the earliest pieces of amber jewellery were not exclusively decorative items but also had magical, ritual and social applications (KOSMOWSKA-CERANOWICZ & KONART 1989, 10 & 20–25; KULICKA 1997, 20; MIERZWIŃSKA 1989, 7–8; 1992, 40). These functions — long-forgotten in urban environments — have been preserved to this day in certain areas of the countryside. This paper attempts to present and analyse the problem of the multifunctional role of amber jewellery in the traditional folk culture of the Kashubia, Kurpie, Pomerania and Kujawy regions. This topic has rarely been mentioned in Polish ethnographic literature, as the issue of jewellery has usually taken second place to detailed descriptions of the cut characteristic of a given region or to the subject of folk costume decoration. With the exception of few articles (CHĘTNIK 1952; 1973; KOSMOWSKA-CERANOWICZ & KONART 1989, 149–153; KWAŚNIEWSKA 1995) describing amber jewellery mainly from the regions of Kurpie and Kashubia there is no monograph dealing with this question. Materials for studies of this problem are to be found scattered among various ethnographic publications and these mostly give purely descriptive data and short notes which rarely discuss the functions and symbolic meaning of jewellery.

Moreover, even though folk costume has often been one of the favourite topics of ethnographic exhibitions, there have never been any exhibitions in Polish museums devoted exclusively to amber folk jewellery, which no doubt has always been an important and meaningful part of folk costume, not only in the north of Poland but also in the Łowicz area and other parts of the country. For example, as Oscar KOLBERG (1963, 170) notes, in the second half of the 19th century strings of "[...] yellow, brown and reddish [...]" amber beads were also worn in the Wielkopolska region (Pleszew district) alongside beads made of natural coral.

I would like this article to be just the first step in my future explorations of the meaning and functions of amber in Polish folk culture. Some of the authors and sources cited in this paper have never previously been mentioned, even in very detailed bibliographies of works on Baltic amber (cf. KOSMOWSKA-CERANOWICZ ed. 1993), and I hope those interested in this topic will find these citations useful.

Adopting the ethnological approach to cultural traits as a means of communication (BARTHES 1970; BOGATYREV 1979), attire, including folk costume and jewellery, can be viewed as a meaningful symbol in a complex cultural system. BOGATYREV (1979, 32–53 & 171) stressed the importance of context for an adequate understanding of the role of a single element of costume, noting that the meaning of the same element in various contexts can be completely different. Thus, the same element may symbolize different values and have different functions. Among the varied functions of peasant amber jewellery, I would like to distinguish six major ones:

1) prestige — amber jewellery as a manifestation of wealth;

2) commercial value — amber jewellery as an investment (accumulation of wealth, such as wedding gifts of amber beads, for example) or used instead of money;
3) social significance — amber jewellery as a sign of social status (brides' necklaces) or emotional attachment;
4) aesthetic value — amber jewellery as an adornment;
5) magical and protective power — amber jewellery used in magic and folk medicine;
6) votive use — amber jewellery as an offering to the Virgin Mary and saints.

All of the above functions are closely related to one another and they often overlap. Thus, it would be impossible to determine the most important one. For example, amber beads were often purposely strung onto red thread to enhance their natural beauty and colour but, at the same time, also to strengthen their protective power through the use of apotropaic red colour. In this paper I would like to pay special attention to the magical and protective function of amber jewellery.

Commercial value and prestige

The quantity and quality of jewellery used as part of a costume, and particularly of one used for a festive occasion, was a manifestation of a person's affluence and social position in folk cultures (and indeed other circles) all over Poland. The most affluent peasants owned items of jewellery, the total value of which often equalled that of more than several heads of cattle.

Amber was generally referred to as a "stone" or "precious stone" in traditional Polish cultures. It was even termed a "yellow stone" in the folk dialects of the Kashubia and Kociewie regions (POPOWSKA-TABORSKA 1964, 91). Amber was considered to be of great value and a sign of affluence classed together with natural coral, pearls and diamonds (KOLBERG 1962, 105). Sometimes it was even compared to gold as this Kurpie proverb demonstrates: "Amber is just as good if you do not have gold" (Polish *Dobry i burśtyn jek nima złota*; CHĘTNIK 1952, 401). Amber is mentioned in folklore as a desirable object, which one would search for (FOLFASIŃSKI 1975, 267) and even steal (SIMONIDES 1977, 141). "Amber vessels" used during a wedding reception, and mentioned in a popular Kurpie wedding song, were yet another symbol of wealth and good luck (CHĘTNIK 1952, 413).

The most valued kind of amber jewellery were beads, commonly worn by peasant women in the 19th century, but already well-known much earlier (cf. CHĘTNIK 1952, 367). Because of their commercial value and prestige, as well as the protective power amber beads were believed to posses, they were an obligatory part of festive apparel, especially that worn for a wedding. Poorer girls would take on additional seasonal work to earn enough money to buy a necklace for their wedding or other festive occasion. In some villages, in order to be allowed to perform an honorary function in church ceremonies, girls had to wear a real amber necklace (CHĘTNIK 1973, 195; FRYŚ *et al.* 1988, 140). When the worse came to the worst and there was no way a woman could afford an amber necklace, she would have tried to get one made of glass beads that resembled real amber as closely as possible. Later on, during the first years of this century, these glass beads became even more fashionable than the older ones made of amber. To increase their prestige and commercial value further still, sometimes silver or even golden coins or medallions were attached to the beads (CHĘTNIK 1952, 389). I think we could agree that the more important functions of different objects are usually reflected in language. Thus, among the different categories of amber distinguished by inhabitants of the Kurpie region we find names such as "dowry" amber. These were often (though not always) amber wedding necklaces prepared by a mother for her daughter (CHĘTNIK 1952, 401; 1973, 196). We also find names such as "currency" or "tax" amber because small pieces of the resin and single beads were sometimes used instead of money, especially in times of unrest, such as periods of warfare and during post-war years (CHĘTNIK 1952, 374 & 396–8).

Social significance

Wedding rings, already recognized in Rome as a symbol of marriage, were not used in Polish villages until the beginning of the 20th century. Amber rings were, however, sometimes given to girls by young men as a token of affection, (CHĘTNIK 1973, 192–3) especially in the period following the First World War. Another manifestation of love or devotion were amber necklaces, which raftsmen would bring their sweethearts or sisters from their trips to Gdańsk (CHĘTNIK 1952, 413). This is probably how amber beads became fashionable in Łowicz costumes at the end of the 19th century. A string of amber beads was even sometimes used in the distant Tatra Mountains region as a gift for a fiancée (KOMOROWSKA 1976, 30). When talking about the social significance of amber jewellery we have to mention special "mourning" black amber beads known from the Kurpie region. These were not, however, usually made of real amber as black amber is very rare indeed (KOSMOWSKA-CERANOWICZ & KONART 1989, 151). The well-known Kashubian proverb "black amber — bad luck" (Kashubian *bursztyn czarny, szczęscé marné*) (KRZYŻANOWSKI ed. 1969, 216) was probably associated with the custom of using black amber beads as a symbol of mourning.

Aesthetic value

The notion of beauty is often influenced by fashion and tends to change fairly often. The aesthetic value of amber jewellery is most evident in the numerous strings of small beads which became fashionable in the Kurpie region at the beginning of this century. Being a relatively cheap product, their commercial worth and value as prestigious items was much reduced, as was their magical power. At about the same time it became fashionable in certain parts of Poland (especially in the Rzeszów and Lublin regions) to wear necklaces made of strings of beads of different kinds, i.e. natural coral, amber, glass, etc., which were all put together to create a multicoloured composition (Fig. 1). Nevertheless, we can still see how perceptive and sensitive to the beauty of their "burning stone" natives of the Kurpie region were. As noted by Adam CHĘTNIK (1982), a keen scholar of their culture, they gave amber many different names according to its degree of transparency, its colour and form. Some of the names are quite poetic, e.g. "feathery" (Polish *pierzasty*), "foamy" (Polish *piankowy*), "glittering" (Polish *mieniący się*) or "nebular" (Polish *obłoczkowaty*). Traditionally, only the best kind of amber, called "a gem" (Polish *cacko*), was used for necklaces. This amber is transparent with reddish overtones and is considered to be the most beautiful of all varieties. The beauty of amber is pointed out by comparing it to the shining eyes of a beloved girl who, according to the Kurpie saying, has "eyes like amber" (Polish *ocy jek burśtyn* — CHĘTNIK 1952, 401).

Magical and protective powers

Red and yellow gems are considered to be "warm" and a good protection against evil spirits all over the world (KOPALIŃSKI 1990, 138). The apotropaic significance of amber jewellery has always been very strong. As amber is often harvested from the sea it is associated with fertility, life and catharsis — all of which water is symbolic of (OESTERREICHER-MALLWO 1992, 179). This symbolic significance is particularly evident in folklore. The most popular wedding song in the Kurpie region states: "A dajciez jej na burśtyny, coby miała śtyry syny" (CHĘTNIK 1952, 413), which in free translation into English reads: "Give her [money] for an amber necklace so she may have four sons". Other versions of this song, recorded by ethnographers in the regions of Podlasie, Lublin, and Łowicz, mention "beautiful sons", whilst in the Mazovian region reference is made to "only sons", with "daughters and sons" being cited in a Łowicz folk song (cf. BARTMIŃSKI ed. 1996, 423). Amber was already associated with fertility in the 16th century, as it was believed that drinking water in which amber had been soaked for three days made child delivery quick and easy (SEWERYN 1947, 280).

Fig. 1. Strings of various beads (amber, natural coral, glass) made into one necklace were popular at the beginning of this century. From the collections of Lublin Museum. Photo. K. SĘDEK. Reproduced from IMIOŁEK 1994.

Until recently it was believed in the Kurpie region that inhaling amber incense would have the same effect (CHĘTNIK 1952, 399). In Kashubia wearing a piece of amber was believed to protect young girls from sterility (KWAŚNIEWSKA 1995, 194). As is often the case with magic, amber was also believed to help in a contrary situation, i.e. to prove a girl's chastity. If a girl who had lost her virginity drank a cup of water in which an amber nodule had been soaked, she would not be able to keep the liquid down. "A virgin would not be bothered by that" states an old adage (SEWERYN 1947, 280). These last beliefs are an example of amber serving a "cathartic" role and are also known from 19th-century Pomeranian folklore (HAAS 1899a)[1].

Another example which shows that amber has always generally been associated in Polish culture with "purity", in both the physical and spiritual sense of the word, is the maxim "pure as amber" (KRZYŻANOWSKI ed. 1969, 397). This belief is expressed in the Kurpie folk legend, claiming that amber came into being at the time of the deluge and was originally made of human tears (both tears and flooding are symbols of catharsis in themselves), and that the more innocent and good the person shedding the tears, the clearer the amber which was created from them (cf. CHĘTNIK 1973, 196). Because of their "cathartic" power amber beads, earrings and pins also helped, according to Kujaw beliefs, to expel "bad juices" and all kinds of illnesses from the body (SZULCZEWSKI 1996, 61 & 114).

Another common belief in the Kujawy region was that amber beads also "protected" a women's breast against the evil eye, curses, illness and all evil forces (SZULCZEWSKI 1996, 103 & 114). Some kinds of amber were even called "healing" (Polish *leczniczy*) or "charlatan's" (Polish *znachorski*) amber. Back in the 16th century amber was believed to "rejoice the heart" and "feed the soul". Drinking water in which a piece of amber had been soaked for three days helped to stop bleeding and alleviate any kind of pain (KOLBERG 1966, 69; SEWERYN 1947, 296). We find that the same convictions existed in 19th-century Pomeranian folklore (HAAS 1899a). Even in the 20th century amber necklaces were valued among the natives of the Kurpie region as a safeguard against headaches (CHĘTNIK 1952, 400; 1973, 194). The tenet that amber relieves headaches was widespread in Pomerania during the 19th century (HAAS 1899a), but was also adhered to in the 1960s in the regions of Wielkopolska and Kashubia (BURSZTA ed. 1967, 413; KWAŚNIEWSKA 1995, 194) and was upheld as late as the year 1984 in the Zamość district (BARTMIŃSKI ed. 1996, 435). In the 16th century, inhaling amber fumes was advocated as a remedy for epilepsy (SEWERYN 1947, 296). This belief survived in Kurpie in a slightly altered form — an amber pendant being believed to protect babies against convulsions (CHĘTNIK 1952, 400). Powdered amber mixed with powdered pearls, emeralds, natural coral and oak mistletoe, which was then added to thin gold leaf, was also believed to help cure epilepsy (CHRÓŚCICKI 1965, 163). Amber in different forms was also used to combat tiredness, asthma, influenza, rheumatism and to reduce enlarged thyroid glands (KWAŚNIEWSKA 1995, 194; PAWLUCZUK 1976, 45–48). In the Konin district peasants wore amber beads to protect themselves against jaundice (MILEWSKA 1891, 423). In the districts of Zamość and Kurpie amber was believed to be helpful in treating eye complaints (CHĘTNIK 1973, 400; BARTMIŃSKI ed. 1996, 435) whilst in Wielkopolska it was considered to help cure ear problems and tuberculosis (BURSZTA ed. 1969, 413 & 426). In the Kurpie region it was also considered a good remedy for lung diseases and runny noses (CHĘTNIK 1973, 196). As we can see, amber used to be utilized extensively in popular medicine and was even proclaimed to be one of the six most effective medications (HACZEWSKI

Fig. 2. A string of amber beads with holy medallions. From the collections of the Ethnographic Museum, Department of the National Museum in Poznań, Poland. Photo by P. JANASIK.

[1] I would like to thank Dr hab. Wojciech ŁYSIAK for supplying me with unique materials about folk beliefs in 19th-century Pomerania from his own collection.

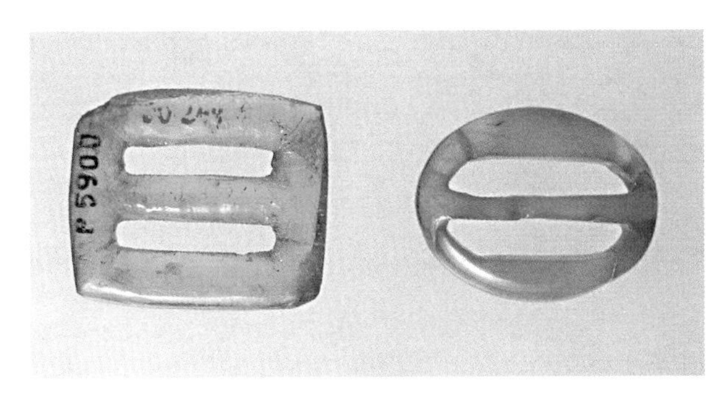

Fig. 3. Amber buckles made in the village of Surowe, Kurpie region: the oval one by Jan CHMIEL in 1923 and the rectangular one by Wiktor DEPTUŁA in 1962. They were attached to hunters' hats as a "good luck" charm. From the collections of the Łomża District Museum, Poland. Photo. B. DEPTUŁA.

after KULICKA 1997, 21; cf. also MIERZWIŃSKA 1989, 311 & footnote no. 26).

In the 16th century amber kept its owner safe from black magic — it cast out devils and kept snakes away (which is understandable if we take into consideration that the devil was often represented in Christian culture as a snake). There was a widespread belief that an eagle always puts a piece of amber in its nest to protect it from snakes (SEWERYN 1947, 290). During the 19th century, in the regions of Wielkopolska and Kujawy, lumps of amber were blessed in church on the Epiphany and then used as incense. On Midsummer's Night (Polish *Noc świętojańska* — 23/24 June: the shortest night of the year, which used to be a great holiday in Slavic folk culture but also a day of extraordinary activity of evil forces) people incensed their houses and cow-sheds to cast out evil spirits and so that witches would not harm their livestock (SZULCZEWSKI 1996, 103 & 115). Similar beliefs were noted in the Gniezno district (KNOOP 1894, 764; 1895, 497). This is probably why, in the 16th century, amber fumes were believed to help guard against thunder and abate storms raised by evil spirits (SEWERYN 1947, 290). In 19th-century sources we find an account relating how people inhabiting the Baltic shore would throw lumps of amber into a rough sea in order to calm it (HAAS 1899b, 160).

Three strings of amber beads (three being a magical, lucky number) were used to adorn a woman's festive costume in the Kurpie region and particularly a bride's outfit. Amber necklaces had special magic powers, especially if they included a pendant in the form of little axe or dart (CHĘTNIK 1952, 360) attached in the middle of a short string of smaller beads tied to the longest string. Later on, under the influence of Christianity, the pendants changed their form to that of sacred medallions (Fig. 2) but the heart-shaped pendant on which a figurine of the crucified Christ was carved still very much resembled the shape of a dart. If an amber wedding necklace was to bring good luck it had to have a special bead in the middle, bigger than all the others and containing inclusions (FRYŚ *et al.* 1988, 169; KOSMOWSKA-CERANOWICZ & KONART 1989, 153). Amber wedding necklaces were also popular in the Pyrzyce region (HAAS 1899a).

Amber was further believed to bring good luck in hunting, which is why hunters from Kurpie ordered amber buckles (Fig. 3) that they attached to their huts together with a red ribbon — another universal symbol of life and good fortune (CHĘTNIK 1952, 393).

Votive use

Amber beads were used in Europe as votive offerings when laying the foundations of a house, in order to ensure the successful completion of its building, as early as the Neolithic period (MIERZWIŃSKA 1989, 8). The votive function of amber in 19th-century Polish folk culture manifested itself in two different ways. Firstly, lumps of amber were burned as incense in churches, and secondly, strings of amber beads were used to decorate icons and altars. Burning amber in a church as incense (which is widely practised throughout Poland to this day) probably further emphasized its aforementioned magical "purifying" properties. Usually, the worst variety of amber (called "church" amber in the Kurpie region) was burned, but quite often peasant women would give their necklaces or individual beads as a church offering to be used as incense during holidays and particularly important religious ceremonies.

A visit to any Polish sanctuary, or quite often even a minor church, will confirm that strings of amber beads are frequently used as votive gifts to the Virgin Mary or saints (CHĘTNIK 1973, 194). In the Chapel of Our Lady of Częstochowa — Poland's foremost sanctuary — there are hundreds of strings of amber beads of all shapes and colours hanging to the left and right of the Black Madonna icon. People from all over Poland have left these offerings at the sanctuary over the last few hundred years as a sign of gratitude. Amber, coral and pearls have been used to make special altar coverings and a decorative "dress" for the icon (Fig. 4). As CHĘTNIK noted (1973, 194), a large statue of the Madonna in the parish church of Czarnia (Kurpie region, near Ostrołęka) had an enormous string of amber beads twisted several times around her neck. Instead of offering necklaces to their local place of worship, the parish women of Mały Płock near Łomża made an interesting modification to this custom by presenting their church with two spectacular chandeliers made of amber beads. Unfortunately, both were lost during the First World War (CHĘTNIK 1973, 194).

The analysed material clearly shows that amber jewellery in folk beliefs was much more than just an adornment. Have these beliefs really been forgotten nowadays? Recently, collecting materials for this paper I talked to my colleague from another museum and she mentioned that she wears an amber bracelet every day as it: "helps her body to function better". There are many people who share the belief that wearing amber jewellery positively influences their vital energies (cf. — BRUSIUS 1997, passim; KOSMOWSKA-CERANOWICZ 1991, 152–162; LORENZO, 57). Different kinds of amber drops or mixtures are still in use in conventional and alternative medicine, and are increasingly widely applied as components of environmentally-friendly cosmetics, which are so fashionable these days (cf. K. B. 1983).

Joanna DANKOWSKA
Muzeum Etnograficzne
Oddział Muzeum Narodowego w Poznaniu
ul. Grobla 25
61-858 Poznań, Poland

CHŁOPSKIE ZŁOTO — BIŻUTERIA BURSZTYNOWA W KULTURZE LUDOWEJ

Joanna DANKOWSKA

Streszczenie

W artykule zaprezentowano wielofunkcyjność biżuterii ludowej w kulturze tradycyjnej wsi na terenie

Fig. 4. Altar and ciborium decorative textiles and other amber offerings to Our Lady of Częstochowa. From the Collection of Votive Art at Jasna Góra, Częstochowa, Poland. Photo. Jasna Góra Archives.

Pomorza, Kaszub, Kurpiowszczyzny i Kujaw na przełomie XIX i XX w. Temat ten nieczęsto poruszany był w polskiej literaturze etnograficznej i nie istnieje monografia poświęcona temu zagadnieniu.

Biżuteria z bursztynu pełniła w kulturze ludowej wiele różnorodnych funkcji:

- funkcję prestiżową — jako wyraz zamożności;
- funkcję ekonomiczną — jako forma inwestycji i akumulacji dóbr (np. bursztynowe naszyjniki jako prezenty ślubne);
- funkcję społeczną — jako oznaka statusu społecznego (np. naszyjnik panny młodej) lub zaangażowania uczuciowego;
- funkcję estetyczną — jako ozdoba;
- funkcję magiczną i ochronną — w magii i medycynie ludowej;
- funkcję wotywną (jako ofiara lub wota dla Matki Boskiej i świętych).

Wszystkie wyżej wymienione funkcje połączone są ze sobą i niemożliwe jest wyznaczenie najważniejszej z nich. W artykule podkreślam szczególnie apotropaiczne znaczenie bursztynowych ozdób, wykorzystując źródła niecytowane dotąd w literaturze przedmiotu. Przedstawiony materiał wyraźnie wskazuje, że biżuteria z bursztynu była w wierzeniach ludowych czymś znacznie więcej, niż tylko ozdobą.

Bibliography

BARTHES R.,
1970 *Mit i znak*, 1–327, Państwowy Instytut Wydawniczy, Warszawa.

BARTMIŃSKI J. (ed.)
1996 *Słownik stereotypów i symboli ludowych, Vol. I: Kosmos*,1–439, Uniwersytet Marii Curie-Skłodowskiej, Lublin.

BOGATYREV P.
1979 Funkcje stroju ludowego na obszarze morawsko-słowackim, 26–96, *Semiotyka kultury ludowej*, 1–367, Państwowy Instytut Wydawniczy, Warszawa.

BRUSIUS H.
1997 *Magia kamieni szlachetnych. Ich kosmiczne znaczenie, działanie i promieniowanie*, 1–176, Astrum, Wrocław.

BURSZTA J. (ed.)
1967 *Kultura ludowa Wielkopolski, Vol. 3*, 1–616, Wydawnictwo Poznańskie, Poznań.

CHĘTNIK A.
1952 Przemysł i sztuka busztyniarska nad Narwią, *Lud*, **39**, 355–415.

1973 Jantar w sztuce kurpiowskiej, *Polska Sztuka Ludowa*, **27** (4), 191–199.

1982 Mały słownik odmian bursztynu, *Prace Muzeum Ziemi*, **34**, Warszawa.

CHRÓŚCICKI J. A.,
1965 Korale, sosulka i szczygieł. (Sztuka gotycka a tradycje i wierzenia ludowe), *Polska Sztuka Ludowa*, **19** (3), 157–166.

FOLFASIŃSKI S. (selected and compiled)
1975 *Polskie zagadki ludowe*, 1–453, Ludowa Spółdzielnia Wydawnicza, Warszawa.

FRYŚ E., IRACKA A. & POKROPEK M.
1988 *Sztuka Ludowa w Polsce*, 1–338, Arkady, Warszawa.

HAAS A.
1899a Der Bernstein im pommerschen Volksglauben, *Blätter für Pommersche Volkskunde*, **VII** (3).

1899b Verwendung von Bernstein, *Blätter für Pommersche Volkskunde*, **VII** (11).

HACZEWSKI J.
1838 O bursztynie, *Sylwan*, **14** (1/2), 191–251 & (3/4), 358–428.

IMIOŁEK M.
1994 *Zdobnictwo i elementy biżuterii w stroju ludowym wybranych regionów Polski Południowej i Wschodniej*, Muzeum Wsi Kieleckiej.

K. B.
1983 Właściwości lecznicze bursztynowych kropli, *Aura*, **12**, 29.

KNOOP O.
1894 Podania i opowiadania z W. Ks. Poznańskiego, Part 1, *Wisła*, **8**, 719–774.

1895 Podania i opowiadania z W. Ks. Poznańskiego, Part 4, *Wisła*, **9**, 470–513.

KOLBERG O.
1962 *Dzieła wszystkie, Vol. **8** Krakowskie, Part 4*, I–XIII & 1–365, Polskie Towarzystwo Ludoznawcze, Wrocław.

1963 *Dzieła wszystkie, Vol. **10** W. Ks. Poznańskie, Part 2*, I–II & 1–386, Polskie Towarzystwo Ludoznawcze, Wrocław.

1966 *Dzieła wszystkie, Vol. **40** Mazury Pruskie* (compiled from manuscripts by Ogrodziński W. & Pawlak D., ed. D. Pawlak), I–XLIII & 1–676, Polskie Towarzystwo Ludoznawcze, Wrocław.

KOMOROWSKA T. (selected)
1976 *Zbójnicki dar. Polskie i słowackie opowiadania tatrzańskie, Part I, Opowiadania polskie*, 1–164, Ludowa Spółdzielnia Wydawnicza, Warszawa.

KOPALIŃSKI W.
1990 *Słownik symboli*, 1–512, Wiedza Powszechna, Warszawa.

KOSMOWSKA-CERANOWICZ B.
1991 Zarys wiadomości o leczeniu bursztynem, [*in:*] *Biomineralizacja i biomateriały*, 152–162, Państwowe Wydawnictwo Naukowe, Warszawa.

KOSMOWSKA-CERANOWICZ B. (ed.)
1993 *Bursztyn bałtycki i inne żywice kopalne. Piśmiennictwo polskie oraz prace autorów polskich w literaturze świa-*

towej. Bibliografia komentowana 1534–1993 [Baltic amber and other fossil resins in Polish literature and works by Polish authors in world literature. An annotated bibliography 1534–1993], *Part I*, Pietrzak T. & Różycka T., *Bursztyn w przyrodzie, kulturze i sztuce [Amber in nature, culture and art]*, 1–164, Muzeum Ziemi PAN, Warszawa.

KOSMOWSKA-CERANOWICZ B. & KORNAT T.
1989 *Tajemnice bursztynu*, 1–231, Sport i Turystyka, Warszawa.

KRZYŻANOWSKI J. (ed.)
1969 *Nowa księga przysłów i wyrażeń przysłowiowych polskich, based on a work by Samuel Adalberg and compiled by Editorial Committee directed by Julian Krzyżanowski, Vol. 1*, I–XXXIX & 1–881, Państwowy Instytut Wydawniczy, Warszawa.

KULICKA R.
1997 Bursztyn w wierzeniach i medycynie ludowej, 20–21, [*in*:] *Bursztyn — skarb dawnych mórz*, 1–28, Sadyba, Warszawa.

KWAŚNIEWSKA A.
1995 Bursztyniarstwo ludowe, 189–194, [*in*:] *Sztuka ludowa Kaszub. Przeszłość i teraźniejszość*, 1–260, Kujawsko-Pomorskie Towarzystwo Kulturalne, Bydgoszcz.

LORENZO L.
[?] *Kamienie szlachetne. Zdobią i leczą*, 1–152, Spar, Warszawa.

MIERZWIŃSKA E.,
1989 *Dzieje bursztynu*, 1–66, Muzeum Zamkowe w Malborku, Malbork.
1992 Wykorzystanie bursztynu w medycynie, 40–41, [*in*:] *Bärnsten — guldet fran Östersjön. Bursztyn — złoto Bałtyku*, 1–79 & 85 pages of colour plates, Excalibur, Bydgoszcz.

MILEWSKA A.
1891 Medycyna ludowa, *Wisła*, **5**, 419–424.

OESTERREICHER-MALLWO M. (compiled)
1992 *Leksykon symboli*, 1–198, ROK Corporation SA., Warszawa.

PAWLUCZUK W.
1976 Rady księdza Podbielskiego — inżyniera-elektryka, *Literatura Ludowa*, **20** (3), 44–52.

POPOWSKA-TABORSKA H.
1964 O "Słowniku gwar kaszubskich na tle kultury ludowej" Bernarda Sychty, *Literatura Ludowa*, **8** (4–6), 84–92.

SEWERYN T.
1947 Ikonografia etnograficzna, *Lud*, **37**, 278–308.

SIMONIDES D. (introduction, collected and compiled by)
1977 *Kumotry diobła. Opowieści ludowe Śląska Opolskiego*, 1–181, Ludowa Spółdzielnia Wydawnicza, Warszawa.

SZULCZEWSKI J. W. (ŁYSIAK W. ed.)
1996 *Pieśń bez końca*, 1–445, PSO, Poznań.

THE PRODUCTION OF JEWELLERY, ARTISTIC AND FANCY GOODS AT THE KALININGRAD AMBER FACTORY (1945–1996)

Zoja KOSTIASHOVA

Abstract

The Kaliningrad Amber Factory was established in 1947 and for the next ten years produced mainly large series of simple jewellery and fancy articles. Throughout the first decade almost all of these goods were very primitive and, as a rule, copied natural forms. During the 1960s Kaliningrad jewellers began to use unprocessed amber nodules, paying attention to their natural colour and texture. This article traces the changes in the assortment of amber goods produced over the last fifty years and outlines the main tendencies in the development of artistic creativity and methods of amber processing.

The Kaliningrad Amber Factory is the largest producer of amber goods in the world. During the fifty years of its existence, 65 million pieces of jewellery and other articles have been manufactured.

The factory was established in 1947 by a decree of the Soviet government. However, the history of amber goods production in the USSR started two years earlier.

After part of Eastern Prussia joined the USSR, Russian people had to master a number of industries in which they did not have enough experience. The most exotic one was that of amber extraction and processing. Nobody knew the true value of amber at that time. It was seen as something similar to colophony and was even used to light stoves. Meanwhile, after the Germans had departed, tens of tons of amber were left in the stores of Palmnicken. The military authorities organised small workshops there, where demobilised soldiers and any remaining German people in the village worked. Later, the workshops were reorganised into a factory, under the control of civilian authorities. At that time seventy-five people worked there, mostly German women. They made beads, bracelets, cigarette-holders and souvenirs. Seventy tons of amber were used within two years for this work (see also two articles: KOSTIASHOVA 1994; 1996).

This approach to the production of amber goods did not suit the new Soviet authorities. In a report to the Ministry of Internal Affairs in 1947, the Head of the Kaliningrad Militia, General TROFIMOV, wrote, that "wasting amber on silly things is an anti-state practice". General TROFIMOV suggested that amber could be used for producing lacquers for military ships and aircraft, filters for gas-masks and insulators for different instruments. Soon after the submission of this report, all amber goods production was transferred under the control of the Gulag system.

Nevertheless, the attempt to change and adapt production for the needs of defence did not succeed.

Further attempts to find other uses for amber, besides jewellery, were made after STALIN's death, when the factory became civilian again. However these were not successful, as there was no market for products manufactured through the chemical treatment of amber, i.e. succinic acid, amber oil and melted amber.

The main products of the factory (up to 90%) are mass-produced art-works and fancy-goods (KOSTIASHOVA 1997). Up until 1951 their quantity was calculated by weight. Later, the goods were divided into two groups. The first of these consisted of ornaments with metalwork (which were classified as "jewellery" and counted per item), the remainder of goods produced being classified under the old heading of "amber articles". This second category was counted by the kilogram until the early 1960s. The dynamics of changes in the quantity of products are shown in Table 1.

1948	1951		1961	1971	1981	1991
1166 kg	5171 kg	100.9	780.3	817.7	2075.7	2223.0

Table 1. Quantity of amber articles produced at the Kaliningrad Amber Factory in 1948–1991 (in thousands of pieces unless otherwise stated).

As can be seen from Table 1, up until 1991 the factory was constantly increasing its production of jewellery and fancy goods. Compared to the early years, it increased more than tenfold. In 1991 (the year of the Soviet Union's break up) the Kaliningrad Amber Factory entered into a crisis period that has not been overcome to this day. The production of amber articles in 1992–1996 decreased by a factor of three.

Data on the assortment of articles produced are available starting from 1955. They are presented in Table 2.

Variety of article	1955	1961	1971	1981	1989**
Brooches	632.0	124.0	125.7	187.8	161.9
Beads	1471.3*	141.2	180.5	314.9	434.2
Bracelets	26.1*	23.9	52.0	58.7	106.0
Ear-rings (pairs)	–	42.1	74.0	119.0	558.5
Finger rings	–	25.1	69.4	787.4	310.0
Pendants	–	6.5	163.0	450.5	187.6
Necklaces	159.4*	2.3	5.1	85.9	27.1
Hairpins, hat-pins	91.7	40.0	–	–	44.9
Cigarette-holders	998.7*	125.0	33.5	17.0	35.0
Cuff-links (pairs)	43.3	123.7	73.9	52.4	43.5
Tie-pins	–	97.0	27.2	–	–
Badges, medals	–	2.6	1.1	1.1	–
Souvenirs	–	26.9	12.3	1.1	14.2
Total	2655.5* 767.0	780.3	817.7	2075.7	1922.9

* kg
** predicted

Table 2. Assortment of jewellery and fancy articles produced at the Kaliningrad Amber Factory in 1955–1989 (in thousands of pieces unless otherwise stated).

From this table the main tendencies in changes of production of different items can be seen. First of all, during the period under discussion, the manufacture of fancy-articles (cigarette-holders, cuff-links, pins, etc.) fell to less than 20% of total production. In the jewellery sector, there was a sharp decrease in the production of brooches, the absolute leaders of the 50s, but the quantity of rings and pendants increased several-fold. In 1981, they made up 60% of all goods produced.

During the first decade of its operation, the factory's range of products did not change very much. The main articles were polished, disk-shaped and faceted beads, brooches, cigarette-holders, bracelets, cuff-links, pins, hairpins and medals. Production of ear-rings started in 1957. The limited use of precious metals in the production of jewellery was introduced in 1952: gold first and then from 1957 — silver.

At that time Russian jewellers, still not aware of the full potential of amber, gave the 'stone' ideal geometric shapes and copied natural forms — flowers, berries, fruit, insects, etc. Typical names for brooches in the 50s were: *Acorn, Flower, Plum, Strawberry, Currant, Two cherries, Five leaves, Two red berries* (Fig. 1). In 1958 the factory's best-selling models were brooches such as the *Oval, Rectangular, Spider, Beetle*, etc.

In the late 1950s different souvenirs came into fashion — statuettes of animals and fish, chess-sets, caskets, powder-cases. Hundreds and thousands of copies of these naïve and, in a way, charming things were produced. In subsequent years the production of souvenirs dropped to one thousand items a year, and those produced acquired visible ideological subjects. The titles of pieces made in the 1980s included names such as *Space exploration* and *Boundary post*.

During the 1950s some massive articles were made by the factory's craftsmen. Their works reflected the tendency towards monumentality, typical of all Soviet fine art of that period. Apart from this, nearly all of the factory's craftsmen were used to carving marble, jasper and other ornamental stones, and very often they copied compositions made of such materials, or used the same methods, while working with amber, which was not always good.

An example of this ostentatious style was the huge vase named *Abundance* — one metre high, made by special order in 1954 (Fig. 2). It was displayed in the Kaliningrad Oblast pavilion of the Moscow Exhibition Centre. The vase was made of natural and pressed amber, decorated with sparkling amber crumbs and medallions with garlands in the form of berries and fruit made of coloured minerals.

Another, and probably the most famous, amberwork of this period was the model of the atomic

Fig. 1. Jewellery and fancy goods produced in mass series in the 1950s. From the collections of the Amber Museum, Kaliningrad. Photo. V. SEMIDIANOV.

powered ice-breaker *Lenin* (Fig. 3). It was made only of natural amber by the leading craftsmen of the factory in twenty days. In creating this model, complicated techniques, carvings and mosaics were used. The ice in this composition was made of white and transparent, blue-tinted pieces of amber, which looked very much like frozen sea-water. A Soviet delegation presented the model to President EISENHOWER on their visit to America in 1960. Later, an accurate copy of this work was made at the factory.

For a long time people were sceptical about works such as the *Abundance* vase or the *Ice-breaker*, even ascribing them to bad taste. However, as time passed so attitudes changed. Today both compositions are on display at the Kaliningrad Amber Museum and always attract visitors' attention.

During the first years of the factory's operation, any amber product made there could be sold. It was the mineral itself, a little known rarity until that point, that generated interest. Soon, though, the market was full of amber goods, all of the same type.

The initial interest in this mineral waned. The crisis in the amber market forced the factory to look for new ways of developing its goods production.

Fig. 2. The Abundance Vase, 1954. From the collections of the Amber Museum, Kaliningrad. Photo: V. SEMIDIANOV.

Fig. 3. Model of the atomic ice-breaker Lenin, 1959. From the collections of the Amber Museum, Kaliningrad. Photo. V. SEMIDIANOV.

This required a different approach to amber and new principles. Such changes came with Chief Artist Alfred MEOS. He was a well-educated man, a graduate of the Institute of Arts in Tallinn, and had a thorough knowledge of gem stone cutting. MEOS brought with him the old Baltic traditions of this art. He helped craftsmen at the factory to see the natural beauty of every piece of amber. They learned to use its natural shape, colour and texture.

In the early 60s Alfred MEOS assembled a team of like-minded young people — Earnest LIS, Aleksej POPOV, Vasylij MITIANIN, Roland BENISLAVSKIJ and Alexander KVASHNIN. These artists, who were unhappy with the monotony of their work, gladly responded to Alfred MEOS's call for originality. They designed models for mass-production. Original ornament designs appeared and artists started trying to emphasize the natural beauty of the material they were working with, occasionally using partially treated amber with an oxidised surface.

In 1967 the Department of Unique and Small Quantities Production was established. Initially, four people worked there, later the number increased to thirty. The articles produced there were more elaborate and difficult to make. The most skilful specialists were employed. Almost all work was done by hand and the number of copies of one model was not more than ten, sometimes only two or three. Only in exceptional cases were the models copied in tens or produced in mass series. At this time the understanding came that amber is not an appropriate material for monumental compositions. Thus, the transition from producing large, bulky vases, caskets and panels to manufacturing small articles began.

Initially, though, this change in style was applied predominantly to individual works. The variety of mass-produced goods changed far more slowly, and in the press the factory was still criticised for its famous *beetles* and *spiders*, and for its naturalistic copying of cherries and beer-mugs. Though more use was made of amber's natural properties, the number of out-of-date, ordinary models was still relatively high.

Apart from the aesthetic qualities of articles produced, updating the existing assortment was also very important. During the 1960s the factory produced, on average, 300–400 items a year. Every year 15–20% of the old designs were replaced with new ones, with the introduction of about sixty new models. Most of them were, however, intended to be produced only in small quantities. Any article in mass-production would normally be produced for a minimum of five years in many thousands of copies. Some of them were out-of-fashion long before their production was stopped. On the other hand, some could be classified as "long-lived". Such was the case with the *Spider* brooch, which was designed in 1958 and remained in production for thirty years. Between 27 to 46 thousand copies of it were made every year, thus the total number amounting to nearly a million.

The majority of jewellery, carved works and fancy-goods — up to 80% — produced in the 60s were made of pressed amber. However, the preferences of buyers were changing. More and more often they wanted to buy goods made of natural amber, with its inimitable shape, colour and texture.

The transition to the preferential use of natural amber in mass-production started around 1962, when a number of new articles were introduced — ear-rings, necklaces and cuff-links, made of small pieces of amber, where only pressed amber had been used previously.

Starting from 1964, the selection of unique pieces with plant and insect inclusions, was instituted. Another important novelty also helped to make goods more attractive — the use of new materials. On the initiative of jewellers A. POPOV, E. LIS, V. MITIANIN and JAROSHENKO materials such as wood, horn, leather, apricot and peach stones came into use alongside amber. At the same time experiments were conducted to combine amber with enamel.

In 1967 the factory widened its range of articles made using amber in combination with silver, whilst in 1968 it started producing articles set in gold. In the same period the first trial models of so-called "sparkling" (or hardened) amber were made. Very quickly the new material became very popular. All these innovations allowed the factory to solve its marketing problems.

The 1970s were the most successful period for the factory. The real "amber boom" started in the second half of the 60s and lasted till 1982. During the whole of this period, the factory never had any problems with finding a market for its products and rapidly increased production. Within ten years its income had increased three and a half times. The 70s were characterized by drastic changes in the ratio between the types of articles manufactured. The production of pendants increased over tenfold, with 36 times more rings being manufactured. This increase was not only due to changing fashions, but also a result of the fact that the factory implemented the methods of centrifugal casting and mechanised soldering during this period. This reduced the cost of running the entire production chain and allowed the factory to its output many times. As for pendants, the technology used to produce them was also very simple, taking into account that the chains for them were supplied by another factory.

At first glance, the assortment of products was very wide — necklaces were produced in 103 different models, bracelets in 68, cigarette-holders — 31, brooches and ear-rings — 115, and rings in 102 different models.

Nevertheless, very often some designs did not differ very much from each other in principle. For example, all models of necklaces were classified by the type of material from which they were made — natural, hardened or pressed — and by the shape of the beads used — round, faceted or oval. Combining these characteristics in a number of different ways and varying the length of the thread gave more than 100 variations. The same was true of many other items.

One of the main reasons for the success of the 70s was that the factory managed to take advantage of the rapid growth in popularity of gold jewellery by expanding the production of articles combining amber with gold.

As can be seen in Table 3, whilst the production of jewellery made of cupro-nickel and other non-precious metals doubled and production of silver jewellery even went down, at the same time the number of gold articles produced increased six-fold. Moreover, the increase in the production of rings and pendants was almost entirely due to gold items. Out of 373.9 thousand pendants, produced in 1980, 334.9 thousand (i.e. 90%) were set in gold, whilst of the 739.6 thousand rings manufactured 764.1 thousand (97%) were made of gold.

Variety of metal	1971		1981		1991	
	No. of pieces (thousands)	% of total production	No. of pieces (thousands)	% of total production	No. of pieces (thousands)	% of total production
Gold	117	14	1130	54	936	42
Silver	264	32	167	8	236	11
Non-precious metals	168	21	387	19	497	22
Without metal	269	33	392	19	554	25
In total	818	100	2076	100	2223	100

Table 3. Amber jewellery production (according to metal used) in 1971–1991.

The "amber boom" lasted till the beginning of the 80s, when, at the end of 1981, the prices for gold jewellery, which gave the factory most of its income, were sharply increased. This resulted in a drop in demand for these goods. Nevertheless, throughout this decade the factory managed to maintain the same levels of production it had previously attained (Fig. 4). The real crisis set in at the beginning of the 1990s. This crisis is usually ascribed to the consequences of radical economic reforms: a slump in production, the severing of former economic links and the disadvantageous reorganisation of the factory into "Russian Amber" — a joint stock company (1993). All this is true, but is not, in my opinion, the main reason for the company's declining fortunes.

A giant factory producing jewellery *en mass* could only exist in a totalitarian country. The low living standards and unpretentious tastes of people who were given no consumer choice guaranteed a market for simple and cheap goods produced by the million. The introduction of a market economy resulted in the formation of a class system and different groups of consumers with different demands. Moreover, within a very short space of time the market became competitive. In Kaliningrad alone, more than one hundred private companies making amber goods were founded. The importation of jewellery also became very widespread. One should add to this the changing motivations of the Russian consumer. This new motivation manifested itself in the urge for originality, in the transition from the wish to look alike to the wish to look different. In such new conditions the factory was doomed to bankruptcy.

Fig. 4. Modern mass products of the Kaliningrad Amber Factory. From author's collections. Photo. V. SEMIDIANOV.

A possible solution may lie in establishing a decentralized chain of small jeweller's workshops, competing against each other and being able to respond to changes in demand and fashion.

Zoja KOSTIASHOVA
Amber Museum
Marshal-Wasiljewskij Square 1
236016 Kaliningrad, Russia

PRODUKCJA WYROBÓW JUBILERSKICH, GALANTERYJNYCH I ARTYSTYCZNYCH W KALININGRADZKIM KOMBINACIE BURSZTYNU (1945–1996)

Zoja KOSTIASHOVA

Streszczenie

Kaliningradzki Kombinat Bursztynu jest największym przedsiębiorstwem rzemiosła bursztynowych ozdób w świecie. W ciągu pięćdziesięcioleciu lat istnienia wyprodukowano tu 10 milionów wyrobów biżuteryjnych. W Kaliningradzie było kilka okresów w rozwoju sztuki jubilerskiej. Kombinat powstał w 1947 r. i przez dziesięć lat produkował na rynek wewnętrzny wielkie serie ozdób i przedmiotów galanteryjnych, jak pierścionki, broszki, wisiorki, naszyjniki, kolczyki, cygarniczki, spinki do mankietów, spinki do włosów i inne. Większość z nich była bardzo prymitywna z punktu widzenia artystycznego. W latach pięćdziesiątych najbardziej popularne wyroby miały wzory geometryczne albo imitowały liście, jagody, owoce, chrząszcze i inne owady.

W latach sześćdziesiątych z pomocą doświadczonych specjalistów przybyłych do Kombinatu z Leningradu i republik bałtyckich nastąpił rozwój nowych stylów i nowych metod artystycznej obróbki bursztynu. Jubilerzy z Kaliningradu uświadomili sobie, że bursztyn jest kamieniem bardzo lekkim i plastycznym. Wymaga on innego obchodzenia się aniżeli twardy kamień, z którym pracowali wcześniej. Artyści zaczęli wykorzystywać naturalne formy bursztynu, zwracać uwagę na jego strukturę i kolor, aby robić różnorodne kompozycje, łączyć bursztyn z nietradycyjnymi materiałami jak drewno, wełna i masa plastyczna. W tym samym czasie rzemieślnicy Kombinatu wykonali kilka wspaniałych prac z bursztynu, jak mozaikową kompozycję z białych i niebieskich kawałków bursztynu wyobrażającą atomowy lodołamacz „Lenin" uwięziony w lodach i gigantyczną wazę „Obfitość".

Biżuteria bursztynowa podlegała wymaganiom zmieniającej się mody. W latach siedemdziesiątych bursztyn osiągnął w ZSRR szczyt popularności. Prawdziwy „bursztynowy boom" był związany z użyciem złota w kompozycjach z bursztynem. Następny okres około piętnastoletni charakteryzował spadek popytu. Obecnie w Rosji bursztyn znów staje się modny.

Bibliography

KOSTIASHOVA Z. V.

1994 Powojenna historia kombinatu eksploatacji i obróbki bursztynu na Sambii, *Przegląd geologiczny*, **43** (4), 364–367, Warszawa.

1996 *Die Nachkriegsgeschichte des Kombinats für Gewinnung und Bearbeitung von Bernstein in Jantarnyi/ Palmnicken*, [*in*:] Ganzelewski M. & Slotta R. (Hrsg.), Bernstein — Tränen der Götter, *Veröffentlichungen aus dem Deutschen Bergbau-Museum Bochum*, **64**, 237–247, Bochum.

1997 *Kaliningradskij jantarnyj kombinat*, IZO-centr. Kaliningrad.

In preparing this article the author has used previously unpublished statistical and other materials from the State Archives in Kaliningrad (Fund 54. Kaliningradsky Jantarny Kombinat) and the Amber Factory Archives in Jantarny. In addition, the author has interviewed more than ten artists and the oldest workers of the Amber Factory (all texts of these interviews are in the Kaliningrad Amber Museum).

SICILIAN AMBER

Piotr SZACKI

Abstract

The leading fossil resin in Europe is succinite. Yet at the far end of the continent, in Sicily — the centre of classical civilizations — a Miocene fossil resin can be found which is called simetite. This name derives from the River Simeto, whose waters wash away and transport amber from the island interior to the sea. Simetite occurs in a variety of shades and frequently exhibits a bluish fluorescence.

Ancient authors do not record the incidence of amber in Sicily. The first mention of it dates from the 16th century. By the 17th century, Sicilian and foreign authors had begun to devote a lot of attention to simetite. As a raw material used in jewellery, simetite was popular and the most highy valued among all fossil resins. Today, however, it is considered somewhat old-fashioned. Nevertheless, it has become an important symbol of the "Sicilian character".

To examine the local amber traditions, an ethnographic research project was carried out in Sicily in 1980 and 1984 as part of the work of the Polish-Italian Interdisciplinary Work Group of Sciences Applied in Archaeology and Protection of the Cultural Heritage.

The characteristics of amber distinguished it from the stock of substances at man's disposal. If it could be found, it was "egalitarian" due to the simplicity with which it was obtained and processed. Its fragrant smoke made it an important means of communicating with the gods. Through its frequent inclusions, especially the animal ones, it was an emphatic reference to the distant past and arrested life. Thus, a complex of knowledge, beliefs and symbolic meanings surrounding amber developed: built, compiled and preserved, it was a specific myth with history of considerable fluctuations resulting, among other things, from amber's changing position in man's system of values. Individual threads of the "amber myth" intermingled and reverberated from the Mediterranean basin to Scandinavia.

In Europe, the leading position among fossil resins is taken by northern succinite. But at the other end of the continent, in Sicily — the centre of classical civilizations — a Tertiary Miocene fossils resin can be found which is called simetite, the name deriving from the "amber-bearing" River Simeto. Palaeoentomological analysis of inclusions indicates its far-reaching convergence with the amber of the

Fig. 1. Amber (simetite?) found during archaeological reaserch at a necropolis site in Marianopoli (central Sicily), dating approximately from the 6th century BC. Photo. P. SZACKI.

Fig. 2. The mouth of the River Simeto in Catania Bay. Photo. P. SZACKI.

Bologna Appenines (SKALSKI 1984[1]). Simetite is distinguished by a specific variety of shades and, primarily, by a frequently encountered bluish fluorescence which significantly influenced the techniques used in its processing.

The occurrence of amber in Sicily was not recorded by ancient authors. It was first mentioned in the 16th century by AGRICOLA. Starting from the 17th century, both Sicilian and foreign authors devoted a lot of attention to this resin. In 1805, F. FERRARA from Catania published a minor systematic study which also recapitulated the observations which had been made thus far. Modern research has extended and clarified our knowledge of the nature and specific physico-chemical features of simetite and the palaeobiological contents of its inclusions.

Old texts written by Sicilian and foreign authors, which deal in greater or lesser detail with simetite, unanimously suggest its relatively abundant occurrence, often in the form of fairly large nodules, in some coastal areas of the island. In their opinion, amber occurred (or was formed due to volcanic activity from bituminous contents and salt) in sedimentary layers of the "ancient sea" in the mountainous areas of central Sicily. From there, it was washed out (especially during the rainy season) by tributaries of the Simeto, and then carried to the mouth of the river near Catania. Coastal currents conveyed amber to the south, scattering it across the coast all the way to the beaches of the south-easterly headland in the area of Pachino. Another amber-bearing river is the Salso. Fed by tributaries, it cuts across central Sicily to the south, between Enna and Caltanissetta. Its mouth lies on the southern coast near Licata; it scatters amber all the way to Porto Empedocle near Agrigento. The occurrence of amber in the vast area around Nicosia is still evidenced to this day by finds of this material during earthworks. As a local peculiarity, simetite products were very popular among collectors, travellers and tourists many of whom visited Sicily starting from the 17th century.

Catania was the major amber craft-centre in Sicily. Amber-craft was also practised to a smaller extent at

Fig. 3. Necklace and lump of simetite (found about 1934 near Nicosia). From the collections of Fratelli Fecarotta jewellers, Catania. Photo. P. SZACKI.

[1] Information from tapescript of the report by A. SKALSKI, 1984; see also bibliography, TABACZYŃSKA, *Amber as a subject of archaeological research: the experiences of Polish-Italian collaboration*, in this volume.

Fig. 4. The purchase of crude amber. Porto Empedocle. Photo. P. SZACKI.

Palermo and Trapani (otherwise important centres of the goldsmith's and jeweller's crafts, as well as of the coral and tortoise-shell processing industry). Later on, Taormina and perhaps also Lentini were added to this list. In the period between the two world wars, Catania and Taormina were still centres of amber-craft. The demand for Sicilian amber exceeded the supply of raw simetite as early as the days of F. FERRARA. This devastated a particular category of relics, i.e. simple folk ornaments. Sold to jewellers on a mass scale, they were subjected to secondary professional processing.

The occurrence of simetite articles among folk ornaments is mentioned i.a. by FERRARA (1805) as a phenomenon which, though obsolete, remains rich in religious connotations. The author especially stresses the symbolic meaning of an amber gift in matrimony. Interviews carried out in the eighties confirmed the persistence of this tradition. Access to such articles — probably still kept somewhere — is made difficult by their rarity and scattered distribution. However, chance finds of single, primitively processed elements of simetite ornaments seem to corroborate the existence of folk amber-craft in the past. Starting from the 19th century, the amount of raw simetite found has gradually been declining — a fact attributed to the slower currents of rivers.

Considerable amounts of Baltic amber were also imported to Sicily. In the 1960s, rare and valuable simetite gave way to imported Dominican, Brazilian and — later on — Sakhalin amber, which looks very

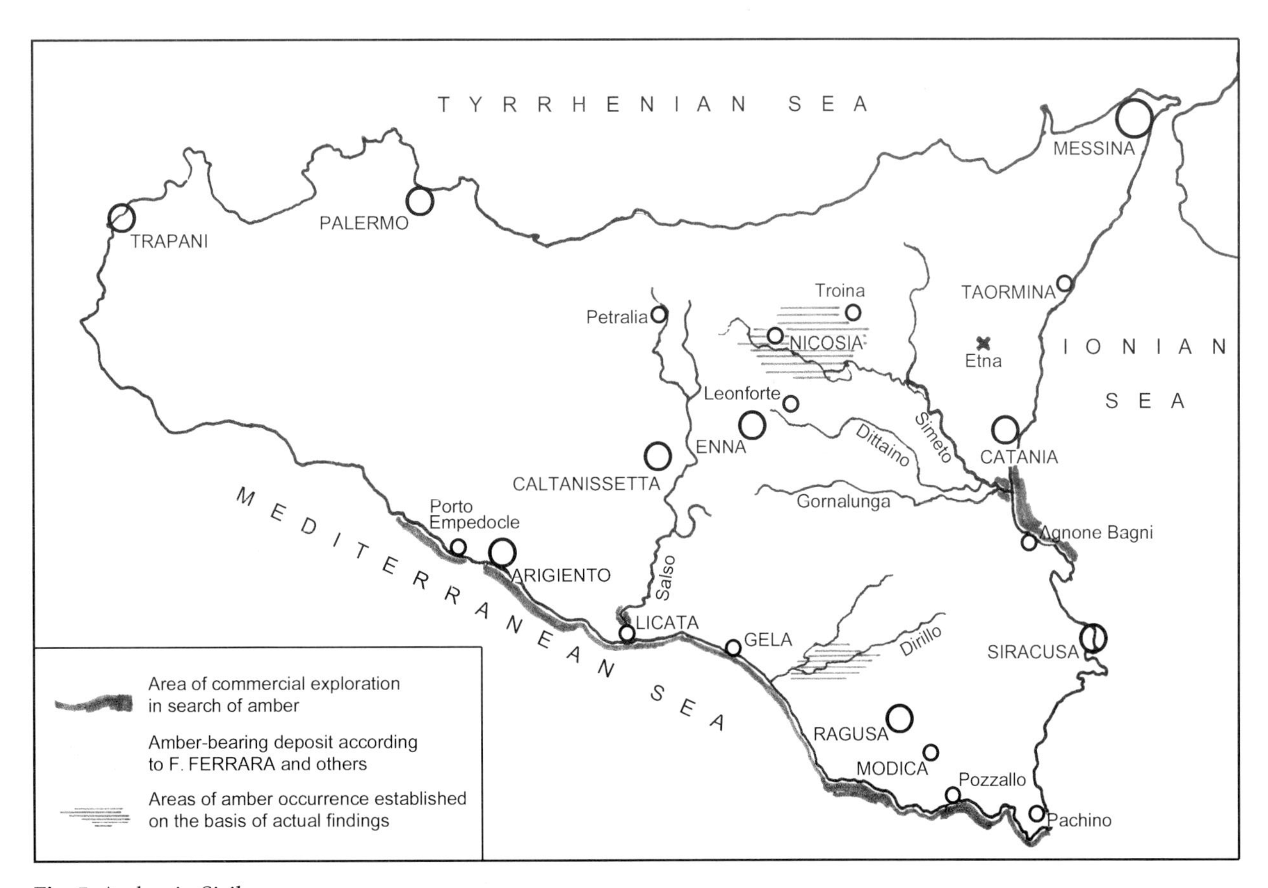

Fig. 5. Amber in Sicily.

much like simetite and frequently displays its fluorescence. These resins were passed off as the Sicilian variety. The imports of "simetite-like" resins attests to the continuing existence of an established canon of beauty based on the features of simetite. As a raw material used in the jeweller's trade, simetite has always been the most highly valued of all fossil resins. Today, however, simetite ornaments tend rather to

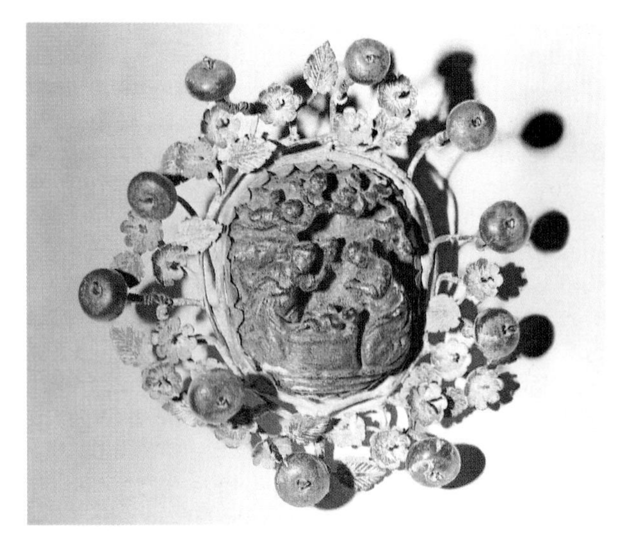

Fig. 6. Nativity scene. Silver setting. Turn of the 17th century. Catania workshop (?). From the private collection of Mario CIANCIO. Catania. Photo. P. SZACKI.

Fig. 7. Elements of ornaments found by amber prospectors. Baltic amber and Simetite. Catania Bay, area of Syracuse, Pachino, Porto Empedocle, 1975–1990. Photo. J. SIELSKI.

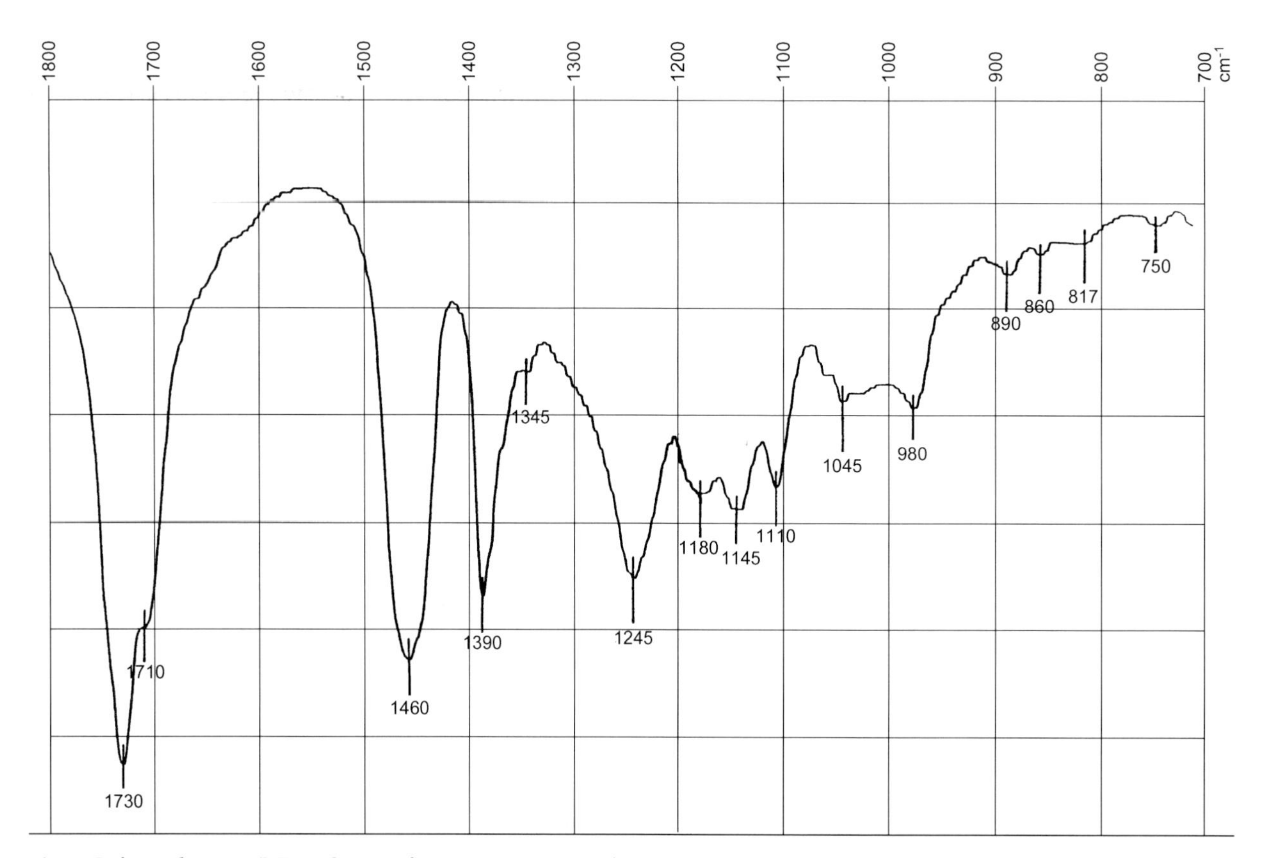

Fig. 8. Infra-red curve (MZ IRS 130) of simetite — a piece from the Simeto river (from the amber collection of the Museum of the Earth, inv. no. 18973).

be associated with the tastes of older generations. This status led to simetite becoming an important symbol of the local "Sicilian character". This is reflected in the opinions and motivations expressed in surveys carried out among jewellers and clients buying jewellery, and also in a relatively large number of publications. One such recent publication which undoubtedly deserves attention was produced in 1996 and consists of a collection of major historical texts (dating from 1639 to 1805) about Sicilian amber, published as a series of anthologies of sources referring to Sicily.

In order to study the local amber-craft traditions, ethnographic research was carried out in Sicily in 1980 and 1984 as part of a project conducted by the Italian-Polish Interdisciplinary Work group of Applied Sciences in Archaeology and Protection of the Cultural Heritage. Researchers sought to contact amber prospectors and jewellers — individuals of the highest professional competence in their fields — and to procure as large as possible a group of respondents to their survey questionnaire. The findings of this survey were modest in terms of quantity but sufficient to ascertain the existence of an "indigenous" amber-craft tradition (admittedly, only a vestigial one in some of its aspects) and to state that the problem of the Mediterranean origin of the "amber myth" remains unresolved.

A relatively extensive body of material was obtained on the particulars of amber prospecting and processing, preferred ornament forms, and the therapeutic and magical uses of this resin.

A serious difficulty is the lack of reliable analyses identifying the raw material used in both ancient and modern artefacts, most of which are dispersed among private collections and church treasuries — posing yet further difficulties in obtaining access to them.

Piotr SZACKI
State Ethnographical Museum in Warsaw
ul. Kredytowa 1
00-056 Warszawa, Poland

BURSZTYN SYCYLIJSKI

Piotr SZACKI

Streszczenie

Wokół bursztynu narósł złożony kompleks wiedzy, przekonań i symbolicznych znaczeń — swoisty mit, którego dzieje odznaczają się znacznymi fluktuacjami wywołanymi także zmiennym miejscem bursztynu w systemie wartości. Wątki „bursztynowego" mitu komplikowały się, odbijając wielokrotnym echem od basenu Morza Śródziemnego po Skandynawię. W Europie wśród żywic kopalnych czołową rolę odgrywa północny sukcynit. Na drugim jednak jej krańcu, na Sycylii, w centrum cywilizacji klasycznej, występuje żywica kopalna, trzeciorzędowa, mioceńska, zwana symetytem od nazwy rzeki Simeto, której wody wypłukują i transportują bursztyn z głębi wyspy do morza. Symetyt wyróżnia się gamą szczególnych odcieni, a przede wszystkim często występującą błękitnawą fluorescencją. Występowanie bursztynu na Sycylii nie jest odnotowane przez autorów starożytnych. Dopiero w XVI w. wzmiankuje go AGRICOLA a poczynając od XVII wieku autorzy sycylijscy i cudzoziemscy poświęcają mu już wiele uwagi. Badania współczesne precyzują poglądy o jego naturze. Jako surowiec jubilerski symetyt był popularny i ceniony najwyżej spośród żywic kopalnych. Obecnie kojarzony jest raczej z gustem minionych generacji. Lokując się jednak w ten sposób w „przestrzeni pamięci", przyobleczony w stosowną ideologię, stał się ważnym symbolem „sycylijskości".

Uchwyceniu miejscowych tradycji bursztyniarskich poświęcono badania etnograficzne przeprowadzone na Sycylii w 1980 i 1984 r., w ramach działań Polsko-Włoskiej Interdyscyplinarnej Grupy Roboczej Nauk Stosowanych w Archeologii i Ochronie Patrymonium Kulturowego. Starano się o dotarcie do środowisk poszukiwaczy bursztynu i jubilerów — najbardziej zawodowo kompetentnych w tym przedmiocie, oraz objęcie możliwie szerokiego kręgu informatorów ankietowym sondażem, by stwierdzić istnienie autochtonicznej tradycji (w niektórych aspektach wprawdzie śladowej) i uznać, iż problem śródziemnomorskiej proweniencji „mitu" bursztynu pozostaje otwarty. Istotnym utrudnieniem jest brak miarodajnych analiz identyfikujących surowiec materiału zabytkowego, zarówno starożytnego, jak nowożytnego, zresztą bardzo rozproszonego i trudno dostępnego.

Bibliography

DELL'AMBRA...
1996 Dell'Ambra siciliana 1639–1805. Testi di antichi autori siciliani 1639–1805, a cura di C. E. Fiore, 1–128, *Quaderni di Bibliotheca*, **1**, Edizioni Boemi.

FERRARA F.
1805 *Memoire sopra il lago di Naftia nella Sicilia meridionale, Sopra l'Ambra siciliana, Sopra il miele ibleo, Sopra Nasso e Callipoli*, 73–159, Reale Stamperia, Palermo.

GDAŃSK'S AMBER-WORKERS, WORKSHOPS AND THE MANAGEMENT OF AMBER RESOURCES IN THE TWENTIETH CENTURY

Wiesław GIERŁOWSKI

Abstract

Gdańsk's position within the context of the entire country's economic situation during the twentieth century proved to be a deciding factor in the success of its amber industry. The peripheral role which the city played under Prussian rule led to the decline of the amber-working trade, which used a total of only 5 tons of raw material in 1914, its products being marketed within a radius of barely several dozen kilometres. Poland's new political system of the 1990s enabled the market for these goods to be expanded across all continents and production to increase forty-fold, reaching an annual turnover rate of 200 tons.

The following paper concerns amber-working not only in the City of Gdańsk but also in the areas surrounding the Vistula estuary, which belong, at present, to the administrative regions of Gdańsk and Elbląg. These were German territories up until 1919, which were divided up between the Free City of Gdańsk, Germany (Eastern Prussia) and Poland during the inter-war period.

The intensity of amber-working in this area was dictated by general trends in the country's economic situation as well as by the degree of state regulation of the economy. The formidable boom in amber-working over the past decade has come about primarily as a result of popularizing research into amber, which has generated wide-scale interest in the subject. The revolution in design of amber goods initiated by artists from Gdańsk's coastal region has also played a significant role in helping the amber-working trade to flourish.

Around 450 qualified artist-craftworkers are employed in the field of amber-working in the Gdańsk region. They not only make designs for mass-produced items, but also create original works of art and carry out reconstructions of historic pieces dating from "Gdańsk's Golden Age".

At the beginning of this century the economy of the Gdańsk region was in a state of stagnation with little hope of any improvement in this situation, constituting as it did a distant province of the Prussian Empire. The traditional craft of amber-working, which concentrated on the production of luxury items for manor houses and members of the nobility had completely died out during the nineteenth century. The amber industry's largest centre for the mass-production of goods was situated near the amber-mining complex in Sambia. In Gdańsk there were only a few outdated workshops duplicating this mass-produced assortment. This situation remained unchanged until the mid-1920s when an economic revival was brought about by the construction of a major sea-port in the neighbouring town of Gdynia (20 km from Gdańsk). This provided open access to the whole Polish market and also stimulated a significant increase in the demand for amber, the amount of operational amber-working centres increasing twofold and the overall value of goods produced rising by approximately one hundred and fifty percent.

This brief spell of prosperity was cut short in 1939 with the onset of the Second World War. The war effort and ensuing resettlement of Gdańsk left not one single amber-workshop in the City nor indeed any masters or apprentices of the amber-working craft.

The first four years immediately following the war saw the rapid establishment of a considerable number of amber-working centres. Finding a market for their products after such a long period of absence proved no problem. By 1949 there were already sixty centres in operation in Gdańsk's coastal area (twice as many as there had been before the war).

Over the years to come, nearly all of them were, however, brought to ruin as a result of the liquidation policies practised on private companies during the Stalinist era. Only six craft centres and one state enterprise created from privately owned property outlived this political regime.

Over the next twenty-five years, right up until the changes introduced into the Polish economy, in 1990, the national production of amber goods grew slowly yet systematically. It was during this period that the foundations of Poland's success in the world amber market were laid down. This success was based on the following facts:

1. The leading role in dictating the range and variety of goods produced was taken on by professional designers, who were not constrained by the same restrictions as were enforced by the communist authorities on other industries.
2. Despite the dearth of amber goods on the local market, the government gave preference to their export.
3. Amber-workers began to make items of jewellery thanks to the fact that they were given unlimited access to supplies of silver at half the price which it was valued at on the stock market.

The growth rate of Gdańsk's amber-working market reached its climax in 1990 following the introduction of a free trade economy. It was at this point that the main barrier preventing further growth and development was lifted, i.e. the state-held monopoly on foreign trade. At present, it is the producers of amber goods themselves who establish retail prices and decide on which transactions to enter into. Trade with even very distant countries is managed by international couriers, who deal with all the customs formalities on behalf of the exporter. This puts the Polish amber-working industry in a much better position than their Russian and Ukrainian counterparts who still have to obtain a licence for every single export transaction undertaken.

The following charts, diagrams and tables illustrate the development of Gdańsk's amber-working trade:

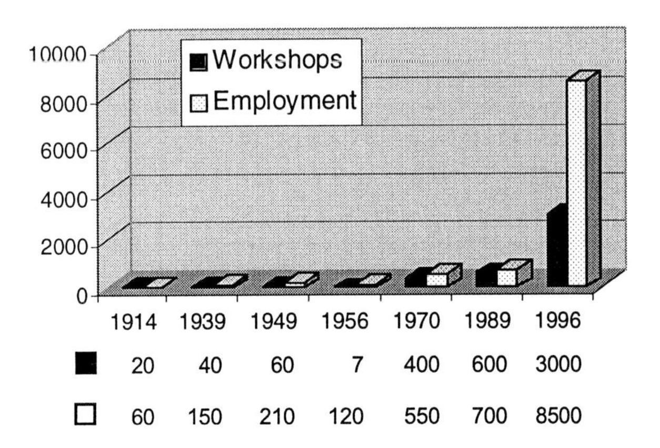

Fig. 1. Amber workshops and employment, 1914–1996.

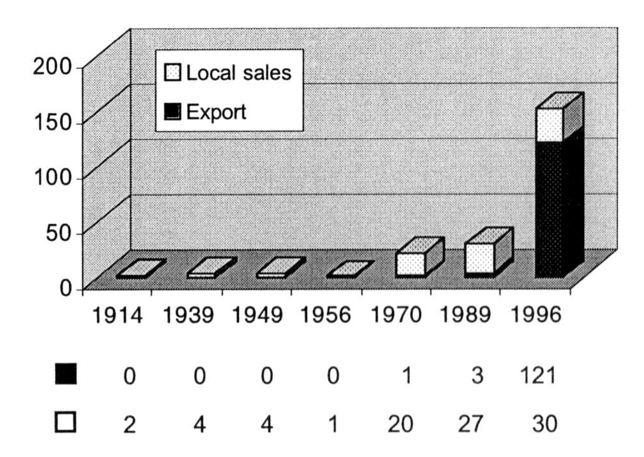

Fig. 2. Structure of amber sales.

Over eighty percent of all amber goods produced in Poland are destined for export. The assortment of these items is increasingly diverse. The most traditional product consists of necklaces made of turned and polished beads of real Baltic amber (succinite). The vast majority of clients choose amber which has been clarified in an autoclave and heat-treated to give it the colour of cognac. The number of jewellery items produced using silver in combination with amber in its natural state is, however, on the increase, as is the production of individual items such as dress accessories, sculptures, trinket boxes and goblets, made either of pure amber or of silver decorated with amber. Fig. 3 shows the scale on which amber goods from the Gdańsk region are exported to various parts of the world.

For the past four years Gdańsk has been the centre of the world's amber trade by virtue of the fact that in mid-March it hosts the annual International Amber Fair — Amberif — organized by the Międzynarodowe Targi Gdańskie S.A. Company.

A permanent solution to supplying amber-working centres with raw material has not yet been found. Amber is already being extracted from known Tertiary deposits in Poland. A certain quantity of material is also acquired through the collection of amber washed up on the beaches of the Bay of Gdańsk, whilst hydraulic extraction methods are used to recover amber from the fossil beaches of the Vistula Delta. However, the majority of raw material comes from the Russian amber mines in Sambia and the Ukrainian mines on Volhynian Upland. The amber extracted here is exported *en mass* and sold at prices lower than the foreseeable cost of recovering material from Polish deposits. The chart shown in Fig. 4 (drawn up for the 1990s) illustrates this fact.

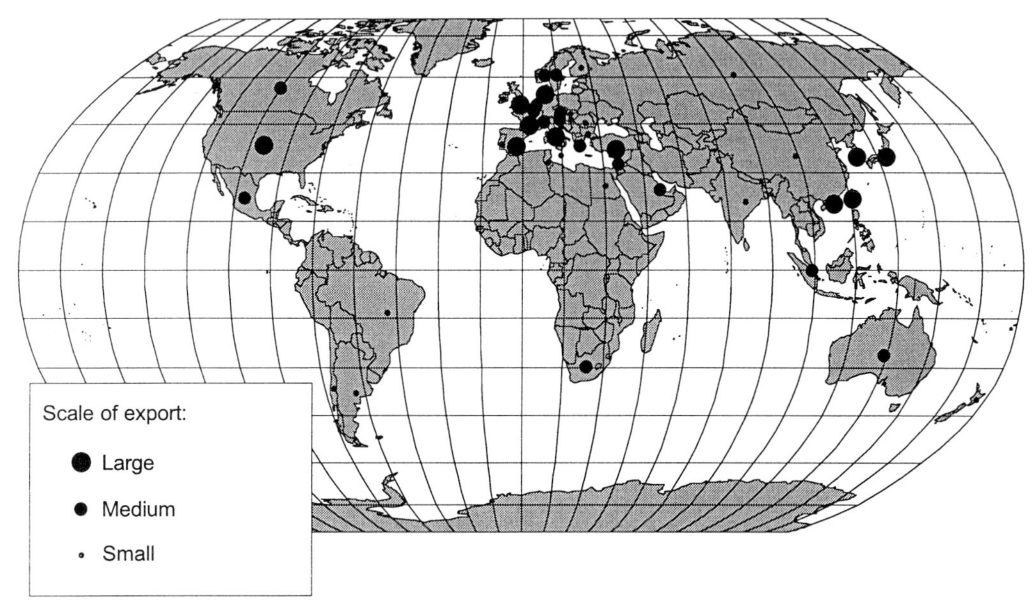

Large	Germany, France, Great Britain, Denmark, Spain, Italy, USA, Japan, Korea, Taiwan, Hong-Kong, Turkey.
Medium	Austria, Belgium, Czech Republic, Greece, Holland, Norway, Sweden, Switzerland, Canada, Mexico, South Africa, Bahrain, Lebanon, Singapore, Australia.
Small	Bulgaria, Slovakia, Slovenia, Croatia, Finland, Ireland, Malta, Portugal, Rumania, Hungary, Russia, Argentina, Brazil, Chile, Egypt, Tunisia, India, China, the Philippines, New Zealand.

(listed according to quantities imported)

Fig. 3. Export of amber items from the Gdańsk region in 1996.

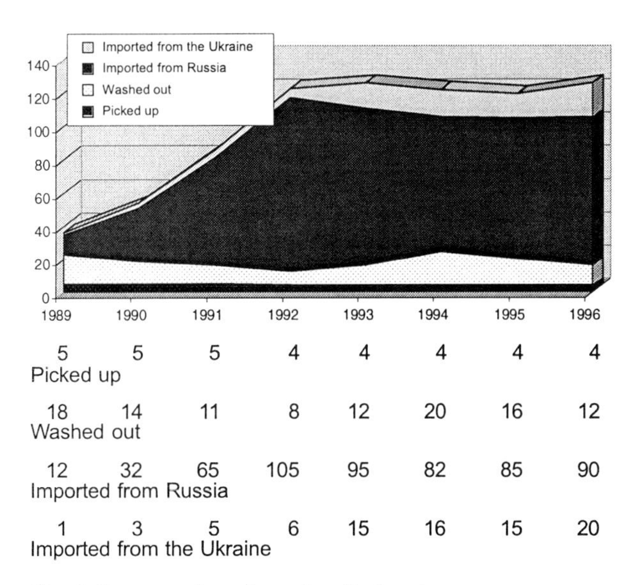

	1989	1990	1991	1992	1993	1994	1995	1996
Picked up	5	5	5	4	4	4	4	4
Washed out	18	14	11	8	12	20	16	12
Imported from Russia	12	32	65	105	95	82	85	90
Imported from the Ukraine	1	3	5	6	15	16	15	20

Fig. 4. Sources of crude amber (in tons).

A chronological overview of the production of amber goods in the period from 1900 to 1996

1. 1900–1918. In this period Gdańsk began to experience a decline in the artistic handicraft of amber-working. Only the turners' guild continued to manufacture round and oval beads for the production of necklaces. Following in the footsteps of one of the leaders in the amber trade in Sambia, Gdańsk introduced the standardization of different varieties of amber by heat-treating the raw material at very high temperatures. The number of shops selling amber goods did not exceed twenty. The "Moritz STUMPF" company was exceptional at this time in that it attempted to uphold contemporary traditions, when Gdańsk was one of the key figures in the field of artistic ambercraft. This company linked the production of silverware (plate, cutlery and jewellery) combined with the production of amber pieces for decorative use and for the needs of conservation of historical works in amber. From 1911 to 1913 they undertook the conservation of all the non-fixed furnishings from the Amber Room in Tsarskoye Selo, thus greatly pleasing the executive board of the Tsar's Summer Palace. Conservation work on large caskets and even items of furniture took place in the Gdańsk workshops of the M. STUMPF company.

2. 1919–1939. During the inter-war period Gdańsk remained in the same customs zone as Poland. This situation caused some complications in the import of raw material from Sambia, although, conversely, it helped to gain access to a whole new market in and around the newly founded town of Gdynia and the fashionable seaside resorts of the Hel Peninsula. The influx of Polish tourists revived Gdańsk's amber trade. The number of specialist shops in Gdańsk reached forty — double the amount which had existed before the First World War. The "Moritz STUMPF" company alone owned a total of five shops in various districts of Gdańsk and in Sopot. However, the export of amber outside Polish territory practically came to a halt. Over a dozen factories producing amber goods were established in the vicinity of Gdynia, the majority of the raw materials used by them coming from the fossil beaches surrounding the new port which was under construction at the time. Among the best known of these companies was that belonging to Piotr TRZEŚNIAK.

3. 1946–1956. During the 1940s high levels of unemployment in Poland's war damaged coastal towns led to the rapid founding of factories producing amber goods, even though few people remained who were familiar with the craft of amber-working. By 1949 there were already sixty factories in existence, five of which exported their wares to the West. The political restrictions imposed by the nationalizing of industry over the next five years meant that by 1956 only seven of these factories were still extant. Some of the machines and personnel from the liquidated plants were taken over by the state-owned amber-working factory, which employed over one hundred people and received a regular supply of 3000 kg of Sambian amber each year.

4. 1957–1970. Following a "political thaw", from 1956 onwards the number of private factories producing amber goods again began to increase. By 1970 there were four hundred workshops in the Gdańsk region and over fifty in the rest of the country. The authorities were not, however, prepared to allow larger companies to come into being. Instead, they gave preference to individual artists and craft-workers whose designs had to be approved by state appointed commissions. As a result, employment in the amber-working sector was relatively low, totalling approximately six hundred workers. In 1968 a stop was put to the import of supplies of crude amber from Sambia, as the development work being carried out at that time in the port of Gdańsk meant that twenty to forty tons of amber were being extracted annually from the Vistula estuary. This quantity by far exceeded that then needed for the production of amber goods in Poland.

5. 1971–1989. These two decades saw a slow but steady increase in production. Only a limited number of new establishments were founded. Poland's authorities did, however, permit the use of hired-labour in amber-working factories, which in turn enabled companies to expand their workforces sufficiently to be able to undertake production destined for export. Up until the fall of socialism export sales remained monopolized by state-owned firms, although a policy was introduced allowing producers a gradually increasing share of the foreign capital gained through export. By the late 1980s the share allotted producers of artistic handicrafts had reached nearly fifty percent.

6. 1990–1996. The radical changes in Poland's economy introduced in 1990 allowed the country's workshops to take advantage of the favourable economic situation on the world's amber market and become one of the leading suppliers of ready-made amber goods across all continents. Currently, Poland's exports account for seventy percent of amber goods supplied worldwide, the value of these exports being estimated at around three hundred million US dollars. Approximately eighty percent of export sales are dealt with by companies based in the Gdańsk region, which has also seen a marked expansion of its local network of retail outlets and the range of goods which they offer. Growth factors in the amber economy have reached an unprecedented scale: the number of factories has risen fivefold (from six hundred to three thousand) with employment rising twelvefold (from seven hundred to eight thousand five hundred). In the city of Gdańsk alone, there are three hundred shops and galleries selling ready-made goods and works of art produced in amber, with a turnover of two hundred million Polish zlotys. This vast progress has been achieved through free enterprise, which, at the moment, is not an option for our eastern neighbours (Russia and the Ukraine), and through combining the use of silver and amber in a wide range of jewellery and items of everyday use.

At present the biggest problem on the Polish amber market is the battle against synthetic amber. Identifying fake amber which has been produced using natural sub-fossil resins is particularly difficult because of the deceptively authentic appearance of this product. In order to protect both those clients making their purchases at retail outlets and those entering into large-scale export transactions, a system of monitoring and recommending reliable companies by the Amber Association of Poland. Those companies who meet with the approval of the Association are awarded the right to use its quality assurance badge, the design of which is shown below:

Wiesław GIERŁOWSKI
ul. Szara 9 m. 50
80-116 Gdańsk, Poland
e-mail: gierlow@fs-samba.com.pl

GDAŃSCY TWÓRCY, WARSZTATY I GOSPODARKA BURSZTYNEM W DWUDZIESTYM STULECIU

Wiesław Gierłowski

Streszczenie

Obszar wokół ujścia Wisły jest u schyłku XX wieku centrum wytwórczości bursztynowej, zaspokajającym około 65% światowego popytu w tej dziedzinie. Ta dominująca pozycja została osiągnięta po radykalnej zmianie ustroju gospodarczego w Polsce w ostatnich ośmiu latach.

Na początku stulecia region gdański przeżywał upadek gospodarczy, który spowodował likwidację prawie wszystkich warsztatów, tak charakterystycznych dla rzemiosła artystycznego w okresie nowożytnym.

Przez cały wiek XX gdańskie bursztynnictwo ledwie wegetowało. Najpierw wskutek sprowadzenia go do funkcji czysto lokalnej, a następnie niszczone wojnami, migracją całej ludności i dławiącą gospodarkę doktryną komunistyczną.

Dopiero ostatnia dekada stulecia stworzyła dynamiczny ośrodek gospodarczy i artystyczny, powiązany z odbiorcami we wszystkich częściach świata. Dalszy rozwój uzależniony jest od dobrego wykorzystania własnych złóż surowca, dotąd prawie nie eksploatowanych.

THE AMBER ASSOCIATION OF POLAND
GENERAL MANAGEMENT IN GDAŃSK

The Amber Association of Poland was registered in June 1996 at the Provincial Court in Gdańsk. The organisation affiliates the owners of production, service and trade companies involved in the mining, processing and trading of amber products. The Association also includes among its members scientists, collectors and artists who design and create works of art in amber.

The aim of the Association is to promote Baltic amber, safeguard against its forgery and maintain its status of prime importance both in Poland and worldwide by spreading knowledge about its qualities: beauty, healing properties, scientific value and cultural significance. To this end the Association deals with the distribution of magazines and books related to the amber industry.

One of the Association's many objectives is to represent its members before both state and local authorities and organise co-operative ventures between associates in the field of design, engineering, work training and the exchange of information relating to the amber market situation. The Association provides help for collectors of natural specimens, crude amber, floral and faunal inclusions. It also offers assistance to individuals conducting scientific research by organising meetings with those involved in the mining, processing and trading of amber. The Association initiates work aimed at searching for, recording and exploiting amber deposits and evaluating their potential. Another guideline of the Association is to consolidate amber producing companies in order to create a lobby and develop joint trade policies and strategies. We invite you to co-operate with members of our Association.

Certificates granted by the Association give the bearer the right to use the Amber Association's trademark. All goods sporting this trademark are guaranteed to have been made from genuine amber which has been correctly processed to obtain the highest quality finished product.

The Association produces a quarterly publication — *Bursztynisko* — and circulates a broad range of articles devoted to amber on the Internet at: http://www.sbp.org.pl. These internet publications come complete with colour illustrations. Professional literature about amber (in Polish, German and English) is also available from the Association.

THE AMBER ASSOCIATION OF POLAND
Address: ul. Beniowskiego 5, hall 03, room 116, 80-382 Gdańsk, POLAND
Tel/fax +48 (58) 5549-223
Bank details: Bank Śląski S.A. O/Gdańsk-Oliwa; Account No. 10501764-2210702292
e-mail: spb@sbp.org.pl
Web site: http://www.sbp.org.pl

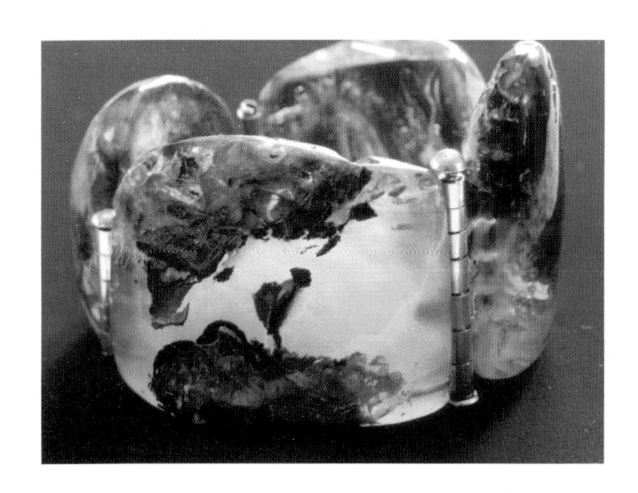

Druk: Drukarnia Wydawnictwa Diecezji Pelplińskiej „Bernardinum",
ul. Bpa Dominika 11, 83-130 Pelplin